Four Centuries
of Clinical Chemistry

Frontispiece. *The Medical Alchemist*, by Matheus van Hellemont, from The Fisher Collection, Fisher Scientific, Pittsburgh, Pennsylvania, USA.

Four Centuries of Clinical Chemistry

Louis Rosenfeld
New York University School of Medicine
New York, NY, USA

Gordon and Breach Science Publishers
Australia • Canada • China • France • Germany • India
Japan • Luxembourg • Malaysia • The Netherlands
Russia • Singapore • Switzerland

Copyright © 1999 Louis Rosenfeld. Published by license under the Gordon and Breach Science Publishers imprint.

All rights reserved.

No part of this book may be reproduced or utilized in any form or by any means, electronic or mechanical, including photocopying and recording, or by any information storage or retrieval system, without permission in writing from the publisher. Printed in Singapore.

Amsteldijk 166
1st Floor
1079 LH Amsterdam
The Netherlands

British Library Cataloguing in Publication Data

Rosenfeld, Louis, 1925 –
   Four centuries of clinical chemistry
   1. Clinical chemistry – History
   I. Title
   616'.0756

ISBN 90-5699-645-2

To

*the many thousands of clinical laboratory
technicians and technologists who practiced
the art and science of quantitative analysis
before it became random assay.*

When I matriculated in the fall of 1902, I had a hazy idea that chemistry might be interesting and approached Prof. F. J. Alway and asked him what chemistry was like. I shall never forget how he looked up at me, smiled and replied, "If you stick at it long enough, some day you will be able to do something that no one ever did before." Then and there I resolved to try to be a chemist.

—Ross Aiken Gortner (1885–1942)

—Samuel Colville Lind
National Academy of Sciences
Biographical Memoirs, 1945

Contents

Preface   xv

I   Introduction. Anatomy, Disease, and Therapy. Transition of Alchemy to Chemistry. Paracelsus. Uroscopy. Van Helmont and Discovery of Gas. Sylvius and Iatrochemistry. Iatrophysicists (Iatromechanists).   1

II   Introduction. Robert Boyle. Color Test and Chemical Reaction. Analysis of Blood. The Gaseous State and Pneumatic Chemistry. The New Chemistry of Combustion and Respiration. Nomenclature.   23

III   Introduction. Urinary Calculi and Discovery of Uric Acid. Cystine and Cystinuria. Preparation and Analysis of Urea. Analysis of Urine. Chemistry vs Physiology. William Prout Discovers the Acid of Gastric Juice. Alexander Marcet, Golding Bird, and Analysis of Calculi.   37

IV   Introduction. Protein and Food Shortage. Analysis of Nitrogen: Methods of Dumas and Will–Varrentrapp. Kjeldahl Analysis. Nessler's Reagent. Berzelius: International Authority. "Vital Force." Wöhler's Synthesis of Urea. Organic Synthesis by Berthelot. The Analytical Balance. Volumetric (Titrimetric) Analysis. Karl Mohr.   55

V   Introduction. Library Medicine, Bedside Medicine, and the Stigma of Dissection. The Stethoscope and the Changing Patient–Physician Relationship. The Changing Role of Therapy. Measurement and the Focus on Specific Diseases. Bright's Disease, Albuminuria, and Scientific Medicine. Body Fluids and the Examination of Blood. Chemical Analysis: Voices Pro and Con.   77

CONTENTS

VI   Rees and the Estimation of Urea and Sugar in Blood.   99
Bence Jones Protein. A Theory of Diabetes. Alkaline Tide.
Medical Education. Liebig's Concept of Stone Formation.
Liebig's Laboratory at Giessen. The Impact of "Animal
Chemistry." Animal Chemistry Society. Chemistry in the
Service of Society, Technology, and Medicine. Chemistry
Separates from Medicine. Chemistry and the Apothecaries'
Act of 1815. Clinical Chemistry—A False Start for a New
Identity.

VII   Textbooks on Urine and Blood Analysis (1863–1899).   131
Albuminuria and Insurance Companies. Esbach's Method
for Urine Protein. Detection of Protein by Dipstick.

VIII   Cholera, Acidosis, and Fluid-Electrolyte Therapy in 1832.   151
Chemical Studies and Therapeutic Responses When
Cholera Returns. The Early History of Polyuria,
Glycosuria, and Hyperglycemia. Sugar Analysis and
Alkaline Copper Sulfate. Fehling's Solution.
Fermentation. Polariscope–Saccharimeter. Claude Bernard
Discovers the Glycogenic Function of the Liver. Chemical
Findings in Diabetic and Other Forms of Acidosis.
Van Slyke's Apparatus for Measuring Bicarbonate
Concentration and Detection of Acidosis. Milliequivalents
and Cation–Anion Balance. Discovery of Insulin.

IX   Respiration and Combustion. Early Studies on Blood   185
Gases. Hoppe–Seyler Isolates Hemoglobin. Physiological
Chemistry and *Zeitschrift für Physiologische Chemie*.
Quantitation of Red Cells and Hemoglobin. Hematocrit,
Urine Sediments, and the Centrifuge. Hydrogen
Electrode. Sørensen, pH, and Buffers. Henderson–
Hasselbalch Equation and New Definitions for Acid
and Base. Glass Electrode. Beckman pH Meter.

X   Early Development and Use of the Microscope. Clinical   215
Microscopy and the Evolution of Clinical Laboratories
in Great Britain. Emergence of Clinical Laboratories
in the United States. Medicine and the Scientific

CONTENTS   xi

Method. The $300 Laboratory. Clinical Laboratory
Testing (1900–1920). Medical Technologists, Commercial
Laboratories, and World War I. Laboratory Supplies and
Suppliers. Origins of the Chemical Glassware Industry
in America. Bausch & Lomb Optical Co.

XI   Introduction to Physiological Chemistry. Chittenden and   245
     the Sheffield Scientific School. Societies, Journals, and
     the Diversity of Biochemistry. Analysis by Color
     Comparison. The Duboscq Colorimeter and Its Early
     Uses. Modern Colorimeter. The Optical Laws of
     Beer–Lambert–Bouguer. Duboscq's Optical Instruments.

XII Introduction. The Flexner Report and Medical School   269
     Reform. Medical Chemistry, Physiological Chemistry,
     and Biological Chemistry. Otto Folin. Metabolic Studies
     at McLean Hospital. Colorimetric Methods for Blood
     Analysis. Nonprotein Nitrogen (NPN). Folin Joins
     Harvard Medical School. The Folin–Wu Protein-Free
     Filtrate and Other Protein Precipitants. Stanley Benedict.
     The Rockefeller Institute for Medical Research. Donald
     Dexter Van Slyke and the Rockefeller Hospital. Ivar Bang
     and Micromethods. Venipuncture and Blood Analysis.
     Evolution of the Hypodermic Syringe. Becton Dickinson
     and Company (1897–1997) and the Vacutainer Tube.

XIII Introduction. Benedict's Reagent for Urine Glucose.   319
     Analysis of Blood Glucose. Method Modifications by
     Folin and Benedict. A New Glucose Reagent—Alkaline
     Ferricyanide. Somogyi Protein-Free Filtrate. Tablets,
     Powders, and Dipsticks. Another New Glucose
     Reagent—Ortho-Toluidine. Enzymatic Methods for
     "True" Glucose. Creatine, Creatinine, and Creatinine
     Clearance. Uric Acid. Urea and Urease. Urea Clearance.
     Phenolsulphonephthalein Excretion. Bromsulphalein
     Retention. Blood Volume.

XIV Chloride, Calcium, Phosphorus, Sodium and Potassium: 351
     Introduction. Gravimetric, Titrimetric, and Colorimetric
     Methods. Spectroscopy. Bunsen and Kirchhoff Apply

Spectral Analysis. Arc and Spark Analysis. Flame Photometry. Ion-Selective Electrodes. Atomic Absorption Spectrometry.

XV Cholesterol: Introduction. Color Reactions. Reaction With Digitonin. Blood Analysis: Esterified and Free Cholesterol. Reference Methods. Enzymatic Methods. Atherosclerosis, Coronary Heart Disease, and the National Cholesterol Education Program. Bilirubin: Introduction. Icterus Index. Diazo, Direct, and Indirect Reaction. Standards: Artificial and Certified. Urine Analysis and Tablet Test. Kernicterus in the Newborn. Direct Spectrophotometric Assay. 377

XVI Enzymes: Introduction. The Catalytic Force. Ferments: Organized and Unorganized. Chemical vs Vitalist Fermentation. Zymase and the Cell-Free Extract of Yeast. Enzyme Theory of Life. Protein Identity of Enzymes. Amylase. Lipase. Alkaline Phosphatase. Acid Phosphatase. Aspartate Aminotransferase and Other Enzymes. SI Units. 395

XVII Proteins: Introduction. Classical Separation of Albumin and Globulin. Salt Precipitation. Sodium Sulfate Fractionation: Euglobulin and Pseudoglobulin I and II. Improved Salt Precipitation Methods. Albumin Binding Reagents. Biuret Color Reaction. Moving Boundary Electrophoresis. Separation on Filter Paper. Visualization of Discrete Bands. Other Support Media. Refractive Index. Specific Gravity. Clot Formation: Introduction. Separation and Quantitation of Fibrinogen (Fibrin). 417

XVIII Victor Myers and the New York Post-Graduate Medical School and Hospital. *Practical Chemical Analysis of Blood*. St. Luke's and St. Thomas's. Hawk's *Practical Physiological Chemistry*. Peters and Van Slyke: *Quantitative Clinical Chemistry*. Photometry: Introduction. The Photoelectric Effect. Photoelectric Colorimeters (Photometers). Hoffman's *Photoelectric Clinical Chemistry*. Spectrophotometry: Visible and Ultraviolet. 443

CONTENTS

XIX Clinical Chemistry Laboratory (1925–1960). Proficiency Testing. Biochemistry Sheds Its Clinical Connection. American Association of Clinical Chemists. Foreign Societies and the International Federation. Bio-Science: The Era of Referral Chemistry. Sigma and Kit Methods. Berson, Yalow, and Radioimmunoassay. Tool–Instrument–Machine–Automation. Skeggs and the AutoAnalyzer. Factory Teaching Facility. Analyzers: Discrete Sample and Random Access. ... 465

XX Pitfalls of Publication. Dangers of Laboratory Diagnosis. The Laboratory: Possibilities and an Image Problem. Scientific Medicine and Overuse of the Laboratory. A New Partnership—A New Vocabulary. Looking to the Future—Cost Containment and Managed Care. New Career Opportunities for the Clinical Chemist. Conclusion. ... 503

Bibliography ... 519

Index ... 533

Preface

Clinical chemistry has been an indistinguishable part of medical practice since earliest times, when the local witch doctor or medicine man examined body fluids as part of a mystical ritual of diagnosis and treatment of physical and mental ailments. Despite its importance today, chemistry in the service of medical practice was relegated to the very periphery of medical science until well into the nineteenth century. Analysis of body fluids was of limited medical value because so little was known about metabolic processes in health and disease.

There existed a vast history of information, beliefs, speculations and groping experiments on the chemistry of living things. The contributions eventually emerging from this mix did not suddenly materialize out of thin air. They had been growing underground, and when the roots subsequently surfaced, they came up a long way from their origins in the sub-soil of antiquity.

This history unravels the origins of clinical chemistry in the sixteenth century by tracing a path through the erratic advances in medical and chemical practices and the parallel developments of physiology and organic chemistry three centuries later. Biochemistry took shape near the close of the nineteenth century, when the emphasis on living systems shifted from physiology to chemistry. The subsequent flowering of clinical chemistry in the opening years of the twentieth century, as the medical application of analytical biochemistry, was inevitable. Clinical chemistry did not emerge as a consequence of biochemistry but was practiced in most places long before the advent of biochemistry.

The development of clinical chemistry and allied sciences is described here in a connected and systematic way without excessive detail or repetition. Emphasis is placed on those discoveries and speculations that contributed to its gradual though uneven growth. This history was written to fill the many gaps in the scientific story of clinical chemistry and to reconstruct the often obscure path of its development to the present. It will interest chemists, pathologists, technologists and others who, in hospitals and industry, in steadily increasing numbers, are participating in the phenomenal late-twentieth-century expansion and metamorphosis of clinical chemistry. This previously unexplored sector of medical history will also interest academic historians by providing informative and useful references to the development of clinical chemistry.

The origins and early years of any rapidly changing scientific discipline run the risk of being forgotten and consigned to oblivion unless a record of its past—recent as well as distant—is preserved. This history aims to inform the new, and remind the not-so-new, members of the profession of the long and illustrious history of clinical chemistry—to tell them who and what went before, as well as where and when. Its development over the centuries illustrates the heterogeneous roots from which modern interdisciplinary clinical chemistry has grown. Important landmarks reveal the international character of the scientific work and the interaction of its participants.

History is usually presented in the order it happens. This sequence is followed here until the end of the eighteenth century. During this period major ideas and events did not have serious competition for attention and can be readily described in chronological order, more or less, without diverting the reader's attention by varying periods of time. Modern chemistry, however, beginning with the start of the nineteenth century, is more suited to thematic treatment because of the progressively numerous lines of investigation that developed in parallel from 1800 on. The inherent disadvantage is that every theme has its own time frame and may cover several decades or even a century; also, a theme usually intersects other themes, resulting in some repetition in the telling. The terminology and units are often those in use at the time. There is frequent use of direct quotes so that the reader may experience the style of the scientific discourse without the overlay of paraphrasing or interpretation by a third party.

All literature references were reviewed in the original to avoid reproducing errors of citation and misrepresentation that sometimes enter the mainstream of review articles and books. Secondary sources are attributed wherever appropriate. Reference to work done more than a century ago is sometimes made only by author and year.

I wish to thank the following for reviewing various chapters of the manuscript and for their valuable comments, suggestions, and helpful criticisms: Theodor Benfey, Johannes Büttner, Wendell T. Caraway, Basil T. Doumas, Joseph S. Fruton, Frederick L. Holmes, Samuel Meites, Leonard T. Skeggs, and Bennie Zak.

The staff of the NYU Medical Center Library was generous with help and cooperation in obtaining books, journals, and reprints of articles through interlibrary loan services. I especially wish to acknowledge the assistance of Joan Himmel and Eileen Brown.

"But thou didst order all things by measure and number and weight."
Solomon, 11:20
Apocrypha: The Book of Wisdom

CHAPTER I

INTRODUCTION

With the beginning of the seventeenth century, a new relation developed between man and nature as the divine will was replaced by the human mind. The point of view which made disease a consequence of sin gave way to the Hellenic idea, that disease is a lack of harmony which nature should cure. The question no longer was, how nature was created, but how does it work [1]. Scientists began to experiment and measure what they observed, and to turn away from scholastic preoccupation with ancient texts and authorities. Specialized anatomic research revealed lesions in particular organs at autopsies, and anatomists were beginning to suspect that disease processes were really localized.

ANATOMY, DISEASE, AND THERAPY

The modern understanding of disease through anatomy began with *De Sedibus* (1761) [2]. This five-volume treatise by Giovanni Battista Morgagni (1682–1771) introduced the anatomic concept of the organ as the seat of disease. For the first time a detailed analysis of postmortem findings was correlated with clinical symptoms and case histories. Disease processes were explained in terms of localized pathological anatomy, rather than as being diffused throughout the system. The treatise did not have an immediate impact. Most doctors of Morgagni's time and for some years afterwards were ignorant of morbid anatomy and did not think in terms of a localized pathologic process or of specific diseases.

This new view was long ignored or opposed by most practicing physicians who believed that the ultimate explanation for pathologic symptoms was not to be found in the observation of gross anatomy. John Locke (1632–1704), the philosopher–physician, did not believe that anatomists could distinguish subtle underlying conditions:

> "Now it is certaine and beyond controversy that nature performs all her operations on the body by parts so minute and insensible that I thinke noe body will ever hope or pretend, even by the assistance of glasses or any other invention, to come to a sight of them, ... and though we cut into these inside,

we see but the outside of things and make but a new superficies for ourselves to stare at" [3].

One explanation for this state of affairs was that medical men lacked sufficient knowledge about normal anatomy to be able to distinguish the changes causing death, from changes occurring after death. To a pragmatic practitioner, postmortems were of no use, since therapy was not guided by knowledge of the cause of death. Anatomical lesions were considered to be the effects, not the causes of disease. The prevailing attitude of the physicians and surgeons was that the cause of death was too obvious to need confirmation and that inspection need be done only if the diagnosis was in doubt. Physicians were convinced that they could determine the nature of each "clinical picture" by the use of their senses alone. They did not understand that the knowledge gained from postmortems would improve their ability to diagnose and to avoid harmful therapies.

Patients were to be observed rather than examined. This approach was reinforced when Thomas Sydenham (1624–1689) revived the methods of Hippocrates (460–370 B.C.E.)—reliance on observation at the patient's bedside and on past experience. The central doctrine of Hippocratic healing was based on a humoral pathology which considered disease to be a general affection and not limited to any one organ. The physician's function was to assist the body's natural ability to respond to disease and restore normal balance.

There was no role for instruments, experiments, and measurements. Sydenham rejected use of the microscope and anatomical investigations for the understanding of disease. He believed that God had not intended for the physician to examine the minute processes in nature's inner parts [4].

Medical practitioners distinguished diseases according to symptoms and the anatomical location of the complaint. They believed that essentially there was one disease state, a humoral derangement, and that different patterns of illness were merely changes in the status of the humoral system. Therapy was directed to restoring harmony and balance to body fluids, whatever the name given to the disease [5]. Unsatisfactory fluids were removed by bleeding, purging, sweating, blistering, and vomiting. Depletion of body fluids appeared to have a beneficial action. By exhausting the patient, it exercised a sedating or calming effect. If a deficiency was believed to exist, the patient was treated with diet and drugs that were administered without a definite knowledge of their action or therapeutic use.

TRANSITION OF ALCHEMY TO CHEMISTRY

Meanwhile, on a parallel track of chronology, chemistry was undergoing its own radical metamorphosis from alchemy. Sometime between the second and third century C.E. there developed an interest to produce gold and silver from base materials. There was nothing inherently absurd in this practice known as alchemy. Numerous examples of changes in material were available since it is the essential nature of chemical reactions that one substance is made to disappear as it is replaced by another with different properties. Since there was no knowledge of compounds, elements, or solutions, the only conclusion from such phenomena was that new substances had been formed [6].

Everything in the world of the alchemist was a creation of a fundamental substance—the quintessential spirit—which alchemists tried to isolate and purify by various processes, usually involving fire. Since a source of heat was central to everything the alchemist did, they transformed the kitchen or the adjacent area into the first chemical laboratories. Their apparatus, derived from cooking technology, was used for fermentation, distillation, ashing, sublimation, filtration, evaporation, extraction, crystallization, and solution. Amidst the confusion and disorder of their laboratories, the alchemists were the first to undertake the methodical experimental investigation of the chemical nature of substances [7].

Paintings of an alchemist's "laboratory" by Renaissance artists usually show a picturesque but dimly lighted room with an oven or furnace, large distilling apparatus, alembics, retorts, flasks, bottles, tongs, and crucibles, on the floor and shelves, clutter everywhere competing for space with books, thick and worn (see Frontispiece). There was total disarray with grimy apparatus and an occasional skull or anatomical model nearby. Often the small figure of an assistant can be seen working the bellows at the furnace. Smoke from a failed experiment gave the impression of mystical or magical goings-on.

Because of their obsession with secrecy, the alchemists were reluctant to use intelligible language to describe their experiments. Their meaning is almost impossible to decipher, and we are unable to trace any rational sequence of progress to their procedures [8]. Deliberate obscurity retarded progress because no one could profit by another's mistakes or learn from another's success. Their obscure writings also served as a cover for their own ignorance and chemical failures.

PARACELSUS

Chemistry was elevated to a new level of importance in the sixteenth century by Philippus Aureolus Theophrastus Bombastus von Hohenheim, the controversial Swiss physician, oculist, astrologer, and alchemist. Calling himself Paracelsus (1493–1541), he claimed parity with the great Roman compiler of medical information, Aulus Cornelius Celsus, who lived in the first century of the current era. Since Paracelsus had no accurate instruments, his rejection of the ancient traditions was not based on new observations or experiments, but rather on philosophical reasoning. His break with the classical ideas gradually led to an awareness that mechanistic views could explain physiological and pathological processes.

Paracelsus urged chemists and physicians to discontinue the search for the means of making gold in the laboratory and to devote their skills to the preparation and purification of chemical substances for use as drugs. His new movement was named iatrochemistry because it was chemistry in the service of medicine. For example, mercury, antimony or gold, was dissolved in mineral acid (sulfuric), then recovered as inorganic salts (sulfate) and used therapeutically. He used mercurous chloride (calomel) as a diuretic, a practice continued into the twentieth century.

Paracelsus declared that there were three elementary principles of the body: sulfur (soul), mercury (spirit), and salt (body). Diseases were caused by disturbances in the natural interaction of these principles. In this manner, chemistry was introduced into medicine. Paracelsus saw the body as a chemical laboratory governed by a conscious entity, the *archeus* (an internal chemist), which performed the chemical operations of the organism. The theory resembled Hippocrates's doctrine of the self-healing powers of the body. The purpose of drugs was to aid the *archeus* in resisting disease.

Paracelsus was unpopular with the establishment practitioners wherever he traveled, because of his caustic style, his lectures in German instead of in Latin, and his unrelenting attacks against orthodox medicine—stagnant for the previous thirteen centuries. Nevertheless, the Paracelsian system was a stimulus to the lethargic and unchanging state of medical theory. It marked the beginning of investigation and explanation of the functions of the human body and the occurrence of disease in chemical terms—thus separating chemistry from alchemy. Whereas alchemy used biological processes as models for the interpretation of inorganic phenomena, modern chemistry seeks to explain biological activity in terms of chemical reactions.

UROSCOPY

As chemistry slowly began to emerge from alchemy about the middle of the sixteenth century, attempts were made to analyze that plentiful body fluid—the urine. This was the first body substance to be examined in relation to disease and consisted primarily of visual observation of color, consistency, and volume. The practice of associating changes in the urine with disease evolved in the early cultures of the Egyptians, Persians, East Indians, and Chinese, and undoubtedly in other early cultures not so well documented. The Sumerian word for doctor in ancient Babylon was *asu*, meaning "one who knows water." What may have been the earliest "chemical" test on urine was one to determine pregnancy and was recorded in the Berlin Medical Papyrus dated about 1350 B.C.E. Egyptian women watered seeds of wheat and other cereals daily with their urine. If the seeds sprouted, the woman was judged to be pregnant. The sex of the expected infant was predicted according to which cereal seeds grew best [9].

Although references to the appearance of the urine occur in ancient records of Egypt and Mesopotamia, the first important descriptions are found in the Hippocratic writings (*circa* 400 B.C.E.), where a rational attempt was made to relate observations on the urine to the clinical situation of the patient for purposes of prognosis. These writings were carried over into Roman medicine and can be seen in the works of Celsus and Galen (130–200 C.E.). This pattern of transmission was continued by Arabian physicians who copied the opinions of the ancient writers without significant alterations. The School of Salerno (eleventh century), the first independent medical school in western Europe, derived its medical practices from the Latin translations of Arabic and other ancient manuscripts.

By the year 1000, observations by Persian physicians included transparency, sediment, odor, and froth. In his medical aphorisms regarding urine—based mostly on the works of Galen—Moses Maimonides (1135–1208), the Spanish-born Jewish philosopher and physician practicing in Cairo, wrote of the necessity and importance of examining the urine during a febrile illness. He described the color, odor, and particulate matter of urine in various disease states and described the clinical symptoms of diabetes mellitus [10].

During the Middle Ages there was very little advance in techniques or procedures used for urinary observations. Visual examinations known as uroscopy were very popular and were used in the diagnosis and prognosis of disease. For more than 500 years, the physician was invariably depicted inspecting a urine glass (matula) (Fig. 1.1)—a round, bladder-shaped,

FIG. 1.1. "The Physician" by Franz Christophe Janneck (1703–1761). (Photo courtesy of Fisher Scientific, Pittsburgh, Pennsylvania.)

wide-mouth, transparent flask carried in a matching cylindrical wicker basket of reeds or straw fitted with a carrying strap, threaded through the lid on either side for extra security. The traditional shape of the urinal is credited to Johannes Actuarius, the author of an elaborate seven-volume treatise on uroscopy, and senior physician at the Byzantine court in the thirteenth century. He borrowed widely from Greek and Arab authors and intended his work to be a complete practice of medicine based on uroscopy. The word *actuarius* was a title denoting his official medical position [11].

The matula actually became the physician's professional emblem for his home or office. In some interpretations, four regions, marked off from top to bottom in the container, were correlated with the head, chest, abdomen, and uro-genital system. Urinary changes of color or movement of sediment occurring in any of these regions indicated the location and nature of the illness. Actuarius divided the matula into three regions subdivided into eleven sections, each with different significance.

Paracelsus was critical of diagnosing disease from the visible appearance of the urine. He added a chemical dimension—examination by extraction or distillation—to separate its parts. He deduced the nature of the ailment from the quality and quantity of distillates and residue left. Paracelsus's followers elaborated on this idea and developed a quantitative evaluation by means of the balance and accurate measurement of volume. This proto-scientific procedure, described in a 1577 publication falsely attributed to Paracelsus, but probably by Leonhardt Thurneisser (1530–1595), remained bound up with the "anatomizing" of the urine, i.e., the idea that in the urine the whole anatomy of man is somehow represented, and that the urine carries the characteristics of all parts of the body, healthy or diseased [12].

In the chemical analysis by "distillation," the still is given the proportions of a human figure, in the belief that the urine mirrors human anatomy. It seemed possible therefore, to locate the seat of the disease process by carefully observing the sequence in which the vapors and spattering appeared during boiling and in which parts of the "anatomical furnace" (Fig. 1.2) they were condensed. In this way the body was "chemically dissected." Though no more rational than earlier systems, directions specified that the urine should be kept in a vessel of glass or stone, not of any other material (metals) which might alter its chemical composition, and should be accurately weighed and its volume measured. The assessment of the urine was based on specific gravity, the weight of gold being used as the standard for comparison [13].

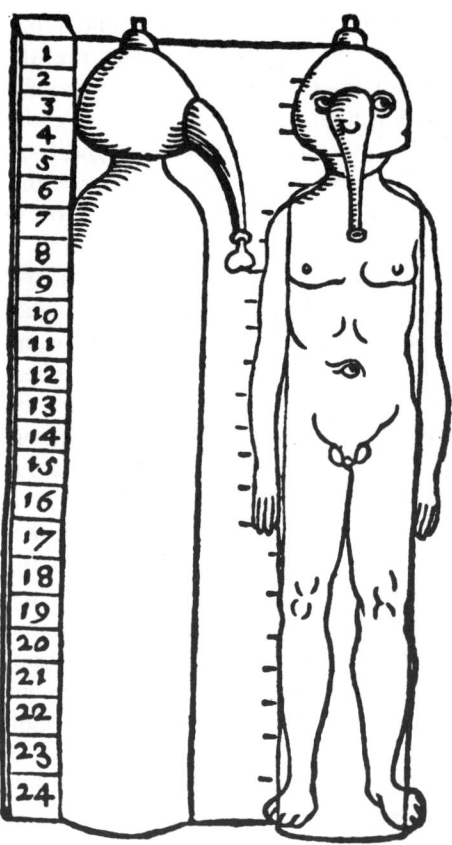

FIG. 1.2. The "Anatomical Furnace" for the distillation of urine and diagnosis of the locus morbi. Originally in *Aurora Thesaurusque Philosophorum Paracelsi* with "Anatomia Viva Paracelsi" (Basel: 1577). Reproduced from Walter Pagel, *Paracelsus*, 2nd ed., 1982, p. 193. (Courtesy of Karger, Basel.)

Another diagnostic aid described in late fifteenth and early sixteenth-century texts on urine were the drawings of many urine containers arranged in a circular design in various colors ranging from white to yellow to red to purple to green to black (Fig. 1.3). Each color was subdivided into various shades as reference points for different clinical conditions. When matched with sediment, suspended matter or turbidity, and other appearances, the resulting combinations provided data for a wide assortment of potential diagnoses. All the descriptions and explanations in books

FIG. 1.3. Disc exhibiting the colors of urine. Originally in *Epiphanie Medicorum* by Ulric Binder (Nurnberg: 1506). Reproduced from Walter Pagel, *Paracelsus*, 2nd ed., 1982, p. 190. (Courtesy of Karger, Basel.)

of this period were taken from preceding works. Used by serious physicians, uroscopy was for centuries a logical procedure based on humoral physiology and pathology and the doctrine of balance of the four humors: blood, phlegm, yellow bile, and black bile.

But there were the unscrupulous practitioners, (barber) surgeons, charlatans, apothecaries, and traveling troubadours, who preyed on the ignorance and gullibility of the masses. Mysticism, quackery, and astrology entered into these procedures, and uroscopy became the "hocus-pocus" of uromancy. These "uromancers" diagnosed pregnancy, sex, all kinds of disease, and predicted future events—all from examination of the urine.

Their influence tended to discourage and discredit the honest practitioners. Urine specimens often made the rounds of these uromancers until a diagnosis acceptable to the patient was obtained [14].

One of the earliest to recognize this practice as unsound was Thomas Linacre (1460–1524), physician to Henry VIII. As a founder of the College of Physicians of London, he tried to prohibit the doctors from offering diagnosis and treatment on the basis of urine examination alone. During the late sixteenth and seventeenth century, abuses of uroscopy drew the ridicule and scorn of several writers. In "The Pisse-Prophet, or, Certaine Pisse-Pot Lectures," published in London in 1637, Thomas Brian vigorously attacked the wide-spread abuses of uroscopy (Fig. 1.4). He stated that it was fallacious to judge disease and prescribe for it by the sight of urine alone. Brian described how the physician or "water-caster," while pretending to discover signs and symptoms in the urine, actually drew his conclusions from his general knowledge of disease and information unwittingly provided by the messenger who brought the urine. Unfortunately, so much confidence was placed in diagnosis by observation of the urine, that it was not thought necessary to see the patient. Brian advised the sick to choose a doctor who is accredited by the universities or the College of Physicians of London. The literature and art of the time reflected these practices and their abuses. Even Shakespeare made several allusions to uroscopists in his comedies, and scenes depicting uroscopy appear often in the art of Flemish painters.

Practices of uroscopy carried beyond logic or reason were not limited to the late Middle Ages and Renaissance. It appears they were in use even earlier according to Isaac Judaeus (Isaac Israeli) (880?–932?), a physician of early tenth century Tunisia, who wrote "A Guide to Physicians." The Arabic original is lost, but a Hebrew version has survived and has been translated. In this compilation of aphorisms, the author castigates charlatans and offers sensible advice to physicians on their behavior in the presence of the patient. He also reminds the reader that "The physician does not make the cure. He merely prepares and clears the way for Nature, who is the real healer." He cautions that the urine provides information only in matters relating to the liver and urinary passages, and only if judged in all its conditions. "But in our time there are fools who would base prophesies on it, without seeing the patient, and determine what disease is present, and whether the patient will die, and other foolishness" [15].

The English physician and leading iatrochemist, Thomas Willis (1621–1675), advocated chemical diagnosis and tried to put uroscopy—or uromancy, as he called it—on a more scientific basis. In a treatise

THE
PISSE-PROPHET,
OR,
CERTAINE PISSE-POT LECTURES.

Wherein are newly difcovered the old fallacies, deceit, and jugling of the Piffe-pot *Science, ufed by all thofe (whether Quacks and* Empiricks, or other methodicall Phyficians) who pretend knowledge of Difeafes, by the Urine, in giving judgement of the fame.

By T H O. B R I A N, M. P. lately in the Citie of *London*, and now in *Colchefter* in E S S E X.

Never heretofore publifhed by any man in the *Englifh* Tongue.

Si populus vult decipi, decipiatur.

LONDON,
Printed by *E. P.* for *R. Thrale*, and are to be fold at his fhop at the figne of the Croffe-Keyes, at *Pauls* gate.
1 6 3 7.

FIG. 1.4. Title page of Thomas Brian's book "The Pisse Prophet" (1637). (National Library of Medicine, Bethesda, Maryland.)

"Of Urines," he criticized those who drew conclusions from the color and consistency of the urine. Although he considered it reasonable to form an opinion of the nature of an illness from an examination of the urine, he did not believe that a complete diagnosis could be made by uroscopy alone. Willis wrote that "the Medicasters and Quacks for the most part behold the Urine sent in a Glass, shake it a little, and presently give Judgment" [16]. By only viewing the urine "there is scarce credit to be given to the single testimony of the Urine, unless there be other signs agreeable;..."[17] (see also Chapter 8).

Besides mere inspection, there were other ways to uncover what lies hidden in the urines and out of sight, namely, evaporation, distillation, and precipitation. Distillation of urine "plainly shows, that the Elements of which its liquor is composed, are a great deal of Water and Salt, and a little of Sulphur and Earth, and a very little of Spirit." It was the variations in these components that produced the changes in "Quantity, Colour, Consistency, and Contents" found in pathological conditions [18].

His treatise has the merit of recognizing the wide range of appearances of normal urine and that even the very sick patient might have normal appearing urine. Willis's treatise is thoroughly scientific in its outlook, in sharp contrast to the bizarre pronouncements made by medieval writers. Although the science was crude, it was a beginning for chemical analysis of urine.

Uroscopy gradually became outmoded and discredited, but managed to hang on until the early nineteenth century. It came to be identified with the ignorant and poorly trained doctors and irregular practitioners with the hit-and-run tactics of "traveling medicine men."

VAN HELMONT AND DISCOVERY OF GAS

Johannes Baptista Van Helmont (1577–1644) (Fig. 1.5) of Brussels, chemist, physician, philosopher, and mystic, represents the transition from alchemy to chemistry in a century of renaissance for chemistry [19]. Van Helmont was a complex personality who embodies curious contradictions. Although attached to the supernatural and fanciful alchemical doctrines of his time, he nevertheless rejected the classical Aristotelian theory of the four elements (air, earth, fire, and water), as well as the three principles (elements) of Paracelsus: mercury (spirit), sulfur (soul), and salt (body). He regarded air and water as the true elements.

According to Van Helmont, vegetable, animal, and mineral were all derived from water, and he directed attention to the large number of

FIG. 1.5. Johann Baptista van Helmont. From the frontispiece to the 1683 German translation of the *Ortus Medicinae* (1648). (Library of the New York Academy of Medicine.)

substances (organic and inorganic) that yield water when heated strongly. As an example, he neutralized acids with chalk and then distilled off water. He considered water as the primary substance of all material things, a conclusion he drew from his tree experiment. Van Helmont planted a 5-lb willow tree in 200 lb of dried earth and regularly added only water. After five years, he found that the weight of the tree had increased to 169 lb, whereas the earth in which it was planted, reweighed after drying, had lost no weight. This was an exceptionally good scientific experiment for the times.

Van Helmont argued that the wood, bark, and roots had been formed from water alone. His conclusion about the tree is mainly correct, since the tree is about 50% water. The irony is that he did not know the part played by the carbon dioxide in the air, although he was the first to recognize the existence of this gas.

Van Helmont vigorously opposed the traditional medicine of the Scholastics with their contemplation of Nature and elaborate demonstrations of deductive logic as a means of obtaining scientific truth. The properties and reactions of matter should be learned, said Van Helmont, not by discourse, but by hands-on demonstrations and experiment. His utilization of chemical medicines and application of chemical theory added to the prestige of chemistry so that medical men soon accepted it as a necessary part of their education. Medical chemistry smoothed the way for the reasonable practice of medicine; it also cleared the way for a less ambiguous chemistry.

Van Helmont criticized the conventional medicine of his time and described Galenical herbal medicine as full of deceit. He rejected traditional therapy directed against humoral imbalance as a whole, notably bloodletting and purging, practices that persisted into the nineteenth century. He also distinguished between the sick patient and the illness as an entity, and treated the cause of the disease, not the symptoms. Disputations about disease, he thought, did not produce cures, but chemistry offered better possibilities. An important feature of Van Helmont's scientific method and experimental chemical work was its quantitative character. He made extensive use of the balance and other measuring devices. Van Helmont ridiculed the "chemical anatomy" of urine, but utilized the gravimetric idea in the analysis and recommended determination of specific gravity. He weighed 24-hour specimens, but there was no practical outcome from this study. He conducted many chemical experiments on urinary calculi, but was not successful with their analysis. As a leader in the new movement to explain vital phenomena in chemical terms, Van Helmont was one of the earliest biochemists.

Van Helmont understood the law of conservation of matter when he stated that metals dissolved in acid are not thereby destroyed or transmuted, but are recoverable in their original quantity and continue to exist throughout a series of chemical changes. He was the first to understand that chemical reactions occasionally liberated air-like substances, and that these gases are distinct from atmospheric air and condensible vapor seen over water in cold air, and that they had different properties. Van Helmont is remembered today as the discoverer of "gas." He claimed more than once that he was the "inventor" of gas, and may have derived the term from the Greek word *chaos*, indicating without fixed order, volume or shape—the original material out of which the Universe was created (according to Greek mythology). Its phonetic sound in Flemish is *gas*.

He mentions gas formed by effervescence when vinegar is added to limestone (calcium carbonate); gas in mineral waters; gas forming from fermenting wine; and a combustible gas formed on dry distillation of organic matter (mixture of hydrogen, methane, and carbon monoxide). He distinguished the poisonous gas that collects in mines and extinguishes the flame of a candle; also the *gas carbonum* formed by burning charcoal. Inasmuch as he could not capture his gas, his only test was to determine if the gas would support combustion or was itself flammable.

Van Helmont's great contribution to chemistry was the discovery of carbonic acid gas (carbon dioxide) which he often called *gas sylvestre*, meaning "of the woods," or "wild." He sometimes used this expression as a general term for gas since he could not devise a method to collect or isolate it in a vessel or reduce it to a visible body. The discovery was entirely overlooked by succeeding chemists until Joseph Black (see Chapter 2) rediscovered it in 1756 and named it "fixed air" because it could be combined (fixed) in such a way as to form part of a solid substance, e.g., calcium carbonate. Until Van Helmont's time, all gaseous substances were regarded as being mere varieties of air.

Van Helmont's main biochemical discovery was the significant chemical role of acid in gastric digestion, which up to this time, and for long after, was attributed to heat and trituration. He understood that heat is a general factor whose action is the same wherever applied, but reasoned that gastric digestion is not due to acidity as such, because neither vinegar nor lemon juice will digest food. He attributed digestion to a specific "vital" acidity which, to emphasize the special nature of this transforming agent, he called "ferment." Van Helmont gave his chemical concept of digestion a vitalistic identity by stating repeatedly that the acid in the stomach is not the "vital ferment" itself, but made its action possible.

SYLVIUS AND IATROCHEMISTRY

The iatrochemical school which succeeded Van Helmont explained the greater part of physiology and pathology in terms of a balance in production of acids and alkalies, and administered therapy based on these principles. Although there was no proof, the concept of a balance in the body was one that physicians and laymen could understand. However, acid and alkali conveyed a very different meaning from that of today. They were considered to be two fundamental and opposing principles of matter, a duality to supplement or replace the older *tria prima* of Paracelsus and the ancient Aristotelian quartet of elements. They were the essence of all things, antagonistic by nature and inevitably reacting when brought together [20].

The iatrochemical school explained all vital phenomena in essentially chemical terms. Though their efforts with the scientific method were filled with errors in calculation and conclusion, they represented the first attempts at quantitative and objective observation in medical research.

The leading exponent of the iatrochemical school in the seventeenth century was a Flemish physician, Franciscus Sylvius de le Boë (1614–1672) (Fig. 1.6). He went beyond Van Helmont and explained all body functions, healthy or diseased, as being determined by chemical reactions without involvement or direction by any mystical or spiritual force. He also maintained that the physiological and pathological activities in the body must be analogous to those that could be acomplished experimentally in the laboratory, and can be explained in terms of fermentation, effervescence, and putrefaction. Sylvius compared respiration with combustion and regarded gastric digestion as a chemical fermentation, with saliva playing an important initial role. He was of the opinion that in the second stage of digestion, pancreatic juice (believed to be acid) effervesced with the alkaline bile in the duodenum. He attached much importance to acid and alkali which he considered the fundamental principles in the animal body, and was especially interested in their equilibrium and interaction as an indicator of health. Sylvius distinguished acids and alkalis by taste (acids were sour; alkalis, bitter) or by effervescence on mixing with anything commonly regarded as belonging to the opposite category. In his therapeutics he preferred the new chemical medicines over the Galenicals which frequently contained dozens of ingredients.

In 1669 Sylvius persuaded the Curators of the University of Leyden to build a "Laboratorium"—what seems to have been the first university chemical laboratory for teaching medical students [21]. He also established

FIG. 1.6. Franciscus Sylvius de le Boë. (National Library of Medicine, Bethesda, Maryland.)

bedside instruction as a regular part of the medical curriculum at Leyden.

The list of lectures offered at the young University of Leyden, which rose to leadership in medicine under the chemically oriented Sylvius, illustrates the turning away from Scholasticism in general and the teaching of Galen in particular. In 1601 lectures were still offered on individual texts, including one by Galen. In 1654 professors lectured on Celsus, "the Latin Hippocrates," and on various specific medical subjects. Two professors gave instruction to the medical students at the public hospital in bedside medicine and the treatment of diseases, and demonstrated the causes of death on the dissected cadavers. By the winter of 1681, all lectures were on medical subjects—none on individual authors; clinical instruction was given every weekday; and there were postmortem dissections. In short, the scholastic system had been replaced by medical instruction of a modern type [22].

IATROPHYSICISTS (IATROMECHANISTS)

The iatrophysical school regarded all physiological events as rigid consequences of the laws of physics. An exponent of this school was Giovanni Alfonso Borelli (1608–1679), a Neapolitan mathematician. According to him, the human organism should be regarded as a machine and the circulation, digestion, respiration, and body locomotion, as mechanical processes subject to physical laws. The extremes to which the iatrophysical school in Italy later proceeded is represented by Giorgio Baglivi (1668–1706). He divided the body machine into many smaller machines, e.g., heart and blood vessels (water works), thorax (bellows), teeth (scissors), etc. Chemists and mechanists both argued essentially by analogy from inanimate model systems to events in living bodies. Whereas the mechanists could see matter and motion in animals, the chemists had to imagine their effervescence when they accepted the acid–alkali hypothesis.

With progressing mechanization of theoretical medicine at the hands of the iatromechanical school in the early eighteenth century, Van Helmont's vitalistic doctrine of acid digestion in the stomach was set aside by a theory of liquefaction resulting from agitation. The reasoning behind this was that a chemical substance in the stomach capable of converting solid food, notably meat, into a fluid, would also dissolve the fleshy wall of the stomach [23].

The iatromechanical theory of gastric digestion dominated until the latter half of the eighteenth century. At this time a new line of investigation was opened up by the French physicist, René de Réaumur (1683–1757), when he persuaded a kite (a species of buzzard that easily disgorges what it does not digest) to swallow open-ended tubes containing food. His experiment in 1752 demonstrated the solvent action of gastric fluid on foods.

These findings were confirmed and greatly extended by the Italian physiologist, Lazzaro Spallanzani (1729–1799), who administered food samples in perforated metallic tubes to a variety of animals. The containers were removed by regurgitation, by recovery in the feces, or by sacrificing the animal. Spallanzani experimented on himself with foodstuffs swallowed in linen bags which he recovered for analysis. He also obtained samples of his own gastric fluid by inducing himself to regurgitate on an empty stomach. In 1782 he studied the solvent action of gastric fluid on foodstuffs outside the body and concluded that the basic factor in digestion is the solvent property of the "gastric juice"—a term he introduced. He proved that trituration was not involved in digestion other than to make food particles more accessible to this juice. By showing that the solvent action of gastric fluid can act outside the body, Spallanzani disposed of previously held mechanisms of digestion attributed to concoction (mixing aided by heat), putrefaction, trituration, and fermentation, in favor of a chemical theory of solution. However, he failed to recognize that the solvent action of the grastric juice is due to its acidity [24].

A new advance came from an unexpected source. John Richardson Young (1782–1804), in his M.D. thesis for the University of Pennsylvania, "An Experimental Inquiry Into the Principles of Nutrition and the Digestive Process" (1803), showed that the solvent principle of the gastric juice is an acid and is part of the normal gastric secretion. Young's results were far ahead of anything previously done for physiology in the United States. Experimenting with starving frogs and using litmus paper, Young demonstrated the acidity of gastric fluid but wrongly concluded that it was phosphoric acid [25]. In 1824, analysis by William Prout showed that the acid of the gastric juice was hydrochloric acid (see Chapter 3).

NOTES AND REFERENCES

1. FRANÇOIS Jacob, *The Logic of Life. A History of Heredity* (New York: Random House (Pantheon Books), 1973), pp. 28–29.
2. GIOVANNI BATTISTA MORGAGNI, *De Sedibus et Causis Morborum per Anatomen Indagatis, libri quinque*, (On the Seats and Causes of Disease Investigated by Anatomy, five volumes), Venice, 1761.

3. KENNETH DEWHURST, "Locke and Sydenham on the Teaching of Anatomy," *Medical History*, 1958, 2: 1–12, pp. 4, 5.
4. DAVID E. WOLFE, "Sydenham and Locke on the Limits of Anatomy," *Bulletin of the History of Medicine*, 1961, 35: 193–220, pp. 194–196, 210.
5. RICHARD HARRISON SHRYOCK, *The Development of Modern Medicine. An Interpretation of the Social and Scientific Factors Involved* (Philadelphia: University of Pennsylvania Press, 1936), p. 12; STANLEY JOEL REISER, *Medicine and the Reign of Technology* (New York: Cambridge University Press, 1978), p. 8.
6. *Comprehensive Biochemistry*, Marcel Florkin and Elmer H. Stotz, eds., vol. 30, *A History of Biochemistry* (Marcel Florkin), (Amsterdam: Elsevier Publishing Company, 1972), p. 61.
7. *Ibidem*, p. 63.
8. F. J. MOORE, *A History of Chemistry* (New York: McGraw-Hill Book Co., Inc., 1918), pp. 14–15.
9. H. P. BAYON, "Ancient Pregnancy Tests in the Light of Contemporary Knowledge," *Proceedings of the Royal Society of Medicine*, 1939, 32: 1527–1538; THOMAS R. FORBES, "Early Pregnancy and Fertility Tests," *Yale Journal of Biology and Medicine*, 1957–58, 30: 16–29; P. GHALIOUNGUI, SH. KHALIL, and A. R. AMMAR, "On An Ancient Egyptian Method of Diagnosing Pregnancy and Determining Foetal Sex," *Medical History*, 1963, 7: 241–246.
10. FRED ROSNER and SUSSMAN MUNTNER, "Moses Maimonides' Aphorisms Regarding Analysis of Urine," *Annals of Internal Medicine*, 1969, 71: 217–220.
11. LEONARD, J. T. MURPHY, "The Art of Uroscopy," *Medical Journal of Australia*, 1967, 2: 879–886.
12. WALTER PAGEL, *Paracelsus*, 2nd ed., rev. (Basel: Karger, 1982), p. 192; see also REISER (1978), pp. 122–123.
13. PAGEL (1982), pp 189–198.
14. JOSEPH H. KIEFER, "Uroscopy: The Clinical Laboratory of the Past," *Transactions of the American Association of Genito-Urinary Surgeons*, 1958, 50: 161–172; "Uroscopy: The Artist's Portrayal of the Physician," *Bulletin of the New York Academy of Medicine*, 1964, 40 [2]: 759–766; RONNIE BETH BUSH, "Urine Is an Harlot, or a Lier," *Journal of the American Medical Association*, 1969, 208: 131–134.
15. SAUL JARCHO, "Guide for Physicians (Musar Harofim) by Isaac Judaeus (880?–932?)," *Bulletin of the History of Medicine*, 1944, 15: 180–188. For an independent translation and biographical sketch, see ARIEL BAR-SELA and HEBBEL E. HOFF, "Isaac Israeli's Fifty Admonitions to the Physicians," *Journal of the History of Medicine and Allied Sciences*, 1962, 17: 245–257. Also see WILLIAM I. WHITE, "A New Look at the Role of Urinalysis in the History of Diagnostic Medicine," *Clinical Chemistry*, 1991, 37: 119–125; P. GARCIA-WEBB, "Urinalysis in the 10th Century," *Clinical Chemistry*, 1991, 37: 1660.
16. THOMAS WILLIS, "Of Urines," Treatise III, in *Practice of Physick*, translated from Latin by S. Pordage (London: T. Dring, C. Harper & J. Leigh, 1684), p. 17.
17. *Ibidem*, p. 15.
18. *Ibidem*, pp. 1, 2.
19. The sections on Van Helmont are derived largely from the following: FLORKIN (1972), pp. 73–77; J. R. PARTINGTON, *A History of Chemistry*, (4 vols.), vol. 2 (London: Macmillan & Co. Ltd, 1961), pp. 209–241; "Joan Baptista Van Helmont," *Annals of Science*, 1936, 1: 359–384; *Dictionary of Scientific Biography* (*DSB*), Charles Coulston Gillispie, ed. (New York: Charles Scribner's Sons, 1972), 6: 253–259; WALTER PAGEL, "Van Helmont's

Ideas on Gastric Digestion and the Gastric Acid," *Bulletin of the History of Medicine*, 1956, 30: 524–536. For a review see LOUIS ROSENFELD, "The Last Alchemist—The First Biochemist: J. B. van Helmont (1577–1644)," *Clinical Chemistry*, 1985, 31: 1755–1760.
20. WALTER PAGEL, "J. B. Van Helmont's Reformation of the Galenic Doctrine of Digestion—and Paracelsus," *Bulletin of the History of Medicine*, 1955, 29: 563–568, pp. 566–567; ROBERT P. MULTHAUF, "J. B. Van Helmont's Reformation of the Galenic Doctrine of Digestion," *Bulletin of the History of Medicine*, 1955, 29: 154–163, pp. 161, 163.
21. J. R. PARTINGTON, *A Short History of Chemistry*, 2nd ed. (London: Macmillan and Co., Limited, 1951), p. 54; also see E. ASHWORTH UNDERWOOD, "Franciscus Sylvius and his Iatrochemical School," *Endeavour*, 1972, 31: 73–76.
22. OWSEI TEMKIN, *Galenism. Rise and Decline of a Medical Philosophy* (Ithaca: Cornell University Press, 1973), pp. 173–174.
23. PAGEL (1955), p. 567.
24. *DSB* (1975), 12: 553–567, p. 558.
25. *Dictionary of American Medical Biography*, Howard A. Kelly and Walter L. Burrage, eds. (New York and London: D. Appleton and Company, 1928), pp. 1352–1353; *DSB* (1976), 14: 558–559.

CHAPTER II

INTRODUCTION

By the start of the seventeenth century, chemical methods, though essentially qualitative in character, were being described in great detail. However, chemistry still had to take a back seat to medicine, mining, and metallurgy. Until this time, most men with a serious interest in chemistry had been physicians, but now with increasing frequency, pharmacists with their practical knowledge as compounders of the physician's prescriptions, were taking on a progressively important role in the development of chemistry. During the next two centuries, a great many fundamental chemical discoveries were made by men with pharmaceutical training. This was especially true in continental Europe, whereas in England, chemical advances were made by the scientific amateur.

ROBERT BOYLE

Robert Boyle (1627–1691) (Fig. 2.1) was the prototype of the self-taught amateur scientist-investigator that thrived in England during the next century and a half. Major advances in England resulted from the work of men who pursued science as an avocation. Boyle was independently wealthy and could equip his own laboratory and hire a team of "lab assistants." These consisted of secretaries, talented technicians, mechanics, glassblowers, and apothecaries, who performed the less exacting and supporting tasks for Boyle's experiments. Other amateur scientists held positions that gave them ample time for investigation and theorizing. These amateurs tended to contribute to theoretical science while the continental pharmacists were discovering new substances and reactions [1].

There were three types of chemical workers in that period: alchemists, iatrochemists, and chemical technologists. Alchemists engaged in a futile pursuit of the "philosopher's stone" which was supposed to bring about the transmutation of metals. Iatrochemists made chemical medicines for physicians and apothecary shops. The chemical technologists of the time were the smelters, assayers, glass-workers, dyers, soap-boilers, etc. Chemical theory meant nothing to them. Knowledge and information was passed

FIG. 2.1. Robert Boyle. (National Library of Medicine, Bethesda, Maryland.)

on from master to apprentice. Books stressed the practical side of the chemical process and the reactions were explained in terms of the classical Aristotelian four elements and Paracelsian three principles theory.

Privately tutored, Boyle [2] avoided the restricting Aristotelian education of a university. He spent his life in the experimental study of various branches of natural science and in meditating and writing on theological subjects. For Boyle, carefully designed and executed simple experiments, not logic, were the essential ingredients for the acquisition of knowledge. This became the guideline for the development of science and medicine for the next three hundred years.

Boyle dismissed the idea that matter could be resolved into its basic units by fire, that is, by destructive distillation. His experiments indicated that many substances may be characterized and identified by chemical reactions. He did not accept as elements, the earth, air, fire, and water, of the Aristotelians, or the three principles, salt, sulfur, and mercury of the Paracelsians. In this, Boyle was undoubtedly influenced by the works of Van Helmont to whom he frequently referred as an authority.

COLOR TEST AND CHEMICAL REACTION

Boyle was the first to develop a systematic procedure for identifying substances by means of physical characteristics such as solubility, specific gravity, crystal form, and color in a flame, in addition to chemical characteristics. Boyle and his assistants mixed dry, wet, and in solution, every chemical available, and heated, baked, and otherwise manipulated them to see what would happen.

Color changes occurring during chemical reactions interested Boyle, and there are references to vegetable extracts throughout *Experimenta et Considerationes de Coloribus* ("The Experimental History of Colours") (1665). Medieval dyers and painters had found that plants were a source of a wide range of colors depending on the season of the year, but also on how they were brought into solution. It was long known that some acids turned blue syrup of violets to red. Boyle claimed to be the first to note that alkalies turned syrup of violets green. Boyle observed that the blue opalescence of the yellow solution of *lignum nephriticum* (a South American wood) was destroyed when the solution was acidified and could be restored by addition of alkali. Other vegetable dyes similarly exhibited color changes. Substances that caused no color change were neither acid nor alkaline. Boyle reasoned that if different colors could be prepared from a single plant extract by adding acid or alkali, it would be possible to use such extracts as "indicators" of acidity or alkalinity in solutions of unknown substances. This led to the litmus test, using the natural dye from the lichen *Roccella*. Medieval painters soaked small strips of linen cloth in the juice of certain plants. When dried, the cloths were soaked in water to form solutions of water color. This technique probably led to the preparation of dye-impregnated litmus paper as an indicator [3].

Boyle identified copper by the blue color of its solutions and by the green color imparted to a flame by copper salts. He described the white precipitate formed by calcium salts with sulfuric acid and the precipitation of silver by its formation of silver chloride. Conversely, he demonstrated sodium chloride by reaction with silver nitrate in 1684, making this reaction perhaps one of the oldest tests still in regular laboratory use. Boyle collected a gas (hydrogen) obtained by reacting iron filings and hydrochloric acid and described the flame of the burning hydrogen. Some of his reactions were new, others had been known for years or even centuries, such as the black color produced with iron salts by tincture of gallnuts, known to Pliny the Elder (23–79 C.E.) (see Chapter 11).

These tests enabled Boyle to discuss the composition of substances in positive terms of empirically determined components. Once chemists saw the advantage of distinguishing between empirically verifiable components, they turned their attention from the "why" of cause and effect, to the "how" of what actually happened in a chemical reaction.

The concept of the chemical reaction and the word *analysis*—in the sense that we know it today—are attributed to Boyle. His extensive and systematic chemical experimentation helped establish experimental chemistry as an independent science worthy of study in an age when chemistry was not accepted as a respectable "natural philosophy," but was regarded as either a mystic or pseudo science related to heretical alchemy or a practical art useful to metallurgy or to medicine.

ANALYSIS OF BLOOD

Since antiquity there has existed a vast history of information, speculations, and groping experiments on the composition of living things. Although phlebotomy had been freely practiced as a therapeutic measure since the dawn of medicine, the inspection or analysis of the shed blood was never developed into a system of diagnosis as it was in the case of urine.

During the Middle Ages, blood, distillations of blood, along with other natural products from biological sources, were used to make medicines. In spite of the knowledge gained about blood from these pharmaceutical preparations, not until the middle of the seventeenth century did it occur to anyone to examine blood with physical or chemical methods to learn more about its properties and constituents. This was due in part to the vitalistic theories of the day which represented blood as a living entity. It could not be analyzed outside the body where it had no warmth or spirit.

The discovery of the circulation of the blood by William Harvey (1578–1657) in 1628 drew the attention of many scientists to the blood. Now the appearance of blood as it flowed from the vein and in its coagulated state was regarded as a valuable sign. Especially useful was the rapid sedimentation of the red cells before clotting occurred. The result was a creamy mass of fibrin and white blood cells above the red clot. In this case, the blood was said to have a "buffy layer" or "inflammatory crust" and indicated an inflammation somewhere in the body.

Sometime during the latter half of the seventeenth century, it occurred to John Locke that chemical analysis of blood might have greater diagnostic value than mere inspection. He asked his friend Robert Boyle to undertake the preliminary work of blood analysis. The resulting study of the blood was probably the earliest attempt to apply analytical chemistry to medicine. Boyle published his results in *Memoirs For The Natural History of Humane Blood, Especially The Spirit of that Liquor* in 1684, and addressed the preface "To the very Ingenious and Learned Doctor J. L."

In the preface Boyle observes that "what is generally known of Humane Blood, is as yet imperfect enough, and consists much more of Observations than Experiments;..." He suggests that for anatomists to concentrate, as too many have done, on "the Solid parts of the Body, and overlook Enquiries into the Fluids, and Especially the Blood, were little less improper in a Physician, than it would be in a Vintner to be very solicitous about the Structure of his Cask, and neglect the consideration of the Wine contain'd in it." [4].

Boyle followed Locke's outline of investigation in his listing of the headings under which whole blood should be examined: color, specific gravity, clotting time, agents that cause or delay clotting, and those that dissolve or preserve the clot, odors, heat when freshly emitted, inflammability, and taste, to name but a few; also the differences and respective quantities of "the two obvious Parts of Humane Blood, the Red (and Fibrous) and the Serous" [5]. For quantitative chemical analysis, he subjected the dried blood to fractional distillation, weighing at intervals and examining the residues. Boyle noted the reactions to alcohol, mineral acids, and other chemical reagents. In this manner, apart from its water content, blood was shown to contain volatile substances, salt, oil, phlegm, earthy products, residue when the dry distillation was complete, and fixed salt which was the final residue following prolonged calcination (ashing). Boyle considered medicinal uses of human blood and differences between it and the blood of other animals. He presented a similar list of thirty-one topics for the examination of urine, but did not describe any studies. In addition to color, taste, odor, and analysis by distillation, he included viscosity and differences resulting from medications, diet, or season.

Being a chemist and not a physician, Boyle worked with normal blood (clot and serum) and left it to others with medical training and access to people who were sick to analyze blood in various pathologic states. He stressed that "Knowledge of the Nature of the Blood, when 'tis rightly conditioned, is necessary to those that would discern, in what particulars, and how far it deviates in the Sick,... For having compared the Qualities and Accidents of this vitiated Blood, with those of the Blood of Sound Men... 'twill not be difficult for a Physician to find, to what heads he is to refer those things that considerably recede from such as belong to Healthy Blood. And... in all other Points the Blood of Persons sick of that Disease is not unlike that of those that are Healthy" [6].

Fifty years later, Browne Langrish (d. 1759), a country physician, advocated chemical analysis of the blood to determine the nature and causes of disease. He weighed the serum and clot, noted their relative proportion, and measured the cohesion of the clot.

> "By proper Distillations, and the Force of Fire, we may compel Nature to an Account; and though the Bulk and Figuration of the saline and sulphureous Parts are undoubtedly much altered and commuted by the Action of Fire; yet the Proportions of the several Principles of the Blood are not increased or diminished thereby; and consequently by carefully separating and weighing them, and seeing the several Proportions they bear to each other, we may arrive at a Knowledge very useful in accounting for some of the *Phaenomena* of Diseases, and directing us to a right Method of Cure. It is satisfying and useful as well as curious, to reduce to Measure and Weight the constituent Parts of the Blood; and I am persuaded, no inquisitive Person will judge it a vain Undertaking" [7].

His "chemical analysis" (by dry distillation) of the blood yielded lymph, volatile salt, oil, residue before ashing, residue after ashing, and fixed salt, each of which he weighed.

Langrish urged the daily observation of the three sections of the urine—top, middle, and bottom—as well as chemical analysis during the illness to detect changes in the proportions of its principal components and to learn the state and progress of the disease.

> "If therefore a bare Inspection of Urines is of such advantage towards investigating the Nature, State, Progress, and Cure of Diseases; most certainly the *natural History* of it, or a more curious Search into the Contents of the Urine, in every Period of the Disease, will be of more moment in discovering the several *Dyscrasies* of the Blood, and in indicating the *Method of Cure*, than what we can

meet with in the *Urinal* only. For this Reason I thought it worth while to make the following Experiments, that by an exact *Analysis* we might see the different Contents of the Urine, and the various Proportions of its Principles, in the several Stages of this Disease" [8].

Langrish presents "the Reader with a *chemical Analysis* of the *solid Parts* of an *animal Body*, in order to investigate the Proportions and Qualities of their several constituent Parts; whereby we may probably receive some Light towards finding out which Principles contribute most towards the Cohesion of the Fibres, and which actuate and invigorate their elastic, contractile Property" [9]. Here too, analysis was by dry distillation.

In 1760 Richard Davies (d. 1762) wrote: "The Human Blood is by a Chemical Analysis resolved into Phlegm, Salt, Oil, and Earth; but since almost all bodies, both Animal and Vegetable, are by the action of Fire reducible to the same Principles, we can gain no particular information from this method of Enquiry" [10]. Davies recognized the limitations of blood analysis and advocated the simpler approach of describing the appearance of the blood and the proportion of serum, red cells, and fibrin in various diseases. There were no further advances in blood chemistry until the next century.

THE GASEOUS STATE AND PNEUMATIC CHEMISTRY

As late as the middle of the eighteenth century, chemistry remained disoriented, complicated, and puzzling. There were no criteria for purity or how to define and identify an element, and most important, there was no concept of the gaseous state of matter. Chemistry remained a two-dimensional science, with equipment and apparatus to handle only solids and liquids. What made the modern era of chemistry possible was the eventual understanding that the gaseous state was a new dimension in chemistry; and that the balance sheet of chemical analysis always had to account for this new state [11].

The new methodology that led to advance in chemical individuality was the use of chemical analysis to characterize a substance by its reactions with other chemicals. Instead of understanding the world as a whole, what evolved was an understanding that we can only investigate its parts. Quantitative methods came to be accepted as essential to chemical investigation, and the whole new field of the chemistry of the gases was opened up.

The revolution in chemistry at the end of the eighteenth century resulted from the synthesis of two currents of scientific research: the discovery in Britain of a whole new class of substances, the gaseous elements and compounds, and the quantitative analytical techniques developed in France. The isolation and identification of gases as chemical individuals were the great laboratory advances of the eighteenth century. It made possible for gases to join solids and liquids on the chemical balance sheet.

After this, new developments came rapidly. The discoveries in pneumatic chemistry made it possible to explain and generalize the nature of respiration and the cause of animal heat, problems which had long baffled physicians and physiologists. The modern theory of respiration and combustion was based upon an understanding of the chemical role of gases and the existence of gaseous species chemically distinct from air, and that they participate in chemical reactions according to the same laws of chemistry as elements in the solid state.

Until the discovery of the gaseous elements and compounds, eighteenth-century investigators were primarily interested in solids and liquids because gases were so difficult to capture, confine, and study. The characterization of the different gases was not possible until 1727 when the English chemist and clergyman, Stephan Hales (1677–1761), developed a practical form of the "pneumatic trough" (Fig. 2.2) which allowed the isolation and collection of gases over water. Previously, generation and collection of the gas had taken place in the same container.

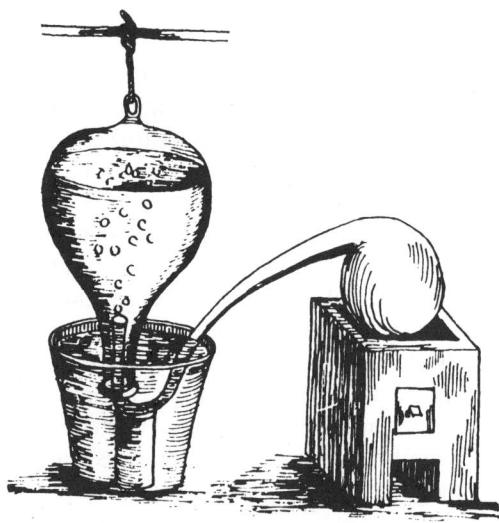

FIG. 2.2. Stephen Hales's gasometer. From his book *Vegetable Staticks* (1727).

The quantitative phase of chemistry got its start in 1754 in the M.D. dissertation of Joseph Black (1728–1799) (Fig. 2.3). He had transferred to Edinburgh from Glasgow to investigate the dissolution of urinary calculi for his doctoral dissertation. Instead of exploring the presumed effectiveness of lime water (calcium hydroxide solution), Black decided to examine other absorbent earths in the hope of finding a more powerful dissolving agent. He chose a white powder, *magnesia alba* (magnesium carbonate), recently in vogue as a mild purgative. When heated or treated with acids, *magnesia alba* emitted a gas (carbon dioxide) and lost weight and became *magnesia usta* (magnesium oxide). Since the gas had previously been

FIG. 2.3. Joseph Black. (National Library of Medicine, Bethesda, Maryland.)

combined as part of a solid substance, Black called it "fixed air." This gas was identical with Van Helmont's *gas sylvestre*. Black's findings demonstrated a connection between the inanimate, e.g., magnesium carbonate and the animate, e.g., wood. Both gave off carbon dioxide upon heating or burning.

Black used the balance—hardly more than a plain apothecary scale—more systematically than any chemist before him. He demonstrated the difference between alkali carbonates and hydroxides and the methods of their interconversion. In a classic paper (1756) he expanded on his dissertation and demonstrated that fixed air was a separate quantitative constituent of alkaline substances such as magnesium carbonate and calcium carbonate. Fundamentally different in its properties from ordinary atmospheric air, this gas did not support combustion or life (respiration). Black noted that fixed air turned lime water milky and caused precipitation of calcium carbonate. He used this test to prove that the air exhaled in respiration contained fixed air, and that it was given off in alcoholic fermentation.

The general view in the eighteenth century, in spite of the work of Van Helmont, was that all gases are composed of air and that this air could not react chemically. This idea was completely discredited by Black's demonstration that a chemical reaction between a solid and a gas must have taken place. What could be more unique than air existing in the form of a hard stone. The significance of Black's discovery was that carbon dioxide, though present in air, is quite different from air. This was the breakthrough that led to the discovery of the gaseous elements [12].

THE NEW CHEMISTRY OF COMBUSTION AND RESPIRATION

The isolated findings with gases by others were summarized, correlated, and constructed into a comprehensive theory by Antoine Laurent Lavoisier (1743–1794). Lavoisier established that oxygen is needed for burning. He also made the first clear distinction between elements and compounds. Lavoisier's work is based on the principle that matter was neither created nor destroyed, but was merely moved from one substance to another. Consequently, the quantitative relationship between the components of reacting and formed substances can be expressed by algebraic equations. The law of the conservation of mass—a philosophy that had been expressed in ancient times—became the keystone of nineteenth-century chemistry. However, Lavoisier was not the first to state the principle of conservation of matter, nor did he introduce quantitative measurement in chemical reactions. But he did emphasize that it was on "this principle that the whole art of making experiments is founded" [13]. Now, for the first time, Lavoisier could explain combustion heat and animal heat as the result of the same chemical process—the chemistry behind both fire and animal heat. Respiration was a *slow combustion* process producing heat from the union of carbon and hydrogen of food with oxygen, forming the carbon dioxide exhaled in breath, and *water*, respectively. The chemical process was the same, and the same amount of carbon yielded the same quantity of heat upon combustion in fire and in the animal body (1783). Oxygen consumption was proportional to heat production. The non-respirable part of air, mofette or azote, later called nitrogen, was exhaled unchanged. Lavoisier revealed the dominant role played by oxygen *in vivo* and *in vitro*.

NOMENCLATURE

In 1787, believing that vocabulary was the key to identification, order, and analysis, Lavoisier and several collaborators developed a new and logical system of chemical terminology that was published as *Méthode de Nomenclature chimique*. What emerged was a complete break with the past. Compounds were named to indicate their constituents.

The chemical revolution, dominated by the work of Lavoisier [14], was essentially completed with the publication of his *Traité Élémentaire de Chimie* (1789) which set down the foundation of modern chemistry. It discarded the heterogeneous terminology and the enigmatic and outmoded alchemical symbols and iatrochemical names of the Middle Ages and presented a unified picture of chemical knowledge based on the new analytical language of chemistry—in accordance with the accepted known evidence. It is understandable today. It was the first modern chemical textbook.

The book was extensively illustrated with engravings of chemical apparatus drawn by Lavoisier's wife, Marie-Anne, who took lessons from the noted artist Louis David in order to accomplish this. David, in appreciation, portrayed the Lavoisiers together (Fig. 2.4).

Chemistry was transformed into an independent and comprehensive science with its own techniques, language, and concepts. Mystery was replaced by careful observation and exact record keeping. It was possible now to begin to understand in a rational manner chemical changes which were seen to occur in nature as well as those produced artificially.

FIG. 2.4. Antoine Laurent Lavoisier and his wife (1788) by Jacques Louis David. (The Metropolitan Museum of Art, Purchase, Mr. and Mrs. Charles Wrightsman Gift, in honor of Everett Fahy, 1977. All rights reserved, The Metropolitan Museum of Art.)

NOTES AND REFERENCES

1. For an interesting article on the role and status of the assistant in Boyle's laboratory see Steven Shapin, "The Invisible Technician," *American Scientist*, 1989, 77: 554–563; see also KENNETH DEWHURST, "Locke's Contribution to Boyle's Researches on the Air and on Human Blood," *Notes and Records of the Royal Society of London*, 1962, 17: 198–206, p. 199.
2. *Dictionary of Scientific Biography (DSB)*, Charles Coulston Gillispie, ed. (New York: Charles Scribner's Sons, 1970), 2: 377–382.
3. WILLIAM H. BROCK, *The Norton History of Chemistry* (New York: W. W. Norton & Company, 1993), p. 178. Also see Ferenc Szabadváry, *History of Analytical Chemistry*, translated from Hungarian by Gyula Svehla (London and New York: Pergamon Press Ltd, 1966). Reprinted in 1992 (Langhorne, Pennsylvania: Gordon and Breach Science Publishers S.A.), pp. 257–258.
4. ROBERT BOYLE, *Memoirs For The Natural History of Humane Blood, Especially The Spirit of That Liquor* (London: Samuel Smith, 1684), preface, np (pp. 2–3).
5. *Ibidem*, pp. 11–14.
6. *Ibidem*, preface, np (pp. 11–13).
7. BROWNE LANGRISH, *The Modern Theory and Practice of Physic* (London: A. Bettesworth and C. Hitch, 1735), pp. 79–80.
8. *Ibidem*, pp. 90–91.
9. *Ibidem*, p. 51.
10. RICHARD DAVIES, *Essays to Promote The Experimental Analysis of the Human Blood* (Bath: J. Leake, 1760), pp. 1–2.
11. BROCK (1993), pp. 42, 84.
12. MARY ELVIRA WEEKS, "Daniel Rutherford and the Discovery of Nitrogen," *Journal of Chemical Education*, 1935, 11: 101–107; THOMAS S. KUHN, "Historical Structure of Scientific Discovery," *Science*, 1962, 136: 760–764, and footnote 7; E. A. Underwood, "Lavoisier and the History of Respiration," *Proceedings of the Royal Society of Medicine*, 1943–44, 37: 247–262; BROCK (1993), pp. 101–108, 112, 115–116; J. R. PARTINGTON, "The Discovery of Oxygen," *Journal of Chemical Education*, 1962, 39: 123–125; SZABADVÁRY (1992), pp. 60–70, 93–96; Mary Elvira Weeks, "The Discovery of the Elements. IV. Three Important Gases," *Journal of Chemical Education*, 1932, 9: 215–235.
13. FREDERIC L. HOLMES, "Antoine Lavoisier & The Conservation of Matter," *Chemical & Engineering News*, 1994 (September 12), 72: 38–45.
14. The literature on Lavoisier is vast and controversial. For many references, see *DSB* (1973), 8: 66–91. BROCK (1993), pp. 682–684, discusses the mythology of Lavoisier as *the* founder of modern chemistry. Lavoisier sought distinction and recognition for his achievements. For comments on his vanity and ambitious personality, see also SZABADVÁRY (1992), pp. 93–94. For an interesting analysis of Lavoisier's work and a critique of historical treatments by others, see HOLMES (1994).

CHAPTER III

INTRODUCTION

Bladder stone is one of the oldest clinical conditions described in medical history [1] and was the first to be treated by an elective surgical procedure. For centuries, patients submitted to the agony of entry into their bladder in order to escape the tortures of the stone. Of the three "elective" surgical procedures first performed by mankind—circumcision, trephination of the skull, and cutting for bladder stone—only the latter had no religious or ritual association, and so may be called the oldest operation performed for the relief of a specific surgical condition. The oldest bladder stone known was found in the grave of a boy about 16 years old in the prehistoric cemetery at El Amrah in Upper Egypt and was dated at about 4800 B.C.E. About 6.5 cm in diameter, it was made up of calcium phosphate and uric acid. The specimen, housed in the Hunterian Museum of the Royal College of Surgeons of England, was destroyed during the bombing in 1941. The Hippocratic physicians of the fifth and fourth centuries B.C.E. referred bladder stone patients to specialists in this work.

Despite the frequency of bladder stone throughout medical history, especially in children, even infants, it is rare today in England and northern Europe, although occasionally seen in southern parts, such as Sicily and Greece. It remains a problem in Turkey, India, Thailand, and China. There is evidence that bladder stones are related to diet deficient in animal products but with an excess of vegetable proteins. The art of surgery evolved from this pathology. No respecter of royalty or intellectual achievement, this ailment has afflicted many of the great figures of history, e.g., Isaac Newton, William Harvey, Peter the Great, Louis XIV, Oliver Cromwell, Samuel Pepys, Thomas Sydenham, Napoleon I, Napoleon III, Benjamin Franklin, and Leopold I of Belgium.

URINARY CALCULI AND DISCOVERY OF URIC ACID

Uric acid—named "acide urique" by Fourcroy and Vauquelin in 1799—was originally called the acid of calculus or lithic acid (Greek: *lithos*: stone) by the Swedish pharmacist Carl Wilhelm Scheele (1742–1786), who first detected it in a bladder calculus in 1776. He also showed that it was present in urine; and he believed that all urinary concretions consisted solely of that substance and were all fundamentally the same. Scheele noted the red color when the stone is moistened with concentrated nitric acid. When this is evaporated to dryness, the addition of dilute ammonium hydroxide produces the characteristic purple-red color of the ammonium salt of purpuric acid. Purpurate of ammonia was prepared by Prout and later renamed murexide by Liebig and Wöhler [2] (see Chapters 4 and 6).

Bladder stone was much more prevalent in the eighteenth and nineteenth century than it is today and attracted much attention from the medical profession. Every hospital had a collection of urinary calculi. Beginning in 1798, Antoine François de Fourcroy (1755–1809) (see Chapter 5) and his frequent collaborator, Nicholas Louis Vauquelin (1763–1829), analyzed hundreds of urinary stones and classified them according to chemical composition: uric acid, urate of ammonia, phosphate of lime, ammonium magnesium phosphate, oxalate of lime, silica (very rare), and animal matter. About two-thirds of the examined stones consisted of the two "earthy" phosphates and oxalate of lime. About one-third were uric acid calculi. Fourcroy hoped that the analysis of urinary calculi would lead to the discovery of solvents suitable for dissolving them by injection into the bladder, e.g., nitric acid and muriatic acid (hydrochloric acid) for the phosphate and oxalate stones. But this procedure was too inconvenient and never went beyond the experimental stage. Chemical analysis of voided gravel and sand had to precede any possible chemical treatment.

The only real solution was the knife because once the calculus is formed, "further enlargement is probably a common chemical process, and will proceed whether the urine be healthy or not, for all urine naturally contains the ingredients most commonly met with in calculi" [3].

Even before the chemical analysis of urinary stones provided a new basis for treating this condition, substances and compound remedies known as lithontriptics [4] (Greek: *tribo*, to grind, rub, waste), with an alleged capacity to dissolve urinary stones, formed a major field of interest of eighteenth-century pharmacology and therapeutics. Remedies for bladder stones were frequently listed in contemporary domestic manuals and were obviously preferred by the public as an alternative to the dreaded lithotomy operation.

The long tradition of lithontriptic remedies can be traced back at least to Graeco-Roman antiquity when the power to break bladder stones was attributed to several vegetable substances. Medicines against urinary stones, mostly of vegetable origin and arranged in rather complex recipes, figured largely in the materia medica of medieval Arabic medicine. Substances repeatedly mentioned were melon seeds, wild carrot seeds, parsley, and the ashes of scorpions. However, by the start of the eighteenth century, the actual existence of lithontriptic remedies and whether they were possible at all, was questioned and debated. There were some trials at the Edinburgh Royal Infirmary in 1745, with injection of lime water directly into the bladder via the urethra of patients. Obviously too inconvenient and unpleasant, this mode of application did not gain ground. But, taken orally, lime water remained the predominant remedy (see also Chapter 6, reference 21).

Drinking the waters of some mineral springs was a common treatment against urinary stones. When Joseph Black, hoping to find a better lithontriptic than lime water, discovered "fixed air" (carbon dioxide) in magnesium carbonate, it was suggested that the "air" contained in the waters of some mineral springs was the same as Black's fixed air. In that case, water artificially impregnated with fixed air could be a convenient remedy. The technical method of performing this process was described by Joseph Priestley (1733–1804) in 1772. The new impregnated water—pleasant to take compared to the nauseous lime water—now became *the* new lithontriptic and a means of self-medication for sufferers of the stone. According to the theory, the stones were supposed to be dissolved by a "superaddition" of fixed air. Machines for producing carbonated water were soon commercialized.

Anesthesia eventually removed the horror of pain from lithotomy, and wound infections were better avoided after the introduction of antisepsis by Joseph Lister (1827–1912).

CYSTINE AND CYSTINURIA

By 1810, when William Hyde Wollaston (1766–1828) had described cystine [5] (the first amino acid to be discovered) in a new and rare type of urinary calculus, he had already characterized five chemically distinct principle constituents of urinary calculi in humans. These were lithic or uric acid, ammonium magnesium phosphate (triple phosphate), calcium oxalate, calcium carbonate, and sodium urate which he had also discovered in gouty joints in 1797 [6]. These calculi were known and their analyses described in books on urine analysis and disorders. Various other designations were used, e.g., bone earth (principally calcium phosphate); fusible calculus (mixture of triple phosphate and calcium phosphate which fuses to a black enamel-like mass under the blowpipe);

mulberry calculus (calcium oxalate). Fourcroy and Vauquelin reported similar investigations but gave no recognition to Wollaston.

Wollaston suggested that the cystine stones resulted from an increased amount of cystine in the urine. He did not recognize the relationship of cystine to protein or that it contained sulfur. Wollaston believed that the substance was an oxide, and since the calculus came from the bladder, he named it cystic oxide, to distinguish it from other calculi. Berzelius, the authority on nomenclature (see Chapter 4), rejected "oxide" as inappropriate for organic substances since most of them contain oxygen; he named it *cystine* (Greek: *kystis*, bladder) in 1833.

Cystinuria was later identified by Archibald Edward Garrod (1857–1936) as one of several "inborn errors of metabolism," a consequence of a genetic defect. He identified four syndromes: alcaptonuria, albinism, cystinuria, and pentosuria [7]. These were "chemical analogues of structural malformations," which is the concept of a "chemical disease." Experimental proof that these inherited diseases of metabolism are caused by the lack of a particular enzyme, did not come until the 1950s. A. E. Garrod, physician and lecturer on chemical pathology at St. Bartholomew's Hospital in London, was the son of Alfred Baring Garrod (1819–1907) who, in 1848, found that uric acid was increased in the serum of patients with gout (see Chapter 13).

PREPARATION AND ANALYSIS OF UREA

In seeking the cause of urinary calculi, Fourcroy and Vauquelin investigated the composition of urine and developed procedures for the isolation and study of urea. There had been earlier descriptions. Van Helmont had isolated two salts from urine; one was sea salt, taken with food; the other, salt of urine, was of different crystalline form and unlike the sea salt, was volatile when heated (probably urea). Sometime before 1727, Hermann Boerhaave (1668–1738), the renowned Dutch chemist-physician, described a crystalline residue obtained from urine that he had concentrated by heating, filtering, washing, and evaporation, which he called "the native salt of urine" and distinguished from sea salt (sodium chloride) also present in urine. Hilaire Marin Rouelle (1718–1779) in 1773 prepared an impure urea from the alcoholic extract of an evaporated urine residue. Rouelle called it *matière savonneuse* (soapy matter). In 1797 in England, William Cruickshank (see Chapter 5) obtained crystalline urea nitrate following the addition of concentrated nitric acid to evaporated urine. In 1799 Fourcroy and Vauquelin prepared a nearly pure urea which they named urée [8].

They prepared a much purer urea in 1808 when the sparingly soluble urea nitrate was neutralized by aqueous potassium carbonate. After evaporation to dryness, the residue was extracted with alcohol to separate the urea from the potassium nitrate. Evaporation of the alcoholic solution yielded crystals of urea. Because aqueous solution of urea decomposes on boiling, to carbonic acid, acetic acid, and ammonia, they speculated that ammonia-containing calculi might be formed by the partial fermentation of urea in the bladder.

Finally, preparation of a pure urea product, its properties, appearance, and chemical reactions, was described in 1817 by William Prout (1785–1850) (first exhibited, he says, at his lectures three years earlier). Prout introduced a purification step with animal charcoal before extracting with boiling alcohol. This procedure became the textbook method of choice for the preparation of urea. Prout's analysis (by combustion) of the percentage composition of the component elements of urea was virtually identical with the values calculated from its formula. A pure urea had been prepared by Berzelius in 1808, by way of the oxalate, but the work was not known in England until the English translation of his book appeared in 1813 [9].

Prout used Gay-Lussac's method of completely oxidizing the urea with black oxide of copper which, at a suitable temperature, readily gives up its oxygen to hydrogen and carbon, but not to nitrogen. The nitrogen, uncombined, is collected as a gas and accurately measured. Prout's apparatus was a simplified modification of that used by Berzelius. Prout calibrated his weights with platinum standards and calibrated the graduations of the gasometer. All substances were dried in a vacuum over sulfuric acid at about 200°F [10].

Years later, Prout acknowledged that "from its composition I was satisfied that it might be formed artificially. I made numerous attempts to form it, but did not succeed; *and the honour of forming the first organic product artificially is due* to Wöhler." Prout also claimed to have found urea (or a substance having most of its properties) in blood in 1816. Believing it was accidental, he did not pursue the inquiry, but made a note of it [11].

Chemical science had a slow and difficult beginning. Nearly a century elapsed between the first isolation of urea by Boerhaave and the preparation and analysis of the first pure specimen by Prout. Only a few years later, in 1828, the laboratory synthesis of urea by Wöhler played the pivotal role that ushered in the era of modern experimental organic chemistry.

ANALYSIS OF URINE

Entering the nineteenth century and the era of atomic weights and atomic theory, chemistry began to disengage from its qualitative and descriptive character in the previous century. "Chemistry being a science of observation," William Prout (Fig. 3.1) looked forward to the time "when chemistry

FIG. 3.1. William Prout. From a miniature by Henry Wyndham Phillips. (Courtesy of the Royal College of Physicians, London.)

shall be brought more under the control of the laws of quantity...." [12]. Applications of chemistry to medicine at the beginning of the nineteenth century were directed to the understanding of disease rather than to its relief. Vitalists denied chemistry a role in physiology. However, Prout [13]—himself a vitalist—was an early and consistent advocate of the benefits to be derived from the application of chemistry to physiology in the treatment of disease. He also favored the study of physics and chemistry by medical students. One of Prout's admirers was Henry Bence Jones (see Chapter 6) who, in 1850, credited him with being first to make the true connection between chemistry and medical practice [14].

Urine, as a readily available body fluid, was subjected to systematic and scientific examination by Prout. His major work, *An Inquiry Into the Nature and Treatment of Gravel, Calculus, and Other Diseases Connected with a Deranged Operation of the Urinary Organs* (1821), was very popular and helped establish his reputation in Great Britain and Europe as Britain's most distinguished chemist. Prout's goal was to establish a coherent connection between the chemical processes of metabolism and excretion as manifest in the urine, with the observed changes in the patient's clinical status.

By 1825, when the second edition of Prout's text was published, now renamed, *An Inquiry Into the Nature and Treatment of Diabetes, Calculus, and Other Affections of The Urinary Organs*, most of our present day knowledge of the composition of stones had been discovered. This book is one of the earliest to contain a list of "Tests, Apparatus, &c. required in making Experiments on the Urine;" and this included litmus paper (blue and red), tumeric paper, a specific gravity bottle or a small portable hydrometer that Prout designed, blowpipe, forceps, two small discs of plate glass for discriminating pus from mucus, and a watch glass for detecting an excess of urea on addition of nitric acid. "These, with one or two small test tubes, and small stoppered phials, containing solutions of pure ammonia, potash, and nitric acid, can be readily packed into a small portable case, or pocket book, and will be sufficient, by the aid of a common taper or candle, to perform all the experiments on the urine, and urinary productions, that are commonly necessary in a practical point of view" [15]. Barely a year later, Richard Bright's studies of renal disease would add a spoon to this portable laboratory, for revealing the presence of albumin in heated urine.

Prout's routine for testing urine began with the 24-hour volume, the color, and the transparency. Some of the imaginative notions of the uroscopists of the past were still in evidence here. The belief persisted that the urine's volume, color, and appearance, was indicative of certain personality traits and was characteristic of particular diseases. Specific gravity was measured and if 1.030 or greater, was considered diagnostic of diabetes. The reaction of urine, known to be acid normally, was taken with litmus paper. Albuminous urine (protein) was detected by heat coagulation, and bile by the yellow staining of linen. Sugar in the urine was still recognized by taste—there was no other test—and was "not found in the blood even of individuals labouring under diabetes, in whose urine it exists in the greatest abundance;...." An excess of urea—regarded as abnormal but of vague significance—was inferred from the length of time for crystallization to occur when nitric acid was added to the urine on a watch glass. Prout claimed that in diabetes and some other diseases of the urine, very little urea is sometimes present. The color and appearance of any sediment settling out on standing were noted, but no microscopic examination was made [16]. Some of the tests may have been crude, but they were chemical.

Prout's book appeared in five editions and several name changes, appearing finally in 1848 as *On the Nature and Treatment of Stomach and Renal Diseases; Being an Inquiry Into the Connexion of Diabetes, Calculus, and Other Affections of the Kidney and Bladder, with Indigestion*. By then, the book's lack of chemical formulae, which Prout dismissed as unphilosophical expedients because they did not represent true compositions, and his omission of continental discoveries and advances, showed an inability to keep up with the newer developments of science. His inertia and conservatism were sharply criticized by *The Lancet* [17].

Prout was very skeptical of so-called chemical remedies because of the possibility of side-effects and their potential for ultimately aggravating the disease. Since "the *object* of the chemical practitioner is at best ... to prevent the effects of disease rather than to remove it," he considered "chemical remedies as palliatives only," and attributed "their acknowledged good effects" to "their general than their chemical operation;...." [18].

In 1827 Prout classified foodstuffs into saccharine (carbohydrates), oleaginous (fats), albuminous (proteins), and water, and urged that a satisfactory diet should include all four [19]. As a vitalist, he maintained that organized (organic) substances contained independent principles which transformed the four foodstuffs into blood and tissues.

CHEMISTRY VS PHYSIOLOGY

In his Gulstonian lecture on "The Application of Chemistry to Physiology, Pathology and Practice," at the Royal College of Physicians in 1831, Prout cited the lack of progress in animal chemistry. He attributed this partly to the inherent difficulty of the subject, but also to the lack of understanding by the pure, i.e., inorganic, chemist of the unfamiliar field of biology. Prout's remedy was for physiologists to become chemists [20]. This was reminiscent of an earlier appeal in 1816 when he stated: "Chemistry, however, in the hands of the physiologist, who knows how to avail himself of its means, will, doubtless, prove one of the most powerful instruments he can possess;...." Furthermore, cautioned Prout, "Organic substances should be compared with one another, and not with inorganic ones, with which they have little or no analogy." He advised the physiological chemist to pay attention only to what is actually observed and to avoid superfluous experiment and visionary hypothesis [21].

Prout's Gulstonian lecture got him embroiled in a running acrimonious debate with Wilson Philip in the pages of the *London Medical Gazette* [22]. Philip, a physiologist and vitalist, could see no value in applying chemical methods and reasoning to the problems of physiology. He resented Prout's claim that almost no progress had been made in physiology in twenty years. Philip charged: "Chemistry, and the science of the vital functions, are of so different a nature, that if they be pursued with ardour, and without this nothing can be done in such subjects, the one will tend constantly to abstract the mind from, and perhaps in some degree to unfit it for, the other;...." [23]. By the 1840s, the vitalist Philip did a complete turnaround to Prout's viewpoint, even claiming that the nervous system was essentially chemical. Prout, on the other hand, became more committed to vitalism, but avoided a new confrontation.

Vitalists found much to criticize in Prout's views. It was inconceivable to them that bodily functions could be explained by chemistry; at best it offered only an incomplete explanation. The composition and workings of the body required a study of the vital functions. In any event, Prout was a vitalist, for it always remained a shortcoming of his scientific method to use lack of knowledge as an argument for vitalism. Prout believed that there exists "in all living organised bodies some power or agency, whose operation is altogether different from the operation of the common agencies of matter, and on which the peculiarities of organised bodies depend,...." Of the various hypotheses explaining these differences, Prout favored that of "independent existing vital principles or 'agents,' superior to, and capable of controlling and directing, the forces operating in inorganic matters; on the presence and influence of which the phenomena of organisation and of life depend" [24].

WILLIAM PROUT DISCOVERS THE ACID OF GASTRIC JUICE

William Prout is known for his remarkable discovery "On the nature of the acid and saline matters usually existing in the stomachs of animals." The report was read before the Royal Society of London on December 11, 1823 and appeared in the *Philosophical Transactions of the Royal Society* the following year [25]. Prout identified free muriatic acid in the gastric juice of various animals and man after a meal and suggested it was derived from the common salt of the blood by the force of galvanism (electricity). Prior to this finding Prout favored phosphoric acid as responsible for the acidity of gastric juice. He expressed concentration of acid as grains "to one pint."

Prout's results received confirmation from U.S. Army surgeon William Beaumont's (1785–1853) classic research on Alexis St. Martin, a Canadian with a permanent gastric fistula that resulted after the wound from an accidental gunshot in 1822 had healed. Beaumont studied the appearance and function of the exposed living stomach over a number of years and provided new information about the nature of the gastric juice and the process of digestion in the stomach. He published his findings in *Experiments and Observations on the Gastric Juice, and the Physiology of Digestion* (1833). Beaumont recognized the acid character of the gastric juice which he studied in response to the stimulus of food and alcohol (see Chapter 10, reference 40). Its identity as muriatic acid was verified at the University of Virginia and at Yale [26]. But old ideas die hard. Writers on gastric physiology continued to deny the presence of hydrochloric acid in gastric juice or mentioned it merely as one of the acids present.

In 1817, François Magendie (1783–1855), French experimental physiologist, had also established the acidity of gastric juice, but he attributed this to lactic acid. Because of his prominence, the acidity question remained a problem for many years, despite Prout's discovery. As late as 1856, Claude Bernard, the famous French physiologist (see Chapter 8), who had been a student of Magendie, still accepted his teacher's view that gastric acidity was due to lactic acid.

In their two-volume publication on digestion, *Die Verdauung nach Versuchen* (1826–27) (Experiments in Digestion), Friedrich Tiedemann (1781–1861), professor of anatomy and physiology at Heidelberg, and Leopold Gmelin (1788–1853), member of a distinguished dynasty of physician-chemists [27], gave Prout credit for the discovery of hydrochloric acid, even though they had found the acid in gastric fluid by a different method in 1824. However, they claimed also to have found butyric and acetic acid in the gastric juice.

Investigators considered that the hydrochloric acid was produced secondarily. They believed that the primary acid secreted was lactic acid, which then acted upon the sodium chloride present to produce the hydrochloric acid which Prout and others had found. Still others attributed gastric acidity primarily to the presence of acid phosphates in the juice. There was general reluctance to accept Prout's conclusions because no one had actually handled pure gastric juice, except Beaumont, and the material examined was usually mixed with other juices and food residues. Sometimes the juice stood for a while before it was analyzed, long enough for some lactic acid to be produced by fermentation of carbohydrates, especially with gastric contents of low acidity. Finally, it was difficult for some physiologists to accept the idea that acid as strong as hydrochloric acid could be secreted by living cells [28]. In 1850, Henry Bence Jones wrote: "This gastric juice, then, is a highly acid fluid secreted by the stomach...What acid it is has not yet been determined. Hydrochloric, phosphoric, acetic, lactic, and butyric acids, have each been said to exist in the gastric juice" [29].

All doubt was finally dispelled by the publication in 1852 of "Gastric Juice and Metabolism. A Physiological–Chemical Investigation," by Friedrich Bidder (1810–1894) and Carl Schmidt of the University of Dorpat. From their quantitative analyses of the gastric juice collected by means of a fistula created in different species of live fasting animals, they proved that the acid of gastric juice is exclusively hydrochloric acid [30].

The production of hydrochloric acid by the gastric tissue represents one of the most interesting phenomena of nature. Physiologic processes are generally very sensitive to minute changes in the reaction of body fluids. Even in disease, variations beyond the narrow range of pH 7.1 to 7.7 are incompatible with life. Nevertheless, the parietal cells of the stomach actually secrete a fluid whose acidity reaches a pH of 0.9 or less [31].

Although the awareness of fluid metabolism and its delicate mechanisms for the preservation of acid–base balance is a modern development, physiologists of the previous century undoubtedly were aware that such extremes of acidity were not compatible with life. One can well imagine the scientific sensation when Prout claimed to have found "free" hydrochloric acid in the gastric juice. Although his scientific standing carried weight, production of a strong free acid by gastric glands in the mammalian body was so contrary to physiological probability that many years passed before it was generally accepted. Like other medical discoveries, this finding was the culmination of years of investigation on the part of many workers in many countries. Prout's discovery of hydrochloric acid in the gastric juice and that none other was present under normal circumstances—a discovery of fundamental importance in the explanation of the chemistry of digestion—was not even mentioned in his obituaries.

ALEXANDER MARCET, GOLDING BIRD, AND ANALYSIS OF CALCULI

Prout discovered a new acid in 1822 in urine that turned black soon after being voided and exposed to the air. The specimen had been sent to him for analysis by Alexander Marcet (1770–1822) (Fig. 3.2) [32]. Prout reported that the black color was due to an unknown principle combined with ammonia which he appropriately named *melanic acid*. The substance

FIG. 3.2. Alexander Marcet. Engraving by Henry Meyer from original painting by Sir Henry Raeburn. (Courtesy of the Royal College of Physicians, London.)

involved was called *alkapton* by C. H. D. Boedeker (1859). It is homogentisic acid (2,5-dihydroxyphenylacetic acid), a product of protein metabolism. The fact that Marcet, a physician at Guy's Hospital, who gave lectures in chemistry to the medical students, was quite capable of making the analysis himself, indicates the high regard Prout enjoyed as an expert in urine analysis.

Marcet was an authority on urinary calculi and their identification by chemical reaction and attempted to correlate chemical composition of the calculi with diagnosis of the pathology causing their formation. In a popular work titled *An Essay on The Chemical History and Medical Treatment of Calculous Disorders*, he described his discovery of a new substance in a urinary calculus. He named it xanthic oxide (Greek: *xanthos*, yellow) because it forms a lemon-yellow colored compound when treated with nitric acid. Liebig and Wöhler correctly determined that xanthic oxide contains one atom of oxygen less than in uric acid (1838). Although Berzelius called it urous acid (1840), it is now known as xanthine.

Marcet dedicated the book to Wollaston whom he credits for many of the discoveries and remarks which it contains. He notes that Fourcroy, in his history of urinary calculi and in various papers on this subject, has

"in a most unaccountable manner, entirely overlooked Dr. Wollaston's labours, and in describing results exactly similar to those previously obtained and published by the English chemist, has claimed them as his own discoveries. Yet Dr. Wollaston's paper was printed in our Philosophical Transactions in 1797, that is about two years before Fourcroy published his Memoir in the 'Annales de Chimie,'....

"It is extremely painful to be compelled by justice to notice such an apparent want of fairness and candour, in a philosopher, who devoted a long and brilliant career to the advancement of science" [33].

However, Marcet does not find fault with Vauquelin, often a co-author with Fourcroy, because he was "not conversant with the English language" and it was well known "that the task of publishing the results of their common labours" always fell to Fourcroy [34].

Marcet noted that "Physicians and chemists from Galen to Paracelsus, and from Paracelsus to Van Helmont and Boerhaave" were unable "to form any rational conjectures on the composition of urinary calculi" [35]. Marcet described and illustrated calculi by means of colored drawings and listed a protocol for analysis of the calculus. He also drew line illustrations of a set of simple and portable chemical apparatus that could be used for analysis at the bedside [36]. Marcet wanted the average medical practitioner to have a reliable method of diagnosis requiring very little chemical skill or knowledge to easily distinguish the various kinds of urinary calculi. He was confident that physiological processes would ultimately be explained by chemistry.

Marcet's reagents were dilute mineral acids, acetic acid, caustic alkalies, alkali carbonates, ammonium hydroxide, ammonium oxalate, and potassium cyanide. The blowpipe in conjunction with a small spirit (alcohol) lamp was also useful. His procedures were designed on a very small scale using only a few drops of solution and pinhead pieces of calculi, a microtechnique he credited to Wollaston and considered a development of primary importance.

Diabetes, readily detected by examination of the urine, also interested Marcet. Initially, in 1797, following the views of William Cruickshank, he believed that sugar was present in the blood of diabetics, but by 1811 he agreed with Wollaston that sugar was not to be found in diabetic blood [37].

Golding Bird (1814–1854) was another (assistant) physician at Guy's Hospital with an interest in medical chemistry. His detailed studies on urinary sediments and renal calculi were more sophisticated than any up to that time. He regarded the nucleus as the key ingredient leading to formation of a calculus. His "Observations on Urinary Concretions and Deposits" was expanded into a monograph titled *Urinary Deposits, Their Diagnosis, Pathology, and Therapeutical Indications* (1844) [38].

For the first time, chemistry was related to organs rather than to the total animal. In a large folding table, Bird lists the chemical composition of the constituents of the blood and some of their most important chemical modifications that are eliminated by the liver and by the kidneys. He also identifies the authoritative sources for the respective empirical formulas of these products. Like Marcet and Prout, Bird devoted much of his chemical work to a study of urine analysis and urinary calculi. He made extensive use of the microscope to illustrate the various urinary deposits and recognized that casts, which had been described by Henry Bence Jones, were diagnostic of Bright's disease. Twenty-four-hour urine collections were weighed and specific gravity was measured. From these values and a formula, he calculated the quantity of solids in 1000 grains of urine. He also used nitric acid to test for bile in urine (a green color), and a simple polariscope or alkaline copper reagent for glycosuria [39]. A fifth edition of Bird's book appeared in 1857.

After completion of his training at Guy's Medical School in 1836, Bird was awarded the licence of Apothecaries' Hall without examination, based on his reputation as a chemist while a student. In 1838, on payment of the fees and armed with the appropriate letters of recommendation and testimonials, but without residence or course work—a common practice—he was awarded an M.D. degree from the University of St. Andrews.

NOTES AND REFERENCES

1. HAROLD ELLIS, *A History of Bladder Stone* (Oxford: Blackwell Scientific Publications, 1969), preface, pp. 1, 66, 71.
2. J. R. PARTINGTON, *A History of Chemistry*, (4 vols.) vol. 3 (London: Macmillan & Co Ltd, 1962), pp. 549, 713–714.
3. WILLIAM PROUT, "Observations on the Nature of Some of the Proximate Principles of the Urine; with a Few Remarks Upon the Means of Preventing Those Diseases, Connected with a Morbid State of That Fluid," *Medico-Chirurgical Transactions*, 1817, 8: 526–549, p. 546.
4. This section is derived from "Dissolving the Stone: The Search for Lithontriptic Remedies in the Eighteenth Century," reported by Andreas-Holger Maehle to the Pybus Society for the History and Bibliography of Medicine, Newcastle upon Tyne, 13 October 1994. See also ARTHUR J. VISELTEAR, "Joanna Stephens and the Eighteenth Century Lithontriptics; A Misplaced Chapter in the History of Therapeutics," *Bulletin of the History of Medicine*, 1968, 42: 199–220.
5. WILLIAM HYDE WOLLASTON, "On Cystic Oxide, a New Species of Urinary Calculus," *Philosophical Transactions of the Royal Society of London*, 1810, 100: 223–230.
6. WILLIAM HYDE WOLLASTON, "On Gouty and Urinary Concretions," *Philosophical Transactions of the Royal Society of London*, 1797, 87: 386–400.
7. ARCHIBALD E. GARROD, "The Incidence of Alkaptonuria: A Study in Chemical Individuality," *Lancet*, 1902, 2: 1616–1620; "The Croonian Lectures on Inborn Errors of Metabolism," *Lancet*, 1908, 2: 1–7, 73–79, 142–148, 214–220.
8. PARTINGTON (1962), vol. 3, pp. 78, 549.
9. PROUT (1817), pp. 528–530; PARTINGTON (1962), p. 549; FREDERICK KURZER and PHYLLIS M. SANDERSON, "Urea in the History of Organic Chemistry," *Journal of Chemical Education*, 1956, 33: 452–459.
10. PROUT (1817), pp. 530–536.
11. WILLIAM PROUT, *On the Nature and Treatment of Stomach and Renal Diseases; Being an Inquiry Into the Connexion of Diabetes, Calculus, and Other Affections of the Kidney and Bladder, With Indigestion*, 5th ed. (London: John Churchill, 1848), pp. 529–530, footnote p. 531.
12. WILLIAM PROUT, *Chemistry, Meteorology, and the Function of Digestion, Considered with Reference to Natural Theology* (Philadelphia: Carey, Lea & Blanchard, 1834), p. 292.
13. W. H. BROCK, "The Life and Work of William Prout," *Medical History*, 1965, 9: 101–126.
14. H. BENCE JONES, *On Animal Chemistry in its Application to Stomach and Renal Diseases* (London: John Churchill, 1850), preface.
15. WILLIAM PROUT, *An Inquiry Into the Nature and Treatment of Diabetes, Calculus, and Other Affections of The Urinary Organs*, 2nd ed. (Philadelphia: Towar and Hogan, 1826), p. 300. The 2nd ed. was published in England in 1825.
16. *Ibidem*, pp. 6, 16, 23–25, 258–263.
17. Editorial, *Lancet*, 1843–44, 1: 486–490.

18. PROUT (1817), pp. 548–549. Prout did have a chemical remedy. After iodine salts had been found in certain marine life forms, it occurred to him that burnt sponge (a well known remedy for goiter) might owe its properties to the presence of iodine. In 1816, after trying hydriodate of potash (potassium iodate) on himself in small doses and experiencing no ill effects, Prout suggested iodine treatment for goiter. This therapy was successfully adopted by Dr. John Elliotson (1791–1868) at St. Thomas's Hospital early in 1819. Prout (1834), footnote pp. 75–76.
19. WILLIAM PROUT, "On the Ultimate Composition of Simple Alimentary Substances with Some Preliminary Remarks on the Analysis of Organized Bodies in General," *Philosophical Transactions of the Royal Society of London*, 1827, 117: 355–388, p. 357. Also see WM. PROUT, "Observations on the Application of Chemistry to Physiology, Pathology, and Practice," *London Medical Gazette*, 1831, 8: 321–327.
20. WM. PROUT, "Observations on the Application of Chemistry to Physiology, Pathology, and Practice," *London Medical Gazette*, 1831, 8: 257–265, p. 258.
21. WILLIAM PROUT, "Inquiry Into the Origin and Properties of the Blood," *Annals of Medicine and Surgery*, 1816, 1: 10–26, 133–157, 277–289, p. 289.
22. *London Medical Gazette*, 1831, 8: 641–, 705–, 737–, 769–770, 770–776, 802–, 843–; 1831–32, 9: 38–, 69–, 73–; see "Dr. Prout's Rejoinder to Dr. W. Philip's 'Reply,'" pp. 769–770.
23. A. P. W. PHILIP, "Some Observations on An Abstract of Dr. Prout's Gulstonian Lectures," *London Medical Gazette*, 1831, 8: 770–776, p. 772.
24. PROUT (1848), p. 452. Prout's views on vitalism that sparked the exchange with Philip are reviewed by Brock (1965), pp. 113, 115–116, 120–121. A strong prejudice existed in those days against chemical ideas, and quarrels among medical men were frequently virulent and vituperative. Personal antagonisms were aired in the medical journals, most notably *The Lancet*, whose founder and editor, Thomas Wakley (1795–1862), was adept at provoking dissension. The letters to the editor were often lengthy and frequently signed by a pseudonym. Accusations of bias or incompetence against individuals or institutions usually generated a response and counter-response which went on from issue to issue and helped stimulate circulation of the journal.
25. WILLIAM PROUT, "On the Nature of the Acid and Saline Matters Usually Existing in the Stomachs of Animals," *Philosophical Transactions of the Royal Society of London*, 1824, 114: 45–49.
26. *Dictionary of Scientific Biography* (*DSB*), Charles Coulston Gillispie, ed. (New York: Charles Scribner's Sons, 1976), 14: 558–559; (*DSB*) (1970), 1: 542–545.
27. PAUL WALDEN, "The Gmelin Chemical Dynasty," *Journal of Chemical Education*, 1954, 31: 534–541. During this period, chemistry first separated as a discipline independent of the medical faculty (see Chapter 6).
28. FRANKLIN C. BING, "Friedrich Bidder (1810–1894) and Carl Schmidt (1822–1894)—A Biographical Sketch," *Journal of Nutrition*, 1973, 103: 637–648, p. 642.
29. H. BENCE JONES, "Diagnosis and Treatment of Stomach and Renal Diseases," *Lancet*, 1850, 1: 69–71, p. 70.
30. BING (1973), p. 642.
31. ANTHONY M. KASICH, "William Prout and the Discovery of Hydrochloric Acid in the Gastric Juice," *Bulletin of the History of Medicine*, 1946, 20: 340–358.
32. ALEX. MARCET, "Account of a Singular Variety of Urine, Which Turned Black Soon After Being Discharged," *Medico-Chirurgical Transactions*, 1822–23, 12: 37–45. The report by Prout is on pp. 43–45.

ANALYSIS OF CALCULI 53

33. ALEXANDER MARCET, *An Essay on The Chemical History and Medical Treatment of Calculous Disorders*, 2nd ed. (London: Longman, Hurst, Rees, Orme, and Brown, 1819), pp. 4, 67–68; also see N. G. COLEY, "Alexander Marcet (1770–1822), Physician and Animal Chemist," *Medical History*, 1968, 12: 394–402; ARCHIBALD GARROD, "Alexander John Gaspard Marcet. Physician to Guy's Hospital, 1804–1819," *Guy's Hospital Reports*, 1925, 75: 373–387. Mrs. Marcet (Jane Haldimand) (1769–1858), born in London of Swiss parentage, was a celebrated author who wrote the very popular *Conversations on Chemistry* (1806, 2 vols; 18 editions). Published anonymously at first—it was a badly-kept secret and most people knew she had written it—her name first appeared as author in the 13th edition (1837). American publishers made 23 impressions of various editions of the work. The book first interested the young Michael Faraday in chemistry when he was binding a copy as a bookbinder's apprentice. Partington (1962, vol. 3), p. 708. Mrs. Marcet's book was directed at beginners and women in particular. For a discussion on the teaching of chemistry and science at college-level women's schools in the first half of the nineteenth century in the United States, see M. SUSAN LINDEE, "The American Career of Jane Marcet's *Conversations on Chemistry*, 1806–1853," *Isis*, 1991, 82: 8–23. For details about Mrs. Marcet's life and excerpts from the book, see EVA V. ARMSTRONG, "Jane Marcet and Her 'Conversations on Chemistry,'" *Journal of Chemical Education*, 1938, 15: 53–57. She also wrote *Conversations on Political Economy, Conversations on Natural Philosophy, Conversations on Vegetable Physiology*, and books for small children.
34. MARCET (1819), footnote pp. 68–69.
35. *Ibidem*, p. 63.
36. *Ibidem*, plate X.
37. WILLIAM HYDE WOLLASTON, "On the Non-existence of Sugar in the Blood of Persons Labouring Under Diabetes Mellitus. In a Letter to Alexander Marcet," *Philosophical Transactions of the Royal Society of London*, 1811, 101: 96–105; Marcet's reply, pp. 106–109.
38. GOLDING BIRD, "Observations on Urinary Concretions and Deposits; With an Account of the Calculi in the Museum of Guy's Hospital," *Guy's Hospital Reports*, 1842, 7: 175–232; *Urinary Deposits, Their Diagnosis, Pathology, and Therapeutical Indications*, 2nd ed. (London: John Churchill, 1846); see also N. G. COLEY, "The Collateral Sciences in the Work of Golding Bird (1814–1854)," *Medical History*, 1969, 13: 363–376.
39. BIRD (1846), pp. 36–40, 291–293.

CHAPTER IV

INTRODUCTION

Organic chemistry has been distinguished from inorganic chemistry since the seventeenth century. The French chemist Nicolas Lemery (1645–1715) separated naturally occurring substances according to their mineral or animal/plant origin. There was little chemical knowledge about the latter group.

In an effort to reduce the empirical element in medicine, the "animal chemistry" of the late eighteenth and early nineteenth centuries had sought to isolate chemical substances from plant and animal sources, in a state unchanged by the process of isolation. Procedures for isolating organic material from plants or animal matter had to avoid destructive calcination (distillation until dry), a common procedure in both inorganic and organic chemistry for centuries, that yielded much the same poorly characterized substances from everything studied. Aside from the commercial value of pure substances with reproducible drug potency, there was the belief that the identification of these materials would lead to an understanding of biological organization and physiological function [1].

With the emergence of modern chemistry at the end of the eighteenth century, and during the early years of the nineteenth century, organic chemistry, which was concerned with nutrition and the phenomenon of living things—not with the chemistry of carbon—began gradually to separate from inorganic chemistry. At this stage, organic chemistry was mostly a descriptive science engaged in isolating, identifying and, with newly developed methods, techniques, and improved facilities, in analyzing a great number of compounds from plant and animal sources. These compounds were made up chiefly of carbon, hydrogen, oxygen, and nitrogen. Some, in addition, had phosphorus and sulfur.

PROTEIN AND FOOD SHORTAGE

Protein chemistry began in the mid-eighteenth century with investigators who extracted proteins from vegetable and animal sources. The study of animal and vegetable substances was a subject of increasing activity throughout the eighteenth century. Some of the materials investigated were commercially important for dyeing and tanning, agriculture, and the manufacture of soaps and glues, and for supplying the pharmacy with much of its materia medica. It also seemed that the conversion of vegetable foods into animal substances, i.e., *animalization*, would provide new insights into animal physiology, and could be investigated by comparative chemical examination of animal and vegetable substances [2].

The most elaborate of the early investigations of animal and vegetable materials were those by Fourcroy. In 1789 he discovered a nitrogenous material in cruciferous plants with the same properties as the albumin of egg white. Other investigators obtained nitrogenous extracts from numerous plants and vegetables whose appearance, solubility, and color reactions, resembled products of animal origin. As a result, the terms albumin, fibrin, and casein were soon applied to these nitrogenous plant substances. The presence of nitrogen in many vegetables could account for its presence in herbivores without having to assume that the animal absorbed it from the atmosphere. Fourcroy concluded that animal substances had a more complex composition than vegetables, but he could find no chemical tests which would distinguish between them. Although this attempt by him and others to bring chemistry into biological classification was not successful it was an indication of the growing importance of chemistry in the eighteenth century [3].

The impetus given to "animal chemistry" by Fourcroy was later reinforced by its inclusion in the teaching programs of the new university chemical laboratories. With the growing specialization in chemistry, animal chemistry emerged as a separate subject in the medical school curriculum. The same happened to agricultural chemistry as a result of increased official concern about human nutrition [4].

Food shortages and high prices caused by the wars between France and England and the political upheavals at the end of the eighteenth and beginning of the nineteenth century led to the study of the chemical constituents of plants and animals used for human food. These foods were extracted with water, acids, alkalies, alcohol, and ether, and the extracts subjected to various methods of precipitation in an attempt to isolate pure substances. Many of these techniques, thought to be universally applicable, had adverse effects on protein material. The solubility of these compounds did not improve with heating. They coagulated irreversibly at moderately elevated temperatures and were damaged by acid or alkali.

By the early nineteenth century, solvent methods for separation and characterization of the "animal" materials—fibrin, albumin, casein, and similar substances—became so highly developed that they emerged as fundamental tools for determining the composition of the fluids and solids of animals and plants. Solvent separations were a prerequisite for all of the efforts during the nineteenth century to illuminate the chemical processes which occur in living organisms [5].

Another factor making food an important social problem in England at the end of the eighteenth century was the enclosure movement. Consolidation of smaller holdings, reduction of tillage, and increase of pasture lands, tended to swell the number of migrant workers and greatly impoverish their diet in quantity and variety. In addition, during the first half of the nineteenth century, the Industrial Revolution was drawing increasing numbers of laborers to jobs in the cities. With fewer agricultural workers to feed them, there was a need for increased food production.

In England, the government was urged to develop policies to put farming techniques on a scientific basis. Studies to find food substitutes appeared in several countries in Europe. As chemistry made inroads into the sugar beet and fermentation industries, there developed an increased need to understand chemical changes during digestion and assimilation in living systems under physiological and pathological conditions [6]. However, many chemists believed that the same proteins were present in plant and animal food, and that their preformed nitrogenous constituents are assimilated almost unchanged (see Chapter 12).

In 1816 Magendie investigated the source of organic nitrogen in the animal—was it the air or food? He discovered from feeding experiments that nitrogenous foods are needed for life. Dogs fed on a nitrogen-free diet—pure sugar, olive oil or butter, and distilled water exclusively—survived only for 32 to 36 days. The analysis of nitrogen became important because of the role of protein nitrogen in physiological processes and for determining the nutritional value of the protein content of animal and plant products and of plant fertilizers. Such determinations were also valuable in the control of industrial processes utilizing nitrogenous raw materials. However, nitrogen analysis was infrequently made at this time because there was no simple and accurate method. Whereas hydrogen and carbon were oxidized to water and carbonic acid, nitrogen combined with oxygen in many different proportions; consequently, reproducible results were not obtained.

ANALYSIS OF NITROGEN: METHODS OF DUMAS AND WILL–VARRENTRAPP

For nitrogen-containing organic substances, the method of Jean Baptiste André Dumas (1800–1884) (Fig. 4.1), devised in 1831, was highly accurate.

FIG. 4.1. Jean Baptiste André Dumas. (National Library of Medicine, Bethesda, Maryland.)

The method demonstrated decided differences in the elementary composition of many of the known proteins and eventually helped dispose of the notion of their identity. However, the procedure was too time-consuming and complicated for large-scale routine use. The test substance was completely combusted in a closed system with copper oxide as oxidizing agent in the presence of a carrier stream of carbon dioxide to move the combustion products along. Dumas used metallic copper for reduction of nitrogen oxides to elemental nitrogen after first activating the copper by heating in a stream of hydrogen. The nitrogen gas is collected in a eudiometer tube filled with alkali hydroxide thus eliminating the need for a separate absorption of carbon dioxide formed during the combustion. The weight of the liberated nitrogen gas is calculated after correcting its volume for pressure and temperature to standard conditions. Continued modifications improved the technique and shortened the analysis time to a few minutes. The Dumas method is still used by organic chemists as the standard procedure.

When it was first introduced, difficulties with the dry combustion technique caused investigators to consider the possibility of a wet method of analysis. Methods of this type are based on the principle that when certain organic nitrogen-containing substances are ignited either alone or in the presence of alkalies, the nitrogen is split off in the form of ammonia which can be determined to give the nitrogen content. This method is easier and more reliable than the complicated gasometric method [7].

In 1841, such a procedure—also utilizing dry combustion—was devised by Heinrich Will (1812–1890) and Franz Varrentrapp (1815–1877). In this method organic nitrogen was converted into ammonia by ignition with barium hydroxide. Originally, the liberated ammonia was passed into hydrochloric acid and determined gravimetrically by precipitation as ammonium hexachloroplatinate. In 1847 the method was modified by Eugéne Melchior Péligot (1811–1890), an analyst at the mint. The ammonia was absorbed in a known amount of hydrochloric acid and the excess acid back-titrated with a standard solution of lime dissolved in water containing sugar to increase the lime's solubility. The iron and tannic acid indicator turned violet when the alkali was present in excess [8]. Although inexact and not applicable to all nitrogen compounds, the Will–Varrentrapp method played a useful part during the development of organic chemistry until replaced near the end of the century by the simpler and quicker Kjeldahl digestion method which was more suited to large-scale demands.

KJELDAHL ANALYSIS

The Kjeldahl [9] method for organic nitrogen was very important for the development of organic chemistry during the closing years of the nineteenth century. Concepts about the structure and reactions of organic compounds depended on accurate determination of the elementary composition of carefully purified compounds and of the products of their chemical reactions. The method was of great value to physiological chemistry because it was readily adapted to the determination of protein in biological fluids, namely, plasma, serum, and urine.

Johan Gustav Christoffer Thorsager Kjeldahl (1849–1900) (Fig. 4.2) [10], a chemist at the Carlsberg Brewery in Copenhagen, was investigating the changes in protein content during barley germination and alcohol fermentation. His research was slowed by lack of a quick, accurate, and reliable method for determining nitrogen. Wet methods developed to avoid the constant attention necessary when heating dry mixtures, showed that complete ammonification followed preliminary destruction of the organic substance. An easy technique used by James Alfred Wanklyn (1834–1906) and other English chemists in 1877 for the determination of "albuminoid ammonia" in drinking water, appeared suitable for Kjeldahl's purpose. Wanklyn improved the Will–Varrentrapp method for estimating organic nitrogen as ammonia by adding alkaline potassium permanganate to assist in the decomposition. The sample of water so treated was distilled and the ammonia in successive portions of the distillate was determined with Nessler's reagent. Although the method gave incomplete decomposition and inconsistent results, it was widely used to determine the protein content of vegetable material.

Kjeldahl tried the method on plant juices, but could not obtain complete conversion of the protein nitrogen into ammonia. Reasoning that ammonia would be more easily formed in acid than in alkaline solution, he boiled his samples in dilute sulfuric acid and excess permanganate. Alkali was added and the liberated ammonia was distilled into an acid receiver. Higher yields were obtained immediately, but the conversion was still incomplete and results fluctuated. Kjeldahl eventually found that heating with concentrated sulfuric acid close to the boiling point destroyed the organic matter and brought the material completely into solution. Subsequent addition of an excess of dry potassium permanganate to the hot solution completed the oxidation of the nitrogen to ammonium sulfate. After making the solution alkaline, the ammonia was distilled and determined by iodometric titration which Kjeldahl preferred but which did not find general use.

FIG. 4.2. Johan Kjeldahl. (Carlsberg Foundation Picture Archives, Copenhagen, Denmark.)

The great advantage of the new method was its speed. Numerous modifications and improvements in the composition of the digestion reagent soon appeared. Various catalysts and additives were introduced, e.g., copper sulfate to accelerate the digestion process; phosphoric acid or potassium sulfate to raise the boiling point; and potassium persulfate or 30% hydrogen peroxide as oxidizing agents. The oxidation takes place at the expense of the oxygen in the sulfuric acid which is consequently reduced. Because the sulfurous acid fumes which result are very irritating to the nose and throat, the digestion must be performed in a hood with a good exhaust.

The titrimetric assay was simplified in 1913 by Lajos Winkler (1863–1939) who absorbed the ammonia in boric acid solution. The boric acid need not be accurately measured because a borate ion is liberated for each molecule of ammonia absorbed. The borate is titrated with standard hydrochloric or sulfuric acid in the presence of methyl red indicator.

The ammonia produced by Kjeldahl digestion can be quantitated colorimetrically either by direct nesslerization of the digest or, for greater accuracy and dependability, by distillation into an acid receiver and back-titrating with standard alkali or adding Nessler's reagent and measuring colorimetrically.

NESSLER'S REAGENT

Nessler's reagent was first developed in 1856 by Julius Nessler (1827–1905), a German agricultural chemist and authority on wine growing, for the analysis of ammonia in water. It is a strongly alkaline solution of HgI_2 and KI as a double iodide, whose exact composition is as unknown as is the colloidal complex reaction product with ammonium ions [11]. Because of its simplicity, direct nesslerization was a very popular technique, despite the tendency for the final color to develop turbidity. This led to many modifications of the Kjeldahl digestion mixture, the Nessler reagent, and to the use of stabilizing additives, e.g., gum ghatti, to prevent, minimize, or delay clouding following addition of Nessler's reagent. The Kjeldahl method was later adapted for the determination of plasma proteins and of urea nitrogen and nonprotein nitrogen in protein-free filtrates. Although the Kjeldahl method does not give quantitative results with many types of nitrogen compounds, these are not encountered in the animal system. Consequently, for body fluids, tissues, and excretions, the Kjeldahl method determines total nitrogen [12].

Transfer of liberated ammonia to the acid receiver by steam distillation is now the method of choice for all Kjeldahl analyses. Micro apparatus is commercially available, easy to use, almost automatic in operation, and requires only a few minutes for each sample. The Kjeldahl method is used as a reference against which all other protein methods are calibrated because it is very precise and accurate. Only eight years after the method was published, an investigator wrote: "In the history of analytical chemistry, no method has been so universally adopted, in so short a time, as the 'Kjeldahl method' for the estimation of nitrogen." [13].

Meanwhile, by the 1830s, the determination of the elementary composition and empirical formulas of many organic compounds isolated from the tissues and fluids of the human body could be made with reasonable accuracy and had become routine. However, only sketchy speculations about the internal mechanisms could be made without more complete information about the many compounds which are found in the animal body. It was not until after the mid-nineteenth-century advances in organic synthetic chemistry that the structural relationships among the substances could be used in the interpretation of their formation or alteration in health and disease [14].

BERZELIUS: INTERNATIONAL AUTHORITY

The analytical methods and determinations of atomic weights by Jöns Jakob Berzelius (1779–1848) (Fig. 4.3) played a key role in building a foundation for continued advances in chemistry. His 1818 table of atomic weights listed values for 45 of the 49 elements then known. Thirty-nine of the determinations were his; six were by his students. Berzelius discovered selenium and thorium and was co-discoverer of cerium. The abundance of rare minerals and ore deposits in Sweden led to the development of mineral analysis to a high degree in that country and explains in part why as many as twenty-three elements were discovered by Swedes. Lack of coal and petroleum, on the other hand, probably accounts for the absence of a tradition in synthetic organic chemistry.

In 1813 Berzelius suggested that each element (and a single atom) be represented by the first letter of its Latin or English name. In the case of duplicate initials, a second letter from the name is added. He wrote the formula of a compound with numerical superscripts to indicate the number of atoms in the molecule. Later, the numbers were written as subscripts [15].

Berzelius, a physician, was the leading chemist in Europe during the first half of the nineteenth century. With no adequate Swedish textbook on chemical subjects, he published a book which summarized the knowledge

FIG. 4.3. Jöns Jakob Berzelius. Portrait by O. J. Södermark, 1843. (Swedish Information Service, New York, NY.)

of that time on animal chemistry (1806) and included the results of numerous analyses he had made on animal tissues and fluids. The first of its kind, the book directed the interest of physiologists and chemists to the chemical composition and processes taking place in the animal body. He defined organic chemistry as "the part of physiology that describes the composition of living bodies, together with the chemical processes that occur in them" [16].

In 1808 Berzelius published the first volume of a general textbook of chemistry, which became the most authoritative chemical text of its day. Beginning in 1821, he also published an annual survey of the advances of physics and chemistry. Translated into German and widely distributed, these *Berichte* gave critical reports, not just abstracts. Berzelius had to design and build almost all his own scientific apparatus and to synthesize most of his own reagents because what was available in Sweden was inadequate. His improved analytical techniques and the new apparatus for studying animal fluids and tissues became standard. He was especially skillful with the blowpipe [17] and popularized its use abroad.

For many years Berzelius was the international authority for new chemical terms. In addition to the modern system of chemical symbols for the elements, he also introduced isomerism, catalysis, polymer, amino acids, glycine, cystine, and protein [18]. Berzelius's influence on the chemists of his time came not only from the skill of his laboratory techniques and experimental discoveries but also from his power to collect diverse facts to produce important generalizations, critiques, and summaries which he published. During the eighteenth century, the continuing demands of mining, metallurgy, the textile industry, and the importance attached to the medicinal value of mineral waters, had furthered the analytical chemistry of inorganic compounds. More than anyone else, it was Berzelius who combined analytical skill and theoretical insight to convince most of his contemporaries that the molecules of organic compounds were, like those of inorganic compounds, composed of elements present in fixed and multiple proportions [19]. As he grew older he resisted the new developments which contradicted some of his own theories. Unable to accept any modification of the ideas of his most active years, he spent much time trying to discredit the new ideas in organic chemistry.

"VITAL FORCE"

Organic compounds were regarded as puzzling substances which would not behave according to the then current laws of chemistry. It was widely believed that they were only produced by living organisms. Although chemists, many with degrees in pharmacy or medicine, could easily decompose organic substances to the inorganic state by heating or other harsh treatment, they did not yet know how to accomplish the reverse and long rejected the idea that Nature could be duplicated in the laboratory. The changes accompanying the passage of matter through a living organism, they contended, ran contrary to the laws of inorganic chemistry, which did not apply to living matter.

To move atoms and radicals around so precisely and exactly to a specific place in the molecule and produce a new compound would seem to require more than the known laws of chemistry could account for. These compounds had radically different properties from those of mineral origin studied in inorganic chemistry. Furthermore, combinations of the same elements were often found to exhibit different properties, depending on whether they occurred in organic or inorganic substances. Consequently, most chemists of that time thought that a special "vital force," operating only within organized living things, was required to synthesize organic substances and convert inorganic material into organic compounds. As a result, only organic compounds were identified or associated with living forms. This concept of vitalism developed chiefly from the theory of animism preached a century earlier by Georg Ernst Stahl (1660–1734), professor of medicine and chemistry at the University of Halle. In his system, all the regulatory directive forces functioning in living organisms were combined into an aggregate he named *anima*. This intelligent agent had the power to organize and rule matter with purposeful goals, and to use the human body as the instrument for achieving these ends. The doctrine of vitalism had a profound influence on the thinking of chemists and physicians for many years after Stahl's death.

After the discovery of fire, it was inevitable that mankind would divide the world into matter that burned and that which did not. From this recognition it was a short step to characterize these two states (combustible and non-combustible) as originating from the world of the living and the non-living, respectively, and furthermore, that a peculiar force must be responsible for the difference in their behavior. It was clear that non-living substances could bear up under harsh treatment, whereas substances from living or once-living matter were easily altered and could not be reconstituted as before. The differences seemed irreconcilable.

Invoked by the organic chemists, vital force was considered an impenetrable barrier between the processes of living systems and the man-made reactions of the laboratory. According to the vitalists, there were two distinct sets of natural law: one for the living (no matter how primitive) and one for the inanimate. Ordinary laboratory methods were believed to be inapplicable to living substances, and it was thought improbable that they would ever be prepared in the laboratory.

WÖHLER'S SYNTHESIS OF UREA

The doctrine of vitalism experienced a major challenge in 1828. Friedrich Wöhler (1800–1882) (Fig. 4.4), a German chemist, reported that the evaporation of an aqueous solution of ammonium cyanate, regarded as an inorganic substance and unrelated to living matter, resulted in the production of urea, $NH_4CNO \rightarrow NH_2$—CO—NH_2, a natural waste product in man and many animals. Wöhler's paper, "The Artificial Formation of Urea," was only four pages long. He had reacted ammonia with cyanic acid (HCNO) and expected to obtain ammonium cyanate. This was formed, but upon evaporation it underwent an internal rearrangement into urea. Experimenting further, he found other ways to prepare urea. The best way was to react ammonium chloride with silver cyanate, or ammonia and lead cyanate:

$$Pb(CNO)_2 + NH_3 \rightarrow NH_4CNO \rightarrow NH_2\text{—CO—}NH_2$$

Wöhler made no resounding claims, but merely remarked that the unexpected formation of urea is so far "noteworthy in that it furnishes an example of the artificial production of an organic, indeed a so-called animal substance, from inorganic materials." In a letter to Berzelius in 1828 he wrote: "I must now tell you that I can make urea without calling on my kidneys, and indeed without the aid of any animal, be it man or dog" [20]. Wöhler was aware of the change that occurred and added that it had been observed in the case of cyanic and fulminic acids, two distinctly different substances that he and Justus Liebig (1803–1873) (see Chapter 6) discovered. Berzelius congratulated his former pupil on the discovery and later identified it as another example of isomerism (Greek: *isomeros*, composed of equal parts), a term he proposed in 1832.

Following the new chemistry of Lavoisier, analyses and syntheses revealed a great variety of new compounds. It was generally assumed that each of these compounds must have its own unique composition. Yet here were two compounds, urea and ammonium cyanate, totally different in their chemical properties, but yielding an identical analysis, i.e., composed of the same elements in the same proportions. The realization slowly

FIG. 4.4. Friedrich Wöhler. (National Library of Medicine, Bethesda, Maryland.)

developed that structure was as important in organic chemistry as composition.

Basic to vitalistic thought was the belief that only living organisms—plant or animal—could produce organic compounds and that these could never be made in the laboratory, and that therefore, there was an impassable gulf between the organic and the inorganic world. Nevertheless, the synthesis had no immediate effect on the vitalistic outlook of the chemical community. It was generally recognized by Wöhler's contemporaries that he had made an important discovery. But the artificial preparation of urea in itself did not overthrow vitalistic doctrine. Chemical textbooks published during the decade following Wöhler's synthesis made little of the anti-vitalistic significance. This came to be emphasized by more modern treatises written after the breach in vitalistic organic chemistry had been widened by later discoveries [21].

The transition from a doctrine of vital forces to a unified scheme of chemistry that applies to organic and inorganic compounds was not sudden and dramatic. Science does not advance in that way. It was a function of time and the steady accumulation of contradictory evidence. It could be argued that Wöhler's synthesis of urea was merely the result of an internal rearrangement of the atoms of ammonium cyanate, and that this was not an inorganic compound. Nevertheless, the achievement was certainly one of the events at the start of the synthesizing activities of the chemists which was finally to expel "vital force" from organic chemistry. The intrinsic historic importance of Wöhler's synthesis was not in the synthesis itself but in directing attention to the possibility of synthesizing other organic compounds. Chemists now felt encouraged to attempt these syntheses instead of pursuing established areas of research based mainly on analysis.

ORGANIC SYNTHESIS BY BERTHELOT

Then, in 1845, Adolph Wilhelm Hermann Kolbe (1818–1884), a pupil of Wöhler's, synthesized acetic acid, without doubt an organic compound. However, Kolbe's synthesis of acetic acid and later salicylic acid did not settle the question of vital force, because the starting material here, as in the case of urea, was a carbon derivative. The inability to link elemental carbon and hydrogen was believed to be an insuperable barrier between the organic and inorganic that could only be overcome by a vital force.

Finally, during the 1850s, the French chemist Pierre Eugène Marcellin Berthelot (1827–1907) (Fig. 4.5) went beyond the synthesis of compounds by special reactions and developed a protocol for the general method of synthesizing the entire range of an almost limitless number of organic compounds. Acetylene became the key starting point in his system since by oxidation or reduction he obtained precursor materials that were clearly

FIG. 4.5. Marcellin Berthelot. (*Comprehensive Biochemistry*, M. Florkin and E. H. Stotz, eds., vol. 30, *A History of Biochemistry* (Marcel Florkin), Amsterdam: Elsevier Publishing Co., 1972.)

inorganic for synthesis of a series of other organic compounds, including methyl alcohol, ethyl alcohol (traditionally the product of fermentation of sugars with yeast), methane, and benzene. As for acetylene, he achieved this monumental synthesis by direct combination of carbon and hydrogen by passing hydrogen through an electric arc formed between carbon electrodes. This established the final link between organic and inorganic chemistry and overcame the mental barrier set up by the chemists. What further proof could the vitalists want that there were not two chemistries?

In 1854 Berthelot heated glycerol with stearic acid and produced tristearin, the triglyceryl ester of stearic acid, that was identical with tristearin from natural fats. The tristearin of Berthelot was the most complex natural product to be synthesized up to that time. He repeated this procedure with acids similar to stearic, but not obtained from natural fats and produced substances strongly resembling ordinary fats, but not like any fats known in nature. It soon became routine to synthesize organic-type compounds unknown in a living organism. The artificial division into organic and inorganic on the basis of production by living tissue was losing its rationale. However, there still were lingering doubts.

The organic compounds prepared by Wöhler, Kolbe, and Berthelot were relatively small molecules. Carbohydrates, fats, and proteins were more characteristic of life and much more complex. The proteins especially were not easy to handle. They were large molecules, unstable, difficult to isolate, to analyze, and to characterize. Throughout the nineteenth century the vitalists hoped that the proteins would help them to redeem their views. Unfortunately, neither the techniques nor the concepts needed for protein investigation were available at that time. The relatively small molecules of sugars and fats were amenable to the existing methods of organic chemistry. They could be purified, analyzed, their properties studied, and their transformation followed during metabolism. In an effort to bring common ground to the rival camps, Friedrich August Kekulé (1829–1896) stated in 1859 that there was no difference between organic and inorganic compounds and defined organic chemistry as merely the chemistry of carbon compounds. Notwithstanding the new definition, the controversy between chemical and vitalist philosophy was not resolved until the end of the century when chemistry and biology were reconciled.

THE ANALYTICAL BALANCE

With the introduction of quantitative scientific chemistry by Lavoisier and his contemporaries during the last quarter of the eighteenth century, synthesis could now be followed by analysis, and gravimetric methods came into prominence. Titration procedures were developed later.

The art and science of weighing was known in Egypt as early as 3000 B.C.E. There is evidence of a system of weights and measures in Babylon, about 2300 B.C.E. The earliest written reference to an equal arm balance dates back to approximately 1300 B.C.E. as described in a papyrus that depicts the soul of the deceased being judged against the weight of a feather on the opposite pan [22].

Analytical balances before the first quarter of the nineteenth century were custom made by craftsmen who specialized in the construction of scientific instruments. The first balances with standardized design for commercial use were made by a London instrument maker named Robinson, shortly before 1825. These balances were the first of the precision type to be used in America. In the United States such balances were first manufactured and sold about 1855 by Christopher Becker, a Dutch instrument maker who had settled in New York City. His success with the balance led Becker to devote the entire business to their manufacture in association with his sons [23].

In the analytical balance in use in the first half of the nineteenth century, a pointer attached to the center of the beam moved over a graduated scale placed low on the pillar supporting the beam. The pans were suspended by thin metal wires or silk threads from hooks at the ends of the beam. Weighing was tedious because of the slow movement of the long beam, the absence of pan arrests, and because all weights, even the smallest, had to be added and removed from the pan by forceps. The sensitivity of these early balances was no better than about 0.5 mg.

The graduated beam with rider was introduced at a London exhibition in 1851 by L. Oertling. Becker's balances in the third quarter of the nineteenth century had pan arrests and agate knife edges for supporting the pans. The chain weight device was invented in France in 1890 for an industrial use. The first successful chainomatic balance—which bypasses the use of a rider—was developed in 1915 by Christopher A. Becker, grandson of the company founder. In 1945, Erhard Mettler departed from the equal arm construction and introduced the substitution principle.

VOLUMETRIC (TITRIMETRIC) ANALYSIS

French chemists made many of the basic contributions to volumetric analysis that were started in the late eighteenth century, but were largely perfected and developed in the nineteenth century. Pioneer work was carried out by François Antoine Henri Descroizilles (1751-1825) who devised the earliest burette in 1791 for controlling the concentration of chlorine water in the bleaching industry. The burette was more like a graduated cylinder. The closed lower end fitted into a stand, while the open upper end had a flared rim. The chlorine water to be assayed was placed at the bottom, and the indigo solution was added until it was no longer decolorized by the fixed volume of chlorine water [24].

The subsequent work of Joseph Louis Gay-Lussac (1778-1850) was more influential in bringing the concept of titration into chemical practice. His 1824 paper on the standardization of indigo solution for the determination of chlorine contains the first use of the terms "pipette" and "burette" for those pieces of apparatus. He later introduced the verb "to titrate." The pipette had essentially the form of its modern counterpart. The burette was more like a graduated cylinder with a connecting side arm and was an improvement on that used earlier by Descroizilles. Gay-Lussac showed that volumetric analysis was convenient, rapid, and accurate. However, the establishment of a general system of volumetric analysis had to wait until the next generation of chemists.

KARL MOHR

The breakthrough of methodology in titration was largely the work of Karl Friedrich Mohr (1806-1879) (Fig. 4.6) [25]. His inventive genius was responsible for many innovations that propelled analytical chemistry forward. He devised the burette named after him, as well as various types of pipettes, a cork borer, specific gravity balance, automatic pipette washer, goggles for dangerous experiments, fume hood for domestic laboratories, gas generating apparatus, and gasometers. He devised many new titration procedures and often improved the older ones. The burettes of his time, which were modifications of Gay-Lussac's design, were inaccurate. The glass stopcocks tended to stick and the controlled addition of standard solution was difficult. Mohr redesigned the burette by inventing the pinchclamp and a burette tip which could deliver fractions of a drop and provide a more accurate end point.

Mohr's greatest contribution to the new field of analytical chemistry was his textbook on volumetric analysis, *Lehrbuch der Chemisch-analytischen Titrirmethode* (1855). It went through numerous editions and translations into the next century. There are very few references in the book, and Mohr often received the credit which was due to others. The book's popularity,

FIG. 4.6. Karl Friedrich Mohr. (Ralph E. Oesper, *Journal of Chemical Education*, 1927, 4: 1357–1363.)

more than any other factor, was responsible for the general adoption of titration methods and the *normal* system of standard solutions which was already in use by the late 1840s. The use of equivalent weights was a challenge to those chemists who still preferred to employ solutions measured in percentages.

Mohr's other outstanding contributions to analytical chemistry include his method for volumetric determination of chloride with standard silver solution and potassium chromate as indicator; the Mohr test for iron;

introduction of oxalic acid as a primary standard in alkalimetry; ferrous ammonium sulfate (Mohr's salt) as primary standard for potassium permanganate; protection of solutions from air by a layer of mineral oil; and many analytical procedures which, if no longer in use, often formed the basis of present-day methods. Titrimetry did not come into its own until indicators such as phenolphthalein and methyl orange were synthesized and a theoretical explanation of their function and the theory of hydrogen ion concentration was provided by the physical chemists [26].

The application of volumetric methods in the growing chemical industry probably furnished some of the impetus which gave Germany world leadership in chemistry. Mohr had a critical and combative nature which made many enemies, but he maintained an intimate friendship with Liebig, who also had an adversarial personality, for more than thirty-five years.

NOTES AND REFERENCES

1. JOSEPH S. FRUTON, "The Emergence of Biochemistry," *Science*, 1976, 192: 327–334.
2. D. C. GOODMAN, "The Application of Chemical Criteria to Biological Classification in the Eighteenth Century," *Medical History*, 1971, 15: 23–44, p. 23.
3. *Ibidem*, pp. 29–34, 44.
4. JOSEPH S. FRUTON, *Molecules and Life. Historical Essays on the Interplay of Chemistry and Biology* (New York: Wiley-Interscience, 1972), pp. 1–21, p. 7.
5. FREDERIC L. HOLMES, "Analysis by Fire and Solvent Extractions: The Metamorphosis of a Tradition," *Isis*, 1971, 62: 129–148, p. 148.
6. MIKULÁŠ TEICH, "On the Historical Foundations of Modern Biochemistry," *Clio Medica*, 1965, 1: 41–57, pp. 47, 52, 53.
7. FERENC SZABADVÁRY, *History of Analytical Chemistry*, translated from Hungarian by Gyula Svehla (London and New York: Pergamon Press Ltd., 1966). Reprinted in 1992 (Langhorne, Pennsylvania: Gordon and Breach Science Publishers S.A.), pp. 295–297.
8. *Ibidem*, pp. 232, 297.
9. J. KJELDAHL, "Neue Methode zur Bestimmung des Stickstoffs in organischen Körpern," *Zeitschrift für Analytische Chemie*, 1883, 22: 366–382.
10. RALPH E. OESPER, "Kjeldahl and the Determination of Nitrogen," *Journal of Chemical Education*, 1934, 11: 457–462; S. VEIBEL, "Johan Kjeldahl (1849–1900)," *Journal of Chemical Education*, 1949, 26: 459–461; H. Holter and K. Max Møller, eds., *Carlsberg Laboratory, 1876/1976* (Copenhagen, Denmark: Carlsberg Foundation, Rhodos Publishing House, 1976), pp. 50–62. Also see LOUIS ROSENFELD, *Origins of Clinical Chemistry. The Evolution of Protein Analysis* (New York: Academic Press, 1982), pp. 49–64.
11. L. F. WICKS, "A Cheaper Nessler's Reagent by the Use of Mercuric Oxide," *Journal of Laboratory and Clinical Medicine*, 1941, 27: 118–122.
12. JOHN P. PETERS and DONALD D. VAN SLYKE, "Total and Non-protein Nitrogen," in *Quantitative Clinical Chemistry*, vol. 2, *Methods* (Baltimore, Maryland: Williams & Wilkins Company, 1932), pp. 516–538; "Proteins of Urine, Blood Plasma, and Body Fluids," *Ibidem* (1932), pp. 678–700; FREDERICK C. KOCH and MARTIN E. HANKE, "The Quantitative Analysis of Urine," in *Practical Methods in Biochemistry*, 4th ed. (Baltimore,

Maryland: Williams & Wilkins Company, 1943), pp. 222–251; OTTO FOLIN and HSIEN WU, "A System of Blood Analysis," *Journal of Biological Chemistry*, 1919, 38: 81–110; GLENN E. CULLEN and DONALD D. VAN SLYKE, "Determination of Fibrin, Globulin and Albumin Nitrogen of Blood Plasma," *Journal of Biological Chemistry*, 1920, 41: 587–597.

13. HUBERT BRADFORD VICKERY, "The Early Years of the Kjeldahl Method to Determine Nitrogen," *Yale Journal of Biology and Medicine*, 1946, 18: 473–516 and reference to Kebler; R. B. BRADSTREET, "A Review of the Kjeldahl Determination of Organic Nitrogen," *Chemical Reviews*, 1940, 27: 331–350.
14. JOSEPH S. FRUTON, "Biochemistry and Clinical Chemistry—A Retrospect," *Journal of Clinical Chemistry and Clinical Biochemistry*, 1982, 20: 243–252, p. 243.
15. WILLIAM H. BROCK, *The Norton History of Chemistry* (New York: W. W. Norton & Company, Inc., 1993), pp. 154–155.
16. J. R. PARTINGTON, *A History of Chemistry*, vol. 3 (London: Macmillan & Co Ltd, 1962), p. 198; FRUTON (1976), p. 328.
17. The object of the blowpipe is to admit (blow) air into the flame and produce a high temperature. It was used by goldsmiths since antiquity, and by the seventeenth century, in the glass industry, to build up the fire of small furnaces. It was used by Swedish analysts at least as early as the 1740s, but was first described for systematic chemical analysis in 1758 by Alexander Frederik von Cronstedt (1702–1765), a Swedish Master of Mines, and the discoverer of nickel. Initially, the flame of wax and tallow candles was used; later on, lamps were designed to burn different oils. SZABADVÁRY (1992), pp. 50–53; ERNEST CHILD, *The Tools of the Chemist. Their Ancestry and American Evolution* (New York: Reinhold Publishing Corporation, 1940), p. 135.
18. HUBERT BRADFORD VICKERY, "The Origin of the Word Protein," *Yale Journal of Biology and Medicine*, 1950, 22: 387–393.
19. FRUTON (1976), p. 328 and reference 17 (p. 333).
20. FREDERICK GOWLAND HOPKINS, "The Centenary of Wöhler's Synthesis of Urea (1828–1928)," *Biochemical Journal*, 1928, 22: 1341–1348, p. 1342.
21. *Ibidem*, pp. 1345, 1341; W. H. WARREN, "Contemporary Reception of Wöhler's Discovery of the Synthesis of Urea," *Journal of Chemical Education*, 1928, 5: 1539–1553; TIMOTHY O. LIPMAN, "Wöhler's Preparation of Urea and the Fate of Vitalism," *Journal of Chemical Education*, 1964, 41: 452–458; for a critical and negative interpretation, see DOUGLAS MCKIE, "Wöhler's 'Synthetic' Urea and the Rejection of Vitalism: A Chemical Legend," *Nature* (London) 1944, 153: 608–610.
22. SZABADVÁRY (1966), p. 5; *A History of Analytical Chemistry*, Herbert A. Laitinen and Galen W. Ewing, eds. (*American Chemical Society*, 1977), p. 8.
23. CHILD (1940), pp. 75–91; LAITINEN and EWING (1977), p. 9.
24. CLÉMENT DUVAL, "François Descroizilles, the Inventor of Volumetric Analysis," *Journal of Chemical Education*, 1951, 28: 508–519, p. 514; see also SZABADVÁRY (1992), pp. 208–215, 220–222. The burette was known initially as *berthollimètre* to identify it with the Berthollet bleaching process for cloth. Descroizilles's first publication on the method and the use of this instrument was in 1795.
25. RALPH E. OESPER, "Karl Friedrich Mohr," *Journal of Chemical Education*, 1927, 4: 1357–1363; JOHN MARK SCOTT, *Chymia*, "Karl Friedrich Mohr, 1806–1879, Father of Volumetric Analysis," 1950, 3: 191–203. Also see *Dictionary of Scientific Biography* (*DSB*), Charles Coulston Gillispie, ed. (New York: Charles Scribner's Sons, 1974), 9: 445–446; SZABADVÁRY (1992), pp. 241–247 ff.
26. BROCK (1993), pp. 183–184; also see SZABADVÁRY (1992), pp. 260–263 ff.

CHAPTER V

INTRODUCTION

The French Revolution of 1789 that overthrew the monarchy produced a medical revolution that swept away the "systems" that had ruled medicine during the seventeenth and eighteenth centuries. The old regime's academies and institutions of medicine and education, anchored in medieval traditions, were abolished. In the reorganization that followed, there developed a very new kind of medical teaching that was centered in the clinic and the hospital and made the theoretical and speculative discourses of the past obsolete.

French medicine at the time of the Revolution had reached a low point and one goal of the revolutionaries was a radical reform of the health care system. In 1791, Fourcroy (Fig. 5.1), a chemist and non-practicing physician who was interested in the application of chemistry to medicine and in the changes that accompany the pathologic state—devised what probably was the first plan for establishing clinical laboratories in hospitals. It came as part of the reforms that swept away the archaic institutions of the French monarchy in the wake of the Revolution. He proposed under the title *Idées sur un nouveau moyen de rechercher la nature des maladies* that a chemical laboratory be located near the wards that had twenty or thirty beds, where chemical analysis of urine, excretions, and other discharges of the sick could be carried out by young physicians who were well trained in modern natural sciences. Fourcroy believed such investigations represented a new means of studying diseases. Inasmuch as the chemical laboratory of the seventeenth and eighteenth century served primarily to prepare chemical medicines for the physician, the idea of chemical examinations on hospital patients in a hospital laboratory was quite a new concept [1].

Although Fourcroy's proposal to the National Convention resulted in the establishment of a large teaching hospital in Paris, with laboratories and equipment, most hospitals were not able to set up their own clinical laboratories. As a result, during the nineteenth century, clinical chemical analyses were done in the pharmacist's laboratory. This arrangement suited French clinicians who were less interested in chemical than in

FIG. 5.1. Antoine François de Fourcroy. From an oil painting by François Gérard in the Versailles Museum. (© Photo RMN-Gérard Blot.)

physiological methods [2]. In any event, the practical implementation of Fourcroy's concept was destined to fail at this early stage since adequate chemical methods were not yet available. However, there were initial successes with the analysis of urinary calculi (see Chapter 3) aided by the chemical expertise developed in assays of ores.

Fourcroy served on several reform committees of the Revolution. Appointed as a councilor of state by Napoleon in 1799, Fourcroy played a large part in drafting a new educational system, from primary schools to advanced colleges; and in 1802 Napoleon chose him as director-general of public instruction to implement the proposals.

LIBRARY MEDICINE, BEDSIDE MEDICINE, AND THE STIGMA OF DISSECTION

There were no facilities anywhere in England for the systematic teaching of medicine or surgery. Instruction was mainly by lectures illustrated by plates, tables, and colored drawings. Adoption of bedside teaching in the London hospitals was slow, for as late as 1834, clinical lectures were described to a Parliamentary Medical Committee as *"quite a new thing"* [3]. Complaints from medical students of not hearing any clinical lectures continued to reach *The Lancet* as late as 1841 [4].

What took place at Oxford and Cambridge, "where Physic was either not taught at all, or taught in such a manner as to make it a byeword and a jest," [5] consisted only of reading the ancient Greek and Latin texts and in the exposition of theoretical works. The *practice* of medicine and surgery had to be learned by apprenticeship. Consequently, students sought their medical *education* at Edinburgh or Leyden, Padua, Montpellier, Bologna, and other foreign universities. After graduating there in arts and medicine with the M.D. degree (sometimes after only a very short period of attendance), on their return to England their own university, Oxford or Cambridge, usually conferred its degree of M.D., "by incorporation" *ad eundem*. This signified the honorary granting of a degree or academic standing to one whose actual work for the degree or academic standing was done elsewhere.

In contrast to the unadventurous spirit of Oxford and Cambridge universities, where institutional inertia and all kinds of barriers blocked the introduction of scientific work, the Scottish universities were not asleep and were tuned to the currents of change on the continent. Edinburgh and Glasgow had flourishing medical schools where the scientific prerequisites to medicine were taught. In any event, medicine, as a profession, did not enjoy great prestige in England during the first 60 years of the nineteenth century. Society preferred the Church, the Law, and Government or the Armed Forces—in that order—as suitable occupations for a gentleman [6].

There was another serious disadvantage to British medical education. Students of anatomy and surgery in England, Scotland, and Ireland faced a shortage of legally-obtained cadavers for dissection or demonstration.

In Great Britain, the law, dating back to 1540, allowed a limited number of dissections of executed criminals. This produced a deep-seated identification of dissection with the commission of serious crimes during life, and strengthened the public feeling against dissection, even of unclaimed bodies of paupers who died in workhouses, prisons, and hospitals. Although legal medical postmortems were tolerated—these were perceived as distinct from *dissection*—they were viewed uneasily and were not frequently requested [7]. The public opposition to legalized dissection was finally overcome by the Warburton Anatomy Act of 1832. But first, Parliament had to repeal the law mandating the dissection of all executed murderers, to remove the connection in the public's mind between dissection and executed criminals.

There was no such problem in Paris where a large percentage of hospitalized patients died and were legally available for a small fee for dissection and research, if not claimed for burial by friends or relatives within 24-hours [8]. The teaching of bedside medicine in the Paris clinic and hospital was in sharp contrast with the situation in England.

The reforms and reorganization in the wake of the Revolution produced a new doctor–patient relationship. By the 1820s, the Paris physician no longer merely "observed" the patient, now he "examined" him. The new practice of medicine was based on the patient's history, physical examination by hand (percussion) and ear (auscultation), pathological anatomy, and statistics. Students followed physicians and surgeons from ward to lecture room to dissecting theater and attempted to correlate before-death external symptoms and patient's complaints and appearances, with the postmortem pathologic appearances of tissues and organs. Hundreds of students and doctors from all over the world were drawn to Paris by the new teaching of pathological anatomy—no longer to read, but to see and to do [9].

THE STETHOSCOPE AND THE CHANGING PATIENT–PHYSICIAN RELATIONSHIP

An additional attraction drawing foreign students with clinical interests to Paris was the new technique of listening with the stethoscope. It followed the publication in 1821 of John Forbes's (1787–1861) English translation of parts of Laennec's *Traité de l'Auscultation Médiate*. The stethoscope, invented by René-Théophile-Hyacinthe Laennec (1781–1826), was the first major diagnostic tool available to clinical medicine. It was a momentous advance and marked a new period in clinical medicine. This instrument enabled the physician to visualize pathological changes while the patient was still alive, thereby changing both the physician's perceptions of disease and his relation to the patient. The physician no longer merely observed his patients and listened to their complaints. Now he examined them and formed a diagnosis primarily from objective findings in a more detached relationship. The heart and lung sounds were independent physical evidence hidden from the patient and which the patient could not simulate, conceal, exaggerate, or minimize. The dislike of close physical contact and use of the hands, because it might damage the professional and social status of the physician and cause him to be classed with the surgeon as mere craftsman, was lessened by placing an instrument between doctor and patient [10]. Unfortunately, as so often happens when something new is introduced, the older men tended to ridicule what they did not understand and found difficult to learn. Besides, as many pointed out, there was no advantage to this new technique since the treatment was usually the same anyway.

Few diseases had been sufficiently differentiated to be easily described and about fewer still were the causes and pathological processes known. Diagnosis was in general terms, e.g., "fever," "liver disease," "stomach trouble," "kidney problems." The practitioner had to proceed empirically, as he had always done, hoping that his treatment this time would be as successful as it had been previously, in what seemed to be a *similar* case. He used a "shot-gun" pharmacy, sometimes helpful, oftentimes not. Working largely by guess and hope, the doctor could not be sure whether recovery was due to the passage of time, the remedies, the nature of the disease, or family prayers [11].

THE CHANGING ROLE OF THERAPY

Between the 1820s and the 1880s, medical therapeutics in America was fundamentally altered [12]. Traditional medical practices, founded on relieving the symptoms of sick individuals, began to be displaced by strategies that minimized differences among patients. Previously, medical theory had stressed the principle of specificity—the notion that treatment had to be matched to the particular characteristics and environments of individual patients. By the last third of the century, the objective of therapy changed from restoring the individual patient's natural state of health to correcting the abnormal state. This would be done by bringing disordered processes into the reference ranges for a normal population as defined by laboratory science.

The shift from the physician's judgment to knowledge derived from experimental science resulted from a change in understanding of the meaning of restoration of health. It involved a subtle change in medical language. Whereas the term *natural* was commonly used throughout the first half of the century to describe the desired state of well-being, by the mid-1870s, *normal* had almost completely replaced *natural* when used in this way.

For much of the nineteenth century, physicians viewed disease as essentially a systemic imbalance. The fundamental objective of therapy was to restore the body's natural state of balance. This was defined separately for each individual and was shaped by factors such as ethnicity, gender, and family background. What was considered natural changed with season, age, altered social position, and geographical location.

With only a few exceptions (smallpox and syphilis being the most notable), etiology in the early nineteenth century was nonspecific. Diseases were thought to be generated not by discrete causative agents, but by a variety of factors acting singly or more often as a group to destabilize the system. By mid-century, physicians began to think of disorders less as systemic imbalances in the body's natural harmony, and more as complexes of discrete signs and symptoms that could be identified, analyzed, separated, and measured. The signs of bodily order and disorder were reduced to objectively measured and quantified values. With the developing concepts of specific diseases and specific causes, specific therapy came into play and diagnosis attained a new therapeutic importance.

MEASUREMENT AND THE FOCUS ON SPECIFIC DISEASES

Quantitative advances in physics and astronomy spurred an interest in experiment and the use of numbers by medical men. However, quantitation did not meet with immediate success or acceptance by medical practitioners probably because of the belief that body functions were best understood by study of morphology. Furthermore, the idea of "numbers" was new to the physicians, and in biologic phenomena—which were complex to begin with—measurement appeared to deprive human phenomena of its mystery and beauty. Physicians were convinced that they could determine the nature of each "clinical picture" by the use of their senses alone. Even easily made medical measurements such as pulse and temperature—advocated before 1700—were largely ignored until the second half of the nineteenth century. The antipathy to quantification by physicians, as late as the mid-nineteenth century, stemmed from their dislike of submitting their insights and cumulative wisdom to numerical tests. They prided themselves on their intuitive skills in making a diagnosis. Such feeling usually appears within any discipline when it is first threatened, as this one was by quantitation [13].

Clinical science came into being in the early part of the nineteenth century when interest shifted from a generalized (humoral) to a localized (structural) pathology, and to a focus on specific diseases. Clinical medicine abandoned the age-old and almost fruitless search for cures and began to concentrate upon a search for the cause and mechanisms of disease. It gradually became recognized that only after various forms of disease could be distinguished was it possible to pursue their prevention and cure in a scientific manner. Once it was possible to identify diseases, then quantitative studies could be undertaken to reveal the incidence of the disease and show the effectiveness of a particular treatment.

This newer approach in the early 1800s not only called for careful study at the bedside and autopsy table, but also for the application of the methods of the physical sciences. This approach brought about the utilization of physical instruments such as the microscope and stethoscope, and later on, timing devices to record pulse and respiration. In addition to physical methods, chemical procedures were gradually introduced and developed for the study of the urine.

Not all doctors accommodated to change and continued to treat patients according to the authoritative judgments of past knowledge, often bestowing onto their suffering patients the mistakes of the past. Others, attracted by the novelty of change, readily tried anything new despite doubtful evidence and false claims. Change became a way of life without the necessity of a personal commitment. It was an uneasy alliance with modern science. The rapid changes in medicine were also occurring in society at large. A combination of forces was generating a new social structure, and change and contradiction were accepted as the natural course of events [14].

The emphasis upon morbid anatomy and the possibility of a correlation between the anatomical changes found at autopsy with clinical findings of disease before death was the great medical advance of French medicine in the early nineteenth century. The large Paris hospitals that were built in the wake of the Revolution presented previously unavailable opportunities for comparative studies of observation and autopsies on a mass scale. The large number of similar cases gave greater certainty to medical diagnosis. Gradually, instead of thinking in terms of the sick individual as a single entity, the profession now began to think of *diseases*, and to classify them [15].

During the first four decades of the nineteenth century, the French clinical school brought about a complete change in the character of scientific medicine. By mid-century, however, a stagnant routine set in, and it was obvious that statistical compilations, physical examination, and gross pathological anatomy had produced as much progress as it could. When new directions were called for, the French clinicians were unable to adapt. French "hospital medicine" had maneuvered itself into a dead-end. The French were simply unable to assimilate foreign contributions—not due to ignorance, but to a "provincialism of conceit." The French school of anatomic pathology was primarily interested in diagnosis, but the real problem was finding the cause of disease. This was considered to be unsolvable. That was because the entrenched hospital-based clinicians had neglected and ignored the basic sciences and their own great French scientists. Medical leadership passed to Germany where "hospital medicine," having displaced the "library" and "bedside medicine" of its predecessors, was now succeeded by the new "laboratory medicine" which featured chemistry, microscopy (histology), and animal experimentation (physiology). The English stopped studying in Paris and began going to Germany. The Americans soon followed in large numbers [16].

BRIGHT'S DISEASE, ALBUMINURIA, AND SCIENTIFIC MEDICINE

Three outstanding practitioners of the new pathology were Richard Bright (1789–1858), Thomas Addison (1793–1860), and Thomas Hodgkin (1798–1866). They were medical graduates from the University of Edinburgh who began to teach at Guy's Hospital in London in the early 1820s. They followed the new scientific approach to medicine of careful and detailed observation to relate postmortem findings to the clinical picture seen in the wards. Addison and Bright were both excellent clinical observers, and in Hodgkin they had an outstanding morbid anatomist (pathologist) with an equally keen interest in the correlation of disease and symptoms.

In the diseases named for Addison and Bright, physiological processes during life are correlated with an anatomical condition found after death. Addison observed the peculiar pigmentation or "bronzing" of the skin along with pallor and lack of energy during life and noted their association with the diseased adrenal glands (then called supra-renal capsules) found after death. Bright correlated the presence of edema in the tissues, albumin in the urine, and renal failure during life, with diseased kidneys found at autopsy. However, Bright was not the first to describe edema associated with albuminuria or kidney disease. An association between dropsy and hardened kidneys had been pointed out by the Italian surgeon William of Saliceto (Guglielmo Saliceti) (1210–1277) [17].

The discovery of albuminuria is generally credited to Domenico Cotugno (1736–1822) of Naples in 1764, but it was Fredericus Dekkers (1648–1720) of Leyden who first observed the precipitation of albumin in urine in 1694. Chemistry made its entry into clinical diagnosis when Dekkers thought that wasting diseases might be caused by bodily substances leaking away in the urine. He examined the urine from such patients by using heat and acetic acid. Dekkers was impressed by the similarity in appearance of boiled urine and milk, and he probably added the acid to confirm the notion that this fluid was milk. Apparently, he did not examine many specimens, nor did he associate this albuminuria with dropsy, but he regarded it as a frequent occurrence in wasting diseases. Although his explanation was incorrect, the procedure involving heat and acetic acid survived and is still in use three centuries later.

Cotugno performed the heat and acetic acid test on urine from an edematous patient and observed the same large white precipitate as when serum or edema fluid was tested. However, he completely missed the connection of albuminuria with edema and he too made no further observations on this reaction. It should be noted that the term "albuminuria," introduced by Fernand Martin-Solon (1795–1856) in 1838, is not strictly accurate, for the usual coagulable urine contains not only albumin, but globulins and at times other proteins. "Proteinuria" is a preferable term, but albuminuria has become sanctioned by more than a century of use. Proteins were originally called "albuminous substances" because a good example was egg white (Latin: *albus*, white).

William Cruickshank (1745–1800) appreciated the importance of chemical analysis to medical practice. In three cases in 1798 he demonstrated the diagnostic significance of chemical changes in urine in dropsy. He described the coagulation of dropsical urine as the formation of a whitish precipitate and milkiness upon addition of nitrous acid or corrosive muriate of mercury (mercuric chloride), and even by heat. This did not occur in cases of dropsy from diseased liver and other morbid states [18]. Cruickshank offered "these imperfect hints, merely with a view to induce others to pay some attention to a subject, which has of late been much neglected, but which, in our opinion, is capable of affording great assistance in the investigation, and cure of many diseases" [19].

Although several European chemists had published compilations of chemical findings in the analysis of animal substances by those who preceded them, their limited approach and brief descriptions caused many of the recorded phenomena to be omitted, and some of the most interesting particulars to be forgotten. Few European physicians at the end of the eighteenth century used chemistry to evaluate disease; most disregarded it. In a three-volume work on the *State of Animal Chemistry* (1803), W. B. Johnson attempted to present the subject on a larger scale and with a more connected and systematic arrangement [20]. In examining the sources of indifference he noted that: "The analysis of animal substances is the most difficult and least advanced of any part of chemistry." He attributed this neglect to several causes, viz., "the disagreeable products" that result from experiments with biological material; the absence of "proper means and methods" to separate or synthesize these substances; "and lastly, the small degree of interest which the chemist who was not a professed physician took in the results of their analysis. Such have been the principal circumstances that have arrested the progress of this part of the science."..."It was owing to these causes that the old chemists contented themselves with the mere distillation of animal matters in the open fire; an operation which, it is now well known, entirely changes their nature and composition" [21].

The usefulness of testing for albumin in the urine was not widely appreciated until the independent observations of two British physicians early in the nineteenth century. In 1811, William Charles Wells (1757–1817), an American colonial loyalist who had migrated to England in 1775, showed a correlation between dropsy and albuminous urine in his report "On the Presence of the Red Matter and Serum of Blood in the Urine of Dropsy, Which Has not Originated from Scarlet Fever." He tested the urine with heat and acid. John Blackall (1771–1860) made systematic studies of albuminuria. He noted that the coagulation of the urine occurred long before boiling heat was reached. In his *Observations on the Nature and Cure of Dropsies, and Particularly on the Presence of the Coagulable Part of the Blood in Dropsical Urine* (1813), he noted that albuminuria was present throughout the course of the dropsy. Not until 1827, however, was the overall correlation of edema, albumin in the urine, and diseased kidneys observed after death, clearly established by Richard Bright (Fig. 5.2). Having no particular knowledge of chemistry himself, Bright left the urinalysis to John Bostock (1773–1846), an Edinburgh M. D. graduate who succeeded Alexander Marcet as lecturer

FIG. 5.2. Richard Bright. Engraving from the painting by F. R. Say. (National Library of Medicine, Bethesda, Maryland.)

in chemistry at Guy's Hospital Medical School. Bostock heated urine in a spoon over the flame of a lamp or candle and noticed the large amounts of coagulated albumin that appeared before the fluid reached the boiling point in urine of patients with renal disease.

Although Bright could not claim priority in connecting dropsy and kidney disease, he was the first to point out the frequent association of the two and to accurately describe the symptoms and autopsy appearance in these cases. And most important, he showed that this form of kidney disease, which became known as Bright's disease, could be detected during life by a simple chemical reaction. Here was an example of chemistry serving clinical practice, rather than being expected to clarify physiology.

In his two-volume *Reports of Medical Cases....* (1827, 1831), a landmark in the history of medicine because it helped distinguish between cardiac and renal dropsy (accumulation of fluid), Bright described and illustrated the renal disease (chronic nephritis) that still bears his name. In describing the coagulation of urine by heat, he wrote [22]:

> "...most commonly when the urine has been exposed to the heat of a candle in a spoon, before it rises quite to the boiling point it becomes clouded, sometimes simply opalescent, at other times almost milky, beginning at the edges of the spoon and quickly meeting in the middle. In a short time the coagulating particles break up into a flocculent or a curdled form, and the quantity of this flocculent matter varies from a quantity scarcely perceptible floating in the fluid, to so much as converts the whole into the appearance of curdled milk. Sometimes it rises to the surface in the form of a fine scum, which still remains after the boiled fluid has completely cooled."

Both volumes were beautifully illustrated by hand-colored plates. The drawing and engraving were carried out under Bright's immediate supervision. Bright's work is an excellent example of the perceptive internist who utilizes data from both pathologic anatomy and chemical pathology to characterize a new disease. This is where chemistry, with the first really useful diagnostic laboratory test, made its first great impact on clinical medicine. It was the starting point of modern clinical pathology.

Bostock analyzed multiple urine and serum samples from many of Bright's 23 cases of kidney disease, with the emphasis on specific gravity and gross appearance of the urine after coagulation by heat or chemical reagents. He also carried out rough quantitative estimates of urinary albumin, urea, and solids, by using simple procedures of extraction and weighing. Bostock showed that the specific gravity of dropsical urine is less than that of urine from healthy individuals and deficient in natural constituents (urea and salts), while at the same time containing extraneous matter (albumin) that coagulates on application of heat. He found that the blood serum of several edematous patients "contained less albumen than in health, although I am not able to state precisely the amount of this difference" [23]. Bostock credits Cruickshanks [sic] with originally noticing that in certain species of dropsy, the coagulation of urine in a greater or less degree by heat, must be due to the presence of albumin [24].

Wishing to determine the quantity of albumin and its proportion to the urea and salts, he estimated its quantity by measuring the specific gravity of the urine before and after removing the albumin by heat, or by heat and bichloride of mercury. These results, he acknowledged, "although by no means perfectly accurate, were sufficiently so for the object in view." As the result of his experiments, he concluded "that the quantity of albumen in the urine bore no exact relation to the total amount of its solid contents, or to that of the urea in particular" [25]. The data is presented in the narrative style of letters to Bright that were incorporated in volume one of his "Reports." There are no tables, no graphs or other format that would permit comparison or correlation.

Bostock suggests that the presence of albumin in the urine may not necessarily be an indication of disease, since "an albuminous state of the urine is produced by such a variety of circumstances, and many of them of so trifling a nature, as to render it almost a constant occurrence. In a great majority of cases it may be detected in the urine of persons in apparent health, by means of the appropriate tests." His own urine is seldom entirely free of albumin and is considerably increased by the slightest causes. He refers to the "different states of the albuminous matter in urine" as a result of variable response to heat or acid or chemical agents and wonders whether this is indicative of different stages of the disease [26]. Bostock showed how loss of albumin into the urine was associated with a decrease in the quantity of albumin in the blood [27]. A century later, this was confirmed by one of the earliest moving boundary electrophoretic analyses of the serum and urine of patients with nephrotic disease [28]. It showed that the protein patterns of nephrotic urine are similar to those of normal serum. This was convincing evidence that in proteinuria of any notable degree most of the protein is derived from the plasma. Bostock, Bright's first chemical collaborator, is remembered today for the first complete description in 1819 of hay fever—his own ailment.

Bright was aware of the possibility of false-positive results from either the application of heat or upon the addition of nitric acid. He recommended not to depend on either of these tests singly, but to use them together, because "wherever a decided precipitate occurs from heat which is not re-dissolved by dilute nitric acid, we may fairly infer the presence of albumen." Because urates are occasionally precipitated by acid alone, but are redissolved on application of heat, Bright recommended the use of heat *and* acid as a trusted test for the presence of albumin in the urine [29].

Bright credited one of his assistants, Dr. Rees, for explaining the occasional false-positive test for albumin on the application of heat alone, as being due to the decomposition of urea with formation of ammonia. The calcium and magnesium phosphates are precipitated in the alkaline medium and can be redissolved by addition of acid. During initial boiling of the urine, carbon dioxide is driven off, decreasing the acidity. This also may cause precipitation of phosphates. Addition of acid also increases the acidity to the optimum for protein coagulation by heat.

The circumstances leading to false-positive tests was subsequently reported by Rees in his "Observations on Real and Supposed Pathological Conditions of the Urine," in which he discussed his findings that medications derived from certain plants produce a false-positive test for albumin in urine [30].

In 1842, when Bright was given control of two clinical wards at Guy's Hospital, 42 beds in all, for a six-months intensive study of renal disease, Rees was put in charge of the chemical tests. George Hilaro Barlow (1806–1866) assisted Bright with the patients. A room between the two wards was used for discussion and record keeping, and a small laboratory connected to the middle room was outfitted for this purpose. This was the forerunner of today's metabolic wards and was an innovative team approach to clinical research. Bright considered this project "the first experiment which, as far as I know, has yet been made in this country to turn the ample resources of an hospital to the investigation of a particular disease, by bringing the patients labouring under it into one ward, properly arranged for observation" [31].

In the 1827 study by Bright, Bostock provided mostly descriptive data and specific gravities. Sixteen years later, Rees [32] reported quantitative measurements for a number of parameters: fibrin, corpuscles, serum solids, and water (by difference) in the blood of six renal patients and one healthy individual for comparison. He also gave quantitative values for urea, albumin, and specific gravity for some of the serum and urine specimens, and arranged his data in tabular form. Only six cases were subjected to a chemical workup because the multi-step quantitative analyses were tedious, labor-intensive, and time-consuming procedures.

In his paper titled "Observations on the Blood, with Reference to its Peculiar Condition in the Morbus Brightii," Rees directed attention to several findings: (a) the excessive quantity of water in the blood; (b) the existence in the blood of one of the ingredients (urea) of the urine; (c) the existence of the same ingredient of the urine in the fluids effused into the various serous cavities; (d) the absence or deficiency in the urine of one or more of its natural ingredients (uric acid, urea); (e) the general watery condition of the urine; and (f) the presence of albumin in the urine. He was convinced of the existence of urea in the blood and effusions from these patients: "I have never yet failed to obtain it in sufficient quantity to shew its physical and chemical characters."

The trend from colored engravings to numerical tables was part of the early nineteenth-century medical goal of identification of diseases. But beyond the more precise fact-gathering was a search for better treatment for their patients, as the full title of Bright's book indicates. Bright, Rees, and their associates believed that pathological anatomy and pathological chemistry could guide therapy [33]. Bright expressed his vision in the preface to "Reports of Medical Cases" [34]:

"To connect accurate and faithful observation after death with symptoms displayed during life, must be in some degree to forward the objects of our noble art: and the more extensive the observation, and the more close the connexion which can be traced, the more likely we are to discover the real analogy and dependence which exists, both between functional and organic disease, and between these, and the external symptoms which are alone submitted to our investigation during life."

However, in the middle of the nineteenth century, despite the correlations found by Richard Bright, most medical men still regarded illness as an essentially general phenomenon and did not think it necessary to look for an association between symptoms in the living and structural pathology in the dead.

The application of chemistry to medicine was new when these investigators tried to correlate chemical tests with symptoms of a particular disease, in order to reach a more reliable diagnosis. Although the utility of chemistry in medicine was gradually gaining recognition, the medical profession in general was still indifferent and even hostile to the idea that investigative work in animal chemistry could lead to improved methods of diagnosis, prevention, and cure of diseases.

BODY FLUIDS AND THE EXAMINATION OF BLOOD

By the early 1800s, the anatomical approach that replaced the centuries-old interest in the body fluids—the centerpiece of humoral theories—had become too dominating and exclusive, and there was a renewed interest in examination of body fluids. For example, as early as 1806, Berzelius, in his *Föreläsningar i Djurkemien* (Lectures in Animal Chemistry), presented a protocol for examination of body fluids.

Chemical analysis was seen as a refined type of dissection. The application of chemistry to determine the changes in the body fluids during disease was increasing, and by 1842, according to one book reviewer, "seem to be daily acquiring increased importance in the eyes of medical men. Indeed, this disposition to return to the study of the fluids, forms a very prominent feature of the medical mind at the present time, as evinced by the number of interesting researches to which it has given rise" [35].

The book, by Alfred Becquerel (1814–1862), emphasized the need to collect a 24-hour specimen, in order to minimize variations in urine constituents due to food intake. Becquerel performed analyses for water, urea, uric acid, and other constituents excreted by healthy individuals and compared these values to similar analyses on patients with illnesses. Since normal values varied, Becquerel specified that the mean of his normal measurements must be taken as the healthy standard. However, only large differences from the healthy standards may be judged as abnormal. He hoped to show that chemical changes in the urine could help establish diagnosis and monitor the course of the illness.

The 1840 memoir by Gabriel Andral (1797–1876) and Jules Gavarret (1809–1890), *Recherches sur les modifications de proportion de quelques principes de sang*, did much to further an interest in blood pathology. Andral, an eclectic, recognized the existence of symptoms without lesions in solid organs, and turned his investigations to the blood when examination of the anatomy revealed no changes. In one of his lectures to the medical faculty in Paris, he called for research on the fluids, similar to that done on the solids during the previous forty years. He examined the blood physically, chemically, and microscopically, and predicted that chemical analysis of body liquids altered by disease would play an increasingly important role in the investigation of pathogenesis. His laboratory studies on blood established clinical hematology as a separate discipline in internal medicine. A reviewer of Andral's book concluded that "it is chiefly by an accurate observation of the changes of the blood and of the secretions that we are to look for any important advancement in our knowledge of the nature of diseases, and of the most successful method of treating them" [36].

Contrast this to what Berzelius said in *Lehrbuch der Chemie* (1840): There was a long way to go before chemical examination could differentiate between normal and diseased blood beyond the variations occurring in healthy individuals [37].

In *Essai d'Hematologie Pathologique* (1843), Andral studied the normal composition of the blood and followed the abnormal variations of some constituents during disease. He calculated the proportions in which the major elements of the blood—which he called globules, fibrin, solids, and water—existed in health and in illness, and described circumstances which changed the quantitative relationship of these elements to each other. He recognized that the absolute amounts and relative amounts in which they existed, could vary considerably within the healthy population. Accordingly, he set out to determine the average proportions of each element: blood fibrin (2 to 4 parts in 1000), globules in blood (127 parts in 1000), serum solids (albumin) (80 parts in 1000), and water in the blood (790 parts in 1000). Andral also distinguished normal from pathological variations. He noted that in patients with edema, albumin in the urine was associated with reduced levels in the serum. Fibrin was increased in inflammation and was reduced in delayed clotting. He found reduced amounts of red globules in anemia. The separations were carried out by heating, drying, and weighing.

Although one of the reviewers [38] of Andral's *Essai* was critical, interest in blood chemistry continued, especially in the relationship of blood sugar to diabetes and uric acid to gout. The large quantities of blood necessary for testing were readily available owing to the continuing popularity of bloodletting as a therapeutic measure.

CHEMICAL ANALYSIS: VOICES PRO AND CON

In an age when the description of disease was often a mere catalog of symptoms, there were a few medical men who appreciated the usefulness of chemistry in the explanation and treatment of disease. Although investigations in the chemistry of disease were being carried out in the 1830s and 1840s by Bright, Bostock, Rees, Prout, Bence Jones, Marcet, and Bird, its application in routine diagnosis was not very widespread.

In 1848, Alfred B. Garrod, assistant physician at University College Hospital in London, in an introductory lecture, described "the application of chemistry to pathology and therapeutics," as being "of the greatest importance to the medical practitioner." He went on to say, "how very imperfect our knowledge must be, both of the healthy and diseased condition of the body, if we do not call in the aid of chemistry to elucidate its phenomena" [39].

The younger physicians were beginning to carry a few test tubes and reagents along with the stethoscope which by then was no longer an oddity. In his memoirs, Charles J. B. Williams, the first President of the Pathological Society of London, told of a house call early in his career and the favorable impression he had made by his "habit of bringing... not only a stethoscope, but also test-tubes and a few chemical reagents for the examination of the state of the secretions, &c." Writing in 1884, he added that this practice was a matter of course in later years, "but it was not so forty years ago;..." [40]. Even as late as the 1860s, physicians were cautioned not to be ostentatious with their scientific examination of the urine because "abuse of his knowledge in this respect will stamp him in the eyes of his colleagues and of the public as a charlatan" [41]. Nevertheless, when it came to urine analysis, chemical information was well ahead of physiology and pathology and would remain so for the rest of the nineteenth century.

And yet, the belief that chemical studies were not relevant to clinical medicine was widely held and taught as late as the third quarter of the nineteenth century. Armand Trousseau (1801–1867), the last of the great classic clinicians, advised those entering medicine not to lose time "in acquiring too extensive a knowledge of chemistry." Although not "wholly useless," accessory studies are "too unimportant to be pursued at the sacrifice of physiology, clinical instruction, and therapeutics,...." He condemned "an exaggeration of their importance, their pretentiousness, their being mixed up with our art in an inappropriate and impertinent manner." He pleads for "a little less science, and a little more art!" [42].

Trousseau, a convinced vitalist, is critical "of the vanity of the pretensions of the chemists, who believe that they can explain the laws of life and of therapeutics, because, forsooth, they know the nature of some of the reactions which take place in the living body." The laws of living matter "for the present remain autonomous, special, unexplained, inexplicable,...." He agreed "with the majority of physiologists and physicians in believing that the acts of organic life, and... those of animal life, are subject to laws which... ought to be regarded as essentially different from those which govern inorganic matter." "...we are all iatro-chemists, with this distinction,... that in living organisms chemistry is controlled by special powers, which give it a special direction,...." "Chemistry plays its part in both instances, but that part is very different in each: and we must admit that there are special properties in the one case, because in it there are special results." Trousseau is willing to confess his "ignorance as a chemist, but only on condition that chemists admit their ignorance as physiologists and physicians."

Trousseau and others notwithstanding, the value of chemical findings was recognized by others. In fact, as early as 1823, Thomas Hodgkin, in his M.D. dissertation on the absorptive functions within the body, challenged the tradition of vitalism taught by many clinicians. The widespread view that body processes could never be understood was supported by the theology and some of the science of the time. However, Hodgkin insisted that chemical studies were relevant to clinical medicine, and he began his dissertation by saying that he treated the subject "from a chemical point of view." He disagreed with physiologists who denied that digestion and absorption could ever be analyzed in meaningful chemical terms. Such changes, he said, belong to chemical investigation. If chemistry is unable to explain physiological phenomena, it "proves only that science has not yet arrived at the height of its perfection" [43].

Hodgkin believed that the new and important information contributed by research in animal chemistry and by microscopic observations (see Chapter 10) was indicative of a gradual return to a new and scientific humoral pathology. This would add more importance than before to alterations in the fluids in connection with changes unquestionably taking place in the solids. Inasmuch as exchanges are continually occurring between the solid parts and the blood, "it is in the blood that we must look for many important modifications in connection with disease." Hodgkin believed that disease will one day be explained in terms of "molecular movements," more chemical than mechanical, "by which our bodies...are continually changing the elements of which they are composed" [44].

Hodgkin was acquainted with Andral whose inspections at the Charité he had observed during his student year in Paris and was undoubtedly influenced by his later work. Hodgkin and Andral were correct with their predictions. What has happened in the second half of the twentieth century is that the practice of medicine, whether centered in the doctor's office, the clinic, or the hospital, has become oriented around the analysis of blood and other body fluids.

NOTES AND REFERENCES

1. J. BÜTTNER, "The Origin of Clinical Laboratories," *European Journal of Clinical Chemistry and Clinical Biochemistry*, 1992, 30: 585–593; "The Programme Devised in 1791 by Fourcroy for the Establishment of Clinical Laboratories," *Tractrix*, 1991, 4: 39–48.
2. *Idem*.

3. "Abstract of Evidence Relating to the London Apothecaries' Company, Taken Before the Parliamentary Medical Committee in 1834. Published in June 1836," *Lancet*, 1835–36, 2: 486–490, p. 489 (paragraph 133).
4. *Lancet*, 1841–42, 1: 266–267, p. 266.
5. "The Question of Titles," *Lancet*, 1862, 2: 626.
6. M. JEANNE PETERSON, *The Medical Profession in Mid-Victorian London* (Berkeley: University of California Press, 1978), p. 194; also see F. N. L. POYNTER, "Medical Education in England Since 1600," in *The History of Medical Education*, C. D. O'Malley, ed. (Berkeley: University of California Press, 1970), p. 242.
7. *Lancet*, 1827–28, 2: 659–663, pp. 659, 660; RUSSELL C. MAULITZ, "Channel Crossing: The Lure of French Pathology for English Medical Students, 1816–36," *Bulletin of the History of Medicine*, 1981, 55: 475–496, pp. 479–480.
8. MAULITZ (1981), pp. 479–480, 482–483.
9. ERWIN H. ACKERKNECHT, *Medicine at the Paris Hospital 1794–1848* (Baltimore, Maryland: The Johns Hopkins Press, 1967), pp. xi–xii. For an account of the evolution of pathological anatomy in France and the transport of this new science to Great Britain, see RUSSELL C. MAULITZ, *Morbid Appearances. The Anatomy of Pathology in the Early Nineteenth Century* (New York: Cambridge University Press, 1987).
10. STANLEY JOEL REISER, *Medicine and the Reign of Technology* (New York: Cambridge University Press, 1978), pp. 30, 31, 37, 38. Also see R. A. YOUNG, "The Stethoscope: Past and Present," *Transactions of the Medical Society of London*, 1931, 54: 1–22; P. J. BISHOP, "Evolution of the Stethoscope," *Journal of the Royal Society of Medicine*, 1980, 73: 448–456.
11. HELEN CLAPESATTLE, *The Doctors Mayo* (Minneapolis: The University of Minnesota Press, 1941), p. 22.
12. This section is derived from JOHN HARLEY WARNER, *The Therapeutic Perspective. Medical Practice, Knowledge, and Identity in America, 1820–1885* (Cambridge, Massachusetts: Harvard University Press, 1986), pp. 1, 5, 7, 85–91. The medical case history records for this study were from the Massachusetts General Hospital in Boston and the Commercial Hospital of Cincinnati.
13. E. ASHWORTH UNDERWOOD, "The History of the Quantitative Approach in Medicine," *British Medical Bulletin*, 1950–51, 7: 265–274; RICHARD H. SHRYOCK, "The History of Quantification in Medical Science," *Isis*, 1961, 52: 215–237, pp. 218, 223, 230.
14. JOHN S. HALLER, JR., *American Medicine in Transition. 1840–1910* (Urbana: University of Illinois Press, 1981), pp. vii, x–xi.
15. ERWIN H. ACKERKNECHT, "Elisha Bartlett and the Philosophy of the Paris Clinical School," *Bulletin of the History of Medicine*, 1950, 24: 43–60, pp. 49–50.
16. ACKERKNECHT (1967), pp. xiii, 121–127; (1950), pp. 49–50.
17. FIELDING H. GARRISON, *An Introduction to the History of Medicine*, 4th ed. (Philadelphia: W. B. Saunders Company, 1929), pp. 153–154, 421; ARTURO CASTIGLIONI, *A History of Medicine*, 2nd ed., translated from Italian by E. B. Krumbhaar, ed. (New York: Alfred A. Knopf, 1947), pp. 336–337. Also see WILLIAM DOCK, "Some Early Observers of Albuminuria," *Annals of Medical History*, 1922, 4: 287–290; "Proteinuria: The Story of 280 Years of Trials, Errors and Rectifications," *Bulletin of the New York Academy of Medicine*, 1974, [2] 50: 659–666; A. LEVINSON, "Domenico Cotugno," *Annals of Medical History*, 1936, [n.s.] 8: 1–9.
18. JOHN ROLLO, *Cases of the Diabetes Mellitus;...*, 2nd ed. (London: C. Dilly, 1798), pp. 443–448.

19. *Ibidem*, p. 451.
20. W. B. JOHNSON, *History of the Progress and Present State of Animal Chemistry*, (3 vols.), vol. 1 (London: J. Johnson, 1803), preface (pp. v–vi).
21. *Ibidem*, pp. 2–3.
22. RICHARD BRIGHT, *Reports of Medical Cases, Selected With a View of Illustrating The Symptoms and Cure of Diseases By a Reference to Morbid Anatomy* (2 vols.), vol. 1 (London: Longman, Rees, Orme, Brown, and Green, 1827, 1831), p. 3. The section on renal disease is available in facsimile in *Great Men of Guy's*, William B. Ober, ed. (Metuchen, New Jersey: Scarecrow Reprint Corp., 1973), p. 77.
23. JOHN BOSTOCK, "Observations on the Chemical Properties of the Urine in the Foregoing Cases," in Bright (1827, vol. 1), facsimile in *Great Men of Guy's* (1973), pp. 76, 83.
24. *Ibidem*, p. 76.
25. *Ibidem*, pp. 78, 80.
26. *Ibidem*, pp. 80–81.
27. *Ibidem*, p. 85.
28. LEWIS G. LONGSWORTH, THEODORE SHEDLOVSKY, and DONALD A. MACINNES, "Electrophoretic Patterns of Normal and Pathological Human Blood Serum and Plasma," *Journal of Experimental Medicine*, 1939, 70: 399–413; LEWIS G. LONGSWORTH and DONALD A. MACINNES, "An Electrophoretic Study of Nephrotic Sera and Urine," *Journal of Experimental Medicine*, 1940, 71: 77–82; JOHN A. LUETSCHER, JR., "Electrophoretic Analysis of Plasma and Urinary Proteins," *Journal of Clinical Investigation*, 1940, 19: 313–320.
29. RICHARD BRIGHT, "On the Functions of the Abdomen, and Some of the Diagnostic Marks of its Disease," *London Medical Gazette*, 1833, 12: 378–384, pp. 378–379.
30. G. O. REES, "Observations on Real and Supposed Pathological Conditions of the Urine," *Guy's Hospital Reports*, 1841, 1st series, 6: 121–130.
31. RICHARD BRIGHT, Preliminary Note (pp. 189–190) in George Hilaro Barlow, "Account of Observations Made Under the Superintendence of Dr. Bright, on Patients Whose Urine Was Albuminous: with a Chemical Examination of the Blood and Secretions, by G. O. Rees, M. D.," *Guy's Hospital Reports*, 1843, series 2, 1: 189–316.
32. GEORGE OWEN REES, "Observations on the Blood, with Reference to its Peculiar Condition in the Morbus Brightii," *Guy's Hospital Reports*, 1843, series 2, 1: 317–330.
33. STEVEN J. PEITZMAN, "Bright's Disease and Bright's Generation—Toward Exact Medicine at Guy's Hospital," *Bulletin of the History of Medicine*, 1981, 55: 307–321.
34. BRIGHT (1827, vol. 1), preface (p. vii).
35. Book Review. *Semeiology of the Urine, or a Treatise on the Alterations of the Urine in Diseases; followed by a Treatise on Bright's Disease, &c. &c.* by ALFRED BECQUEREL (Paris: 1841), in *American Journal of Medical Sciences*, 1842 (new series vol. 3), 29: 155–169, p. 155.
36. Lecture Review. "M. Andral on the Modern Doctrine of Humoral Pathology," *Medico-Chirurgical Review, and Journal of Practical Medicine*, 1841, 35: 177–178; Review. "M. Andral on the Changes of the Blood in Disease," *Medico-Chirurgical Review, and Journal of Practical Medicine*, 1841, 34: 196–201, p. 201.
37. JOHANNES BÜTTNER, "Clinical Chemistry as Scientific Discipline: Historical Perspectives," *Clinica Chimica Acta*, 1994, 232: 1–9, p. 3.
38. Book Review. *Essai d'Hématologie Pathologique*, by G. ANDRAL (Paris, 1843), in *Dublin Journal of Medical Science*, 1843, 23: 288–299; Book Review. *British and Foreign Medical Review*, 1844, 17: 136–152.

39. ALFRED B. GARROD, "Application of the Science of Chemistry to the Discovery, Treatment, and Cure of Disease," *Lancet*, 1848, 1: 353–355, p. 353.
40. CHARLES J. B. WILLIAMS, *Memoirs of Life and Work* (London: Smith, Elder, & Co., 1884), p. 323.
41. CARL NEUBAUER and JULIUS VOGEL, *A Guide to the Qualitative and Quantitative Analysis of the Urine*, translated from German by WILLIAM ORLANDO MARKHAM (London: New Sydenham Society, 1863), p. 282.
42. A. TROUSSEAU, *Lectures on Clinical Medicine*, 3rd ed., translated from French by John Rose Cormack and P. Victor Bazire (2 vols.), vol. 1 (Philadelphia: Lindsay & Blakiston, 1873), Introduction (pp. 33–61), pp. 34–36, 60.
43. AMALIE M. KASS and EDWARD H. KASS, *Perfecting the World. The Life and Times of Dr. Thomas Hodgkin, 1798–1866* (Boston: Harcourt Brace Jovanovich, 1988), pp. 109–110.
44. SAMUEL WILKS, "An Account of Some Unpublished Papers of the Late Dr. Hodgkin," *Guy's Hospital Reports*, 1878, 3rd series, 23: 55–127, pp. 112–113.

CHAPTER VI

REES AND THE ESTIMATION OF UREA AND
SUGAR IN BLOOD

George Owen Rees (1813–1889) (Fig. 6.1) [1] was one of those enterprising physicians and chemists who were early in the field in studying the chemical changes in the blood in disease. In an age when medicine was beginning to be dominated by morbid anatomy, he realized the fundamental importance of biochemistry. As a student at Guy's Hospital, before the age of twenty, he attracted the attention of Richard Bright who requested his assistance in the analysis of urinary calculi and urine in diseases of the kidney. Rees also made quantitative analyses for albumin and urea in the serum and urine of patients with kidney disease, and Bright often mentioned his help.

Rees's association with Bright led to a lifelong friendship, and in 1845, Rees dedicated the second edition of his book, *On the Analysis of the Blood And Urine, in Health and Disease; and on The Treatment of Urinary Diseases*, to Bright as an expression of "gratitude for the kind encouragement received from you when, as a mere boy, I first entered on the study of pathological chemistry." He also mentions Bright's confidence in his observations and results, and his "friendly aid and council" [2].

The first edition, published in 1836, was one of the earliest such books in English on animal chemistry. He wrote it in response to "The increased desire for a more intimate acquaintance with animal chemistry, which has lately been evinced by the medical profession, ..." Rees hoped it would stimulate study of animal analysis, as applied to disease. He deserves credit for "directing attention to a subject which, in all probability, is no less rich in discovery, than it is neglected and uninvestigated by the great body of the medical profession." The book was written as a manual for the medical practitioner who might wish to do his own analysis. "Since chemists are not physicians, we shall scarcely benefit by their art, except by making the physician a chemist" [3].

Rees listed "the following animal ingredients, capable of separation" from serum: water; albumin; extractive matter soluble in water and alcohol; albumin combined with soda; crystalline fatty matter; animal oily matter;

FIG. 6.1. George Owen Rees. (National Library of Medicine, Bethesda, Maryland.)

chlorides of potassium and sodium; alkaline subcarbonate, phosphate, and sulfate; earthy phosphate and subcarbonate; subphosphate of iron; and oxide of iron. For the analysis of blood he included fibrin and coloring matter (red particles) [4]. From this list we see the beginning of a search for analytical methods for the determination of specific compounds.

Rees's analyses are quantitative and starting materials were carefully weighed.

The kidney's role in disposal of waste nitrogen by removing urea from the blood was not yet a settled matter or even accepted at this time. Consequently, the search was on for evidence of increased levels of urea in the blood of patients with diseased kidneys. In 1823 Jean Louis Prévost (1790–1850) and J. B. A. Dumas, removed the kidneys of dogs. The animals died several days later. In the blood they detected urea by means of crystallization with nitric acid. By comparing analyses before and after removal of the kidneys, they concluded that kidneys remove urea from the blood, but do not produce it. But, what about humans? At Guy's Hospital and in the pages of the *London Medical Gazette* of 1833, there was an ongoing debate concerning the presence of urea in blood of renal dropsy patients. George Owen Rees pointed to his experiments as support for the presence of urea, and Bright referred to a patient with renal disease whose urea concentration in the blood serum was at least 15 parts per 1000 (1500 mg/dL) [5]. M'Gregor, a few years later, was unable to find urea in the blood of healthy individuals, even though he used "several pounds of serum in each experiment" [6].

In 1840 Rees published "On the Proportion of Urea in Certain Diseased Fluids." There was no longer any doubt, he says, of the existence of urea in the blood in several forms of disease. He was not aware, however, "that the *proportion* in which that substance exists in morbid blood or secretions, has yet been very accurately determined" [7]. He describes an ether extraction procedure that is a "more perfect method" than the nitric acid precipitation technique generally used. His analysis of serum urea from two patients with Bright's disease were 0.2096 and 0.5 parts per 1000 (21 and 50 mg/dL). Undoubtedly, the true values were really much higher than those he found. Having found no urea in serum of healthy venous blood, he tried his new method on serum from healthy arterial blood. He found no evidence of urea and apparently concluded that urea probably occurred in the blood only in disease. His report also included quantitative analyses for albumin and "extractives and salts."

The method in brief, for urea, is as follows: dried serum is digested with water over an open steam bath to maintain a near, but below boiling temperature, and the mixture is filtered. The filtrate and washings from the residue in the filter are combined and evaporated, and the residue is extracted with several portions of ether, which selectively removes only the urea. The ether is evaporated and the urea residue is directly estimated by weight. According to Rees, this process yields urea that is quite pure and colorless. The new method did not appear in the second edition of *On the Analysis of the Blood and Urine,*....(1845). Instead, he reprinted the procedure from the first edition, a procedure that involved an extraction with alcohol and the familiar precipitation of urea with nitric acid as urea nitrate crystals. In neither edition did he give any data on urea in blood in health or in disease.

As for the existence of urea in the blood of healthy individuals, not only was there considerable doubt, but the same doubt existed for sugar (glucose). Even the occurrence of hyperglycemia in diabetes was doubted until Rees, in 1838, clearly demonstrated it [8].

In Rees's process 12 ounces of blood is evaporated to dryness over a water-bath. The dried mass is ground up and digested for several hours in boiling water, then filtered and the filtrate evaporated. The dry residue is then digested in alcohol of specific gravity 0.825. The resulting alcoholic solution is filtered or carefully decanted and evaporated to dryness. The dry mass is treated with several portions of purified ether, to dissolve out urea and some fatty matter. The sugar remains behind in the mixture with osmazome and sodium chloride. Solution of this mass in alcohol, followed by spontaneous evaporation in a flat glass dish, yields crystals of alkaline chloride and diabetic sugar, which are easily distinguishable and may be separated by shaking with alcohol. The chloride settles, and the sugar, which accumulates on top of the surface of the solution, may be carefully spooned out with a spatula and examined—without delay, before the crystals redissolve in the alcohol. The evaporated alcoholic extract has a sweetish taste and syrupy smell.

His analysis was the first quantitative estimation of sugar in the blood. The analysis yielded 1.8 parts per 1000 (180 mg/dL), which he considered to be only a close approximation, owing to impurities and loss on handling. He rightly attributed his success to the extraction step with ether, which separated the sugar from the urea and fatty material. However, rather than evaporate whole blood at the start, a slow procedure, Rees should have coagulated the proteins and filtered them off before commencing to evaporate the solution. But the chief difficulty of Rees's method was in the final step, when the upper layer of sugar was removed with a spatula.

BENCE JONES PROTEIN

Several years after Richard Bright showed that heat-coagulable albumin in the urine was an indication of serious renal disease, Henry Bence Jones (1813–1873) (Fig. 6.2) of St. George's Hospital observed a protein with

FIG. 6.2. Henry Bence Jones. (Library of the New York Academy of Medicine.)

uniquely different properties of heat coagulability [9]. The protein, which now bears his name, is differentiated qualitatively in urine by a characteristic and unusual reaction. It begins to coagulate and precipitate at temperatures (40°–60°C) considerably lower than that for other proteins and, on continued heating, dissolves at or just below boiling, only to reprecipitate as the urine cools to 40°–60°C, and to redissolve below 40°C. Its identification is easy if it is the only protein in the urine, but becomes more difficult when other proteins are also present, which is often the case. Therefore, modifications of this basic procedure were necessary for specific identification and quantitation. This unusual reaction is not mentioned by

Bence Jones in his autobiography nor is it mentioned in any of his obituaries. He did not follow up on this unusual disease and it failed to interest his own generation. At the time of his death, it had not been observed by anyone else and did not appear in textbooks until near the close of the century.

Bence Jones carried out extensive chemical analyses on this unusual heat-precipitable substance and concluded that it was an oxide of albumin—specifically, the hydrated deutoxide of albumin. The ultimate analysis he represented by $C_{48}H_{38}N_6O_{18}$ or $C_{40}H_{31}N_5O_{15}$, depending on whether protein is equal to $C_{48}H_{37}N_6O_{15}$ or $C_{40}H_{30}N_5O_{12}$. According to his estimate, the enormous quantities of this peculiar albuminous substance voided in the urine were in the same concentration as ordinary albumin in the serum. No amount of food could compensate for such a loss [10].

A THEORY OF DIABETES

Bence Jones's researches with diabetics showed that sugar is still found in the urine when sugar-forming food is withheld from the patient, and he suggested that the disease might be due to blockage of the action of oxygen on the non-nitrogenous constituents of the food and tissues. He believed diabetes was due to deficient oxidation, affecting first the non-nitrogenous and ultimately the nitrogenous constituents of the food and tissues. He explained the error of calculating the quantity of sugar in urine from specific gravity tables as follows: "But diabetic urines contain a multitude of other substances besides sugar, each of which is variable, and each of which may cause the specific gravity to vary, whilst the quantity of sugar remains constant." He supported this contention—and the reverse relationship—with many examples of comparisons of specific gravity and sugar content determined directly by measurement with "Soleil's saccharometer" (polariscope). His treatment for diabetes: "Whatever is beneficial for excessive acidity is still more useful in diabetes. Small meals; free from sugar and acid, and the substances which can give rise to sugar and acids, constitute the best diet" [11]. Because of the tendency to acidity in diabetes, he treated this disease with ammonium carbonate and other alkaline solutions to stop the formation of sugar in the urine. However, there was no scientific justification for this general approach.

ALKALINE TIDE

From about 1844 to 1853, in systematic analytical studies of the chemical composition of urine in health and disease, Henry Bence Jones sought data that would make the diagnosis of *chemical diseases* more reliable. It was already well-known that large amounts of phosphates were often present in some diseases. He tried to correlate the relative proportions of the alkaline and earthy phosphates with diet, exercise, and certain medications, but found wide day-to-day variations in healthy individuals also. When he followed the variations in urinary acid in healthy individuals, he noted the 24-hour cycle of ebb and flow. He concluded that the rise in alkalinity of urine after a meal ("alkaline tide") was time-related to the eating of food and the concomitant increase of hydrochloric acid in the stomach. If acid is released from the blood to the stomach, an equivalent amount of alkali must be set free in the blood, the excess of which shows up in the urine. Thus, he noted the reciprocal relationship between variations in urine acidity and the quantitative changes of acid in the stomach. Urine acidity was lowest several hours after meals, and highest before meals. From his experiments Jones drew the significant physiological conclusion that urine composition was determined by dietary intake and not necessarily by any particular disease state [12].

He also observed that the administration of ammonium salts produced an increased acidity in the urine, owing, we now know, to conversion of ammonium ion to urea. But, in applying Liebig's theories of oxidation, Bence Jones explained the effect as due to the oxidation of ammonium salts to nitric acid. Jones also showed that animal foods produced more gastric acid than vegetable foods and consequently, a more alkaline urine; conversely, vegetable foods produced increased urine acidity. The nature of the acid is not important, said Bence Jones, but the variations in the acidity of the urine in health provide practical guidelines for evaluating the reaction of random specimens during the treatment of patients. From his experiments it was clear that the reaction of test paper on a random urine specimen "should never determine the use of acid or alkaline medicines. The different deposits which take place in the urine are far better tests of the state of the urine, and of the necessity for these remedies." If guided by the reaction of the urine, the twenty-four hour volume must be examined [13].

Bence Jones lectured on chemistry at St. George's Hospital and published part of his course as *On Animal Chemistry in its Application to Stomach and Renal Diseases* (1850). The book earned him recognition as an authority on the subject. He stressed the practical diagnostic value of chemistry. "By such examination, both in serious diseases and in slight disorders, I believe that as much or even more useful evidence will be obtained regarding complaints of the stomach, the kidneys, and the system than has been acquired respecting diseases of the lungs and heart by the stethoscope" [14].

MEDICAL EDUCATION

Henry Bence Jones's appointment as President of the Chemical Section of the British Association for the Advancement of Science, the first for a practicing physician, indicated the high regard in which he was held as a chemist and was "evidence of the relationship that exists between chemistry and medicine,...." In addressing the Chemical Section, he took a position in the politics of medical education. He appealed for greater attention to chemistry in the training of doctors, and said: "...whatever sets forth the union of chemistry and medicine tends to promote not only the good of science but also the welfare of mankind." "...the discovery of Dr. Bright has proved that chemistry is absolutely requisite for the detection of a large class of diseases, and that without chemistry the nature of these diseases cannot be understood." "...it is daily becoming more and more certain that...every medical man [must] become a chemist if he wishes to have any clear idea of the action of air, food, and medicine,...." [15].

Bence Jones also urged a revision in the curriculum for all would-be physicians to include a first-rate instruction in English so they "could explain the nature of the disease and the course to be followed in the most idiomatic and unmistakeable [sic] English,...." Medical men would be much better served, he said, if they spent some time in acquiring knowledge about chemistry and physics instead of "learning some Latin and less Greek" [16]. And, he contrasted "the present state of medical education with that reasonable knowledge, which...ought to be possessed by those who attempt to understand and to regulate an apparatus that works only whilst oxygen is going into it and carbonic acid is coming out of it" [17].

Opposition to these chemical ideas came from those who still believed that the chemistry of life was governed by a vital force. Most physiologists were vitalists, as were many of the leading animal chemists—including Liebig—who equated the vital force with gravity, magnetism, and electricity. They believed that, like these forces, the vital force would obey experimentally determinable simple physical laws. Bence Jones postulated a quantitative relationship between physical and chemical forces with a minor contribution from the vital force, but believed, as did other nineteenth-century chemists, that advances in chemical knowledge would eventually obviate the notion of a vital force. This was "but a collective term, including many forces, as the nervous force, the contractile force, the chemical forces, and the formative forces,...." [18].

LIEBIG'S CONCEPT OF STONE FORMATION

Henry Bence Jones had come to Liebig's laboratory in Giessen for six months soon after passing the examination for Licentiate of the College of Physicians in 1841, and became a close friend of his teacher. Jones was greatly impressed with Liebig and his views on animal physiology and based most of his own work on Liebig's concept of the oxidative metamorphosis of tissues, i.e., the interconversion of biological molecules.

In his first book, *On Gravel, Calculus, and Gout; chiefly an Application of Professor Liebig's Physiology to the Prevention and Cure of These Diseases* (1842), Jones used Liebig's concept to explain the causes, treatment, and prevention of formation of bladder stones. Urea was defined as arising from uric acid. Bence Jones's approach to preventing the formation of uric acid stones was to increase the rate of oxidation of uric acid to the more soluble urea by increasing the supply of oxygen. Consequently, he emphasized vigorous exercise and limiting the intake of non-nitrogenous foods, which were thought to inhibit such oxidation, and administered alkaline medication to keep the uric acid in solution, where it was more readily in contact with oxygen. The fault with his book was his uncritical acceptance and reliance on Liebig's theories and the many speculative applications and physiological doctrines offered without explanation or discussion.

According to Liebig's chemistry of physiological processes, substances in the tissues and in food were broken up and their atoms then rearranged into new combinations—by the vital force. If the resulting products matched the living tissues, they were absorbed; if they were markedly different, they were rejected and excreted. Since, in the majority of biochemical reactions, only the initial and final products were known, vital force was an easy explanation for the intermediate steps connected with secretion and animal metabolism. Liebig's approach to converting one substance to another by moving atoms around on paper was convenient, but formulas were often incorrect and did not always balance. The addition and subtraction of atoms looked plausible on paper, but why do the atoms recombine as they do, and what are the mechanisms involved? (See later this chapter for critique by Berzelius.)

Liebig's theories were attractive because they were based on a balanced system of oxidative reactions, all fitting neatly together to explain animal metabolism. They were, of course, obviously incorrect. According to Liebig, uric acid was the first product of the oxidation of tissues and was later converted to urea. How much was converted depended upon the degree of oxidation available in the body. Imperfect oxidation of the tissues led to the formation of uric acid, while a more complete oxidation produced urea, oxalic acid, and carbon dioxide.

Liebig and Wöhler began work on uric acid in 1837 and prepared a series of derivatives by fairly simple oxidation reactions with nitric acid, and by reduction reactions. These derivatives—some of which were identical with organic substances found naturally—were named by Wöhler (1838): allantoin, alloxan, alloxantin, uramil, thionuric acid, etc. Their provisional view was that urea preexisted in uric acid. They assumed a hypothetical body, *uril*, to be combined with urea in uric acid. Accordingly, uric acid was uril + urea. Changes of uric acid occurred by oxidation and combination with atoms of water. From this work, Liebig and Wöhler concluded that the laboratory production of all organic materials was not only probable, but certain—as soon as the precursor compounds became known [19].

Because of the strong impression made by Liebig's work, Bence Jones always preferred the findings of German chemists over those of the French. This led him to ignore or even to reject some important findings in physiological chemistry made during his time. For example, he was not satisfied with Claude Bernard's experiments showing that pancreatic juice saponified neutral fat, and he cited the contrary opinion of a German researcher [20]. The weak point in his and Liebig's work was a too-direct application of the laws of chemistry to the complex phenomena of the human body. Bence Jones's friendship with Emil du Bois-Reymond (1818–1896), founder of modern electrophysiology, led him to attempt to revive an early technique of electrolysis for the *in situ* dissolution of urinary calculi [21]. However, his experiments were limited to *in vitro* tests since there were mechanical difficulties in constructing a perfectly safe device for use in the body.

LIEBIG'S LABORATORY AT GIESSEN

Justus Liebig (Fig. 6.3) [22] was one of the forces making chemistry (in which France had led the way in the eighteenth century) almost a German monopoly in the nineteenth century, as his school of chemistry was attracting students from all over the world. It was the outgrowth of his own post-doctoral study in Paris (1822–24) with Gay-Lussac and Louis

FIG. 6.3. Justus Liebig. (National Library of Medicine, Bethesda, Maryland.)

Jacques Thenard (1777–1857), for which he had obtained a travel grant from Grand Duke Louis I of Hesse.

In Paris, Liebig encountered a rigorous, quantitative, experimental chemistry unlike anything he had found in Germany where "natural philosophy" led to nowhere. Returning to Germany in 1824, he determined to make the opportunities of Paris, where he mastered methods of organic analysis and systematic investigations, available to a larger number of students. Previously, practical laboratory exercises were almost completely neglected in the universities and students were taught only theory. Experiments were almost always limited to demonstrations by the instructors and his assistants.

While in Paris, a research presentation by Liebig attracted the attention of Alexander von Humboldt (1769–1859), the renowned natural scientist, intellectual, and world traveler, who recommended to Louis I of Hesse that he provide Liebig with an academic position. Thus, in 1824, Liebig, only twenty-one years old, was appointed extraordinary professor in the philosophical faculty at the University of Giessen by the Grand Duke and not by election of the faculty. Only a year later Liebig succeeded to the chair for chemistry at the university. This gave Liebig the opportunity to establish the first systematic laboratory course designed expressly to train new chemists. Through a carefully planned program of exercises, students could progress systematically from elementary operations to independent research under the guidance of an established scientist. It was the first physiological chemistry laboratory in the modern sense. Before this time, much of the work in this science had been done in laboratories of general chemistry, physiology, pathology, and clinical medicine.

Many of Germany's leading chemists, after studying in Paris, established important, but relatively modest university centers of chemical training. The decisive change came when Liebig set up his teaching laboratory at Giessen. This was soon followed by others, and by 1840, owing to the growing interest in the applications of chemistry in agriculture and industry, a steady flow of well-trained chemists was emerging in Germany. Whereas chemistry in the previous century emphasized its relation to pharmacy, metallurgy, and mining, it now became a scholarly subject taught in universities to future highly specialized professionals. At the core of these programs was laboratory experimentation. This was in sharp contrast to the teaching of chemistry in the eighteenth century in the universities, where chemistry had been part of the medical curriculum [23].

Liebig's Chemical Institute in Giessen was a remodeled guardhouse of a former military barracks. The lower floor housed the laboratory and the service rooms for the balance, chemical supplies, glassware washing, and Liebig's laboratory assistant. The apartment in the upper floor was occupied by Liebig and later, by his family.

The laboratory was an unventilated room with a large charcoal stove in the center. There was no chemical hood. If necessary, the windows and outside door were opened for ventilation. Chemists at that time had to produce all the necessary reagents themselves or isolate them from commercially available impure raw products. Water for chemical experiments and for rinsing glassware was provided primarily by rain water which was passed through a sand and gravel filter and collected in cisterns. Because of the institute's limited budget, Liebig bought most of the equipment and chemical supplies and paid his assistant, from his own modest salary. A larger analytical laboratory with tables along the walls, cupboards and drawers below, shelves above, and scientists at work back-to-back (Fig. 6.4) was built for him in 1839, after he received and turned down an offer from the University of St. Petersburg. In 1852 he left for the University of Munich which offered better facilities and freed him from teaching responsibilities.

The Giessen building was restored in 1920 and opened as a museum dedicated to the memory of Liebig. Damaged during World War II, it was

FIG. 6.4. Liebig's Analytical Laboratory in Giessen. From a sketch by Trautschold and v. Ritgen (1840). (Courtesy of the Library of the Justus Liebig University, Giessen.)

reopened in 1952 and is today among the world's most important chemistry museums.

Liebig made chemistry teachable and he made chemical research learnable. By 1831 he had established a national and international reputation for his program of research and instruction. It was the beginning of a whole new mode of training scientists, and it became the model for others in Germany and elsewhere. Among Liebig's other lasting contributions to organic chemistry were the quantitative techniques and apparatus he designed to overcome difficult operations and to eliminate sources of error in determining the composition and reactions of physiologically important compounds.

While at Giessen, Liebig was asked by Phillip Lorenz Geiger (1785–1836), a Heidelberg pharmacist, to join him as co-editor of *Magazin für Pharmacie* (renamed *Annalen der Pharmacie* in 1832). Geiger needed Liebig to verify the accuracy of chemical statements in articles submitted for publication. Following Geiger's death, Liebig changed the name of the journal to *Annalen der Chemie und Pharmacie* (1840). It became the leading journal of chemistry of his time. After Liebig's death in 1873, the journal was briefly known as *Justus Liebig's Annalen der Chemie und Pharmacie*, then as *Justus Liebig's Annalen der Chemie*. In 1979 it became *Liebig's Annalen der Chemie*, and in 1995—*Liebig's Annalen*.

Liebig did as much as anyone to bring about the era of large-scale research, in which the ability to organize men became as critical as the ability to conceive and carry out experiments. Between 1830 and 1840, he was at the very center of the rapidly growing field of chemistry. He was clearly one of the chief pioneers in physiological chemistry and had a long lasting influence on the younger generation of scientists and physicians, many of whom studied with him at Giessen and then went out to develop the new discipline at various hospital centers in central Europe. Liebig helped establish the independent and scientific status of the chemist.

THE IMPACT OF "ANIMAL CHEMISTRY"

Liebig appreciated the need to link chemistry, especially organic chemistry, to physiology, and he worked hard to put plant and animal physiological chemistry directly to the service of agriculture and medicine. He expressed this in his book "Organic Chemistry in its Application to Agriculture and Physiology" (1840), written at the request of the British Association for the Advancement of Science. In the companion book, a major work titled *Die Organische Chemie in Ihrer Anwendung auf Physiologie und Pathologie*

(1842) ("Animal Chemistry, or Organic Chemistry in Its Application to Physiology and Pathology"), Liebig developed a new comprehensive concept of metabolism. He tried to predict the chemical transformations which the fats, carbohydrates, and proteins undergo in the body to yield heat, based on his knowledge of organic chemical compounds and their reactions in the laboratory. Liebig treated physiological processes as chemical reactions subject to the laws of chemistry and physics He was confident that the application of quantitative chemical methods would provide solutions to problems that had eluded physiologists. Although "Animal Chemistry" was accepted uncritically by some, others were antagonized by his wishful thinking and speculative excesses that went far beyond the available experimental evidence.

In his critique of "Animal Chemistry" in 1843, Berzelius wrote: "This easy kind of physiological chemistry is created at the writing desk, and is the more dangerous, the more genius goes into its execution, because most readers will not be able to distinguish what is true from mere possibilities and probabilities,...." [24].

Though frequently in error in his scientific discoveries, Liebig's commanding personality and vigorous literary style exerted much influence, but also brought him into bitter conflicts with others. Engaging in every controversy and rarely able to preserve a distinction between intellectual disagreements and personal attacks, Liebig became embroiled in furious literary polemics. He often used his publications to publicize his own views and to denounce and discredit other chemists, German and foreign, with whom he disagreed—a practice which earned him many enemies. Quick to charge plagiarism, he was not inclined to credit the work of others in his own investigations and conclusions. He rarely underrated the importance and novelty of his own contributions. Liebig's disputes with Dumas were part of the rivalry between German and French chemists to dominate organic chemistry [25].

Many details of Liebig's hypotheses were later shown to be in error. He drew most of his arguments from the work of others and from strong deductive reasoning, and relied on his own findings to decide between rival theories. And, although he had never performed an experiment on living animals, Liebig provided one of the first comprehensive pictures of chemical exchanges of the vital processes. "Animal Chemistry" had the most significant single impact upon the future course of physiological thought and investigation. Perhaps, the most important result of Liebig's work was the discussion and new research by others that he helped to stimulate [26].

ANIMAL CHEMISTRY SOCIETY

Animal chemistry had generated interest among chemists, physicians, and surgeons, long before Liebig's work. In England in 1808, a group of Fellows of the Royal Society with shared interests formed an affiliated Society for the Improvement of Animal Chemistry which was designated as an Assistant Society to the Royal Society. Since the Royal Society by itself could not provide the facilities for in-depth discussion of all the specialized branches of science that interested its members, various groups had separated and established independent organizations. Affiliation offered benefits but gave up independence. This new designation was an effort by the parent society to control the proliferation of break-away groups that were seen as rivals and a threat to its monolithic structure. Without an independent status, the newly-formed society was destined to remain small in numbers—never more than about ten or twelve members—and to be relatively ineffective. Its meetings took the form of after-dinner discussions held in the homes of the members every three months [27].

The six founding members were the chemists Humphry Davy, Charles Hatchett, and William Brande, the physician William Babington, and the surgeons Everard Home and Benjamin Brodie. It is surprising that William Wollaston, John Bostock, Alexander Marcet, Thomas Thomson, and William Prout were not members, since they all made important contributions to the study of animal chemistry about this time. Although sixteen papers were submitted to the Royal Society by the Animal Chemistry Society and were published in the *Philosophical Transactions*, the members seem to have worked mainly on their own and probably would have produced their work in any case. The society went out of existence in 1825.

CHEMISTRY IN THE SERVICE OF SOCIETY, TECHNOLOGY, AND MEDICINE

Philosophical chemists in the mid-eighteenth century tried to explain chemical phenomena in terms of particles and forces, ether, phlogiston, and affinities. They persisted in spite of an awareness that their explanations were not satisfactory. As late as 1803, Joseph Black noted: Chemistry is not yet a science.

Notwithstanding the uncertainties, chemistry was remarkably popular in the eighteenth century. It was seen as being of potential benefit to medicine and technology. Students crowded into lectures in Glasgow, Edinburgh, and Paris. In London at the end of the century, the Royal Institution's first professorship was in chemistry. It had been founded in 1799 as a forum for publicizing science to the rising middle class. Other platforms of public education, beginning in 1823, were the numerous Mechanics' Institutions that were formed throughout England and Scotland. They were the result of a widespread movement to make scientific and technical information available to the working class in a popular format of instruction during evening hours. It was an educational by-product of the Industrial Revolution. This was not a sudden development. There had been extensive provision of public lectures on scientific subjects, both in London and in many provincial centers of England since early in the eighteenth century.

In actual practice, chemical technology rejected the traditional views of historians and economists that the Industrial Revolution had little or nothing to do with science on the grounds that science was too primitive and inaccurate to be applicable to technological problems. Even before Lavoisier had stabilized chemical theory with a new language and a new foundation, chemistry was experiencing a technological revolution producing benefits for agriculture, mining, bleaching, dyeing, and leather production. Britain, more attuned to commerce than other nations, was especially aware of the need for chemicals to prop up its growing textile industry [28].

During the latter part of the eighteenth and early nineteenth century, the study of chemistry in Great Britain at the university or hospital medical school was considered a branch of medicine, and the professor of medicine frequently held both chairs. In 1747 William Cullen (1710–1790) was appointed the first independent lecturer in chemistry in Britain at the University of Glasgow. In 1756 he accepted the chair of chemistry at the University of Edinburgh, where he lectured until 1766 before transferring to the chair of Institutes of Medicine (physiology). He was succeeded in this position at both locations by his former student Joseph Black, who later picked Thomas Charles Hope (1766–1844) to succeed him as lecturer in chemistry.

Chemistry was becoming increasingly important in the education of the medical profession. At Guy's Hospital Medical School in London, chemistry had been the first subject to be given up to a specialist by the clinicians. A course in chemistry was given at Guy's by William Babington (1756–1833), Alexander Marcet, and William Allen (1770–1843). Allen, a prominent Quaker pharmacist, was the first nonphysician to lecture in the medical school (1802–1826), with frequent absences to attend to religious missions in Europe. He was a well-known man of science and collaborated with Humphry Davy (1778–1829) (see Chapter 14) and other prominent scientists. Allen was often consulted for chemical analysis and delicate and difficult experiments at his pharmaceutical firm in Plough Court. In 1803, Davy asked Allen [29] to give the same course of lectures at the Royal Institution that he was giving at Guy's.

The Royal Institution, founded by Count Rumford (Benjamin Thompson) (1753–1814), was outfitted for the research of gifted investigators like Davy and later, Michael Faraday* (1791–1867). However, its principal educational function was to provide public enlightenment through popular lectures. Davy was appointed to a lectureship in chemistry at the Royal Institution in 1801. The following year he was promoted to professor of chemistry. The lectures on science attracted large and fashionable audiences. Under Davy this trend grew stronger, until the Royal Institution became a center for advanced research and for polished demonstration lectures. Davy was a showman and lectured to capacity audiences. By about 1810, as many as a thousand people came to see and hear him, including the aristocracy. His lectures on agricultural chemistry were the first serious attempt to apply chemistry to agriculture, and it remained the standard work until Liebig's publications a generation later [30].

Early American medical colleges recognized chemistry as a branch of medical study and, at the end of the eighteenth century, added chemistry to their science curricula. The College of New Jersey (known as Princeton since 1896) was the only American institution of learning other than medical schools to teach chemistry from a separate department. John Maclean (1771–1814) in 1795 was the college's first professor of chemistry. Benjamin Rush (1745–1813), a signer of the Declaration of Independence, and a graduate of the College of New Jersey (1760) and Edinburgh University (M.D., 1768) where he studied under Joseph Black, practiced medicine in Philadelphia. Elected professor of chemistry in 1769 at the Philadelphia Medical College (absorbed by the University of Pennsylvania in 1791), he was the first to hold an independent chair of chemistry in America. Rush's *Syllabus of a Course of Lectures on Chemistry* (1770) was presented in a medical context and contained much material on pharmaceutical chemistry. He wanted chemistry to be useful to the community at large; therefore, he offered a course to the educated public in 1775 and to

---

*Faraday introduced the terms electrode, anode, cathode, anion, cation, ion, and ionization.

the students of the Young Ladies Academy of Philadelphia in 1787. However, he did not appreciate the experimental method and gave no demonstrations.

CHEMISTRY SEPARATES FROM MEDICINE

At the beginning of the nineteenth century, the philosophical and practical dependence of chemistry on medicine, which Paracelsus had set into motion as iatrochemistry three centuries earlier, had not changed. Chemistry served the physician, as chemists (apothecaries) continued to prepare the medications prescribed by the physician. This relationship contradicted the belief that Robert Boyle had endowed chemistry with new goals and fundamentals and was the "founder of modern chemistry" [31]. The crucial separation of chemistry from medicine—in spirit and doctrine— occurred when chemistry moved from the medical faculty to the philosophical faculty as an independent discipline [32].

Chemistry had been a didactic subject taught mainly by physicians in the medical schools of Europe until Liebig changed it into an experimental laboratory discipline in its own right. Even earlier, in 1789, the Weimar minister, scientist, and poet, Johann Wolfgang von Goethe (1749–1832), in agreement with the Duke of Saxony-Weimar—thus bypassing university protocol—arranged for the appointment of the apothecary Johann Friedrich August Göttling (1753–1809) as professor of chemistry in the philosophical faculty of the University of Jena and the designation of the gifted Johann Wolfgang Döbereiner (1780–1849) as his successor in 1810. From 1820 on, predating Liebig in Giessen, Döbereiner gave a practical chemical course of laboratory exercises at Jena [33].

CHEMISTRY AND THE APOTHECARIES' ACT OF 1815

The teaching of chemistry was raised to new importance in England by the Apothecaries' Act of 1815. Intended to regulate the practice of apothecaries, the Act specified new educational requirements as qualification for the diploma of the Society of Apothecaries. The required courses were in anatomy and physiology, theory and practice of medicine, chemistry, and materia medica (medical botany). The Act opened the door to eventual legalization of the practice of medicine by apothecaries, but did so by maintaining the obsolete *status quo*, viz., control and supervision of apothecaries by the Apothecaries' Company—a mere private incorporation in the wholesale drug business—that supporters of the Act had sought to reform. Physicians and surgeons were specifically exempted as were chemists and druggists, who received no medical education whatever [34].

After the establishment of the Royal College of Surgeons of London in 1800, it became a regular but voluntary practice for many apothecaries to apply for the license (diploma) of the College as an additional qualification. Licenses from the Royal College of Surgeons and the Society of Apothecaries, held by the same man, became the best available credentials for the "general practitioner," a term *The Lancet* castigated as clumsy and vulgar. *The Lancet* and the rank and file preferred the term *doctor*, but this did not take hold until late in the century [35].

By 1815 the new class of "surgeon-apothecary" had become the largest part of the profession in town and country and tended to the health of over 95% of the population of England and Wales. They prescribed the medicine, performed minor operations, pulled teeth, dressed wounds, set fractures and reduced their dislocations, and attended women in childbirth. These surgeons and apothecaries often served long apprenticeships and were probably better prepared for their work than the physicians, especially if they were able to add to their vocational training by a course at a private medical school. Besides, the poor and the working-class could not afford the high fees charged by the physicians [36].

By the 1830s, the stratified system of health services, in place for centuries, and established partly by law and largely by custom, was beginning to fall apart. The sharp traditional divisions within the medical profession were disintegrating under the pressures of new societal needs and structures arising from the growth of industrial towns, the spirit of free competition, and rugged individualism. Physician, surgeon, and apothecary, as distinct and separate practitioners, existed in little more than name alone. Their practices overlapped. It was even becoming difficult to distinguish them from a new group, the chemists and druggists. Few physicians could make a living *as physicians* by limiting their practice to examining patients, diagnosing disease, and prescribing medicines for *internal* diseases to be compounded and dispensed by the apothecaries. By 1834 many physicians included surgery, midwifery, and even pharmacy in their practice. However, as respects the public in general, and perhaps even a large proportion of the profession also, these distinctions were very little thought of and very little understood. The wounds of the public were treated with little if any reference to them [37].

CLINICAL CHEMISTRY—A FALSE START FOR A NEW IDENTITY

In Germany in the early nineteenth century, there was a renewal of interest in an essentially speculative philosophy of Nature (Naturphilosophie). It applied the brakes to the use of quantitative chemical and physical methods in biology and medicine. Ideas and principles now became the ultimate goal of science since experimental attempts to explain cause and effect were not providing decisive answers. It was a reaction to the mechanistic rationalism of the Age of Enlightenment. Physicians and natural philosophers returned to the concept that vital phenomena are not mechanical processes obeying the laws of lifeless inorganic matter.

However, it was only a temporary lapse. Opposition to vitalism and philosophic speculation gradually developed, and by the 1840s physiologists and physicians in France and England and particularly in Austria and Germany acknowledged that vital phenomena in health and disease should be explained by physicochemical processes. They wanted to use the methods of the natural sciences in medicine— empirical research and experimentation—and practice "scientific medicine." The change was brought about mainly by Liebig's "Animal Chemistry." The book was significant for the development of clinical chemistry because it introduced a quantitative method of observation into physiological chemistry and thereby encouraged doctors also to apply quantitative analysis to the diagnosis of diseases [38].

This was the scientific environment during the first half of the nineteenth century, in which clinical chemistry emerged from applications of chemistry to medical diagnosis. The discoveries of new substances in the healthy and diseased body that accompanied the beginning of scientific medical research, and the development of organic and physiological chemistry, spawned a wave of interest in clinical chemistry as a recognizable identity, in the late 1830s and 1840s. There followed a systematic search for pathological changes in the chemical composition of body fluids to guide medical diagnosis, follow the course of the disease, and control therapy. A search for chemical explanations for biological phenomena became a major preoccupation of leading scientists during the nineteenth century.

In England, already mentioned, were the books on the chemical composition of urine, urinary deposits and calculi, by Alexander Marcet (1817, 1819), William Prout (1821, 1826), Golding Bird (1844); and on the analysis of urine and blood in health and disease by George Owen Rees (1836, 1845), and Henry Bence Jones (1850). On the Continent, clinical chemists who applied chemistry in a systematic manner to clinical medicine were Johann Franz Simon (1807–1843) in Berlin, Johann Florian Heller (1813–1871) in Vienna, and Johann Joseph Scherer (1814–1869) in Würzburg.

At Berlin's Charité Hospital in 1839, Simon, who started as a pharmacist, then studied chemistry and was awarded a Ph.D., demonstrated chemical and microscopic examinations, while Johann Lucas Schoenlein (1793–1864) gave his clinical lectures at the bedside. Since there was no laboratory space available at the hospital at that time, Simon set up a laboratory in his apartment, where he performed tests for patients and also gave private courses in chemical analysis for the physicians of the clinic. Simon's *Handbuch der Angewandten Medizinischen Chemie* (2 vols., 1840, 1842) appeared later in English translation as *Animal Chemistry with Reference to the Physiology and Pathology of Man* (1845, 1846). In the journal *Beiträge zur Physiologischen und Pathologischen Chemie und Mikroskopie in ihrer Anwendung auf die Praktische Medizin* (1844) which he founded and edited, Simon stressed the importance of urinary pathology. Urine was to be tested for specific gravity, acid or alkaline reaction, albumin and sugar, and the sediment examined under the microscope for inorganic constituents (uric acid, urates, calcium oxalate, magnesium ammonium phosphate) and organic constituents (blood, pus, tubules or cylinders) [39]. Only one volume was published. After Simon's untimely death, the journal was taken over and edited by Heller as *Archiv für Physiologische und Pathologische Chemie und Mikroskopie*.

For Heller (Fig. 6.5) as well, the objectives of the pathological chemist were the chemical and microscopic identification and characterization of pathological products of the bodily humors (fluids) and solids. Heller received his chemical training in Prague and later studied with Liebig and Wöhler. He established a laboratory of pathological chemistry in Vienna's Allgemeines Krankenhaus (General Hospital) in 1844, but his official appointment as "pathological chemist" was delayed until 1855 because some of the medical faculty believed that the head of the laboratory should be a doctor of medicine who was also an expert in organic chemistry. This reflected the basically favorable, but reserved attitude towards the role and significance of pathological chemistry for clinical medicine. Heller's best known test was the so-called "ring test" for detecting albumin in urine by adding nitric acid (1852). Heller's journal commenced publication in 1844 and was the first journal to deal exclusively with pathological chemistry. However, there were not enough original papers or experienced contributors to support it; publication ceased after only six volumes. Besides, there were other journals accepting papers on clinical chemistry,

FIG. 6.5. Johann Florian Heller. (*Roots of Clinical Chemistry*, J. Büttner and C. Habrich, eds., Darmstadt: Git Verlag Gmbh, 1987, p. 83.)

e.g., *Annalen der Chemie und Pharmacie*, *Annales de Chimie*, *Archiv für Physiologische Heilkunde*, and others [40].

Another key location joining in the almost simultaneous beginning of clinical chemistry in the German-speaking countries was Würzburg, where

Scherer was director of the first independent hospital laboratory in the Juliusspital (Julius Hospital) of the University of Würzburg. He was appointed "university reader" in 1842 to teach organic chemistry in connection with the chemical investigations of blood and urine from the patients admitted to the hospital. It was the first academic position devoted entirely to this new discipline. His appointment to a chair was confirmed in 1847 by King Ludwig. Scherer (Fig. 6.6) was the first to use the term "clinical chemical laboratory" (klinisch-chemischen Laboratorium) in the foreword of his monograph *Chemische und Mikroskopische Untersuchungen zur Pathologie* (1843). This was almost 100 years before the term came into general use in Germany. "Pathological chemistry" was the usual term in German-speaking areas at the time. Scherer, a physician, had studied chemistry then spent one year in the laboratory of Liebig in Giessen, where the important groundwork was being built for the application of new chemical knowledge and methodology to problems in medicine—with the emphasis on quantitative methods [41].

Scherer developed methods for the quantitative analysis of serum and blood which were widely used in the middle of the nineteenth century. Water content of serum was determined gravimetrically after drying on a water bath and in an air bath. The "soluble salts" were determined by weighing the residue following ignition with Berzelius's high temperature spirit lamp burner. Serum proteins were coagulated in boiling water, isolated by filtration, dried and weighed. The "extract" is the residue after evaporation of the protein-free filtrate. For whole blood analysis, the fibrin is separated first by pressing the coagulated blood through a linen cloth. In the squeezed-out fluid, protein content, "extract," and "soluble salts," are determined as in the method for serum. Lipid content can be measured by weighing, after extraction with ether and drying [42]. Although the analysis had come a long way from the dry distillation procedures of Boyle and Langrish, the methods were too elaborate and complicated for any practical application by the clinician at the bedside.

An additional stimulus to the idea that the examination of blood and urine might yield information useful for medical diagnosis or prognosis was provided at this time by two pathologists, Andral in Paris and Karl Freiherr von Rokitansky (1804–1878) in Vienna. They emphasized the special role played by blood in pathogenesis and the importance of chemistry in extending the range of pathological methods. Much of this clinical work, however, was directed more toward the identification of diseases than toward their cure. The clinician-pathologists of 1800–1850 were sceptical about therapy [43].

FIG. 6.6. Johann Joseph Scherer. (Picture Archive of the Austrian National Library, Vienna.)

Research in clinical chemistry did not come to a standstill. During the mid-century period, tests were developed for many constituents in urine, and volumetric methods replaced the laborious gravimetric techniques. Characteristic reactions were reported for protein by Heller in 1852, and by Auguste Nicolas Eugène Millon (1812–1867) in 1849; for bile (acids) by Pettenkofer in 1844; for sugar by Fehling in 1848 (see Chapter 8); and for the estimation of urea in urine by Liebig in 1853. One cannot overemphasize the importance of these and other tests and the associated methods and techniques, as the tools of research and the guideposts of progress in the development of biochemistry.

Proteinuria and glycosuria, as well as glucose, urea, and bile pigments in blood, became known as "diagnostic signs." Another such sign was the detection of tyrosine and leucine in urine in cases of acute atrophy of the liver. These signs and the anticipation of finding others heightened interest in applications of chemistry to medical problems. But it was soon realized that there was little benefit to clinical medicine. There were many isolated pieces of chemical information about blood and urine in health and disease that did not hang together because there was not enough known about basic physiology and pathology. Analytical chemistry had outdistanced them. Hence, the concept of a "chemical sign" was premature. Writing to his teacher, Justus Liebig, in 1849, Max Josef von Pettenkofer (1818–1901) (Fig. 6.7) had this to say: "The reagent-case now holds the same position as the crocodile and basilisc used to in the stalls of those itinerant Aesculapian quacks. We must have it, but we can get no use out of it." [44]. Apparently, it was not yet time for clinical chemistry to take on a partnership role to clinical practice.

This first phase of clinical chemistry as an independent science did not last very long. Developments came to a standstill, more or less, about 1860. The burst of interest and activity in the simple chemical examinations of blood, particularly urine, ended abruptly, probably because it failed to produce any significant benefits for the clinician. The majority of practitioners lost interest in chemical analysis of biological fluids [45]. Besides, the knowledge of chemistry and dependable methodology in the middle of the nineteenth century did not allow for rapid further development. During this period, chemical diagnosis was largely absent from medical practice since results of chemical analyses did not fit in anywhere. Proteins presented a special problem. Their structure was unknown and there were no suitable methods to characterize them.

Investigators did not realize how little they knew, so they overlooked the need for qualitative research as a preliminary to quantitation. There was

FIG. 6.7. Max von Pettenkofer. (National Library of Medicine, Bethesda, Maryland.)

no firm biochemical foundation for clinical chemistry applicable to pathology and clinical medicine. Furthermore, the metabolic processes in health and disease were practically unknown. The early clinical chemists, filled with the excitement of exploring new territory, took shortcuts across areas of unrecognized complexity of biological phenomena. They rushed in to attack the most difficult problems and tried to explain pathologic chemical processes before they understood normal physiology. Unsuccessful in these goals after a decade of enthusiasm for pathological chemistry (1840–50), biochemical research turned increasingly to problems of basic physiological chemistry. Analysis of complex systems was replaced by the isolation, purification, and eventually, synthesis, of defined substances. What followed were marked advances in biochemistry and biophysics, and the beginnings of a physiological chemistry with closer ties to medical physiology than to basic chemistry. Modern physiology and pathology arose out of the rejection of the speculative philosophy of nature. It evolved from the advances of physics and organic chemistry applied to life's phenomena in health and disease [46].

Meanwhile, the clinical laboratories continued to provide analytical data with diagnostic applications to clinical medicine, as well as the practical training of physicians and medical students in physiological and pathological chemistry. And, Americans were coming for post-graduate work that they couldn't get at home. The revival of clinical chemistry in Germany and Austria came sometime after 1860. It took place in the clinics and was integrated with internal medicine. The clinicians became chemists and began using chemistry for their experimental research work. The preferred term was "pathological chemistry" or "diagnostic chemistry." The subject was covered in medical school courses as "chemical-microscopical examinations." Similar developments followed in England and the United States— with a lag time of about twenty years [47].

In the training of professional chemists, England trailed both Germany and France. The amateur tradition of the Royal Society and the emphasis in Oxford and Cambridge on liberal, rather than professional education, were factors in the slow growth of British chemistry during the first half of the nineteenth century. By the end of the century, chemistry was well-established at Oxford and Cambridge and the other British universities [48]. However, it was mainly the microscope and bacteriological examinations, not chemistry, that led to the setting-up of clinical laboratories in England—beginning about 1880.

NOTES AND REFERENCES

1. LOUIS ROSENFELD, "George Owen Rees (1813–1889): An Early Clinical Biochemist," *Clinical Chemistry*, 1985, 31: 1068–1070; N. G. COLEY, "George Owen Rees, M.D., F.R.S. (1813–89): Pioneer of Medical Chemistry," *Medical History*, 1986, 30: 173–190.

A FALSE START FOR A NEW IDENTITY 127

2. G. OWEN REES, *On the Analysis of the Blood And Urine, in Health and Disease; and on The Treatment of Urinary Diseases*, 2nd ed. (London: Longman, Brown, Green, and Longmans, 1845), pp. iii–iv.
3. G. O. REES, *On the Analysis of the Blood And Urine, in Health and Disease. With Directions for the Analysis of Urinary Calculi* (London: Longman, Orme, Brown, Green, & Longmans, 1836), pp. iii–v, 35.
4. *Ibidem*, pp. 28–29, 23.
5. RICHARD BRIGHT, "On the Functions of the Abdomen, and Some of the Diagnostic Marks of its Disease," *London Medical Gazette*, 1833, 12: 378–384, p. 380.
6. ROBT. M'GREGOR, "An Experimental Inquiry into the Comparative State of Urea in Healthy and Diseased Urine, and the Seat of the Formation of Sugar in Diabetes Mellitus," *London Medical Gazette*, 1837, 20: 221–224, p. 224.
7. G. O. REES, "On the Proportion of Urea in Certain Diseased Fluids," *Guy's Hospital Reports*, 1840, 5: 162–166, p. 162.
8. G. O. REES, "On Diabetic Blood," *Guy's Hospital Reports*, 1838, 3: 398–400.
9. N. G. COLEY, "Henry Bence-Jones, M.D., F.R.S. (1813–1873)," *Notes and Records of the Royal Society of London*, 1973, 28: 31–56; LOUIS ROSENFELD, "Henry Bence Jones (1813–1873): The Best "Chemical Doctor" in London," *Clinical Chemistry*, 1987, 33: 1687–1692; FRANK W. PUTNAM, "Henry Bence Jones: The Best Chemical Doctor in London," *Perspectives in Biology and Medicine*, 1993 (Summer), 36: 565–579. So poorly developed was the state of clinical research, that when Bence Jones began to investigate albumin in urine, the only chemical facilities at St. George's Hospital were an alcohol lamp and some nitric acid on one of the surgical wards. He subsequently set up a laboratory at home.
10. HENRY BENCE JONES, "On a New Substance Occurring in the Urine of a Patient with Mollities Ossium," *Philosophical Transactions of the Royal Society of London*, 1848, 138: 55–62.
11. H. BENCE JONES, "On Intermitting Diabetes, and on the Diabetes of Old Age," *Medico-Chirurgical Transactions*, 1853 (2nd series, vol. 18), 36: 403–432, pp. 403–404, 430.
12. H. BENCE JONES, "Lecture II. On Digestion," *Lancet*, 1850, 1: 69–71; "Lecture V. On the Quantity and Acidity of the Urine," pp. 163–165.
13. *Ibidem* (1850), "Lecture V," p. 165.
14. HENRY BENCE JONES, *On Animal Chemistry in its Application to Stomach and Renal Diseases* (London: John Churchill, 1850a), p. 138.
15. HENRY BENCE JONES, (Address to Chemical Section) in *Report of the British Association for the Advancement of Science*, 1866, 36: 28–33, p. 28.
16. *Ibidem*, p. 32.
17. *Ibidem*, p. 31.
18. H. BENCE JONES, "Papers on Chemical Pathology; Prefaced by the Gulstonian Lectures, Read at the Royal College of Physicians, 1846," *Lancet*, 1847, 2: 88–92, p. 92; JONES (1850a), p. 139.
19. J. R. PARTINGTON, *A History of Chemistry*, (4 vols.), vol. 4 (London: Macmillan & Co Ltd, 1964), pp. 333–334.
20. HENRY BENCE JONES, "On the Saliva and Pancreatic Juice," *Medical Times*, 1851, (n.s.) 2: 579–582.
21. HENRY BENCE JONES, "On the Dissolution of Urinary Calculi in Dilute Saline Fluids, at the Temperature of the Body, by the Aid of Electricity," *Philosophical Transactions of the Royal Society of London*, 1853, 143: 201–216. See also JEAN-LOUIS PREVOST and

JEAN ANDRÉ DUMAS, "Note sur l'emploi de la pile dans le traitement des calculs de la vessil," *Annales de Chimie et de Physique* (Paris), 1823, 23: 90–97.
22. *Dictionary of Scientific Biography (DSB)*, Charles Coulston Gillispie, ed. (New York: Charles Scribner's Sons, 1973), 8: 329–350; WILLIAM H. BROCK, *The Norton History of Chemistry* (New York: W. W. Norton & Company, Inc., 1993), pp. 194–209. For an informative and illustrated color brochure (58 pages), see SIEGFRIED HEILENZ, *The Liebig-Museum in Giessen. Guide through the Museum and a Liebig-portrait from a Current Viewpoint* (Darmstadt, Germany: E. Merck, 1986). (German and English).
23. JOSEPH S. FRUTON, "Introduction," (pp. 1–21) in *Molecules and Life. Historical Essays on the Interplay of Chemistry and Biology* (New York: Wiley-Interscience, 1972), pp. 4, 3.
24. *Ibidem*, p. 97. Also see FREDERICK L. HOLMES, "Introduction," (pp. vii-cxvi) in *Animal Chemistry or Organic Chemistry in its Application to Physiology and Pathology*, by Justus Liebig (William Gregory, ed.). A Facsimile of the Cambridge Edition of 1842 (New York and London: Johnson Reprint Corporation, 1964), p. lx.
25. HOLMES (1964), pp. x-xxxiii.
26. For analysis and review of "Animal Chemistry," the reception it received, and the personalities involved, see HOLMES (1964), Introduction (pp. vii-cxvi).
27. N. G. COLEY, "The Animal Chemistry Club; Assistant Society to the Royal Society," *Notes and Records of the Royal Society of London*, 1967, 22: 173–185. Between 1788 and 1838, twelve new societies devoted to science specialities were formed in London.
28. BROCK (1993), pp. 271–272.
29. In 1841, William Allen became the first president of the Pharmaceutical Society of Great Britain. This society united for the first time the chemists and druggists of the country into one organization for the protection of their general interests and the improvement and advancement of scientific knowledge.
30. DSB (1971), 3: 598–604, p. 601.
31. J. R. PARTINGTON, *A Short History of Chemistry*, 2nd ed. (London: Macmillan and Co., Limited, 1951), p. 67.
32. PAUL WALDEN, "The Gmelin Chemical Dynasty," *Journal of Chemical Education*, 1954, 31: 534–541.
33. WILHELM PRANDTL, "Johann Wolfgang Dobereiner, Goethe's Chemical Adviser," *Journal of Chemical Education*, 1950, 27: 176–181.
34. S. W. F. HOLLOWAY, "The Apothecaries' Act, 1815: A Reinterpretation. Part I: The Origins of the Act," *Medical History*, 1966, 10: 107–129; "Part II: The Consequences of the Act," pp. 221–236. For an editorial reaction see *Lancet*, 1825–26, 2: 625–627.
35. HOLLOWAY (1966), p. 108; *Lancet*, 1829–30, 2: 693.
36. HOLLOWAY (1966), p. 108; "Medical Education in England, 1830–1858: A Sociological Analysis," *History*, 1964, 49: 299–324, p. 314; A. P. THOMSON, "The Influence of The General Medical Council on Education," *British Medical Journal*, 1958, 2: 1248–1250, p. 1248.
37. *London and Provincial Medical Directory* (1847), pp. xv–xvi, quoted by HOLLOWAY (1964), pp. 313–314; also see pp. 307–312; THOMAS HODGKIN, *Medical Reform. An Address Read to the Harveian Society, at the Opening of its Seventeenth Session, October 2, 1847* (London: John Churchill, 1847), (18 pages), p. 3.
38. N. MANI, "The Historical Background of Clinical Chemistry," *Journal of Clinical Chemistry and Clinical Biochemistry*, 1981, 19: 311–322; J. BÜTTNER, "Evolution of Clinical Enzymology," *Journal of Clinical Chemistry and Clinical Biochemistry*, 1981, 19:

529–538, p. 531; see also RICHARD H. SHRYOCK, "The History of Quantification in Medical Science," *Isis*, 1961, 52: 215–237, p. 229.
39. J. BÜTTNER, "The Origin of Clinical Laboratories," *European Journal of Clinical Chemistry and Clinical Biochemistry*, 1992, 30: 585–593, p. 589; Mani (1981), p. 318.
40. BÜTTNER (1992), p. 588; Mani (1981), pp. 318–319. See also ref. 42, pp. 87–88.
41. JOHANNES BÜTTNER, "Johann Joseph von Scherer (1814–1869). A Commentary on the Early History of Clinical Chemistry," in *History of Clinical Chemistry*, J. Büttner, ed. (Berlin and New York: W. de Gruyter, 1983), pp. 45–50, translated from *Journal of Clinical Chemistry and Clinical Biochemistry*, 1978, 16: 478–483; also see BÜTTNER (1992), p. 588.
42. JOHANNES BÜTTNER and CHRISTA HABRICH, *Roots of Clinical Chemistry* (Darmstadt, Germany: Git Verlag GMBH, 1987), pp. 94–95. This guide was prepared for the Historical Exhibition at the XIII International Congress of Clinical Chemistry, The Hague, The Netherlands, June 28-July 3, 1987.
43. J. BÜTTNER, "From Chemistry of Life to Chemistry of Disease: The Rise of Clinical Biochemistry," *Clinical Biochemistry*, 1980, 13: 232–235, p. 233; SHRYOCK (1961), p. 230.
44. BÜTTNER (1980), p. 233. Quotation is in K. Kisskalt, *Max von Pettenkofer* (Stuttgart: Wis. Verlagsges, 1948), p. 26. In 1844, working in Scherer's clinical chemistry laboratory in Würzburg, Pettenkofer developed his test for bile acids and began studies which led to discovery of creatinine in urine later that year in Liebig's laboratory in Giessen. *Annalen auf Chemie und Pharmacie*, 1844, 52: 90–96; 97–100. Pettenkofer's fame rests chiefly on his pioneering accomplishments in experimental hygiene, a discipline that he founded. He urged study of the origin of infectious diseases and was convinced of the importance of chemistry in the development of hygiene. In German universities, bacteriology became a part of "hygiene" (public health). Pettenkofer suffered from depression and took his own life by gunshot.
45. JOHANNES BÜTTNER, "Interrelationships Between Clinical Medicine and Clinical Chemistry, Illustrated by the Example of the German-Speaking Countries in the Late 19th Century," *Journal of Clinical Chemistry and Clinical Biochemistry*, 1982, 20: 465–471, p. 465; see also "Clinical Chemistry as Scientific Discipline: Historical Perspectives," *Clinica Chimica Acta*, 1994, 232: 1–9.
46. MANI (1981), p. 319; BÜTTNER (1983), p. 49.
47. BÜTTNER (1992), p. 589; (1982), pp. 465, 468, 466.
48. FRUTON (1972), pp. 4–5.

CHAPTER VII

TEXTBOOKS ON URINE AND BLOOD ANALYSIS (1863-1899)

During the second half of the nineteenth century, numerous books on urine analysis and chemical diagnosis began to appear in England, Germany, France, and the United States, and were frequently revised. Directed to practitioners and medical students, they represent the first major application of chemical analysis to aid in diagnosis and treatment. Visual and microscopic examination of the urine sediment and the organized and unorganized crystals along with detailed drawings and photomicrographs received major attention and description.

By century's end, clinical diagnosis had expanded beyond the laboratory examination of urine to include other body fluids, secretions, and waste products. Advances in physiological chemistry added new reliability to the study of the composition of the urine, and provided a better insight into the relationship of the urine to the organism in health and in disease. Medical chemistry was a relatively new subject, a stage in the development of physiological chemistry as it came to be taught in medical schools during the final decades of the century and continuing to as late as the 1960s.

In 1863, Carl Neubauer and Julius Vogel produced the fourth edition of *A Guide to the Qualitative and Quantitative Analysis of the Urine, Designed Especially for the Use of Medical Men*. By then the major constituents of urine were known and their qualitative and quantitative analyses had been developed. However, except in a few instances such as proteinuria and glycosuria, interpretation of variations was not at all clear, and the idea persisted that every particular disease produced a different kind of urine [1].

In their introduction, Neubauer and Vogel assured the reader that the analysis of urine is no longer a lengthy and difficult procedure. "The physician, armed with the simplest and newest methods of analysis, is now able in a short time, and at the bedside of the patient, to test the urine, and thereby to discover in it the presence of abnormal constituents, or to determine the quantity of any of its normal constituents. This method of analysis of the urine, combined with a scientific application of the microscope, enables us to arrive at positive conclusions concerning changes going on in the body" [2]. They list seventeen normal and fifteen abnormal constituents of urine, most of which, though recognizable by name, are not frequently analyzed today.

However, their enthusiastic claims are not justified. The methods, gravimetric and volumetric, are long, complicated, and require a wide range of chemical reagents and calibrators, and the usual facilities of a chemistry laboratory. Many of the procedures require more than 100 mL of urine. The authors do stress the importance of analyzing timed collections of urine and, where great accuracy is required, averaging replicate analyses to minimize error. In some cases, a rapid approximate quantitative estimation was satisfactory [3]. "Most of the quantitative analyses of the urine,... have... been so simplified, that any properly educated physician may readily undertake them." If the physician does not have the time to run the analysis himself, "... a chemist may always be found ready, for a moderate consideration, to undertake the simplified analysis; and, if necessary, any intelligent attendant, or servant, provided he be careful, may, as I know from experience, be taught enough for the purpose in a very short time" [4].

Quantitative colorimetric methods were unknown. Specific gravity was measured by a specially designed hydrometer (urinometer) or by comparative weighings in a weighing bottle (pycnometer). Urinary sediments were examined under the microscope. Chloride in urine was determined by Mohr's volumetric method and was usually reported as milligrams of sodium chloride per 100 cc. No method was given for sodium but the intensely yellow color was noted when its salts are heated to redness on a platinum wire in the inner flame of a blowpipe.

Albumin was precipitated by acidifying with acetic acid and heating—a procedure first described in 1694 by Fredericus Dekkers—then collected, dried, and weighed. Many color detection methods for protein are listed by Neubauer and Vogel, including the xanthoproteic (yellow) reaction with concentrated nitric acid and heat, and Millon's reaction with mercuric nitrate and heat (red precipitate). Each of the methods is accompanied by cautionary instructions for optimum results and caveats about interfering substances.

In a gravimetric procedure, uric acid was precipitated by concentrated hydrochloric acid and, after 24–36 hours, was filtered, dried, and weighed. Creatinine was also determined by weight after precipitation from an alcoholic extract with zinc chloride. The procedure, newly devised by Neubauer in 1861, was a laborious three-day affair.

An English physician, Arthur Hill Hassall (1817–1894), covered a wide range of substances in the second edition (416 pages) of *The Urine in Health and Disease: Being an Exposition of The Composition of The Urine, and of the Pathology and Treatment of Urinary and Renal Disorders* (1863). Procedures are presented for the analyses of urinary constituents as well as a discussion of physiological factors such as sex, age, weight, diet, and exercise, and various pathological states that influence their concentration. There is frequent reference to the work of others, but very few literature citations are given. Many drawings and descriptions of crystalline structures of urinary sediments as observed under the microscope are included. Instead of structural formulas of urinary constituents, the percent composition of its elements are given. Reagents, liquid and solid, are measured by weight—drachms, grains, ounces—and most analytical methods are gravimetric. Quantitative volumetric (titrimetric) and ponderous (gravimetric) methods are given for chlorine, urea, phosphoric and sulfuric acids, and sugar, "which are amongst the more important constituents of the urine" [5]. Directions throughout are difficult to follow today because of the outmoded terminology sometimes used for chemicals, reactions, and equipment.

Numerous and "exceedingly ingenious" methods have been devised for the determination of urea in urine. However, Hassall describes only "the most convenient... which are of acknowledged practical utility." First are the methods based upon the chemical decomposition of urea followed by measurement of the carbonic acid, ammonia, or nitrogen released. In other procedures, urea is reacted with nitric, oxalic, or tartaric acid, to form insoluble salts. After washing, recrystallization, drying, and weighing, the urea fraction may be calculated. The preferred procedure was Liebig's volumetric method based on precipitation of urea by nitrate of protoxide of mercury (mercuric nitrate) [6]. This complicated procedure appeared in manuals during the remaining years of the nineteenth century. Another procedure, the gasometric measurement of nitrogen released from urea by reaction with alkaline hypobromite, continued to appear in laboratory manuals well past the middle of the next century, especially in Great Britain [7], despite the availability since 1916 of the popular urease enzyme method [8].

"A very good general idea of the *quantity* of urea present in any urine may be obtained by adding a little nitric acid to a few drops of the urine on a slip of glass, when a crust,... according to the amount of urea present, will be left on evaporation," and the crystals examined under the microscope will be found to have all the characters of urea nitrate [9].

Emphasizing the need to collect a 24-hour volume, Hassall gives an average of 512.4 grains (33.8 grams) with a range of 286.1 to 688.4 grains of urea for healthy males 20 to 40 years of age, a value in agreement with today's findings. Increases are noted for diseases accompanied by fever and in inflammatory diseases. However, less than normal quantities are observed in urine in kidney diseases since elimination of the excess of urea formed, as in other acute diseases, is impeded "owing to the impaction of the tubules with fibrinous casts." Urea was believed to originate "in the nitrogenized tissues of the body" and possibly also "in the liver and other glandular organs" [10].

Glucose was believed to be normally destroyed by oxidation on its way to, and principally in, the capillaries of the lungs, by conversion into carbonic acid and water. In diabetes, this process is interrupted and sugar appears in the urine. Very little is to be found in arterial blood and it was believed that minute amounts of sugar are present in normal urine. Many reduction tests with salts of metals were described. In the potash test (Moore's test), a brown coloration when urine and potash are boiled is highly satisfactory "when the amount of glucose is at all considerable, ..." However, the author acknowledges, since most urines darken on boiling with potash, the test is unreliable for detection of very minute quantities of glucose. In Trommer's test (see Chapter 8), a solution of caustic potash and a few drops of dilute solution of copper sulfate are added to a urine specimen. If sugar is present, the mixture turns azure-blue in color and precipitation of copper oxide takes place, then disappears on agitation of the liquid. After some hours of standing at rest, the suboxide of copper (cuprous oxide) deposits, showing a more or less yellow or red tint. If the mixture is initially boiled, the suboxide is immediately precipitated. A gravimetric modification is described for measurement of the cuprous oxide precipitate and calculation of the corresponding amount of "anhydrous grape-sugar" [11].

An unusual test for sugar was a fermentation procedure highly valued by the author. Sugar is detected by microscopic observation of certain characteristic sugar-fungus (yeast) that forms after several days of incubation. A contemporary of Hassall criticized the use of Moore's test, the fermentation test, and observation of the growth of yeast-like fungi, as "unreliable or insufficiently delicate." His preference was for Trommer's test which, if properly done, renders all other reagents superfluous [12].

Analytical procedures for blood chemistry appear in the fourth edition of *A Practical Handbook of Medical Chemistry* (1862) by John E. Bowman of King's College, London. Detailed directions for examination and analysis are presented on "healthy" and "morbid" blood and urine, with sections on calculi, milk, bile, mucus, and bone. A large section is devoted to detection of inorganic poisons in organic mixtures. This was of greater concern then as compared to now because of the greater prevalence of adulteration in food, medication, and water supply.

Nearly half of the book's 300 pages deals with urine. Fifty-one pages are assigned to analytical procedures in blood: water, blood corpuscles, albumin, fibrin, alcohol extractive, water extractive, oily fats, crystalline or solid fats, and fixed saline matters. The methods for urine and blood are cumbersome and depend largely on separating and weighing the constituents. A blood analysis usually begins with as much as ten to twelve ounces of blood. "For most purposes, ... and in the majority of cases, a knowledge merely of the proportion of fibrin, the corpuscles, and the solids contained in the serum, is what the medical practitioner chiefly requires" [13].

In *Practical Treatise on Urinary And Renal Diseases Including Urinary Deposits. Illustrated by Numerous Cases And Engravings* (1872), William Roberts, a physician to the Manchester Royal Infirmary, illustrates the apparatus required for examination or analysis of the urine for clinical purposes (Fig. 7.1). Other items required, but not identified in the diagram, are litmus paper, test tubes, nitric acid, acetic acid, liquor potassae, liquid ammonium fortis, prepared copper solution, drop tubes, stirring rods, spirit lamp, and a small flask. And, at a time when examination of urinary sediments and calculi played a large part in the diagnosis of urinary disorders, a microscope was essential.

In *Animal Chemistry or The Relations of Chemistry to Physiology and Pathology* (1878), Charles Thomas Kingzett presents a wide-ranging review of organs, tissues, and body fluids, and directions for separation of many constituents of no clinical relevance today. Blood is treated briefly and superficially with several procedures for sugar, which the author describes as "more or less reliable." No quantitative directions are given.

One method for sugar credited to Claude Bernard (see Chapter 8) starts with a paste of blood and animal charcoal to which is added "a little water," and this is then filtered. Trommer's test is then applied to the colorless filtrate. In another method, blood is precipitated with "much alcohol," the precipitate is extracted with "tolerably strong spirit," the alcohol is distilled off from the combined solutions, and the watery residual solution is tested for sugar. In a fermentation procedure with yeast, the carbonic anhydride (carbon dioxide) released is captured in baryta water, the barium carbonate precipitate is isolated, dried, and weighed. And finally, blood is coagulated, filtered, and the filtrate is boiled with excess

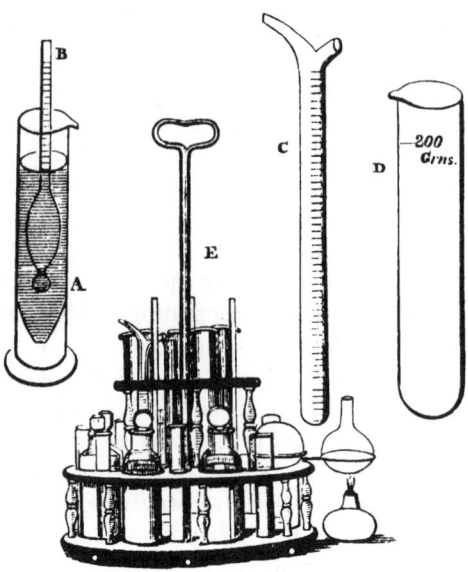

FIG. 7.1. Apparatus for Testing Urine. A, urine glasses; B, urinometer; C, burette, graduated in grains; D, 200 grain measure; E, two-tier circular stand.

potassio-tartrate of copper. The resulting suboxide of copper is oxidized by a few drops of hydrogen peroxide, dissolved in nitric acid, and the copper electrolytically deposited on a platinum spiral and weighed [14].

Kingzett, an English chemist, refers to Claude Bernard who found blood sugar concentrations to range from 1 to 3 parts per 1000 (100 to 300 mg/dL). Bernard also noted the rapid disappearance of sugar from blood after it was drawn. In a dog, a value of 1.07 parts per 1000 in a freshly drawn sample, dropped to 0.44 parts per 1000 after standing 5 hours, and completely disappeared after 24 hours. Bernard also showed that arterial blood (dog) has a higher level of sugar than venous blood.

A review of this 494-page book in *The Lancet* credits the author with successfully compressing "into a volume of reasonable size all the trustworthy work on record in relation to animal chemistry, so far as it concerns the human body" [15].

In a *Hand-Book of Chemical Physiology and Pathology with Lectures upon Normal and Abnormal Urine* (1880) by Victor C. Vaughan, a lecturer on medical chemistry at the University of Michigan, there is a brief section on blood analysis. Although this topic had not received the attention it deserves, says the author, he advises the physician to learn what is already known about the condition of the blood in disease, to be better able to base his treatment upon rational principles. In tabular form, he lists diseases of the liver and kidney, diabetes, puerperal fever, cholera, inflammation, and others, and identifies variations from normal $(+, -, 0)$ found for water, fibrin, corpuscles, albumin, urea, sugar, fat, and salts [16].

Procedures are given for the detection and separation of many constituents of urine, bile, saliva, gastric juice, milk, and feces. Quantitative directions are given for estimation in urine of urea, chloride, sulphuric and phosphoric acids, uric acid, sugar, and albumin. Values are reported in total grams and grains, per kilo and per lb body weight, respectively [17]. Liebig's method for urea is still in vogue. For albumin, urine is boiled with water acidified with acetic acid, the coagulum removed on a filter, washed, dried, and weighed. As would be expected for a book on urine, this 350-page book contains many figures of urine crystals and sediments of various kinds. Vaughan was appointed professor of physiological and pathological chemistry in 1883, the first such position in an American medical school. He later became dean of the medical school.

Clinical Chemistry (1883) by C. H. Ralfe of London Hospital, is subtitled "An account of the analysis of blood, urine, morbid products, etc., with an explanation of some of the chemical changes that occur in the body, in disease." The book is a mix of organic chemistry and clinical biochemistry. Numerous organic and inorganic constituents of the animal body are identified and directions are given for detection and quantitative analysis of several of them.

Five semi-quantitative tests are described for glucose in urine: alkaline copper; fermentation (yeast); indigo carmine; picric acid and liquor potassae; polarized light. At this late date, reagent quantities are still given in drachms, grains, and minims, indicating the endurance of the pharmaceutical terminology. Gmelin's test for bile pigments, Pettenkofer's test for bile acids, the guaiac test for blood in urine (with a warning about false positives from extraneous substances), and the spectroscopic examination of the absorption bands of hemoglobin and its derivatives are also described [18].

Albumin in urine was detected by saturated picric acid solution, concentrated nitric acid, or by dropping a citric acid paper (soaked then dried) into the urine, followed by potassium ferrocyanide, potassio-mercuric iodide, or mercuric chloride. If albumin is present "a delicate haze will diffuse through the fluid." It was convenient for testing at the bedside. Once the presence of a "proteid or albuminous substance" in urine is established, several special tests are applied to determine whether it is albumin, paraglobulin and globulin, fibrin, parapeptone (pro-peptone), or peptone. Precursors of peptone are identified as anti-albumose and hemi-albumose. The latter "is probably equivalent to the so-called c peptone of Meissner, which has been identified with the peculiar form of albumin discovered by Bence Jones in the urine of a case of osteomalacia" [19].

Blood analysis is treated in 36 pages with discussions of specific gravity, acid or alkaline "reaction," fibrin, coloring matter, blood stains, colorless corpuscles, serum, and fatty matters. "Extractives" consist mostly of urea, glucose, kreatin, hypoxanthine, and uric acid. Methods are complex, lengthy, and tedious to perform, and require large volumes of blood. To obtain kreatin in any appreciable amount directions start with 1000 grams of blood. Uric acid cannot be obtained from healthy blood in quantities sufficient for identification, but Garrod's linen thread test can be used to detect it at the higher concentrations occurring in gout [20] (see Chapter 13).

This 308-page pocket-size book is distinguished chiefly as the first of its kind in English to carry the title "Clinical Chemistry." The year was 1883. Little wonder, as one plods through the wordy directions for analysis of body fluids and secretions, that the author acknowledges the lack of advances in determining the changes which blood undergoes in disease. The fault, suggests Ralfe, may be due in part to the difficulty in obtaining sufficient quantity of blood for analysis now that bleeding was no longer part of the ordinary medical treatment. He emphasizes the need to know the daily and hourly variations in the chemical composition of normal blood and the influence of these variations on oxidation and nutrition in the body before judging the effects brought about by changes in blood in disease [21].

In the preface Ralfe states: "In spite of the disparagements of such eminent clinical teachers as Graves and Trousseau, chemistry has become more and more important to the physician as a means of elucidating many pathological conditions, or of determining the character of the morbid changes effected in tissues or secretions. Indeed, it is becoming more and more evident that we must eventually look to Chemistry for information with regard to the primary alterations that occur in fluids and tissues, and which are the first step in every disease." The author does not dwell on instructions for such simple operations as weighing, evaporation, filtering, drying precipitates, etc., since "few medical schools are now without physiological and chemical laboratories, in which students" have not already gone through a course of practical training.

A few years after Ralfe's *Clinical Chemistry* appeared, the publication of *Manuel de Chimie Clinique* in 1891 by the Lausanne pathologist Louis Bourget (1856–1913) put the French imprint on the name.

In 1884, in the United States, Austin Flint (1836–1915) published the 6th edition (87 pages) of his *Manual of chemical examination of the urine in disease; with brief directions for the examination of the most common varieties of urinary calculi, and an appendix containing a diet-table for diabetics.* The frequency of publication was a sure sign of the rapid advances being made in physiological chemistry and of the demand for this information.

The fourth edition of Hermann Sahli's *A Treatise on Diagnostic Methods of Examination* (first published in 1894) translated in 1907 from the fourth German edition (1905), devotes six pages to chemical examination of the blood: total iron, spectroscopic examination of hemoglobin for carbon monoxide and hydrogen sulfide poisoning, methemoglobin, and Garrod's thread test for uric acid. When it came to the reaction of the blood, Sahli concludes that "the method of estimating carbonic acid is too complicated for clinical purposes and requires altogether too much blood." For microscopic examination, his directions are to obtain a few drops of blood from the finger tip or ear lobe with a needle or sharp lancet. For several cubic centimeters of blood, a hypodermic syringe with a large needle is used. Large amounts of blood are best obtained by a very sharp large cannula through which the blood flows directly from the vein. A piece of tubing (less than 5 cm long) is attached to the end of the cannula [22].

This textbook discusses clinical diagnosis and clinical phenomena from physiological as well as pathological perspectives. Color reactions in urine are given for biliary pigments, bile acids, indol, skatol, melanin, urorosein, uroerythrin, urobilin, and diacetic acid. The text of 1008 pages devotes 129 pages to qualitative and quantitative examinations of urine—of which, eight pages deal with the quantitative analysis of urea by gasometric procedures, and includes a method based on specific gravity.

Clinical Diagnosis: The Bacteriological, Chemical, and Microscopical Evidence of Disease (1899) by Rudolf v. Jaksch (1855–1947) lists over 2000 literature references in a 535-page review of analytical procedures in blood, urine, and other body fluids and secretions, and feces. Numerous precipitation and color reactions, most of which have long since been abandoned, are given for the detection of albumin in urine. Several tests that have survived are: biuret reaction, first reported by Rose in 1833 (see Chapter 17); Heller's nitric acid ring test; and Esbach's quantitative estimation of albumin [23] (see later this chapter).

Some may question why the test is named for Heller. Nitric acid had been used earlier by Bright, Rees, Bence Jones, and others. In his frequently cited 1844 paper, Heller discusses, among other things, the use of acetic acid in the heat test, and refers to testing with nitric acid only in passing and makes no other mention of it [24]. But then, in 1852 he illustrates the procedure with several drawings and the use of nitric acid.

The earliest workers with nitric acid must have been familiar with the yellow stain this reagent leaves on the skin. The term "xanthoproteic" (Greek: *xanthos*, yellow), as applied to this color reaction, was used at least as early as 1838 by Mulder (see Chapter 17) for the observation made by Fourcroy and Vauquelin in 1805 when they brought protein into contact with nitric acid. The color with protein and skin is due to reaction with tyrosine and tryptophan.

The nature of the protein molecule was not understood and a hierarchy of related nomenclature was used to describe the intermediate products in the hydration (hydrolysis) of proteids, i.e., deutero-, proto-, and heteroalbumose, the final product being peptone; and there were tests distinguishing "peptone" and "albumose." Peptonuria (distinct from albuminuria) was associated with destruction of white blood cells and with those states attended by the formation and breakdown of pus.

Jaksch treats qualitative and quantitative analyses for glucose, still referred to as grape-sugar, in much detail. He mentions at the outset the long-held view that normal urine contains a trace of sugar, however, "so small, that it may be neglected as a disturbing factor, even in the most sensitive of the tests to be described." In addition to the Trommer and Fehling tests, Jaksch describes Böttger's test (1857). In this procedure, the reagent of sodium carbonate and basic bismuth nitrate turns black or dark gray upon boiling if sugar is present, due to the reduction of bismuth oxide. Nylander (1884) retained the bismuth salt, but used sodium potassium tartrate in strong alkaline solution. Rubner (1884) utilized lead acetate and ammonia and obtained a rose-red precipitate upon heating in the presence of sugar. The Molisch (1886, 1887) reaction detects sugar in urine by reaction with alpha-naphthol and sulfuric acid to form a violet-blue precipitate. This is essentially a furfural reaction that takes place with any carbohydrate. Many detection tests are given, the large number serving to emphasize the defects in all, especially the lack of specificity. Quantitation by difference in density before and after fermentation of 100–200 cc of urine for 24–48 hours is described. For this procedure, hydrometers accurately graduated to four decimal places and provided with a thermometer registering 0.1°C are required. A conversion formula is provided.

According to the author, "Its simplicity renders it very suitable for clinical use" [25]. Analysis by polarimeter (polariscope) is also described. Several color tests for bile pigments in urine, viz., bilirubin, are given. Gmelin's test (1826) layers urine onto nitric acid and a positive appears as a play of colors—a green ring in particular—at the point of contact. In Rosenbach's modification (1876), urine is passed through a filter paper which adsorbs some of the bilirubin. After a drop of nitric acid is placed on the paper, colored rings develop if bilirubin is present. Paul Ehrlich (1883, 1886) mixed urine with an equal volume of dilute acetic acid and added this dropwise to a mixture of sulfanilic acid, hydrochloric acid, and sodium nitrite. This reaction of bilirubin with diazotized sulfanilic acid to form violet azobilirubin has survived to this day [26] (see Chapter 15).

Jaffe's (1886) qualitative test for creatinine is described as follows: "A fairly concentrated solution of picric acid and a little caustic potash are added to the urine. If the fluid be heated, the presence of kreatinin will be shown by the appearance of a beautiful red coloration. Acetone and grape-sugar yield a similar reaction. Picric acid with caustic potash alone gives a slight red colour." A quantitative gravimetric procedure using zinc chloride, first devised by Neubauer, is also described. In a crude gasometric method, urea is decomposed by alkaline hypobromite and the released nitrogen gas is collected and measured after the carbonic acid also formed is removed by caustic soda. Temperature and barometric pressure enter into the calculations. Results are "not entirely accurate, but sufficiently approximate... to determine its variations at different times." Because suitable methods of analysis for urine urea were not available, Jaksch suggested an estimation of total nitrogen by Kjeldahl's method as the best and most satisfactory procedure for following the changes in nitrogen metabolism [27].

The chapter on urine has 154 pages and 793 references. Many of the chemicals for the reagents were not commercially available and had to be prepared, purified, or recrystallized, before use. Grams and cc were the units of measurement and chemical formulas are rarely used. Directions for adding ingredients, especially for qualitative tests, were still being given in tentative "cookbook" style. For example: in Emmo Legal's (1859–1922) "rough test for acetone" (1882, 1889), several cc urine are treated "with a few drops of a freshly made and somewhat concentrated solution of sodium nitroprusside, and with a moderately strong solution of caustic soda or potash. The fluid develops a red colour, which rapidly disappears, and if acetone be present, gives place to purple or violet-red on the addition of a little acetic acid" [28]. The earliest observation of ketone compounds in the urine of diabetics was made by Wilhelm Petters in 1857 and confirmed by a chemical procedure in 1865 with the ferric chloride test for acetoacetic acid by Carl Gerhardt (1833–1902). A Bordeaux-red color indicates a positive reaction.

Legal's test was described as late as 1965 in the 14th and last edition of Philip B. Hawk's (Practical) *Physiological Chemistry*, edited by Bernard L. Oser (see Chapter 18), and has survived in the dry chemistry tablet and dipstick of the Diagnostics Division of the Bayer Corporation (formerly the Ames Division of Miles, Inc., Elkhart, Indiana) and by other companies [29]. Rothera's (1908) modification using ammonium sulfate, alkali, and sodium nitroprusside, was available as a powder from the Denver Chemical Manufacturing Co., New York, as late as 1963. A few drops of urine on a small mound of powder produces a purple color in the presence of acetone. The test is much more sensitive to acetoacetic acid than to acetone, its decomposition product. Both are found in the urine together.

Chemical changes in the blood are described by Jaksch in a short section of only twenty-two pages (the same as that for gastric juice analysis), of which eight pages are given to detection of the various forms of hemoglobin by spectroscopic examination of their absorption bands. Blood is usually collected in cupping glasses following venesection.

Protein nitrogen, determined by Kjeldahl analysis, is multiplied by 6.25 to obtain grams protein, and is reported as albumin. On average, the blood of an adult contains 22.62 grams of albumin in 100 grams of blood; serum contains 8.86 grams albumin. The only mention of globulin is as a proteid in urine that can be removed by full saturation with magnesium sulfate or half saturation with ammonium sulfate. The conversion factor of 6.25 has been traced back to before 1875 and probably originated in the analytical values of protein nitrogen approximating 16.0% in the work of Mulder, Liebig, and Dumas, before 1840 [30].

Citing an 1857 source, Jaksch states that only trace amounts of urea were present in healthy blood. Consequently, even those methods designed to detect urea required large amounts of blood. In one procedure, starting with 200 to 300 cc of blood, the residue of an alcoholic extraction was dried and examined under the microscope for long slender (rhombic) prisms of urea. In a relatively short quantitative procedure, 20 cc of blood are weighed, dried *in vacuo* at low temperature, extracted with alcohol, the extract evaporated *in vacuo*, and the nitrogenous matter, "which consists entirely of urea," is estimated by Kjeldahl analysis [31]. Results were neither precise nor accurate, and of dubious clinical significance because of limited knowledge of what was normal. But it was a beginning.

Uric acid is dealt with in a similar manner. Believed also not to be present in "appreciable quantity" in blood in health, a method for its detection in blood called for 100 to 300 grams of blood. The over-24-hour procedure may yield crystals that may be examined under the microscope and tested with the murexide reaction. Uric acid was "present" in renal disease, severe anemia, and all conditions which induce dyspnoea, notably in heart-disease and pleurisy. Its presence in the blood was not believed characteristic of gout alone and therefore, according to Jaksch [32], has not the diagnostic significance claimed by Garrod (see Chapter 13).

"In health the blood contains a minute quantity of sugar." Its presence was determined by a method of Claude Bernard (1888): a protein-free filtrate is prepared by boiling a mixture of 20 grams of blood and an equal weight of crystallized sodium sulfate until "the froth above the clot becomes white, and the clot itself is free from red specks." Weigh and replace water lost by evaporation. Express the fluid, filter, and place in a burette. Titrate into a mixture of 1.0 cc Fehling's solution (copper sulfate, sodium potassium tartrate, sodium hydroxide) (see Chapter 8) to which has been added "a few small pieces of caustic potash" and distilled water and brought to a boil. The endpoint of titration is the disappearance of every trace of blue color. "As in all sugar estimations, the process must be repeated several times to get accurate results." The weight of sugar is calculated from a calibration formula and is reported as grams per kilogram blood [33].

The filtrate may be tested for the presence of sugar by Trommer's color test. Although very sensitive, the test is not specific and is reliable only in the presence of relatively large concentrations of sugar. Sugar in the filtrate may also be detected as characteristic yellow crystals of phenylglucosazone after reacting with phenylhydrazine hydrochloride. A protein-free filtrate may also be prepared by rubbing the blood in a mortar with solid ammonium sulfate and filtering to remove protein. In addition to grape sugar, the section on carbohydrates gives procedures to detect glycogen and cellulose in the blood.

Fat (fat, lecithin, cholesterin) is estimated gravimetrically after extraction of blood with hot ether for several days and subsequent drying of the extract. Results were too low by far in comparison with levels determined by today's methods. Bilirubin in serum or urine was detected by the froth test—shaking to produce a yellow foam. Uremia was not clearly understood, but the possibility that urea or its breakdown product ammonium carbonate is the poisonous material, was discounted. An increase of urea and, in some cases, "an excess of uric acid," is noted, and the observation is made that "the alkalinity of the blood was greatly less than normal." No mention is made of blood creatinine in this connection. Only passing comment is made of inorganic elements in the blood: for this, the reader is referred to textbooks of physiology and physiological chemistry. Urinary sediments are discussed, but concretions are mentioned only briefly. For their composition, the reader is referred to "the ordinary text-books on the chemistry of urine" [34].

In America, Charles W. Purdy could claim "through uranalysis alone can an almost daily increasing number of diseases be determined, their intensity be gauged, and their progress toward recovery, or their tendency toward a fatal termination be predicted" [35]. The adoption of his book, *Practical Uranalysis and Urinary Diagnosis. A Manual for the Use of Physicians, Surgeons, and Students*, as a text in more than 60 medical colleges in the United States, persuaded the author to publish a fourth edition in 1898.

Purdy echoed the prevailing view that grape sugar existed in normal blood in minute quantity, varying chiefly with the functional activity of the liver. Levels of blood sugar "in the more pronounced diabetic conditions" become markedly increased, reaching a maximum of about 0.1%. He devotes several pages to procedures for the detection of sugar in urine. Directions are usually given in grammes and cubic centimeters and are generally free of the vague language in books of that era and earlier. A chapter on gravel and calculus describes the various urinary stones found in humans and includes a tabular protocol for analysis [36].

Urine urea was a valuable index of the functional capacity of the kidneys. Decreased excretion of urea usually indicated serious forms of kidney disease. Quantitative methods for urine urea are grouped according to: reaction with mercuric nitrate to form an insoluble precipitate (Liebig's method); reaction with hypobromite to release nitrogen gas (Knop–Hufner method); and, specific gravity difference of the urine sample before and after decomposition of urea by hypochlorite [37].

Purdy describes seventeen tests for detection of albumin in the urine, all of which give a reaction with substances in normal and in abnormal urine other than albumin. The increased sensitivity of many of these tests "is nearly always obtained at the expense of trustworthiness." He describes the preparation of albumin test paper strips for use at the bedside. Chemically inert filter paper is saturated with reagent solution. This is dried and cut to convenient size. They are not used as indicator "dip-sticks," but as a source of reagent to be extracted with water—to which the urine is then added [38].

ALBUMINURIA AND INSURANCE COMPANIES

Purdy included an interesting chapter on "Examination of Urine for Life-Insurance," which provides a detailed protocol of testing for albumin, sugar, and an estimate of urea, in addition to questions to ask the applicant. His comments reflected the attitude of insurance companies at the turn of the century toward detectable levels of albuminuria. He warned that the "presence of traces of albumin in the urine, however minute, are often the index of irretrievably-damaged kidneys." These cases *"have always proved so unprofitable to life-insurance associations, through concealed or overlooked disease of the kidneys"* [39]. At this time hypersensitive tests were used and insurance coverage was refused to any person whose urine contained albumin.

However, there was growing evidence that a minute trace of albumin in the urine was frequently without clinical significance. Albuminuria not associated with any disease was first brought to the attention of the medical profession by Moxon and by Dukes (1878) in England, and by Wilhelm Leube in Germany. In 1885 Pavy described a cyclical physiological albuminuria in children and young adults without renal disease. Stirling (1887) noted an apparent relationship by using the expression "postural albuminuria," and Teissier (1899) introduced the general term "orthostatic albuminuria." This benign form of proteinuria may occur in young persons after prolonged standing, especially if at attention [40].

Highly sensitive methods demonstrated a wide margin for safe levels of albumin and changed the attitude of the life insurance companies toward albuminuria. Following ten years of favorable insurance experience during the early years of the twentieth century, with a great number of individuals with "slightest possible trace" of albumin, albuminuria in amounts up to 30 mg/dL no longer prevented a person under 35 years of age and with a negative previous history and otherwise normal findings, from being accepted as a standard risk [41].

The reliable determination of urine protein concentrations required parallel testing of calibration material. To avoid repeated preparation of standards (using sheep serum) with each series of analysis, Kingsbury *et al.* in 1926 [42], working in the laboratories of the Metropolitan Life Insurance Company in New York, developed a set of artificial standards preserved in gelatin and simulating protein concentrations from 10 to 100 mg/dL that were stable for nine months. Their procedure called for 2.5 cc urine diluted to 10 cc with 3% sulfosalicylic acid. After ten minutes, the turbidity was compared visually with the artificial standards to obtain the reading it most closely matched.

These artificial "permanent" standards and the sulfosalicylic acid turbidity procedure for measuring protein in urine were adopted by the medical directors of other life insurance companies and by clinical laboratories throughout the United States. Although the preparation of these standards was complex and highly empirical, at the time, they represented a major contribution of the laboratory to medical practice. The Kingsbury–Clark standards are still available from the R. P. Cargille Laboratories (Cedar Grove, New Jersey) in a new formulation of egg albumin and sulfosalicylic acid in gelatin, preserved by 1% formaldehyde, and have a much longer shelf life than the original preparation.

The availability of photometers and spectrophotometers made it possible to prepare a reproducible turbidimetric calibration curve for analysis of dilute protein solutions such as urine or cerebrospinal fluid, and it became the practice to rely on a one-time calibration for the procedure. However, the relationship between the amount of a substance in suspension and the turbidity or transmittance of the fluid is much more empirical than for substances in solution. Turbidimetric methods exhibit a considerable lack of precision because light absorption is a function of the degree of dispersion of the material, i.e., size, shape, and concentration of suspended particles. Furthermore, the degree of dispersion is affected by the rate and intensity of mixing reagent with specimen, temperature, time delay before reading, and other factors. Consequently, periodic verification of the calibration with standards and/or control specimens of known value is advisable.

ESBACH'S METHOD FOR URINE PROTEIN

There has always existed a need to determine protein in urine in the range from 10 to 100 mg per 100 mL (0.01–0.10%), because small amounts of protein are excreted physiologically. This estimation was usually made by qualitative precipitation tests with heat and/or acid under conditions which were not necessarily optimum. Positive results were reported subjectively as negative, faint trace, trace (+1), small amount (+2), moderate amount (+3), or large amount (+4). Various other descriptive terms were used. Evaluation differed from analyst to analyst. Many physicians still regard urine as a protein-free fluid from a misconception of the laboratory's qualitative report of negative on routine urine analysis of normal specimens. Fortunately for the clinician, the classic screening tests, e.g., heat and acetic acid, nitric acid ring test, and even the relatively recent protein-error indicator of the dipstick, do not detect proteins at the low concentrations normally present in urine, but they are sufficiently sensitive for most clinical purposes. Sulfosalicylic acid is more sensitive and may give a positive reaction in the low but normal range of less than 20 mg/dL.

One of the earliest attempts at quantitation of albumin in urine was made in 1874 by Georges Hubert Esbach (1843–1890) when he introduced a procedure that may still be in use somewhere in modified versions. The Esbach albuminometer was available from laboratory supply distributors as late as 1980. In the Esbach procedure, the urinary protein is precipitated by a reagent containing picric (1%) and citric (2%) acids in the Esbach albuminometer, a specially calibrated test tube with a "U" mark at the 10 cc level for the urine and an "R" at 20 cc for the added reagent. The tube is stoppered and inverted slowly several times to ensure thorough mixing. Measurement is based on the volume occupied by the precipitated protein-picrate after gravity settling for 24 hours at room temperature and is read from the calibrated tube as grams protein per liter of urine [43].

Purdy obtained practically the same readings on the Esbach albuminometer after a few minutes of centrifugation, as was obtained after standing for 24 hours. He criticized the method because the reagent precipitated "peptones, mucin, the proteoses, and practically all the proteids of the urine." In a modification, which he claims "measures *all the albumin*, and practically nothing else," Purdy reacts urine with 10% potassium ferrocyanide and acetic acid in a specially designed graduated 15 cc conical tube. After five minutes of centrifugation, each 0.1 cc of precipitate represents 1% albumin. Modifications were still inaccurate since column height was markedly influenced by particle size, centrifugation, and density of packing [44].

Though able to demonstrate gross changes in proteinuria, Esbach's method was too inaccurate at low levels of protein. As far back as 1914, this test was called "wholly untrustworthy" by Folin and Denis [45]. Magath [46] described it as having "no clinical value." Yet in 1947, the test was described in a British text [47] as the "method used most commonly in clinical work" for quantitative urine protein. Peters and Van Slyke [48], in their classic treatise, observed that Esbach's method, "despite the known fact that it may give results varying from half to twice the true amounts, has held its own through [the] years because of its convenience." And as late as 1974, Freeman and Beeler [49], who read the volume of precipitate after allowing it to settle for only thirty minutes, characterized the Esbach's test as accurate and simple.

In 1914 Folin and Denis [50] used a gravimetric procedure for total protein of urine. They precipitated the protein from 10 cc of urine in a previously weighed centrifuge tube with one cc of 5% acetic acid. The tube was placed in a boiling water bath for 15 minutes then centrifuged. The precipitate was washed with 10 cc of boiling 0.5% acetic acid then with 50% alcohol and dried for two hours at 100°–110°C. The tube was cooled in a desiccator and weighed and the protein determined by difference.

DETECTION OF PROTEIN BY DIPSTICK

In 1957, "Albustix," [51] the dip-and-read colorimetric test for protein in urine, was introduced by the Ames Division of Miles Laboratories (Elkhart, Indiana). This convenient semiquantitative procedure gradually replaced the qualitative heat and/or acid precipitation tests in most parts of the world. The dipstick protein test, whether alone on a strip or as part of a multiple-analyte test strip, is based on the so-called "protein error of indicators" discovered in 1909 by S. P. L. Sørensen, originator of the term pH to indicate the degree of acidity (see Chapter 9). Sørensen noted that protein in a solution containing certain pH indicators caused color changes different from those at the same pH in the absence of protein. This suggested a shift in pH, usually in the alkaline direction, when actually there was hardly any change. This "protein error" made it difficult to measure pH of body fluids with indicators.

The dipstick, impregnated with buffered indicator reagent, utilizes this phenomenon to detect protein. With the pH constant, the indicator has one color in the absence of protein, and another color in its presence. Although dipsticks are not as sensitive as heat and acid or sulfosalicylic acid, they provide clinically useful information on the amount of protein in urine. The sensitivity of the strip is adjusted to read negative at less than 10 mg protein per 100 mL urine, and a "trace" at somewhat greater levels, but still below 30 mg/dL. Since most of the excreted protein is albumin, this dye-binding procedure is designed primarily to detect albumin and usually fails to detect mucoprotein, globulins, hemoglobin, myoglobin, or Bence Jones proteins. Sulfosalicylic acid will precipitate all proteins.

NOTES AND REFERENCES

1. CARL NEUBAUER and JULIUS VOGEL, *A Guide to the Qualitative and Quantitative Analysis of the Urine, Designed Especially for The Use of Medical Men*, 4th ed. Translated from German by William Orlando Markham (London: The New Sydenham Society, 1863), p. 283.
2. *Ibidem*, pp. 1–2.
3. *Ibidem*, pp. 374–376.

URINE AND BLOOD ANALYSIS TEXTBOOKS   149

4. *Ibidem*, p. 373.
5. ARTHUR HILL HASSALL, *The Urine in Health and Disease: Being an Exposition of The Composition of The Urine, and of the Pathology and Treatment of Urinary and Renal Disorders*, 2nd ed. (London: John Churchill and Sons, 1863), pp. 395–396.
6. *Ibidem*, pp. 15–19.
7. HAROLD VARLEY, *Practical Clinical Biochemistry*, 3rd ed. (London: William Heinemann Medical Books, Ltd., 1963), pp. 119–120; G. A. HARRISON, *Chemical Methods in Clinical Medicine. Their Application and Interpretation with Techniques of Simple Tests*, 3rd ed. (New York: Grune and Stratton, 1947), p. 91.
8. DONALD D. VAN SLYKE and GLENN E. CULLEN, "The Determination of Urea by the Urease Method," *Journal of Biological Chemistry*, 1916, 24: 117–122.
9. HASSALL (1863), p. 17.
10. *Ibidem*, pp. 27, 33–35.
11. *Ibidem*, pp. 136–138, 144, 148–150.
12. *Ibidem*, p. 146; WILLIAM ROBERTS, "Lectures on Certain Points in the Clinical Examination of the Urine," *Lancet*, 1862, 1: 507–510.
13. JOHN E. BOWMAN, *A Practical Handbook of Medical Chemistry*, 4th ed. (London: John Churchill, 1862), p. 154.
14. CHARLES THOMAS KINGZETT, *Animal Chemistry or The Relations of Chemistry to Physiology and Pathology. A Manual for Medical Men and Scientific Chemists* (London: Longmans, Green, and Co., 1878), pp. 133–135.
15. Book Review. *Lancet*, 1878, 2: 883–884.
16. VICTOR C. VAUGHAN, *Hand-Book of Chemical Physiology and Pathology With Lectures Upon Normal and Abnormal Urine*, 3rd ed. (Ann Arbor: Ann Arbor Printing and Publishing Company, 1880), pp. 111–112.
17. *Ibidem*, p. 315.
18. CHARLES HENRY RALFE, *Clinical Chemistry* (Philadelphia: Henry C. Lea's Son & Co., 1883), pp. 150–159, 163–164, 76–79.
19. *Ibidem*, pp. 142–147; see also p. 187.
20. *Ibidem*, pp. 86, 92, 93.
21. *Ibidem*, p. 65.
22. HERMANN SAHLI, *A Treatise on Diagnostic Methods of Examination*, 4th ed., translated from German (Philadelphia and London: W. B. Saunders Company, 1907), pp. 613, 610.
23. RUDOLF V. JAKSCH, *Clinical Diagnosis: The Bacteriological, Chemical, and Microscopical Evidence of Disease*, 4th ed., translated from German by James Cagney (London: Charles Griffin and Company, Limited, 1899), pp. 296–301, 304–305.
24. THOMAS B. MAGATH, "Minimal Albuminuria and Tests for Albumin in the Urine" (pp. 440–452) in *The Kidney in Health and Disease*, Hilding Berglund and Grace Medes, eds. (Philadelphia: Lea & Febiger, 1935), p. 442.
25. JAKSCH (1899), pp. 315–316, 317–323, 328, 329–332.
26. *Ibidem*, pp. 336–338.
27. *Ibidem*, pp. 369–370, 363–366.
28. *Ibidem*, p. 354.
29. HELEN M. FREE, ROBERT R. SMEBY, MARION H. COOK, and ALFRED H. FREE, "A Comparative Study of Qualitative Tests for Ketones in Urine and Serum," *Clinical Chemistry*, 1958, 4: 323–330.
30. JAKSCH (1899), pp. 81, 312. Although serum proteins vary in their nitrogen content, it is convenient to use 6.25 for calculating the average protein content. Lipid and

carbohydrate moieties of the protein are excluded from the determination. For studies on the conversion factor, see F. E. KENDALL, "Studies on Human Serum Proteins. II. Crystallization of Human Serum Albumin," *Journal of Biological Chemistry*, 1941, 138: 97–109; ERWIN BRAND, B. KASSELL, and L. J. SAIDEL, "Chemical, Clinical and Immunological Studies on the Products of Human Plasma Fractionation. III. Amino Acid Composition of Plasma Proteins," *Journal of Clinical Investigation*, 1944, 23: 437–444; A. HILLER, J. PLAZIN, and DONALD D. VAN SLYKE, "A Study of Conditions for Kjeldahl Determination of Nitrogen in Proteins," *Journal of Biological Chemistry*, 1948, 176: 1401–1419. For a brief review, see LOUIS ROSENFELD, *Origins of Clinical Chemistry. The Evolution of Protein Analysis* (New York: Academic Press, 1982), pp. 62–63.
31. JAKSCH (1899), pp. 83–84.
32. *Ibidem*, pp. 85–87.
33. *Ibidem*, pp. 87–88.
34. *Ibidem*, pp. 91, 93–94, 290.
35. CHARLES W. PURDY, *Practical Uranalysis and Urinary Diagnosis. A Manual for the Use of Physicians, Surgeons, and Students*, 4th ed. (Philadelphia: The F. A. Davis Company, Publishers, 1898), p. 1.
36. *Ibidem*, pp. 99, 226–230.
37. *Ibidem*, pp. 24–30.
38. *Ibidem*, pp. 71–79.
39. *Ibidem*, pp. 342, 341.
40. ARTHUR M. FISHBERG, "Orthostatic Proteinuria" (pp. 396–408) in *Hypertension and Nephritis*, 5th ed. (Philadelphia: Lea & Febiger, 1954).
41. MAGATH (1935), pp. 450–451.
42. F. B. KINGSBURY, C. P. CLARK, G. WILLIAMS, and A. L. POST, "The Rapid Determination of Albumin in Urine," *Journal of Laboratory and Clinical Medicine*, 1926, 11: 981–989.
43. PHILIP B. HAWK, BERNARD L. OSER, and WILLIAM H. SUMMERSON, *Practical Physiological Chemistry*, 13th ed. (New York: McGraw-Hill Book Company, Inc., 1954), p. 929. For details about Esbach and his method, see W. J. HATCHER and A. G. W. WEBB, "Georges Hubert Esbach and the tube albuminometer," *Medical Laboratory Sciences*, 1979, 36: 185–190.
44. PURDY (1898), pp. 82–84, 64.
45. OTTO FOLIN and W. DENIS, "The Quantitative Determination of Albumin in Urine," *Journal of Biological Chemistry*, 1914, 18: 273–276.
46. MAGATH (1935), p. 444.
47. HARRISON (1947), p. 44.
48. JOHN P. PETERS and DONALD D. VAN SLYKE, *Quantitative Clinical Chemistry*, (2 vols.), vol. II. *Methods* (Baltimore: The Williams & Wilkins Company, 1932), pp. 682–683.
49. JAMES A. FREEMAN and MYRTON F. BEELER, *Laboratory Medicine—Clinical Microscopy* (Philadelphia: Lea & Febiger, 1974), p. 288.
50. FOLIN and DENIS (1914).
51. ALFRED H. FREE, CHAUNCEY O. RUPE, and INGRID METZLER, "Studies with a New Colorimetric Test for Proteinuria," *Clinical Chemistry*, 1957, 3: 716–727. A variation of this colorimetric test, based on the same principle, but in tablet form (Albutest), was introduced by Ames at the same time.

CHAPTER VIII

CHOLERA, ACIDOSIS, AND
FLUID-ELECTROLYTE THERAPY IN 1832

The history of acidosis is a fascinating story of the interplay between medical science and biochemistry. The initial advances in acid–base concepts evolved from clinical observation. Physiological and chemical investigation followed. Very probably the first notice of a relationship such as acid–base balance was reported in a letter to the editor of *The Lancet* on December 29, 1831. It described the acidosis occurring in a cholera patient during an epidemic sweeping England. William B. O'Shaughnessy (1809–1889), who wrote the letter, had read a report by William Stevens (1786–1868), a British physician on St. Croix. Stevens, visiting London shortly before the cholera epidemic struck Britain, was describing his favorable results in the treatment of tropical fevers [1].

Stevens rejected the commonly held belief that fever was an inflammation to be treated by expelling the offending "evil" with strong emetics or by blood-letting. His therapy was to resupply the patient—by mouth—with salts and water lost during the fever. However, since the rapid course of cholera would hardly allow enough time for oral treatment to be effective, O'Shaughnessy believed that intravenous infusion of similar saline solutions would be the better route. But first, to provide the rationale for such therapy, he ran a chemical analysis of cholera blood and feces and reported this in his letter. "Free alkali" was absent from the serum of some cases and "barely a trace in others" due to loss of "carbonate of soda" in the feces. There was also loss of "neutral saline ingredients" (sodium chloride) and water. At that time, chemical analyses were generally conducted on whole blood since the leading French chemists taught that serum not only surrounded the blood cells, but also was inside the cells together with the red pigment. Inorganic substances were determined after the blood sample was dried and incinerated.

By a coincidence of timing, in another letter to *The Lancet* dated only two days later, William Reid Clanny (1776–1850) reported similar findings in a cholera patient. It should be of interest to note that Clanny, by drawing blood into an air-tight flask which he evacuated with an air-pump, was using the "vacutainer" principle a century before the successful commercial introduction of a simpler design by Becton Dickinson and Company, East Rutherford, New Jersey,* in 1949 (see Chapter 12). Using the air pump, Clanny evacuated the carbon dioxide from the blood and made a quantitative estimation of the gas by passing it through lime water (calcium hydroxide) and measuring its increase in weight [2].

Meanwhile, Thomas Latta (d. 1833), a general practitioner in the small city of Leith, Scotland, readily grasped the therapeutic implications of O'Shaughnessy's observations, and became the first to use intravenous injection of aqueous solutions of sodium chloride and sodium bicarbonate to replace losses incurred by diarrhea. His first patient was severely sick and near death when he commenced intravenous injection of six pints of fluid within a 1/2 hour period. Although there was a remarkable improvement in the patient's appearance, she died in 5 1/2 hours. Not discouraged, Latta realized that repeated injections may be required. The next patient received 330 ounces (approximately ten liters) during a twelve hour period. The response was striking and recovery proceeded rapidly. During the epidemic he treated sixteen patients who were near death and rescued eight of them. Some of Latta's cases failed, which he attributed to the quantity injected being too small, or the effects were nullified by extensive organic disease, or its application was too late [3].

Other doctors tried the new treatment and reported "miraculous" recoveries. The solution used by Dr. Latta and his followers contained about 0.4% sodium chloride and 0.3% sodium bicarbonate, so that it was an approximately isotonic solution. They found that in order to save their patients several injections of 2 to 3 liters might be required. It is a remarkable chapter in the history of medicine that the chemical methodology of the time—possibly an inorganic technique following drying and ashing—could determine the salt losses in the blood and their recovery in the feces, and would lead to correct interpretation of the pathophysiology of the disease and appropriate, though novel, therapy.

The reports by Dr. Stevens generated a series of charges by a Dr. Johnson and resulted in exchanges between the two in the pages of *The Lancet* [4]. This medical weekly correctly inferred from O'Shaughnessy's experiments that "water *essentially*, and salts *contingently*, should be added to the blood before it could again discharge its functions." Crediting Latta as the first to carry out this suggestion, the editor expects that "the practice

---

*The company is now located in Franklin Lakes, N.J.

will be repeated all over Great Britain, as a last resource in the desperate cases which have baffled the ordinary methods of treatment, and which would otherwise be abandoned to inevitable death" [5]. There were also those who considered the cholera epidemic as a "*government hoax*, got up for the purpose of producing a counter-revolutionary excitement, and distracting the attention of the people from the reform bill," of 1832 [6].

Though the clinicians were unaware of the relation between water loss and circulatory failure, the therapy inadvertently fulfilled the all-important requirement for the treatment of dehydration. Their concern was to provide enough sodium salts, but bedside observations guided them correctly. Consequently, in 1832, both the rationale and the effectiveness of fluid replacement therapy were roughly but adequately demonstrated.

Since the treatment was used only in advanced stages of the disease and there was often other underlying organic disease, many patients did not recover. The conservative English medical establishment, not understanding supportive therapy, exhibited the traditional prejudices of those in authority to introduction of another system of practice so diametrically opposed to their own and did not accept this bold and innovative procedure. It was, no doubt, even less attractive since the first to carry out chemical analyses on cholera blood, O'Shaughnessy and Clanny, were Irish. *The Lancet*, always crusading for improved medical services for the public, supported the new treatment, but the epidemic ended before intravenous therapy could enter into accepted medical practice. When the next cholera epidemic struck in the late 1840s, the issue was debated all over again.

CHEMICAL STUDIES AND THERAPEUTIC RESPONSES WHEN CHOLERA RETURNS

Carl Ernst Heinrich Schmidt (1822–1894), who studied with Liebig at Giessen in 1843–44, was professor of physiological chemistry at the German language University of Dorpat in Estonia [7] (founded by Czar Alexander I in 1802), when cholera struck the Baltic countries in 1848. Schmidt carried out remarkably complete analyses of whole blood, red cells, plasma, and serum from healthy individuals and from patients with cholera, during the epidemic. The close agreement of Schmidt's determinations on normal specimens with today's average normal values is impressive evidence of the reliability of the laborious methods of those times and his skill in using them. These methods required at least 100 cc of blood, often more, and utilized incineration, resolution, precipitation, and weighing [8].

In cholera, a loss of water and electrolytes, caused by the severe diarrhea, was shown by the increased albumin content of the serum along with reduction of total base. The severe acidosis, evident from the large reduction in bicarbonate, can be derived from Schmidt's other data. Although the basis for corrective therapy in diarrheal dehydration was clearly indicated by Schmidt's data—and corresponded to changes observed by British physicians in the 1831–32 cholera outbreak—it did not lead to other than sporadic attempts at peroral or intravenous water–salt administration in the Baltic countries or elsewhere, in later outbreaks of cholera.

Carl Schmidt's students called him "Wasserschmidt" to distinguish him from "Blutschmidt," Alexander Schmidt (1831–1894), who was also a professor at Dorpat, and described the formation of fibrin in blood as an enzymatic process. Carl Schmidt was the first to demonstrate that the sodium and chlorine content of plasma was considerably higher than in red blood cells; whereas, the reverse was true for potassium and phosphorus. This was contrary to the prevailing conception that the same fluid was present outside and inside of the red blood cells—except for the red pigment in the cell. These findings suggested the selective permeability of cell membranes to inorganic substances. Carl Schmidt is generally credited with originating the term "carbohydrate" (Kohlenhydrate) in 1844 for those sugars containing hydrogen and oxygen in the same ratio as in water. It should be noted that in 1831, William Prout referred to the saccharine group of foodstuffs as "hydrates of carbon" [9].

The effectiveness of parenteral fluid and salt therapy in the treatment of cholera was rediscovered in 1884 by Arnoldo Cantani (1837–1893) during an epidemic in Naples. In this situation, injection was by subcutaneous infusions of 0.4% sodium chloride and 0.3% sodium bicarbonate. Here it was dehydration and the viscosity of the blood that dictated the treatment, not the alkali deficiency. Still, there was no general acceptance of the therapy. In other outbreaks on the Continent in the 1890s, in most cases the amounts of fluid given were too small and had hardly any influence on the mortality rate [10].

There remained much uncertainty concerning the optimal concentration of the bicarbonate solutions employed. There still was a tendency to use comparatively concentrated solutions, as it was not yet understood that it was necessary to treat the accompanying dehydration. Consequently, despite the temporary improvement due to the bicarbonate treatment, the final outcome was often less satisfactory. Whatever the success that may have been achieved over cholera in Europe by about 1900, it was not due to therapeutic measures, but to improved sanitation and water supply, a recognition that cholera was a water-borne disease. However, epidemics still flourished.

Fluid and salt parenteral therapy for cholera was eventually established in the early years of the twentieth century when this disease again drew attention to acid–base balance. In 1910–11 the American physician Andrew Watson Sellards (1884–1941), working in the Philippines, demonstrated the intense acidity of the urine of cholera patients. He found that vastly larger amounts of orally administered sodium bicarbonate were needed to turn the urine of cholera patients alkaline than were needed for normal subjects. He followed the transition from an acid to an alkaline urine reaction with litmus paper during the alkali treatment. He also established the greatly reduced plasma bicarbonate levels in cholera patients—about one-third of normal—by means of acidification, and a reduced pressure and evacuation technique, coupled with gravimetric measurement of the released carbon dioxide absorbed on alkali. When Sellards initiated intravenous infusion of sodium bicarbonate solution there resulted considerable reduction in mortality from cholera and uremia [11].

In India, between 1908 and 1915, Leonard Rogers (1868–1962) lowered the mortality rate from over 60% to 25–30% by administering hypertonic saline solutions. Addition of Sellard's bicarbonate therapy as a supplementary treatment, reduced the mortality from acidotic uremia from 11.1% to 3.2% [12]. During this same period, John Howland (1873–1926) and William McKim Marriott (1885–1936) reported from Johns Hopkins Hospital that infants and children with severe diarrhea suffered from acidosis, characterized by hyperpnea, reduced carbon dioxide tension in alveolar air, decreased alkali reserve of serum, and increased hydrogen ion concentration in serum. Correction of this deficiency required intravenous administration of large amounts of sodium bicarbonate to make the urine alkaline [13]. However, effective parenteral therapy had to wait for development of micromethods of blood analysis to reveal the same changes in diarrheal disease of infants and children, that Schmidt had found in the blood of adult cholera patients.

Finally, in the early 1920s, the rationale of a procedure which seemed obvious to a bedside doctor in 1832 was established on a reliable scientific basis. The life-saving effectiveness of intravenous fluid–electrolyte therapy for cholera and gradually in other forms of diarrhea was dramatically demonstrated by reduction of the mortality of severe diarrheal disease from near 90% to less than 15% [14]. It led to universal adoption of this therapy.

THE EARLY HISTORY OF POLYURIA, GLYCOSURIA, AND HYPERGLYCEMIA

Apart from O'Shaughnessy's observations on cholera, diabetic coma was the first clinical form of acidosis that was recognized. But, before proceeding further into the modern study of diabetes, it would be well to review the early histories of diabetes mellitus, the accompanying acidosis, and the monitoring of blood glucose levels. Each has its own chronology of development. The polyuria of diabetes was known in ancient times, but glycosuria was first reported in Europe in the late seventeenth century, and hyperglycemia not until the early nineteenth century. Clinical acidosis was first recognized in 1832, during a cholera epidemic, and diabetic acidosis late in the same century. The paths of investigation began to converge in 1889 when experimental diabetes was produced in animals. Reliable methods for quantitative determination of blood sugar were developed in 1915, while the determination of carbon dioxide combining power in 1917 provided a simple measurement of acidosis. Everything came together in 1922 with the demonstration that insulin prevented glycosuria, hyperglycemia, and acidosis.

Physicians from ancient times up to the late seventeenth century included all diseases accompanied by an increased quantity of urine under the term diabetes. An ancient Egyptian medical papyrus dating from approximately 1500 B.C.E., 1000 years before the birth of Hippocrates, mentions polyuria, which may have been a reference to diabetes mellitus. The first accurate (and vivid) account of diabetes from ancient times was written in Greek by Aretaeus the Cappadocian (first century C.E.). "Diabetes is a wonderful affection, not very frequent among men, being a melting down of the flesh and limbs into urine." He describes the flow of urine as incessant and thirst as unquenchable, but there is no mention of the taste. Aretaeus explains that the name of the disease, referring to rapid passage, comes from the Greek word for "siphon." Diabetes may have been rare among the ancients since Galen saw only two cases. He described it as "diarrhea of urine." The disease was considered to be a weakness of the kidneys because of the polyuria [15].

Although writings by Hindu and Chinese physicians indicate knowledge of a disorder associated with "honey urine"—so-called because it selectively attracted insects—if it was known in Europe, it was not understood and was rarely recorded. In France, Moliere's comedy "Le Médecin volant" (1650) mentions the sweetness of urine (urine sucre). Thomas Willis—in the late seventeenth century—is credited as being the first European to discover that in polyuria associated with wasting disease the urine was "wonderful sweet water, that tasted as if it had been mixed with Honey;...." [16]—hence, "mellitus" (Greek: *meli*, honey). This adjective, first added by William Cullen, established the basic difference between diabetes mellitus and diabetes insipidus. Either because the old observations had been forgotten, or were now described better, Willis's simple observation introduced the modern era in the study of diabetes. However, he was not the first to associate glycosuria with diabetes. Avicenna (960–1037), a prominent Arab physician, described the disorder and some of its complications, and the honey-like substance in the urine of diabetic patients.

In 1776, more than a century after Willis, Matthew Dobson (1735?–1784) of Liverpool showed the actual presence of sugar in the urine of diabetics. He evaporated two quarts of urine to dryness over gentle heat. The residue was a white cake which broke easily, had the smell and taste of brown sugar, and was capable of vinous, acetous, and putrefactive fermentations. From the sweet smell and taste of diabetic urine and blood serum "it appears that this saccharine matter was not formed in the secretory organ, but previously existed in the serum of the blood" [17].

"This idea of the disease, also, well explains its emaciating effects, from so large a proportion of the alimentary matter being drawn off by the kidneys, before it is perfectly assimilated, and applied to the purposes of nutrition" [18].

"For if it is a disease of the system in general, if it is to be considered as a species of imperfect digestion and assimilation, the obvious indications of cure are, to strengthen the digestive powers, to promote a due sanguification, and establish a perfect assimilation through the whole oeconomy" [19].

In 1780, Francis Home (1719–1813), professor of materia medica at Edinburgh, detected sugar in diabetic urine by reacting with yeast. Undoubtedly, this fermentation test was the first use of enzymes as a specific reagent for a clinical chemistry analysis. In 1798 William Cruickshank reported a difference between urine sugar and milk sugar. In 1815, Chevreul (see Chapter 13) confirmed Thenard's finding in 1806 that the sugar of diabetic urine is fermentable, less sweet than cane sugar, and resembled the sugar in grapes [20].

In 1788 Thomas Cawley may have been the first to associate diabetes with the pancreas. He found multiple calculi and destruction of pancreatic tissue at autopsy of a patient dying of diabetes. In 1683 Johann Conrad Brunner (1653–1727) had observed extreme thirst and polyuria in a dog after removal of the pancreas, but he did not associate this with the clinical symptoms of diabetes [21].

By the start of the nineteenth century, there was no conclusive evidence of the presence of sugar in the blood of a diabetic. Efforts to extract sugar or sugar-like substances from diabetic serum were unsuccessful; the sweet taste of the blood was not sufficient proof, and alcoholic fermentation gave generally negative results, probably because the analyses were performed on old blood, after glycolysis had occurred.

Dobson's work was a stimulus for experiments by the English physician John Rollo (d. 1809) and his colleague, the surgeon and chemist, William Cruickshank, that were first published in 1797. They demonstrated that the chemical composition and weight of the residue of urine from diabetic patients changed during the course of the disease; it sometimes contained sugar and sometimes did not. They were convinced of the importance of chemical examination of the urine for knowing the state of the disease and the response to therapy. Rollo realized that the application of these findings to medical practice was difficult. "But as every medical man may not be significantly chemical, or have the advantages we have had of the co-operation of an expert and intelligent chemist, we would recommend the simple evaporation of the urine at the commencement of the treatment, and of the same process frequently during its progress; for in this way a tolerably accurate state of the complaint, or convalescence from it, may be obtained" [22].

From his studies on two cases, Rollo reported the success of a meat diet in the treatment of diabetes. He concluded that diabetes mellitus is a disease of the stomach and develops from some morbid change in digestion and assimilation, and that the kidneys and other parts are affected secondarily, and generally by sympathy. Since saccharine matter of the disease is formed in the stomach and chiefly from vegetable substance, successful treatment required preventing the formation of sugar by the abstinence from every kind of vegetable matter [23]. The regimen was "a diet, consisting entirely of animal food, being the only one which does not furnish oxygene, and that peculiar, but simple combination of carbone and hydrogene, constituting the basis of sugar, and without which it cannot be produced" [24]. A controlled diet was subsequently the main approach in the treatment of diabetes until the introduction of insulin therapy in 1922.

The question of hyperglycemia in diabetes was a difficult problem for the early workers. Wollaston reported in 1811 that, in spite of known presence of sugar in urine of diabetics, he had been unable to detect it in the blood of diabetics. He accepted the unusual theory proposed by Charles Darwin (1758–1778), uncle of Charles Robert Darwin (1809–1882), the evolutionist, that an unknown pathway between the stomach and bladder allowed the sugar to bypass the blood and enter direcly into the urine. Wollaston failed to detect sugar because he tested old specimens in which glycolysis had occurred. Claude Bernard later showed that it was essential to test fresh serum [25].

In *Die Verdauung nach Versuchen* (1826), Tiedemann and Gmelin demonstrated the presence of fermentable sugar in the intestines and venous blood of healthy dogs after ingestion of starch. In 1835, an Italian chemist, Felice Ambrosioni, was the first to give definite proof of the presence of sugar in the blood of a diabetic person. Using yeast, he demonstrated alcoholic fermentation of blood sugar [26]. In England in 1837, M'Gregor confirmed the observation of Ambrosioni and, by using concentrates of test fluids, showed the presence of sugar "in the blood, saliva, and stool of diabetic patients, and even in the blood of healthy persons who indulge in vegetables." M'Gregor relied on the fermentation test rather than attempting separation of crystals to show the presence of sugar. According to M'Gregor, "There are physicians, in extensive practice too, who state that they have never met with a case of such a disease." [27]. The following year, Rees separated sugar crystals from diabetic blood in a quantitative analysis [28] (see Chapter 6).

SUGAR ANALYSIS AND ALKALINE COPPER SULFATE

Grape sugar in urine has been determined most frequently with an alkaline (potash) copper sulfate solution first formulated in 1841 by Carl August Trommer (1806–1879). This reagent can be traced to ancient Egypt, where the priest-physicians compounded *Unguentum Aegyptiacum* [29], an ointment for wounds and tumors. This old preparation, formerly an official pharmaceutical in Great Britain, was used undiluted or mixed with some mild ointment to destroy fungus granulations or to repress their growth. As an ointment it was a stimulant to flabby, indolent ulcers. It was prepared by boiling down a mixture of verdigris (blue hydrated basic copper acetate), vinegar, and honey to a proper consistence. A red deposit of cuprous oxide results from the interaction of the basic verdigris and the honey (glucose). Many efforts were made in the late eighteenth and early nineteenth centuries to explain the color change.

Trommer [30] differentiated between dextrin, cane sugar, and grape sugar and originated a new test for grape sugar. With his alkaline copper sulfate reagent he obtained a distinctive precipitate of cuprous oxide (Cu_2O) when boiled with dilute solutions of grape sugar. Large amounts of sugar gave the reaction at room temperature after standing for a while. With application of heat the reaction was immediate. If heating was prolonged, the test becomes too sensitive, loses its specificity and shows a positive reaction with non-saccharine substances found in healthy and in morbid urine, such as uric acid, creatine, creatinine, allantoin, hippuric acid, hypoxanthine, glycuronic acid, ether-sulfuric acids—substances known by the mid-nineteenth century—as well as metabolites of ingested medications [31]. Furthermore, the concurrent precipitation of blue $Cu(OH)_2$ and black CuO obscured the color of the red Cu_2O precipitate formed by the reducing action of the sugar.

The next advance in this reagent for the analysis of sugar occurred in 1844, when Louis-Charles Barreswil (1817–1870) stabilized the alkaline solution (potassium hydroxide and sodium carbonate) of copper sulfate by addition of potassium tartrate [32] to prevent precipitation of $Cu(OH)_2$ and CuO and allow reduction of cupric ions to cuprous oxide. Rayer introduced the "blue liquid of Barreswil" into the systematic clinical detection of diabetes [33]. The method was soon used by others.

FEHLING'S SOLUTION

In 1848 Hermann von Fehling (1812–1885) modified Barreswil's reagent with a new formulation (sodium hydroxide, copper sulfate, potassium tartrate) to establish a "quantitative" relationship between the sugar and precipitated copper. Although the method has since been modified numerous times, Fehling's identification with the reagent is based on his three publications on the quantitative determination of sugar in urine [34].

Some historians have credited E. Mitscherlich in 1840 in Berlin, as the first to use the components of "Fehling's solution" in his sugar experiments. However, he was interested in the preparation of cuprous oxide from the reaction of sugar and alkaline cupric sulfate. Trommer, who investigated the reaction with different sugars, was Mitscherlich's student. Earlier still, in 1815, Heinrich August Vogel (1778–1867) showed that cupric acetate, when boiled with different sugars, gave a precipitate that was "*oxide minimum*" (cuprous oxide). He also investigated the reaction of sugar with other copper salts and salts of mercury, silver, and gold. Most oxidations of reducing sugars, especially with heavy metal salts, are not stoichiometric. The errors introduced in this manner make the reagents hardly suitable for accurate quantitative analysis [35], unless the reaction is calibrated.

In Fehling's procedure, the sugar may be determined in two ways—by titration or by weight. In one procedure, the reagent is heated to near boiling and is titrated with the sugar solution (urine) until all the copper is reduced and precipitated as red cuprous oxide. The endpoint is the disappearance of the blue color of the reagent. Calibration is made against a sugar solution of known concentration. Since the blue color diminishes gradually, the endpoint may be difficult to detect. In that case, an excess of reagent may be used and sugar estimated—the reaction is not stoichiometric—from the weight of the precipitated cuprous oxide or the unreduced copper remaining in solution. More than seventy years had passed since Dobson first separated sugar from diabetic urine. It would take that long again for this basic chemical reaction to be adapted to the quantitative analysis of sugar in small amounts of blood.

Fehling's reagent—initially a single solution—was widely used for quantitative determination of glucose in urine by titration. Eventually, it became apparent that the reagent slowly undergoes spontaneous decomposition (autoreduction) of cupric salts in highly alkaline solution. In 1862 Fredrick William Pavy (1829–1911) modified Fehling's formula by substituting potassium hydroxide for sodium hydroxide and storing the copper sulfate and alkaline potassium tartrate solutions separately until they were mixed prior to use. Separate solutions were subsequently adopted as standard procedure for preparation of Fehling's solution. However, the reduction of the cupric salt was still attended by formation of a precipitate which greatly obscured the color at the end of the reaction. To remedy this, Pavy reformulated Fehling's reagent with sodium–potassium tartrate in strong ammonia water. Now, on titration, the blue color disappeared without formation of a cuprous oxide precipitate, the latter being dissolved by the ammonia [36]. Pavy's fluid was much used in England, but despite this improvement, problems in technique persisted and innumerable modifications of this reagent continued to be proposed.

FERMENTATION

Because of the many false positive findings with Fehling's solution due to medications and urates, positive results were often subjected to confirmation by fermentation with yeast for twenty-four hours. The Einhorn tube (see Chapter 10) afforded a quantitative approximation sufficiently accurate for clinical purposes. For rapid determination, reaction with phenylhydrazine could provide a qualitative, but positive indication of any glucose [37].

Sugar in diabetic urine was also estimated by measuring the decrease in specific gravity after fermentation with yeast for twenty-four hours. The calibration relationship was determined by comparison experiments with solutions of known concentrations of sugar added to healthy non-saccharine urine. For every grain of sugar in an ounce of urine, there was a decrease of one degree of specific gravity [38]. Only nine years earlier, Henry Bence Jones had cautioned against making a quantitative determination of sugar in urine from specific gravity tables because of the presence of varying amounts of other substances in the urine [39] (see Chapter 6).

POLARISCOPE—SACCHARIMETER

The polariscope, often referred to as a saccharimeter, was the first optical instrument after the microscope to be used in the chemical laboratory. It was developed in 1840 by the French physicist Jean Baptiste Biot (1774–1862) for the quantitative estimation of sugar in molasses. In 1833 Biot found that a solution of cane sugar, heated with dilute sulfuric acid, underwent a chemical change. Plane polarized light was now rotated to the left instead of to the right. He described the effect as "inversion," a term still used.

This device was first used in America in Philadelphia in 1842 in the analysis of sugar and molasses, probably by James Curtis Booth (1810–1888). Following the Tariff Act of 1883, the polariscope was employed in the classification of sugar for duty assessment. This brought about the establishment of many commercial sugar laboratories. Germany became the dominant manufacturer of polariscopes after the Prussian government instituted a tax on sugar, about 1860 [40].

Booth was one of the first American students to go abroad to extend his knowledge of chemistry under the then growing German influence. After working with Wöhler in 1833, Booth returned to Philadelphia and established a student laboratory which, as college facilities grew, came to specialize in analytical and consulting work. For a long time, systematic instruction in the theory and practice of chemistry received little attention by American academic institutions. Privately owned laboratories like those of Booth in Philadelphia and Charles Thomas Jackson (1805–1880) of Boston offered the best means of learning the methods of chemical analysis [41].

CLAUDE BERNARD DISCOVERS THE GLYCOGENIC FUNCTION OF THE LIVER

In France in 1846 François Magendie demonstrated that the presence of sugar in the blood of normal rabbits and dogs after they were fed on starch is not necessarily a sign of pathology. Consequently, physiologists and physicians now accepted an alimentary origin of sugar and considered glycemia as a physiological phenomenon compatible with health, but inconstant and due to ingestion of special kinds of food. It was Claude Bernard (1813–1878) (Fig. 8.1) who discovered that sugar in blood was a normal and constant phenomenon, independent of whether the food intake was exclusively carbohydrate, meat, or lard—or even, as he discovered to his great surprise in August 1848, if all solid food had been withheld for several days. Where did this sugar come from?

As a young man Bernard had literary aspirations and was especially attracted to the drama. He wrote a vaudeville comedy which was produced and achieved some success. Seeking a literary career, he went to Paris where a critic, knowing the uncertainties of a writer's life, suggested he look for work that could earn him a living. Learning that he had been apprenticed to a pharmacist in Lyons, the critic suggested that Bernard study medicine.

During his medical studies, Bernard assisted Magendie, the foremost experimental physiologist in France at that time. Magendie's major contribution was in the movement to replace anatomical observation with animal vivisection and chemical methods as the principal tools of experimental physiology. He tended to place less emphasis than his German colleagues on the use of physical instruments. It was in Magendie's laboratory that Bernard, who never practiced medicine, learned vivisection and used it in his medical research.

From 1844–49 Bernard worked in the laboratory of chemist Theophile-Jules Pelouze (1808–1867), founder of the leading private laboratory school of chemistry in France. Pelouze trained many students and made his laboratory facilities available for the personal research of Claude Bernard and other French and foreign chemists. In August 1848 Bernard reported to the Académie des Sciences in Paris on the presence of sugar in the liver.

Initially, Bernard was of the opinion that diabetes was a "nervous affection of the lungs." He accepted Liebig's view that sugar was the fuel of life, and he believed that the action of combustion took place either in the lungs, according to Lavoisier's initial hypothesis, or in the general capillaries, the hypothesis of Joseph Louis Lagrange (1736–1813) and

FIG. 8.1. Claude Bernard. (© *Journal of Nutrition*, 1951, 45: 1, American Institute of Nutrition.)

Jean-Henri Hassenfratz (1755–1827). It had not yet been definitely established that sugar was a constant and normal presence in blood. The prevailing theory assumed that sugar present in animals came exclusively as a product from foods, and that it was destroyed in animal organisms by combustion, i.e., respiration [42].

Animals were believed incapable of making sugar, fat, or protein. They might modify, but they could never synthesize compounds of this complexity. Proteins equivalent to each principal nitrogenous constituent of the blood and tissues of animals were received as a preformed plant and animal sustance in food and, prior to being assimilated almost intact, required only minor alterations in form, not in composition. Sugar, fat, and albumin would always originate in plants, and their concentration in the blood was believed to vary and depend essentially on the food consumed. It seemed to chemists like Liebig and Dumas that to search for chemical transformations during the process of digestion and assimilation was to unnecessarily complicate the great simplifying generalization that the substances involved were all alike [43].

In 1855 Bernard discovered by means of perfusion experiments on an isolated liver, that this organ can convert food proteins into a starchlike substance from which sugar is made by the liver. He then proposed the revolutionary theory of a glycogenic function for the liver. Two years later he obtained this substance in a pure state and named it "glycogen" because it was the source of blood sugar. Bernard drew an analogy between the formation of glycogen in the liver and the formation of starch in plants, and linked them to the activity of living cells. He believed that high blood sugar levels in diabetes were due to overproduction of sugar by the liver. There was no reliable blood sugar method, but he was able to recognize in a semiquantitative way the relationship between hyperglycemia and glycosuria in diabetes. Bernard's discovery that the animal body could produce such a complex substance as glycogen helped to discredit the idea that animals cannot synthesize, but can only degrade products built up by plants. This had a profound effect upon prevailing biological theories concerning differences between plant and animal metabolism.

Further progress in carbohydrate metabolism came near the end of the century, with information on the structure of the sugars, chiefly the work of Emil Fischer (1852–1919) (Fig. 8.2) (see also Chapter 16). In 1875 Fischer had discovered phenylhydrazine, the reagent that reacts with the simple sugars to form crystallizable compounds. These can be separated and the sugars characterized. Between 1883 and 1894 he had synthesized and determined structural formulas for most of the common sugars and

FIG. 8.2. Emil Fischer. (Library of the New York Academy of Medicine.)

their isomers. He also synthesized numerous purines, including xanthine, and established the biochemical interrelationships of uric acid. When he developed a hypersensitivity to phenylhydrazine (chronic eczema), Fischer turned from sugar chemistry to the proteins.

In 1902, Fischer, and independently Franz Hofmeister (1850–1922), proposed a new theory of protein structure as complexes of amino acid units joined together by the peptide bond. This made protein synthesis less mysterious. Since the process of peptide formation was a mild hydrolytic condensation involving very small changes of energy, the reaction was within the capability of the animal organism. By 1907 Fischer was

CHEMICAL FINDINGS IN ACIDOSIS 167

synthesizing polypeptides, the largest consisting of fifteen glycyl and three leucyl amino acid residues. It was no longer plausible to believe that an animal could obtain all of its many distinctive constituent proteins intact from the substance of a different organism. Fischer's achievements in the synthesis of sugars and purine compounds were rewarded by the Nobel Prize in Chemistry in 1902.

CHEMICAL FINDINGS IN DIABETIC AND OTHER FORMS OF ACIDOSIS

Meanwhile, other approaches to the problem of acidosis were yielding valuable new information [44]. In 1880 Eugen Hallervorden (1853–1914), working in the laboratory of Bernhard Naunyn (1839–1925) in Königsberg, demonstrated increased urinary excretion of ammonia during severe diabetes. Since phosphate and sulfate excretion were normal, acid poisoning, if present, could not be attributed to abnormal concentrations of phosphoric and sulfuric acids, but to some other as yet unknown acid.

In 1877 investigation of the mechanism for disposal of ingested hydrochloric acid (administered to animals), or acids metabolically generated, led to the discovery by Friedrich Walter (b. 1850) that the total carbon dioxide content of the blood was a useful measure of alkali content—or degree of acidosis—, i.e., the extent to which the alkali of the blood had been neutralized by non-volatile acids. Much of the ingested acid was also neutralized by an increased production of ammonia which was excreted into the urine instead of alkali. Dogs poisoned by acid also exhibited slow and deep respiration suggestive of the deep breathing (hyperpnoea) described by Adolf Kussmaul (1822–1902) in 1874 during diabetic coma.

In 1883 Ernst Stadelmann (1853–1941), one of Naunyn's assistants, discovered large amounts of an organic acid in the urine in diabetes, which he believed to be crotonic acid. Subsequent investigation in Naunyn's laboratory by Oscar Minkowski (1858–1931) in 1884, and independently by Eduard Kulz (1845–1895), identified the "pathological" acid as beta-hydroxybutyric acid. Since alkali therapy for diabetic coma brought about temporary relief, Minkowski proposed that diabetic coma was due to acid poisoning as such, and not to the toxicity of this particular acid. Diabetic coma thus became the second clinical form of acidosis to be recognized— cholera was the first.

Four years later, Minkowski performed the first gasometric analysis of carbon dioxide in blood from a patient in diabetic coma with acidosis. The method, too complicated for the clinical laboratory, used a gas-pump and an alkali absorption apparatus which was weighed before and after the gas was passed through it. The blood contained much less carbon dioxide than is found normally, and the urine contained large amounts of ammonia, as well as the "ketone bodies"—acetone, acetoacetic acid, and beta-hydroxybutyric acid. These changes could be used in making the diagnosis of diabetic acidosis—acid intoxication was the term then in use—and in initiating therapy. However, the analyses—ammonia by distillation, and carbon dioxide with a blood-gas pump—were too cumbersome and difficult for routine use in a clinical setting. A physician's laboratory work was limited to bedside urine analysis for the common pathological components and an occasional blood hemoglobin measurement with a color comparator—tests which they performed in the wards, usually on the window sills. Laboratory facilities were not generally available and blood sampling was a problem until hypodermic needles came into use about 1910.

For some years acidosis was characterized by the finding of acetone and acetoacetic acid in the urine with Gerhardt's (1865) and Legal's (1882) tests, the observance of Kussmaul's slow and deep breathing, and the increased excretion of ammonia which usually paralleled the excretion of ketone bodies fairly closely. The difficulty in distilling ammonia was avoided after S. P. L. Sørensen introduced formalin titration (1908) to determine amino groups released during protein hydrolysis (see Chapter 9). Now urine could be titrated directly for ammonia. Although the formalin titration of ammonia in urine was subject to error from the presence of primary amines, it was simple and convenient and was used in many hospitals to monitor the acidic status of diabetic patients.

Acidosis never occurs as a separate disease, but only as a complication. Its deleterious effects are due to the fact that the increase in the degree of acidity in the tissue fluids brings about changes in the function of the cells and eventually their destruction. These changes may be so serious that acidosis becomes the actual cause of death, probably owing to paralysis of the cells of the respiratory center.

The term "acidosis" was first used by Naunyn (1898) in a discussion of diabetic ketonemia. The concept was later extended to include other forms of the accumulation of acid in the body and such cases in which, owing to loss of alkali (e.g., in severe diarrhea and intestinal fistulas), an increase in the acidity of the blood occurs. Obviously, acidosis may result either from an abnormal formation of acid substances, as in diabetes, or from a decreased elimination of normally formed substances, as in nephritis. Acidosis was used to designate qualitative and/or quantitative changes of acid in the blood, but not the "reaction" of the blood.

Deep breathing was an early but not so easy method for diagnosing the acidosis during diabetic coma. Because of the difficulties in determining carbon dioxide in blood, some attempts were made to determine its tension in quickly expelled alveolar air in equilibrium with arterial blood. There was no means of quantitation until 1905, when John Scott Haldane (1860–1936), an Edinburgh M.D., and John Gillies Priestley (1880–1941) described a practical method for collecting alveolar air for subsequent analysis in Haldane's gas analyzer [45]. The carbon dioxide content in the patient's exhaled air was frequently used for the diagnosis of acidosis because it varies, approximately, with the bicarbonate content of the plasma.

Haldane's method led to design of less complicated devices by others, including a combined sampler and analyzer of alveolar air by Fridericia [46], that was widely used in the early clinical investigations of acidosis. This was the method of choice when the patient can cooperate by blowing into the apparatus as directed. However, this was not always possible with very sick individuals. The results were less accurate than by direct analysis of plasma.

Haldane's laboratory at Oxford played an important early role in developing standard apparatus and techniques for analysis of blood gases. In 1897 he discovered that the small air bubbles that formed when ferricyanide was added to oxyhemoglobin to produce methemoglobin, was oxygen. Haldane used this reaction in 1900 to develop a rapid and accurate method for the oxygen content in blood, which made evacuation by the blood-gas pump and the large sample requirement, unnecessary. He demonstrated a correlation between the "oxygen capacity" of a blood sample and its concentration of hemoglobin.

VAN SLYKE'S APPARATUS FOR MEASURING BICARBONATE CONCENTRATION AND DETECTION OF ACIDOSIS

In 1914, when Lawrence J. Henderson, a Harvard M.D. (see Chapter 9), and Walter W. Palmer, in Otto Folin's department at Harvard (see Chapter 12), defined acidosis as a decrease in the bicarbonate concentration of the blood—a definition applied to metabolic, not respiratory acidosis—there was no method available for early detection of acidosis or for monitoring its correction through alkali therapy. Reasoning that an excess of organic acids in the circulation would decrease the blood levels of salts of weaker acids, notably sodium bicarbonate, Donald Van Slyke (see Chapter 12) devised an apparatus (Figs. 8.3, 8.4) and procedure based

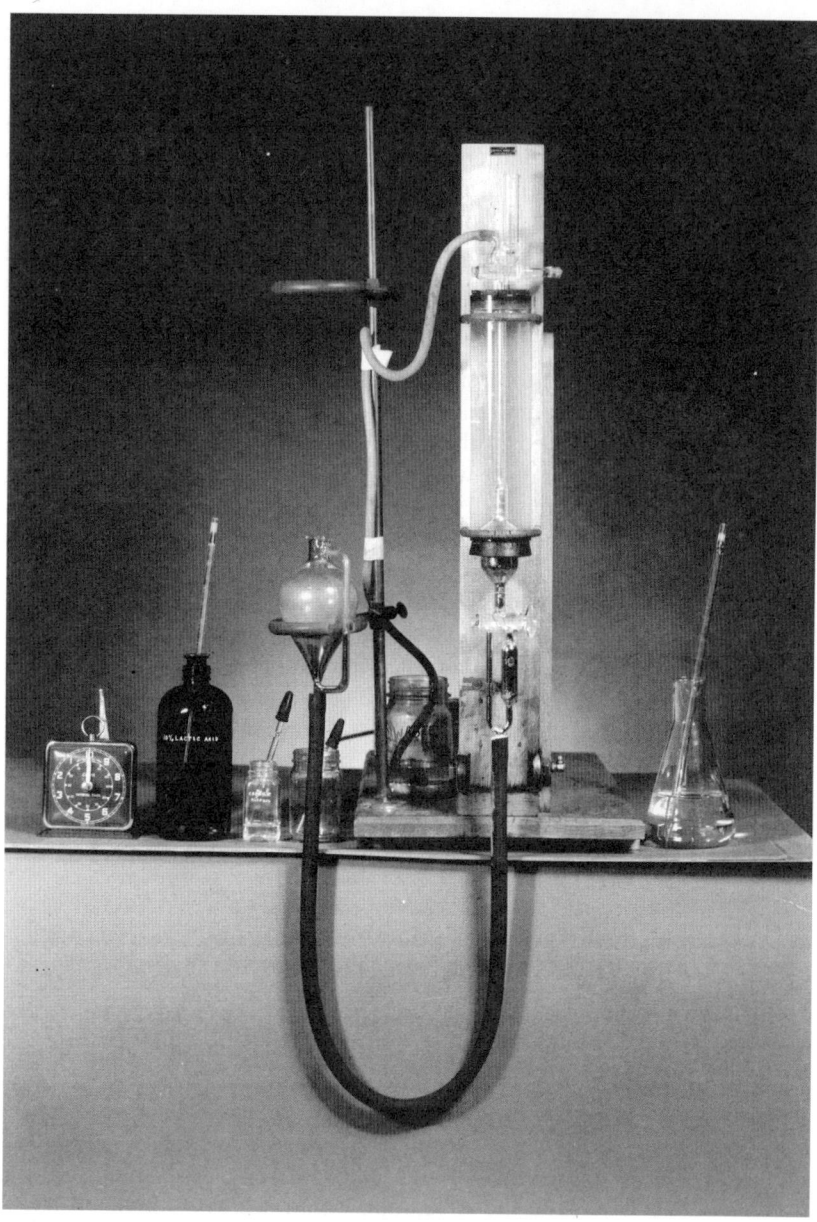

FIG. 8.3. Van Slyke volumetric gas analysis apparatus.

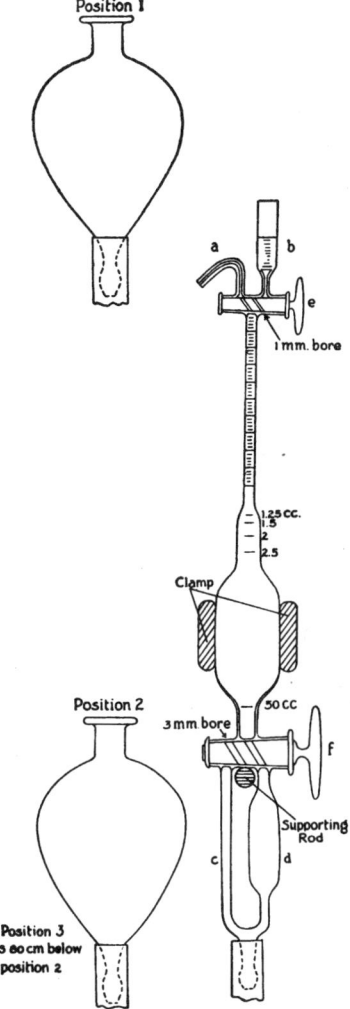

FIG. 8.4. Diagram of Van Slyke volumetric gas analysis apparatus. (*Journal of Biological Chemistry*, 1917, 30: 347–368, p. 349.)

on vacuum extraction of carbon dioxide gas for measuring the bicarbonate concentration in 1.0 mL of blood plasma [47]. It was the first instrument designed primarily for clinical chemistry. In 1920 the complete Van Slyke volumetric gas apparatus—support stand, leveling bulb, pressure tubing, separatory funnel, double-stopcock pipette, and aeration bottle with beads—sold for $27.50.

Introduced in 1917, and refined in 1921 [48], this volumetric gas method was devised for the early detection of acidosis, an inevitable fatal consequence of diabetes in the days before insulin. The simple procedure soon became the most reliable and widely used clinical method for measuring the sharp decrease in plasma bicarbonate levels that occurs in the acidosis of diabetes, nephritis, and other acid–base abnormalities in which bicarbonate is displaced by anions of other acids. The analysis could also be used to assess the severity of alkalosis. The use of this instrument led to fundamental advances in our knowledge of acid–base balance. Between 1917 and 1934, Van Slyke and his colleagues published a series of twenty-three papers under the general heading of "Studies of Acidosis."

In this method, oxalated plasma (or serum) is saturated with carbon dioxide (Fig. 8.5) from the alveolar air of the analyst whose carbon dioxide content normally is approximately 5.5%, or from a tank of compressed air with 5.5% carbon dioxide gas content. This is done to compensate for loss of carbon dioxide by exposure of the blood to the atmosphere during drawing and centrifugation. With anaerobic collection (under oil) equilibration is not necessary and the total carbon dioxide *content* of venous or arterial plasma is about three volumes percent lower (about 5% of the total) than carbon dioxide *capacity*, i.e., bound by the plasma saturated with the alveolar carbon dioxide. Since this relationship is quite constant, no special precautions in blood drawing or plasma storage are needed, and

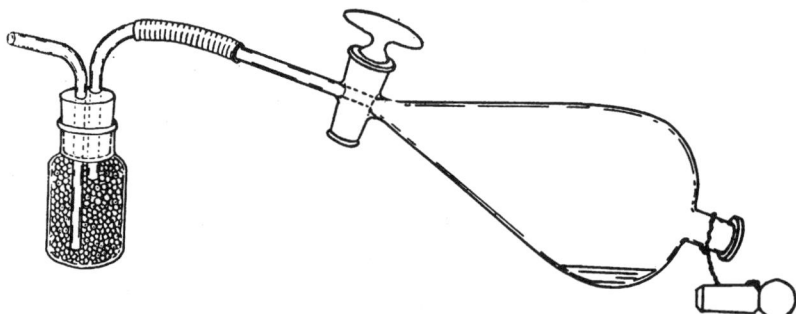

FIG. 8.5. Separatory funnel arranged for saturating plasma with alveolar air. The jar with glass beads removes the moisture from the analyst's breath. (*Journal of Biological Chemistry*, 1917, 30: 289–346, p. 308.)

no serious error is introduced by using the *capacity*, or *combining power*, as it was usually called, for following pathological changes [49].

A sample of saturated plasma is acidified by dilute lactic acid in the upper end of the apparatus which is essentially a 50 mL closed chamber pipette with a stopcock at top and bottom. The lower end is attached to a heavy-walled rubber tube connecting to a mercury leveling bulb. When the mercury bulb is lowered, a Torricellian vacuum is obtained in the pipette and carbon dioxide escapes from the plasma after several minutes of shaking. The pressure of carbon dioxide gas released inside the closed chamber is equilibrated with the outside atmospheric pressure by bringing the reservoir of mercury to the level of mercury in the pipette. The difference in volume of carbon dioxide gas released and gas remaining after subsequent absorption by added alkali, is measured in the graduated upper stem of the pipette at atmospheric pressure and room temperature, then corrected to standard conditions of 0°C and 760 mm pressure, and reported as volumes of carbon dioxide per 100 mL plasma or serum. This represents the total amount of base available for purposes of buffering acid production within physiological ranges in plasma.

The term "bicarbonate level" is preferred to the once frequently employed but less accurate expression "alkali reserve." It was first used in 1892 by Jules Alfred Jaquet (1865–1937) of Basle to denote that part of the carbon dioxide binding to alkali which was not used in the actual transport of carbon dioxide from tissues to lungs [50]. The term "alkaline reserve" was popularized by Van Slyke and Cullen who were unaware of its earlier use. The analysis of plasma carbon dioxide, though it is an indirect indicator of pH, has fewer opportunities for error than the measurement of pH, and is a more convenient basis for the calculation of the amount of alkali needed to treat the acidosis.

In 1924 Van Slyke designed a manometric [51] version (Fig. 8.6) of his gas apparatus that was more accurate than the volumetric method. The released carbon dioxide gas is compressed to a constant volume and is determined from the pressure it exerts on the manometer. In both techniques a correction for temperature is included in the calculations. The manometric method is subject to much less error since gas pressure is more accurately read than gas volume.

Both systems were very versatile and could also be used to determine oxygen capacity as a measure of hemoglobin content, and carbon monoxide content as evidence of poisoning. Van Slyke adapted this new gas technique to many ingenious procedures for the quantitative determination of numerous other constituents of body fluids—inorganic constituents,

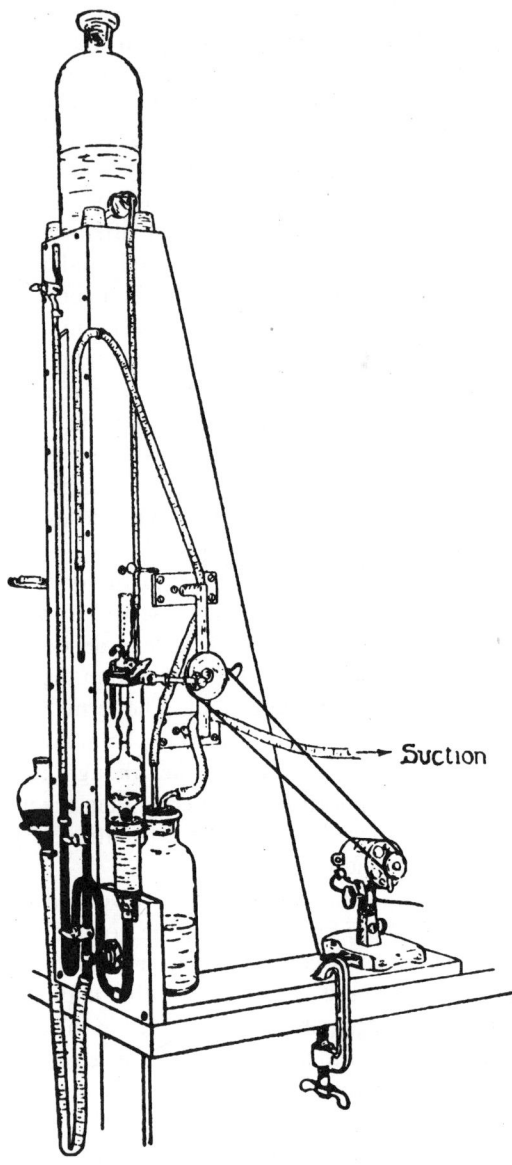

FIG. 8.6. Van Slyke manometric gas analysis apparatus. (*Journal of Biological Chemistry*, 1927, 73: 121–126.)

total nitrogen, urea, amino acids, lipids, sugar, lactic acid, and proteins—that could undergo chemical reactions leading to the production of a gas. In the 1950s he developed microgasometric procedures utilizing specimens of 0.1 mL or less [52]. A popular microgasometer adaptation of Van Slyke's classical manometric method, that required only 0.03 mL serum (Fig. 8.7), was introduced in 1951 by Samuel Natelson (1909–   ) [53]. Van Slyke's volumetric gas apparatus was available from supply catalogs as late as 1980.

MILLIEQUIVALENTS AND CATION–ANION BALANCE

Van Slyke's gasometric method was adopted by most hospital laboratories in the United States and Europe and continued in general use well into the 1960s when it was replaced by automated colorimetric methods for bicarbonate levels in plasma or serum. Reporting carbon dioxide content (or combining power), sodium, potassium, and chloride in terms of milliequivalents per liter—units which can be added for the interpretation of acid–base balance, instead of the volumes/100 mL and mg/100 mL in use as late as the early 1960s—slowly followed the work of James Lawder Gamble (1883–1959) of Harvard Medical School. The rationale was to balance the concentrations of anions and cations for the clinical evaluation of acid–base disorders and electrolyte imbalance. One of Gamble's innovations was the instructive graphs, first published in 1923, which illustrated the acid–base status of the blood. Two side-by-side bar graphs showed the concentration of cations and anions, respectively, in milliequivalents per liter of serum. The cations of serum (sodium, potassium, calcium, and magnesium) are the bases, and the anions (chloride, bicarbonate, phosphate, sulfate, organic acids, and protein) are the acids [54].

In his book *Chemical Anatomy Physiology and Pathology of Extracellular Fluid. A Lecture Syllabus* (1942) and in future editions, Gamble provided the foundation for rational water/salt therapy in hospitals everywhere. By diagrammatic representations of the chemical composition of extracellular fluid (plasma and interstitial fluid), sea water, and of cell fluid, in terms of acid–base equivalence (milliequivalents per liter), he placed an understanding of the interrelationship of anions and cations in health and disease on a scientific basis. His diagrams illustrate how the physicochemical system in extracellular fluid exhibits the necessary resilience to the rapidly changing currents of chemical events, especially in relation to the loss of digestive fluids. Gamble also provided directions for solving the problems of acid–base imbalance in specific clinical situations encountered

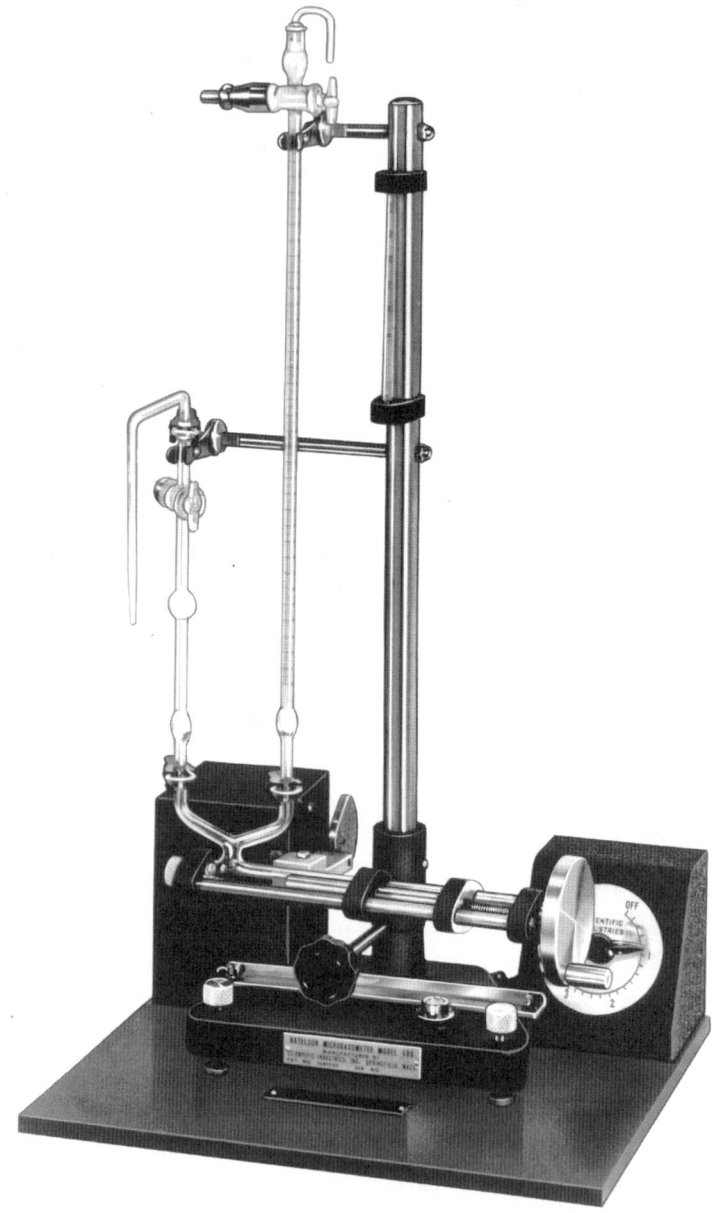

FIG. 8.7. Natelson microgasometer with timer–shaker attachment.

in practice [55]. A century had passed since the trial and error probing of Thomas Latta.

DISCOVERY OF INSULIN

In 1890, Joseph Freiherr von Mering (1849–1908) and Oscar Minkowski successfully removed the entire pancreas from a dog and observed hyperglycemia, glycosuria, and finally death in ketosis and coma. This was the first definite experimental proof that diabetes may be of pancreatic origin and sent investigators in all countries searching for the specific agent in the pancreas responsible for the normal utilization of sugar in the body. Von Mering's interest was the role of the pancreas in fat digestion. Minkowski provided the surgical skill for the experiments and was the one who noticed the polyuria, tested for sugar, and made the right connection between diabetes and the removal of the pancreas. The discovery was made in Naunyn's clinic. At this time, fasting, undernutrition, and perhaps alkali, were the only therapeutic measures that could prolong the life of the diabetic patient.

While an instructor in pathology at Johns Hopkins in 1900, Eugene Lindsay Opie (1873–1971) described hyaline degeneration of the islands of Langerhans in cases of diabetes mellitus. This very important discovery attracted attention to the islets as the probable source of an internal secretion lacking in diabetes. As early as 1894, Sir Edward Albert Sharpey-Schäfer (1850–1935) pointed out on morphological grounds that the islet tissue might be responsible for the internal secretion by which the pancreas produced its effect on the blood sugar level. In 1913, in lectures at Stanford University, he suggested the name "insuline" for the still hypothetical substance. He later acknowledged that he was unaware that the term was introduced by de Meyer in 1909. "Insulin" (Latin: *insula*, island) was independently adopted by the Toronto workers in 1922 [56].

The discovery and introduction of insulin in 1922 by Frederick Grant Banting (1891–1941) and Charles Herbert Best (1899–1978) in Toronto, sharply reduced the mortality from diabetic coma. They succeeded where others had failed in preparing an active extract of the islets of Langerhans—first described in 1869—which were believed to be the source of the internal secretion necessary for the utilization of sugar and prevention of diabetes. Working on the hypothesis that the active principle was destroyed by digestive enzymes also present in such extracts, Banting conceived the idea of preparing an effective extract from the atrophied pancreas of dogs after ligation of the pancreatic ducts for seven to ten weeks. This results in degeneration of the enzyme-producing acinous tissue, but not of the insulin-producing islet tissue, and eliminates the insulin-destroying enzymes of the pancreas. In a preliminary communication, Banting and Best reported that an extract of this degenerated gland, injected intravenously or subcutaneously, invariably reduces the concentration of sugar in the blood and the amount of sugar excreted in the urine in animals [57].

Work on Banting's plan of investigation began in May 1921, in the physiological laboratory of John James Richard Macleod (1876–1935), an authority on carbohydrate metabolism and professor of physiology at the University of Toronto. Macleod provided Banting with laboratory facilities for eight weeks, an undergraduate assistant, and an allotment of ten dogs. What began as a summer research job for a student turned into one of the most exciting medical adventures of modern times [58].

Best had just completed his bachelor's program in physiology and biochemistry and was hired for the summer as a research assistant. He won a coin toss with another student assistant, Edward Clark Noble, as to who would start the project with Banting. Best did all the chemical testing for Banting, measuring blood and urine sugar, and assisting in other ways in the series of experiments on the dogs and the preparation of active extracts. Blood sugar estimations were made by the Myers–Bailey modification of the Lewis–Benedict method [59]. The results with this procedure were confirmed by the recently published Shaffer–Hartmann iodometric titration method at high and low percentages of blood sugar [60]. In a follow-up study on human subjects with diabetes mellitus, blood sugar was estimated at intervals by the revised Folin–Wu method, urine sugar by Benedict's methods, and acetone bodies by Van Slyke's methods [61].

In late July 1921, extracts injected into depancreatized diabetic dogs produced frequent declines in blood sugar and improvement in clinical condition of the animals. These experiments, repeated and extended in subsequent months, were published in the *Journal of Laboratory and Clinical Medicine* (February 1922) [62].

When Macleod returned from a summer in Scotland, he hired James Bertram Collip (1892–1965), a Canadian biochemist, to help with the work of purification. Collip rapidly developed a method for the preparation of insulin from the whole gland of beef pancreas by employing alcohol in varying concentrations to obtain a differential precipitation of impurities. The resulting extract was sufficiently pure to allow the clinical testing of its action in humans.

In January 1922, the first clinical trials began in the Toronto General Hospital on a fourteen year old boy with diabetes, using purified extract prepared by Collip. The antidiabetic effects were as dramatic as they were remarkable. Purification, concentration, and production were then taken over by Eli Lilly and Company. In 1926 John Jacob Abel (1857–1938) of Johns Hopkins University prepared the first crystalline insulin. In 1923 the Nobel Prize in Physiology or Medicine was awarded to Banting and Macleod. Banting immediately divided his share of the prize money with Best. Macleod followed suit for Collip. In February 1941, on his way to England for research liaison with British colleagues, Banting died in an air crash in Newfoundland. Best succeeded Macleod as professor of physiology in 1929. In the late 1930s, Best's laboratory pioneered in the isolation and production of heparin, which found an important clinical application as an anticoagulant in vascular surgery. Best with Norman B. Taylor coauthored a widely used textbook, *The Physiological Basis of Medical Practice* (1937); 10th edition (1979). An 11th (1985) and 12th (1991) edition was edited by John B. West.

NOTES AND REFERENCES

1. W. B. O'SHAUGHNESSY, Letter to the Editor. "Experiments on the Blood in Cholera," *Lancet*, 1831–32 (December 29, 1831), 1: 490; WILLIAM STEVENS, Letter to the Editor. "On the Efficacy of Saline Agents in the Treatment of West India Fevers," *Lancet*, 1831–32 (January 9, 1832), 1: 553–565.
2. WM. REID CLANNY, "Case of Cholera at Sunderland, with an Analysis of the Blood, Taken From the Patient," *Lancet*, 1831–32 (December 31, 1831), 1: 505–508; see also "Description of Apparatus and Experiments for Determining the Composition of the Blood in Health and Disease," *Edinburgh Medical and Surgical Journal*, 1829, 32: 40–43.
3. THOMAS LATTA, Letter to the Secretary of the Central Board of Health, London. "Documents Communicated by the Central Board of Health, London, Relative to the Treatment of Cholera by the Copious Injection of Aqueous and Saline Fluids into the Veins," *Lancet*, 1831–32 (May 23, 1832), 2: 274–277; see also pp. 277–281 for brief case reports of successful treatment by other physicians.
4. Letters to the Editor from Dr. James Johnson and Dr. William Stevens. *Lancet*, 1831–32, 1: 376–377, 412, 591–592, 657–661.
5. *Lancet*, 1831–32 (June 2, 1832), 2: 284–286, p. 286.
6. Letter to the Editor. *Lancet*, 1831–32 (November 15, 1831), 1: 377–378.
7. Nearly all the professors came from German immigrant families who held predominant positions in the Baltic countries, both economically and culturally. Dorpat has undergone many name changes, depending on which foreign power occupied it. The city is now known by its old name of Tartu.
8. JAMES L. GAMBLE, "Early History of Fluid Replacement Therapy," *Pediatrics*, 1953, 11: 554–567. See also EBSEN KIRK, "Acidosis. Clinical Aspects and Treatment With Isotonic

Sodium Bicarbonate Solution," *Acta Medica Scandinavica*, Supplement 183, 1946, pp. 9–11, 48–58, 85–89.

9. C. SCHMIDT, "Ueber Pflanzenschleim und Bassorin," *Justus Liebig's Annalen der Chemie und Pharmacie*, 1844, 51: 29–62, p. 30; WM. PROUT, "Observations on the Application of Chemistry to Physiology, Pathology, and Practice," *London Medical Gazette*, 1831, 8: 321–327, p. 321.
10. POUL ASTRUP and JOHN W. SEVERINGHAUS, *The History of Blood Gases, Acids and Bases*, translated from Danish by Patrick Graham Jørgensen (Copenhagen: Munksgaard, 1986), p. 219.
11. ANDREW WATSON SELLARDS, "Tolerance for Alkalies in Asiatic Cholera," *Philippine Journal of Science*, Section B (Medical Sciences), 1910, 5: 363–390; ANDREW WATSON SELLARDS and A. O. SHAKLEE, "Indications of Acid Intoxication in Asiatic Cholera," *Philippine Journal of Science*, Section B (Medical Sciences), 1911, 6: 53–76.
12. LEONARD ROGERS, "The Results of the Hypertonic and Permanganate Treatment in 1000 Cases of Cholera; With Remarks on the Value of Alkalies in the Prevention of Uraemia and the Role of Atropine," *Lancet*, 1915, 189: 219–223; "Further Work on the Reduction of the Alkalinity of the Blood in Cholera; and Sodium Bicarbonate Injections in the Prevention of Uraemia," *Annals of Tropical Medicine and Parasitology*, 1916, 10: 139–149; "The Mortality From Post-Choleraic Uraemia: A 70 Per Cent. Reduction Through Intravenous Injections of Sodium Bicarbonate," *Lancet*, 1917, 193: 745–746; "The Mortality and Prognosis of Cholera Treated by the Author's Hypertonic Saline Method, Based on 2000 Cases," *Lancet*, 1921, 200: 1079–1081.
13. JOHN HOWLAND and W. MCKIM MARRIOTT, "Acidosis Occurring With Diarrhea," *American Journal of Diseases of Children*, 1916, 11: 309–325; "A Discussion of Acidosis. With Special Reference to that Occurring in Diseases of Children," *Bulletin of the Johns Hopkins Hospital*, 1916, 27: 63–69.
14. GAMBLE (1953), pp. 559–560.
15. RALPH H. MAJOR, *Classic Descriptions of Disease. With Biographical Sketches of the Authors* (Springfield, Illinois: Charles C. Thomas, 1932), pp. 185–188, 192; ADRIAN W. ZORGNIOTTI, "Galenic Urology. Translation of Urologic Portions of De Locis Affectis (Book VI, Chapter 3): Part II," *Urology*, 1979, 13: 701–703. For a historical review, see also CHARLES H. BEST, "Epochs in the History of Diabetes," (pp. 1–13) in *Diabetes*, Robert H. Williams, ed. (New York: Paul B. Hoeber, Inc., 1960).
16. THOMAS WILLIS, Treatise IX, Part I, (Pharmaceutice Rationalis: or, an Exercitation of the Operations of Medicines in Humane Bodies: . . . . ,) in *Practice of Physick*, translated from Latin by Samuel Pordage (London: T. Dring, C. Harper, and J. Leigh, 1684), p. 76; see also MAJOR (1932), pp. 192–194. Willis's *Cerebri Anatome* (1664) was the most complete and accurate account of the nervous system up to that time. He was the first to describe the disease now known as myasthenia gravis (1671), and puerperal (childbed) fever, which he named.
17. MATTHEW DOBSON, "Experiments and Observations on the Urine in a Diabetes," *Medical Observations and Inquiries* (London), 1776, 5: 298–316, p. 307.
18. *Ibidem*, p. 310.
19. *Ibidem*, p. 316; see also MAJOR (1932), pp. 194–198.
20. J. R. PARTINGTON, *A History of Chemistry*, (4 vols.), vol. 4 (London: Macmillan & Co Ltd, 1964), pp. 249, 93.
21. THOMAS CAWLEY, "A Singular Case of Diabetes, Consisting Entirely in the Quality of the Urine; With an Inquiry into the Different Theories of that Disease,"

DISCOVERY OF INSULIN

London Medical Journal, 1788, 9: 286–308; FIELDING H. GARRISON, *An Introduction to the History of Medicine*, 4th ed. (Philadelphia: W. B. Saunders Company, 1929), p. 265.
22. JOHN ROLLO, *Cases of the Diabetes Mellitus;* ...., 2nd ed. (London: C. Dilly, 1798), p. 95. This is an account of two cases of diabetes mellitus, with remarks as they arose during the progress of the cure.
23. *Ibidem*, pp. 138–139.
24. "Experiments on Urine and Sugar by Mr. Cruickshank," (pp. 438–477), in Rollo (1798), p. 477. A review of the comments by earlier European physicians who advanced medical knowledge of diabetes is on pp. 356–379.
25. WILLIAM HYDE WOLLASTON, "On the Non-existence of Sugar in the Blood of Persons Labouring Under Diabetes Mellitus. In a Letter to Alexander Marcet," *Philosophical Transactions of the Royal Society of London*, 1811, 101, 96–105; "Reply of Dr. Marcet on the Same Subject," pp. 106–109; see *Dictionary of Scientific Biography* (*DSB*), Charles Coulston Gillispie, ed. (New York: Charles Scribner's Sons, 1976), 14: 486–494, p. 492; see also N. G. COLEY, "Alexander Marcet (1770–1822), Physician and Animal Chemist," *Medical History*, 1968, 12: 394–402, p. 401.
26. FELICE AMBROSIONI, "Dello Zucchero Nelle Urine, e Nel Sangue dei Diabetici," *Annali Universale di Medicina e Chirurgia* (Milano), 1835, 74: 160–166; see also FELICE AMBROSIONI, "Detection of Sugar in the Blood in Diabetes, and on the Best Mode of Separating the Same Substance From the Urine," *London Medical Gazette*, 1836, 18: 541–542.
27. ROBT. M'GREGOR, "An Experimental Inquiry Into the Comparative State of Urea in Healthy and Diseased Urine, and the Seat of the Formation of Sugar in Diabetes Mellitus," *London Medical Gazette*, 1837, 20: 221–224, 268–272, pp. 272, 268.
28. G. O. REES, "On Diabetic Blood," *Guy's Hospital Reports*, 1838, 3: 398–400.
29. *Dispensatory of the United States of America*, 19th ed., 1907, p. 1458.
30. C. A. TROMMER, "Unterscheidung von Gummi, Dextrin, Traubenzucker und Rohrzucker," *Annalen der Chemie und Pharmacie*, 1841, 39: 360–362.
31. CHARLES W. PURDY, *Practical Uranalysis and Urinary Diagnosis. A Manual for the Use of Physicians, Surgeons, and Students*, 4th ed. (Philadelphia: The F. A. Davis Company, Publishers, 1898), pp. 101–102.
32. L. C. BARRESWIL, *Journal de Pharmacie et Chimie*, 1846, 6: 301–304 (Extract).
33. *Dictionary of Scientific Biography* (*DSB*), Charles Coulston Gillispie, ed. (New York: Charles Scribner's Sons, 1970), 1: 471.
34. H. FEHLING, "Quantitative Bestimmung des Zuckers im Harn," *Archiv für Physiologische Heilkunde*, 1848, 7: 64–73; "Die quantitative Bestimmung von Zucker und Stärkmehl mittelst Kupfervitriol," *Annalen der Chemie und Pharmacie*, 1849, 72: 106–113; "Die quantitative Bestimmung von Zucker," *Ibidem*, 1858, 106: 75–79.
35. PER M. BOLL, "Revisiting Fehling and Discovering Vogel," *Journal of Chemical Education*, 1994, 71: 220–221.
36. PURDY (1898), pp. 110–112.
37. THEODORE C. JANEWAY, "Some Sources of Error in Laboratory Clinical Diagnosis," *Medical News*, 1901, 78: 700–706; GEORGE DOCK, "Clinical Pathology in the Eighties and Nineties," *American Journal of Clinical Pathology*, 1946, 16: 671–680, p. 675; see also MAX EINHORN, "Fermentation as a Practical Qualitative and Quantitative Test for Sugar in the Urine," *Medical Record*, 1887, 31: 91–94, p. 94.
38. WILLIAM ROBERTS, "Lectures on Certain Points in the Clinical Examination of the Urine. On Quantitative Sugar-Testing in the Urine," *Lancet*, 1862, 1: 535–536.

39. H. BENCE JONES, "On Intermitting Diabetes, and on the Diabetes of Old Age," *Medico-Chirurgical Transactions*, 1853, 36: 403–432, pp. 403–404.
40. ERNEST CHILD, *The Tools of the Chemist. Their Ancestry and American Evolution* (New York: Reinhold Publishing Corporation, 1940), pp. 60, 160–161, 162–163.
41. *Ibidem*, pp. 48–49.
42. M. D. GRMEK, "First Steps in Claude Bernard's Discovery of the Glycogenic Function of the Liver," *Journal of the History of Biology*, 1968, 1: 141–154, pp. 145, 142, 143; see also JEAN MAYER, "Claude Bernard," *Journal of Nutrition*, 1951, 45: 3–19.
43. FREDERICK L. HOLMES, "Early Theories of Protein Metabolism," *Annals of the New York Academy of Sciences*, 1979, 325: 171–187; see also JOSEPH S. FRUTON, "The Nature of Proteins," (pp. 87–179) in *Molecules and Life. Historical Essays on the Interplay of Chemistry and Biology* (New York: Wiley-Interscience, 1972), p. 97.
44. ASTRUP and SEVERINGHAUS (1986), pp. 226–248.
45. J. S. HALDANE and J. G. PRIESTLEY, "The Regulation of the Lung-Ventilation," *Journal of Physiology*, 1904–05, 32: 225–266.
46. L. S. FRIDERICIA, "Eine klinische Methode zur Bestimmung der Kohlensäurespannung in der Lungenluft," *Berliner Klinische Wochenschrift*, 1914, 27: 1268–1271.
47. DONALD D. VAN SLYKE and GLENN E. CULLEN, "Studies of Acidosis. I. The Bicarbonate Concentration of the Blood Plasma; Its Significance, and its Determination as a Measure of Acidosis," *Journal of Biological Chemistry*, 1917, 30: 289–346; DONALD D. VAN SLYKE, "Studies of Acidosis. II. A Method for the Determination of Carbon Dioxide and Carbonates in Solution," *Journal of Biological Chemistry*, 1917, 30: 347–368.
48. DONALD D. VAN SLYKE and WILLIAM C. STADIE, "The Determination of the Gases of the Blood," *Journal of Biological Chemistry*, 1921, 49: 1–42; WILLIAM C. STADIE, "A Mechanical Shaker and Other Devices for Use with the Van Slyke Blood Gas Apparatus," 49: 43–46.
49. WILLIAM C. STADIE and DONALD D. VAN SLYKE, "Studies of Acidosis. XV. Carbon Dioxide Content and Capacity in Arterial and Venous Blood Plasma," *Journal of Biological Chemistry*, 1920, 41: 191–194.
50. ASTRUP and SEVERINGHAUS (1986), p. 163.
51. DONALD D. VAN SLYKE and JAMES M. NEILL, "The Determination of Gases in Blood and Other Solutions by Vacuum Extraction and Manometric Measurement. 1.," *Journal of Biological Chemistry*, 1924, 61: 523–573; see also DONALD D. VAN SLYKE, "Note on a Portable Form of the Manometric Gas Apparatus, and on Certain Points in the Technique of its Use," 1927, 73: 121–126.
52. DONALD D. VAN SLYKE and J. PLAZIN, *Micromanometric Analyses* (Baltimore: The Williams & Wilkins Company, 1961).
53. SAMUEL NATELSON, "Routine Use of Ultramicro Methods in the Clinical Laboratory. Estimation of Sodium, Potassium, Chloride, Protein, Hematocrit Value, Sugar, Urea and Nonprotein Nitrogen in Fingertip Blood. Construction of Ultramicro Pipets. A Practical Microgasometer for Estimation of Carbon Dioxide," *American Journal of Clinical Pathology*, 1951, 21: 1153–1172. For a biographical sketch, see WILLARD FAULKNER, "Samuel Natelson, Clinical Chemist," *Clinical Chemistry*, 1986, 32: 216–220.
54. JAMES L. GAMBLE, G. S. ROSS, and F. F. TISDALL, "The Metabolism of Fixed Base During Fasting," *Journal of Biological Chemistry*, 1923, 57: 633–695, p. 636.
55. For the concept of electrolyte balance, see James L. Gamble, *Chemical Anatomy Physiology and Pathology of Extracellular Fluid. A Lecture Syllabus*, 6th ed. (Cambridge, Massachusetts: Harvard University Press, 1954).

56. EDWARD SCHÄFER, *An Introduction to the Study of the Endocrine Glands and Internal Secretions* (California: Stanford University, 1914), pp. 84, 86; E. SHARPEY-SCHAFER, *The Endocrine Organs. An Introduction to the Study of Internal Secretion*, 2nd ed. (2 vols.), vol. 2 (London: Longmans, Green and Co. Ltd., 1926), p. 343; see also *DSB* (1975), 12: 355–357.
57. F. G. BANTING, C. H. BEST and J. J. R. MACLEOD, "The Internal Secretion of the Pancreas," *American Journal of Physiology* (*Proceedings*), 1922 (December 1921), 59: 479.
58. The isolation of insulin was beset by controversy over who made the major contributions to the discovery and by the friction between Banting and Macleod. See "Banting's, Best's, and Collip's Account of the Discovery of Insulin," *Bulletin of the History of Medicine*, 1982, 56: 554–568. See also JOSEPH H. PRATT, "A Reappraisal of Researches Leading to the Discovery of Insulin," *Journal of the History of Medicine and Allied Sciences*, 1954, 9: 281–289; *DSB*, 1970, 1: 440–443; 1971, 3: 351–354; 1990 (supplement 2) 17: 80–81. In a letter dated September 20, 1922, Macleod addresses the questions of priority and credit for the discovery of insulin. "J. J. R. Macleod: History of the Researches Leading to the Discovery of Insulin," *Bulletin of the History of Medicine*, 1978, 52: 295–312.
59. VICTOR C. MYERS and CAMERON V. BAILEY, "The Lewis and Benedict Method for the Estimation of Blood Sugar, With Some Observations Obtained in Disease," *Journal of Biological Chemistry*, 1916, 24: 147–161; ROBERT C. LEWIS and STANLEY R. BENEDICT, "A Method for the Estimation of Sugar in Small Quantities of Blood," *Proceedings of the Society for Experimental Biology and Medicine*, 1913–14, 11: 57–58; *Journal of Biological Chemistry*, 1915, 20: 61–72.
60. P. A. SHAFFER and A. F. HARTMANN, "The Iodometric Determination of Copper and its Use in Sugar Analysis. I. Equilibria in the Reaction Between Copper Sulfate and Potassium Iodide," *Journal of Biological Chemistry*, 1921, 45: 349–364; II. Methods for the Determination of Reducing Sugars in Blood, Urine, Milk, and Other Solutions," 45: 365–390.
61. F. G. BANTING, C. H. BEST, J. B. COLLIP, W. R. CAMPBELL, and A. A. FLETCHER, "Pancreatic Extracts in the Treatment of Diabetes Mellitus," *Canadian Medical Association Journal*, 1922, 12: 141–146.
62. F. G. BANTING and C. H. BEST, "The Internal Secretion of the Pancreas," *Journal of Laboratory and Clinical Medicine*, 1921–22, 7: 251–266. Reprinted in vol. 80, 1972 to mark the fiftieth anniversary of the discovery.

CHAPTER IX

RESPIRATION AND COMBUSTION

The scientific search for the agent of combustion (respiration) began with Robert Boyle's experiments with flames and animals *in vacuo* (1660). Boyle studied the effects of reduced pressure and noted the extinction of a candle flame in an evacuated bell jar. In another experiment he demonstrated that a bird can live only for a limited time in a closed vessel and deduced a connection between respiration and combustion when he found that the volume of air was reduced in both experiments. It was apparent to him that some substance in air is necessary for respiration as well as for combustion, but he could take it no further. He did not connect this substance with the calcination of metals, i.e., the slow oxidation during heating.

Calcination became the main experimental technique during the next century for determining the composition of air. Boyle observed that metals gained weight in this process, a fact known before his time, and if this was conducted in closed vessels, the gas pressure would fall. But he did not attribute this to the consumption of air in the reaction flask. He explained the increase in weight of the resulting calx (ash) as due to particles of fire that had penetrated the glass wall of the reaction vessel during the heating and been absorbed by the metal and concluded that fire had weight. Had he weighed the vessel before opening it he would have found no change in the total weight.

Before William Harvey demonstrated the mechanics and phenomena of the circulation of the blood through the lungs and the heart, it was believed—according to Galen—that the purpose of respiration was to cool the heart, and that the chest's movements served to introduce air. But, just what happened during the change of venous to arterial blood—was a mystery.

Then, Robert Hooke (1635–1703), who assisted Robert Boyle and demonstrated experiments at the Royal Society, showed in 1667 that a dog with an open thorax and exposed lungs motionless could be kept alive by artificial respiration. Using a bellows, he blew in a continuous stream of air intratracheally, without any movement of chest or lungs. During its passage through the inflated lungs the blood turned bright red. This experiment, also performed by Andreas Vesalius (1514–1564), Flemish anatomist and physician, proved that the essential feature of respiration is not in the movement of the lungs, but in certain blood changes the lungs bring about by admitting air to the body. Richard Lower (1631–1691) was first to transfuse blood between two animals (1665); and in 1668 he also observed that dark venous blood turned bright red on passing through lungs kept inflated with air. It was not the movement of the lungs, but the continuous exposure of the blood to air during its transit through the lungs that brought about the color change and sustained life.

John Mayow (1643–1679), chemist and physiologist, first suggested in 1674 that dark venous blood turned bright red by taking up a specific ingredient in air necessary for life, which he termed the nitro-aerial spirit. Although his vague vocabulary was fanciful and confusing, he understood that the principal object of breathing is to bring a supply of "nitrous air" to the blood to maintain the combustion process by which body warmth is produced. Knowledge about atmospheric air and the existence of other gases at this time was only fragmentary. It would take another hundred years before the components of atmospheric gases would be isolated.

EARLY STUDIES ON BLOOD GASES

Despite many experiments, there still was no general agreement and a great deal of uncertainty among physiologists in France, Britain, and Germany, about the site of combustion in the body. To solve this problem, investigators used the air pump to demonstrate the presence of gases in the blood as they searched for the locale of heat generation and to discover whether there was free carbon dioxide in the blood that was capable of passing into the air by way of the lungs.

The first known attempt to evacuate gases from blood had been made by Robert Boyle (1669) when he released large quantities of "air" from sheep's blood in a vacuum created by his air pump. But he did not know how to analyze the released gas. Humphry Davy, at age 19, in his first scientific publication (1799), showed that by heating venous blood on a sand bath while using an air pump to create a partial vacuum, he was able to extract both oxygen and carbon dioxide, becoming thereby the first to establish their presence in blood. He also demonstrated that venous blood was capable of taking up oxygen [1].

Animals forced to inhale oxygen-free gas mixtures continued to exhale carbon dioxide in the expired air, which was hardly expected if the combustion occurred in the lungs. However, this did not reveal the site of internal combustion or whether the carbon dioxide was free to pass from blood to the outside air through the lungs.

The first attempts to quantitate blood gas analysis were reported in 1837 by Heinrich Gustav Magnus (1802–1870), a well-known chemist and physicist. Using a mercurial gas pump of his own design, he produced a Torricellian vacuum, then transferred the released blood gases to a eudiometer for analysis. He demonstrated that oxygen, nitrogen, and carbon dioxide are present in arterial and venous blood and inferred that these gases are simply dissolved in the blood. Since arterial blood contained more oxygen and less carbon dioxide than the venous blood, Magnus concluded that carbon dioxide must be formed in or conducted to the blood during its circulation. In an expanded series of experiments (1845), he concluded that carbon dioxide was not formed in the lungs, but in other locations to which the arterial blood carried its oxygen supply. Oxygen and carbon dioxide could be transported in the blood both from and to the lungs where their exchange with air took place, while the oxidation and generation of body heat occurred elsewhere.

Between 1838 and 1846, while investigating the inefficiencies in the industrial production of cast iron in Germany, and the large loss of valuable by-products and gases accompanying the process, Robert Wilhelm Eberhard Bunsen (see Chapter 14), an inorganic chemist and inventor of the Bunsen burner, developed techniques for the collection, measurement, and study of the escaping gases. His research on the process was published in his only book, *Gasometrische Methoden* (1857). By means of specific chemical reactions, Bunsen eliminated one gas at a time from a mixture of gases. By measuring the volume, the temperature, and the pressure, before and after the reaction, he could determine the quantity of the reacted (i.e., eliminated) gas. Bunsen's carefully developed methods of quantitative measurements brought gas analysis to a level of accuracy and simplicity that matched gravimetric and titrimetric techniques. His methods were a very useful and important contribution to the physiological research in respiration taking place in Germany.

Carl Friedrich Wilhelm Ludwig (1816–1895) and Lothar Meyer (1830–1895), who worked in Bunsen's laboratory in Heidelberg, were pioneers in the quantitative analysis of blood gases. They were among the German physiologists who, in the latter half of the nineteenth century, created modern physiology. Ludwig designed an improved mercurial blood-gas pump (1867) which utilized a Torricellian vacuum to liberate the gases from a blood sample. Rigid pipes in earlier devices were replaced by flexible tubes connected to reservoirs of mercury. This arrangement allowed changing the air pressure within the system by raising or lowering the reservoirs, a principle Van Slyke adopted for his breakthrough volumetric gas apparatus in 1917.

After leaving Heidelberg, Meyer worked in Ludwig's laboratory and, using a method suggested by Bunsen, i.e., boiling in a vacuum to extract the gases, obtained results similar to those of Magnus. Meyer also demonstrated in 1856 what Magnus had failed to observe—oxygen absorption by blood in the lungs occurred independently of pressure. This suggested that apart from a physical binding of oxygen, there was a relatively loose or weak chemical linkage with some constituent. He demonstrated a similar chemical linkage in carbon monoxide poisoning between that gas and a constituent of blood. He also found that the amounts of oxygen and carbon monoxide taken up by the blood were in a simple molecular ratio, and that the carbon monoxide was able to expel volume for volume the oxygen already in the blood. This suggested to Meyer that the same constituent of blood reacted with both gases. However, his search for this constituent was unsuccessful.

It was still not clear whether combustion (oxidation) takes place in the lungs, in the blood, or in the cells. From their work on gasometry of the blood, mechanism of respiration, and oxygen and carbon dioxide interchange, Eduard F. W. Pflüger (1829–1910) and his pupils demonstrated that it was the energy requirement of the cells, not the oxygen content of the blood, that determines the degree of oxygen consumption. Consequently, the tissues, not the blood whose energy needs are comparatively low, was the essential location of respiration. With an improved mercurial gas pump of his own construction (1865), Pflüger and associates showed that the gas exchange between blood and tissue, and between blood and lung-air, did not involve secretion, but could be explained by passive diffusion alone. However, was the uptake of oxygen regulated, and did the red pigment of the blood play a part in such regulation, as Lothar Meyer had intimated? [2].

HOPPE-SEYLER ISOLATES HEMOGLOBIN

Felix Hoppe-Seyler (1825–1895) (Fig. 9.1) was the first head of the chemical laboratory in the new pathological institute organized in Berlin in 1856 by Rudolf Virchow (1821–1902). In 1861 he moved to Tübingen as professor of applied chemistry in the Faculty of Medicine. In 1864 he adopted the hyphenated surname from the husband of his oldest sister, Doctor Seyler, who was his guardian when he was orphaned at a young

FIG. 9.1. Felix Hoppe-Seyler. (*Zeitschrift für Physiologische Chemie*, 1895, 25: facing p. 1.)

age. In 1872, after the Franco-Prussian War, he became professor of physiological chemistry—the first such position in Germany—at Strassbourg, newly occupied by the Germans, and held this post until his death. Hoppe-Seyler was the nineteenth-century epitome of the German professor and natural scientist with many interests. In 1884, a new Institute for Physiological Chemistry was erected in Strassbourg, the first by a German university for the investigation and teaching of the science of physiological chemistry [3]. His laboratory became a leading center of professional training in this discipline. Tübingen and Strassbourg remained the only German universities where *physiological chemistry* was an autonomous subject, until 1928, when an independent institute was set up in Berlin. In the meantime, leading biochemists held posts in physiological or pathological institutes, or hospital clinics.

Hoppe-Seyler's initial researches after receiving the M.D. degree from the University of Berlin in 1851, dealt largely with the improvement of analytical methods for the chemical study of blood, urine, and bile. He borrowed the analytical methods of colorimetry and spectral analysis from physical chemistry for his study of the red pigment of blood which, prior to his work, had been known as "Blutfarbstof," "the colouring matter," and "la matière colorante du sang." Berzelius had split it into a "globulin" and a colored iron-containing component to which he attributed its ability to take up more oxygen than serum could.

In 1862 Hoppe-Seyler noted the characteristic absorption spectrum of the coloring matter of blood; and in 1864 he isolated the pigment as a crystalline compound which he called hemoglobin. He subsequently showed that there was species variation in the crystalline form. With the description of hemoglobin's absorption spectrum, he introduced the new spectroscope of Bunsen and Kirchhoff into medical chemistry (see Chapter 14). Then, also in 1864, using a gas pump that he had constructed, Hoppe-Seyler demonstrated that it was hemoglobin that loosely binds oxygen to form oxyhemoglobin, which then gives up its oxygen to the body tissues. This explained the "chemical binding" of oxygen by the blood postulated by Lothar Meyer a decade earlier. Hoppe-Seyler regarded the loose chemical combination of oxygen and carbon dioxide with constituents of the blood as gas reservoirs from which large quantities can be removed and to which large quantities can be added without major changes in respective gas tensions. He developed a quantitative colorimetric method for hemoglobin by diluting blood until there was a color match with color standards. In 1892 he used stable carboxyhemoglobin for color comparison [4].

PHYSIOLOGICAL CHEMISTRY AND
ZEITSCHRIFT FÜR PHYSIOLOGISCHE CHEMIE

As academic appointments in chemistry became more common in the universities, chemistry was freed to exist as a scientific profession in its own right. No longer was it a sideline for pharmacists or physicians. Separated from medicine and recognized as a "pure" academic science, those parts of chemistry that were oriented towards medical application formed the nucleus of physiological chemistry. However, in Germany, formation as a separate discipline was delayed because many physiologists objected to its withdrawal from physiology [5].

During his years at Strassbourg, Hoppe-Seyler became the leading German proponent for the separation of physiological chemistry from medical physiology. His *Handbuch der Physiologisch und Pathologisch-Chemischen Analyse*, first published in 1858 and later in translations, was a standard text. Together with his four volumes of *Physiologische Chemie* (1877–81), these textbooks contributed greatly to the dissemination of chemistry throughout the medical community. Physiological chemistry, an applied branch of chemistry, became a new discipline in the medical sciences largely through his efforts.

In 1877 Hoppe-Seyler founded the *Zeitschrift für Physiologische Chemie* as a publication outlet for the interests and observations of biochemistry as an active and independent area of science. There was opposition from some physiologists, notably Eduard Pflüger, who resented the inroad of this new publication. To preserve unity in science, he believed physiological chemistry should remain an inseparable part of physiology within the confines of his own journal, *Pflügers Archiv für die gesamte Physiologie des Menschen und der Tiere*, founded in 1868. Hoppe-Seyler's journal was continued after his death as *Hoppe-Seyler's Zeitschrift für Physiologische Chemie*. In 1985 the name was changed to *Biological Chemistry Hoppe-Seyler*. Pflüger's journal is still in existence as *Pflügers Archiv European Journal of Physiology*. At the University of Göttingen, physiological chemistry remained a part of physiology until 1939 when the Institute of Physiological Chemistry was finally founded. This sequence is characteristic for the general establishment of the discipline in Germany [6].

QUANTITATION OF RED CELLS AND HEMOGLOBIN

The attempt to make quantitative measurements had originated in part, from the belief that small alterations in the fluids represented important indices of health and illness. The recognition that the method itself involved small errors helped demonstrate that clinical significance could not be attached to minor deviations from normal standards and that the available techniques of chemistry made absolute precision impossible. In the case of red blood cell counts and hemoglobin values, very few physicians were aware that subjective and environmental factors beyond their control might affect the accuracy of instrumental observations [7]. Qualitative tests for urine also produced difficulties. So many tests for the same constituent were being developed that the ordinary physician could not distinguish good from bad tests. In one report, of eleven tests for sugar, five were judged unreliable [8].

Karl von Vierordt (1818–1884), German physician and professor of physiology at the University of Tübingen, is credited with performing the first blood counts in 1852. Vierordt developed a technique of counting red blood cells by microscopic observation of a certain volume of a measured dilution. Difficult to do and too time-consuming for routine clinical use, and little used, it was followed by other faulty cell counting techniques [9]. At about 1880, Paul Ehrlich distinguished different types of white blood cells by using staining methods.

A marked improvement in red cell counting was made in 1877 by William Richard Gowers (1845–1915), an eminent English neurologist. Gower's device was much easier to use and more accurate than earlier instruments and provided a reliable index of anemia. He adopted Vierordt's value of five million red cells per cubic millimeter as the reference standard for the average of healthy blood. However, he noted that the level may be a little higher in a healthy adult male, and a little lower in an adult female [10].

Gowers believed that this information should be supplemented by the determination of the hemoglobin in the red cells. His reasoning was that the actual quality of blood depends not merely on the number of corpuscles, but also on the amount of hemoglobin in them. At that time several techniques were used to estimate hemoglobin: iron in the red cells; spectroscopic appearance of a solution of hemoglobin; absorption capacity of a given quantity of blood for oxygen. These were difficult procedures not commonly used by physicians. In 1878 Gowers utilized a fourth technique, the determination of the red coloring material of blood by comparison with a standard substance whose color corresponded to the tint of a solution of normal blood. It was the first direct quantitative determination of hemoglobin. His simple apparatus for this estimation was made commercially available in what may have been the first "kit" method [11] (see Chapter 19).

The test was based on color comparison of a blood sample progressively diluted until a color match is observed with a standard. The "Hemoglobinometer" consists of two tubes of the same diameter seated in a small support stand. Because diluted blood is not stable, an artificial standard is used. One tube contains the standard made of glycerine and gelatine colored with carmine and picrocarminate of ammonia to match as closely as possible the tint of a 1 in 100 dilution of "normal blood." The other tube is calibrated in 100 equal divisions, each unit corresponding to 20 cubic mm so that the calibration at 100 corresponds to 2 cc. In the procedure, twenty cubic mm of blood are transferred to the graduated tube partly filled with water; the solution is mixed and the color is compared with the standard by transmitted light. Water is added dropwise with mixing until the color in both tubes is as nearly alike as possible. The final volume read from the scale of the diluted tube corresponds to the percentage of color, i.e., of the "normal" hemoglobin represented by the standard tube [12]. Gowers exhibited his apparatus at a meeting of the Clinical Society of London on December 13, 1878.

In the hemometer of Fleischl-Miescher, the instrument is configured like a microscope and a colored glass wedge of adjustable depth is substituted for the artificial colored liquid [13]. In 1894, the British firm of Lovibond produced a colorimeter (comparator) for the investigation of color. This device employed discs containing graded, permanent colored glass standards for comparison with the test solution (Fig. 9.2). The loss of some precision is usually offset by the convenience and simplicity.

Hermann Sahli (1856–1933), a Berne physician, used the same method and principle of comparison as that of Gowers, but improved the method by replacing the artificial standard with a permanent hemoglobin derivative. He prepared this by adding dilute hydrochloric acid to blood of known hemoglobin concentration to form a brownish-yellow solution of acid hematin—actually a colloidal suspension of minute particles. This solution is then diluted with water to the 2 cc mark, as in the Gowers procedure. The blood to be tested is similarly reacted and diluted with water until there is a color match with the standard solution. The scale reading of the test specimen is the percentage of color—i.e., of hemoglobin—compared to the normal blood represented by the standard. Since comparison can be made more accurately with darker than with lighter shades, Sahli selected for his standard, normal blood with the highest hemoglobin concentration he could find in his "strongest assistants." In Sahli's hemometer, the 100 scale mark corresponded to a hemoglobin level of 17.2 grams per 100 cc. In a variation of the acid hematin procedure,

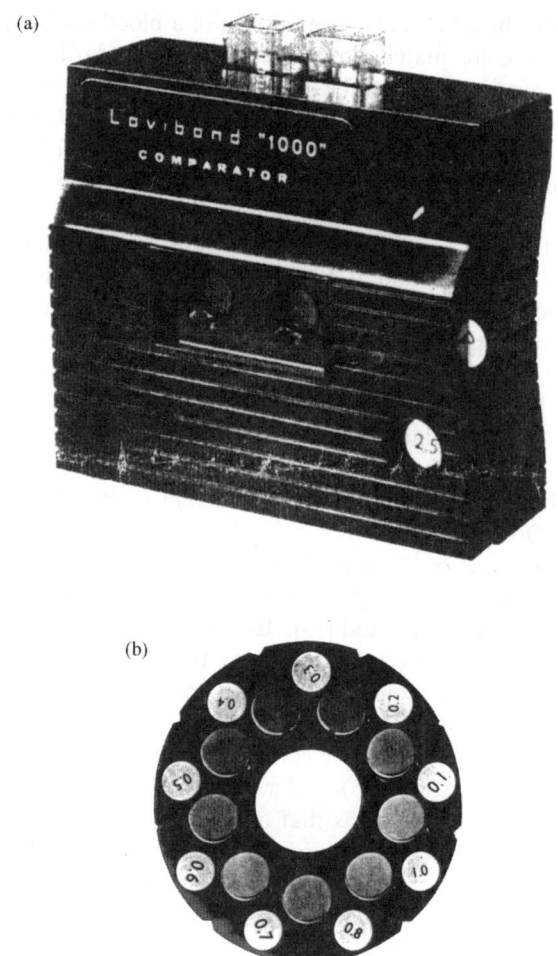

FIG. 9.2. Lovibond comparator and colored discs (*circa* 1959).

comparison was made with a standard colored glass plate adapted to a visual colorimeter [14].

The estimation of hemoglobin was apparently the first chemical determination in the blood to find extensive clinical application. Most of the early estimations were based on dilution and comparison with one or more arbitrary standards randomly set, and an empirical scale with 100 as the normal level. With different colorimetric methods and procedures of standardization, the 100 became a variable factor and the use of "percent of normal," was potentially confusing and misleading. Sometimes the only reference was the blood of one "healthy" person. Consequently, since the hemoglobin content of blood varies in health and disease, clinicians were warned not to draw inferences "from any but gross deviations from the normal standard," and thus outside the possible limits of physiological variation, and instrumental and observer errors [15]. And, as late as 1924, Myers reminded the profession that the hemoglobin test will have greater clinical value when more attention is given to accurate standards and methods and the actual percentage of hemoglobin is calculated, rather than its relation to an indefinite normal in parts per hundred [16].

What was needed for making an accurate colorimetric hemoglobin determination was a calibration standard of known hemoglobin concentration. This became available when the measurement of oxygen capacity was introduced by Haldane and Smith in 1899. Since the oxygen capacity of blood is dependent upon its hemoglobin content, and one gram of hemoglobin combines with 1.34 cc oxygen (0°C, 760 mm), this method provides a correct value for a hemoglobin standard by which other methods for hemoglobin could be calibrated [17]. The relationship had been determined in 1894 by Carl Gustav von Hüfner (1840–1908) who also reported the first accurate value for the iron content of hemoglobin.

For some time, an average oxygen capacity of 18.5 cc per 100 cc of blood obtained on several normal adult males was accepted as "100 per cent" of normal in clinical methods and gave average readings of 100 on the Gowers hemoglobinometer (13.8 grams hemoglobin per 100 cc blood) [18]. These levels of oxygen capacity and hemoglobin are decidedly below that found in normal adults by other investigators, and vary greatly with age and sex [19].

The technique was greatly improved and made practical for the clinical laboratory by Van Slyke's simple gasometric method of determining the oxygen capacity of the blood by using the same gas burette as in his carbon dioxide combining power measurement. This instrument would soon be in use in nearly every clinical laboratory [20].

It was discovered later that normal blood contains about 0.5% inactive hemoglobin (methemoglobin) and about 0.5% carboxyhemoglobin formed by the release of carbon monoxide in the decomposition of hemoglobin. As a result, since 1966 all the hemoglobin variants in blood, except sulfhemoglobin, are converted into cyanmethemoglobin according to the recommendations of the International Committee for Standards in Haematology, and accepted by the International Congress of Haematology in Stockholm (1967). The optical density of the reacted unknown is compared to a known standard solution of cyanmethemoglobin [21].

In 1907 a Boston physician addressing the Congress of American Physicians and Surgeons, said that "the oldest and simplest methods are still the best," and added: "In nine out of ten cases, as I see them in private practice, I make no examination of the blood other than that afforded by direct inspection of the color of the blood when soaked into a slip of paper." Nor did he "regard it as essential, or even of any considerable importance, that every patient suffering from diseases of the heart, lungs or stomach should be examined by the X-ray" [22].

HEMATOCRIT, URINE SEDIMENTS, AND THE CENTRIFUGE

As an alternative to counting red blood cells, Magnus Gustav Blix (1849–1904) in 1885, suggested the use of centrifugal force to pack the red cells together and to estimate their number by measuring their volume. Then, Sven Hedin (1859–1933) developed a machine—the "hämatokrit" (Fig. 9.3)—which he described initially in 1889. He connected a large wheel to a smaller one holding two graduated test tubes (35 × 1 mm). One revolution of the large wheel caused the smaller one to turn 100 times. This whirling motion forced the cells to the bottoms of the tubes where, as a compact mass, their volume could be measured and the number of cells calculated. The hämatokrit required less skill and took much less time than microscopic observation and counting of red cells [23].

In 1891 Daland improved the mechanical construction of the original instrument and simplified the technique. Two graduated capillary glass tubes are quickly filled with blood and rotated in their horizontal axes. A set of cogwheels enclosed in a metal box are geared to rotate a vertical spindle 130 times with each complete turn of the handle (Fig. 9.4). At the rate of 77 manual turns per minute, the spindle and its capillary tubes reach a speed of 10,000 rotations per minute. For the most accurate results, *centrifugalization* is continued for three minutes. In this manner the percentage of red blood cells was determined. The tube holders are not shielded. While in use the instrument is securely attached to the edge of a table or shelf by means of a clamp [24].

The use of centrifugal force for separation of sediments from urine, sputum, and other pathological fluids, for study of bacteria, crystals, and other formed elements, was first described in 1891 by Thor Stenbeck, a

FIG. 9.3. Hedin's "Hämatokrit." (Rudolf v. Jaksch, *Clinical Diagnosis*, 4th ed., London: Charles Griffin and Company, Limited, 1899, p. 29.)

Stockholm medical student. His "Sedimentator" (Fig. 9.5) consisted of a small metal plate from which two or more receptacles were suspended perpendicularly for the purpose of holding the glass tubes containing the fluid to be examined, and so hinged that when the apparatus is in motion they assume a horizontal position. The unit is rapidly rotated by a series

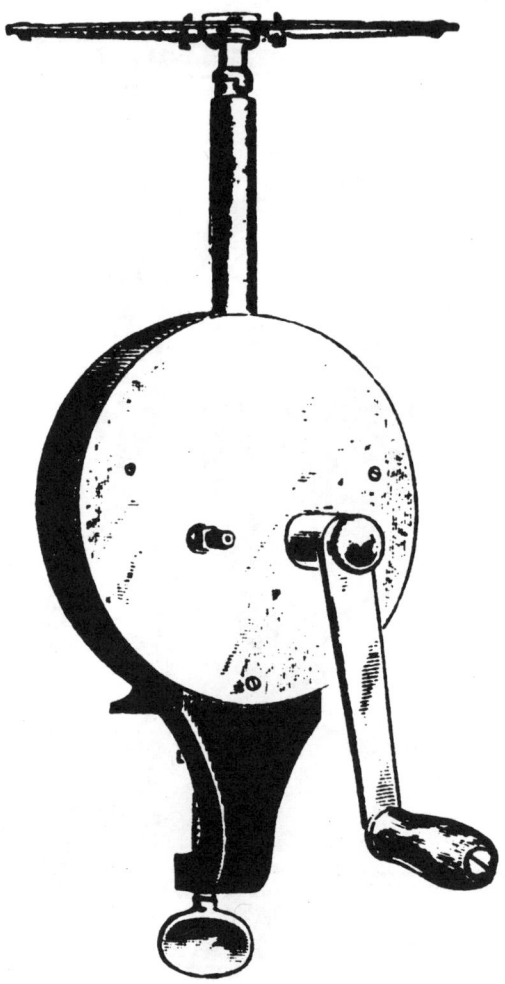

FIG. 9.4. Daland's Hematocrit. (John C. DaCosta, Jr., *Clinical Hematology*, 2nd ed., Philadelphia: P. Blakiston's Son & Co., 1905, p. 92.)

of cogwheels moved by an electric motor, by hand, or a treadle-wheel worked by foot power. Hand power attains a speed of 1200 to 1500 revolutions per minute; by motor, 3000 or more. In the former, 5 to 15 minutes will achieve the desired sediment; by motor, 2 to 3 minutes will suffice [25].

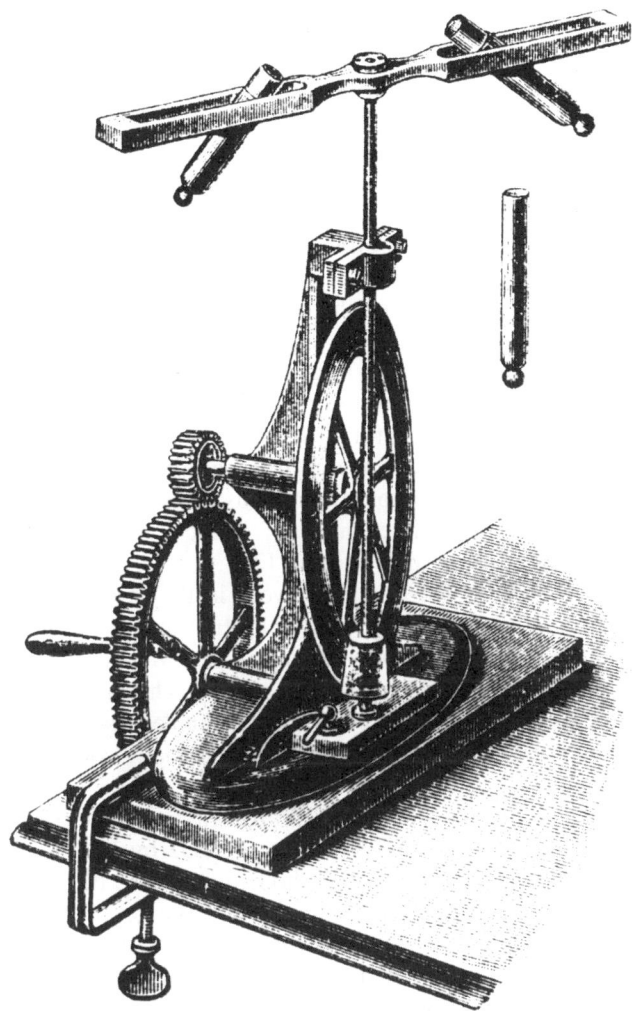

FIG. 9.5. Stenbeck's "Sedimentator." (*Zeitschrift für Klinische Medizin*, 1892, 20: 457–475, p. 462.)

The many advantages of the instrument, in addition to the great savings of time, were very soon apparent. Early enthusiasts of this technique noted that rapid sedimentation of urine by centrifugation yielded a richer and clearer microscopical field before changes in reaction and solubility due to fermentation and bacterial contamination, and before disintegration of some elements had time to develop, as often happened during the many hours of standing in the old procedure [26]. Centrifugation also avoided the delayed precipitation of phosphate or urate crystals which, owing to the changing urine reaction, tended to cover over many of the elements looked for. Previously, if a centrifuge was not available, the urine was allowed to stand at least six hours in a conical glass. William Osler (see Chapter 10) had a unique method for obtaining sediment. When he came from a consultation, he always brought with him a bottle of urine he had carried, cork down, in a coat pocket. In the laboratory he carefully turned the bottle right end up, removed the cork, and spread the sediment from the cork onto a slide [27].

Machines in which centrifugal force is used for separating solids from liquids had been introduced into the sugar, milk, and starch industries, at about the middle of the last century. A German brewer may have been the first to conceive of centrifugal force in the separation of cream from milk, but the practical industrial application of his ideas is credited to Gustav de Laval, a young Swedish engineer of French ancestry. In 1878 he invented the centrifugal cream separator [28].

Centrifuges were adapted to clinical work in the early 1890s and were made in Germany. Speed was generated by friction wheels and was sufficient for the beginning work with the hematocrit. Models driven by water power and electric motor were developed rapidly. The early machines had no protection from spattering or other accidents unless the instrument was enclosed in a wooden case or surrounded by specially constructed metal shields [29].

Centrifugal force had been used for the first time in the chemical laboratory at the University of Karlsruhe in 1852 by Lambert Heinrich Clemens Karl von Babo (1818–1899), who had borrowed the idea from the sugar industry [30]. Gustav Piers Alexander von Bunge (1844–1920), while visiting the laboratory of Oswald Schmiedeberg (1838–1921) in Strassbourg in 1876, may have been the first to use a centriguge in a physiological chemistry laboratory to separate serum from blood cells and to avoid the slow filtration in isolating precipitates. Bunge's apparatus was operated by the propulsive force of hot air generated by a machine. After four hours of centrifugation at 1000–1400 rpm, he could separate blood cells from serum. Through subsequent analyses, he confirmed Carl Schmidt's findings of substantial concentration differences between the content of inorganic components in red blood cells and in serum [31].

The electric centrifuge (Fig. 9.6) designed by Charles Purdy and manufactured in Philadelphia was in response to the 24-hour delay in obtaining

FIG. 9.6. Purdy's Centrifuge. (Charles W. Purdy, *Practical Uranalysis and Urinary Diagnosis*, 4th ed., Philadelphia: The F. A. Davis Company, Publishers, 1898, p. 149.)

urinary sediments for satisfactory microscopic examinations and the changes in the urine during this long period of standing at room temperature. Purdy recommended that it replace the hand-operated centrifuges that did not achieve the necessary speeds of 1500–2000 rpm. He had found few of the very large number of "centrifugals" on the market capable of efficient practical work, which accounted, he said, for the large number of discarded instruments in medical offices [32].

Purdy's model was easily capable of speeds from 500 to 10,000 revolutions per minute. Designed to meet the entire range of medical and bacteriological work, it operated "on the interrupted...[and] on the constant incandescent illuminating current" without overheating. Purdy identifies the manufacturer and assures the reader that since he "has no commercial interest in the centrifugal that bears his name," and never did have, he "feels entirely free to speak of its merits." Centrifugal speed was indicated by a hand-held dial applied to the top of the motor's central axis [33].

With the increasing use of electric power, came rapid developments in methods using centrifugal separation, a great time saver over older methods of gravity settling and suction filtration. Arthur Kendricks, a Harvard physics professor who founded the International Instrument Company, Cambridge, Massachusetts in 1901, was the American pioneer in the development of modern laboratory centrifuges. Kendricks designed the first fully enclosed and compact motor-driven centrifuge, revolutionary in design and utility. It met with such success that the firm extended the line to include a large variety of interchangeable accessories, adapted to a wide range of uses [34]. In 1921 the company produced the first bench-model electric powered centrifuge. In 1929 the firm was reorganized as the International Equipment Company (IEC), indicating an expanded production line. In 1937 IEC developed and began marketing the world's first refrigerated centrifuge. IEC was acquired by the Damon Corporation in 1968.

Centrifuges were slow to find general use in the clinical laboratory. The only mention of a centrifuge in the first edition of Hawk's *Practical Physiological Chemistry* (1907) was for the separation of urine sediments and was illustrated with Purdy's electric centrifuge. As late as the thirteenth edition (1954), precipitates formed in urine or blood analyses were almost always removed by filtration. The expression "protein-free filtrate" was routinely used in the laboratory, even when precipitates were separated by centrifugation. Nevertheless, the potential of the centrifuge was recognized as early as 1907 by Ivar Bang (see Chapter 12) who, in a preliminary step for estimating blood sugar, precipitated the proteins with alcohol and removed the precipitate by centrifugation [35].

HYDROGEN ELECTRODE

Another late nineteenth century innovation in laboratory instrumentation provided a valuable tool that expanded analytical parameters and led to new capability and methodology in clinical chemistry. The ability to measure hydrogen ion concentration was always a major concern because of the practical application not only in biological animal fluids but also in biological industrial processes such as fermentation of grapes for wine production, brewing of beer, and growth of bacteria in a culture medium, where a constant hydrogen ion concentration had to be maintained. Titration to an indicator endpoint, usually litmus, was not sensitive enough to quantitate and could not distinguish a weak acid from a strong acid. An important change in the concept of acids was introduced with the theory of electrolytic dissociation in 1887 by Svante August Arrhenius (1859–1927). An understanding of the role of the hydrogen ion followed in the wake of the dissociation theory, and electrometric means for measuring hydrogen ion concentration were soon developed using gas electrodes.

The most direct method of determining hydrogen ion concentration in solution is by means of a hydrogen electrode. This is formed by bubbling pure hydrogen gas over a wire or small foil, whose surface can establish equilibrium between hydrogen molecules (or atoms) and the hydrogen ions (or protons) in the solution in which the metal electrode is immersed. The most useful surface catalysts are platinum and palladium. The hydrogen electrode is the ultimate standard for the determination of hydrogen ion concentration but, owing to the experimental difficulties associated with it, other electrodes referable to hydrogen ion are commonly used for routine measurements. The performance of the practical secondary electrodes, e.g., the glass electrode, is always evaluated in terms of the hydrogen electrode [36].

The hydrogen electrode was introduced into physical chemistry in 1893 by Wilhelm Ostwald (1853–1932). In 1897 Wilhelm Carl Böttger (1871–1949) first reported the use of a hydrogen electrode to follow changes in acidity during titration. In 1900 Rudolf Otto Anselm Höber (1873–1953) was the first to attempt a direct determination of hydrogen ion concentration in blood with this electrode after equilibration with hydrogen. His results were incorrect because when he bubbled hydrogen through the sample he did not allow for the loss of carbon dioxide when the blood became saturated with hydrogen [37].

Hildebrand, in 1913, brought the hydrogen electrode to the attention of scientists outside the field of physical chemistry by extending its application to potentiometric titrations and other types of reactions [38]. It soon became possible to select a suitable indicator for a given titration by this method. However, the procedures at this time were either inaccurate or too difficult for hospital laboratory use.

SØRENSEN, pH, AND BUFFERS

Søren Peter Lauritz Sørensen (1868–1939) (Fig. 9.7), who succeeded Kjeldahl as head of the chemical department of the Carlsberg Laboratory in Copenhagen, introduced the hydrogen electrode into biochemistry in 1909. His studies of the enzymatic breakdown of proteins resulted in comprehensive publications in 1908 and 1909. In the first, he described the formalin titration of amino acids [39], which was the earliest quantitative method for following the hydrolysis of proteins. In the second work, he introduced and defined the concept of pH as the negative logarithm—an expression of hydrogen ion concentration that facilitated calculations—and developed the principles and technique of colorimetric pH determination. In addition, he reviewed the electrometric determination of pH by the hydrogen electrode and identified the many sources of error with this technique [40].

Sørensen demonstrated the importance of pH control for the enzymatic process and emphasized the key role of hydrogen ion concentration in all biological processes. He also introduced the concept of buffer systems—solutions of weak acids and their respective salts with strong bases. He gave directions for preparation of standard buffer solutions to cover most of the pH scale. The pH values of these buffer solutions change only slightly on addition of acid or base. Therefore, they are ideally suited for controlling the medium of enzymatic processes which release acid or alkaline products that could challenge the pH optimum for the enzyme activity. He illustrated the role of pH by plotting the relationship between pH and enzyme activity for the enzymes invertase, catalase, and pepsin.

HENDERSON–HASSELBALCH EQUATION AND NEW DEFINITIONS FOR ACID AND BASE

The factors involved in acid–base balance were clarified during the early part of the twentieth century by Lawrence Joseph Henderson (1878–1942). He had studied the application of physico-chemical principles and methods to physiological problems under Franz Hofmeister and became intrigued by the extraordinary ability of blood to neutralize large amounts of acids or bases without losing its neutral reaction. As an admirer of Claude Bernard's theories on the maintenance of a constant "milieu interieur," Henderson was interested in discovering uniformities and generalizations and wanted to describe biological systems in mathematical and physico-chemical terms.

In 1909, in a monograph on equilibrium between acids and bases in the animal organism, and the variables affecting the neutrality of blood,

FIG. 9.7. S. P. L. Sørensen. (Carlsberg Foundation Picture Archives, Copenhagen, Denmark.)

Henderson pointed to the significance of bicarbonate as a reserve of alkali in excess of acids. He applied the law of mass action to a weak acid and its salt with a strong base and developed the "Henderson equation" to investigate the buffer systems of blood and its acid–base regulation [41].

$$[H^+] = K \frac{[H_2CO_3]}{[HCO_3^-]}$$

From this mass law equation describing carbonic acid equilibrium in the blood, Henderson calculated blood pH as 7.4. He concluded that the most important variables for maintaining the neutrality of blood were the proteins in blood, including the hemoglobin in the red blood cells, and the buffer systems of carbonic acid–bicarbonate and primary–secondary phosphates.

In 1910, Karl Albert Hasselbalch (1874–1962), a University of Copenhagen M.D., devised a hydrogen electrode that could be used in the presence of carbon dioxide [42]. It provided the first reliable blood pH values with a hydrogen electrode. With it, Hasselbalch and Lundsgaard determined the pH of blood to be 7.4 ± 0.2 [43]. In 1915 Hasselbalch and Gammeltoft demonstrated the influence of respiration on blood pH. They introduced the term "compensated acidosis" to indicate the state in pregnancy where the blood bicarbonate concentration is diminished, but respiration is increased so that carbon dioxide tension had a parallel fall and blood pH remained normal at 7.40 [44]. Hasselbalch used his pH measurements to define clinical acid–base disturbances which he characterized as either "compensated" or "non-compensated." The technique was too difficult for hospital laboratory use. Instead, some clinicians tried indicators directly on serum or diluted serum.

The computational advantages of expressing hydrogen ion concentration as the negative logarithm became obvious when the dissociation constant K was also expressed by its negative logarithm, a practice introduced by Niels Bjerrum (1879– 1958) and adopted by Hasselbalch. In 1916, Hasselbalch converted the Henderson equation to logarithmic form—now known as the Henderson–Hasselbalch equation—making it more convenient and accessible for simple, though indirect, calculation of pH [45].

$$pH = pK + \log \frac{[HCO_3^-]}{[H_2CO_3]}$$

Important contributions were made in 1923 by Johannes Nicolaus Brønsted (1879–1947) in Denmark and Thomas Martin Lowry (1874–1936) in England when they broadened the concept of acids and bases by defining acids as hydrogen ion (proton) donors and bases as acceptors. Earlier in 1916, Gilbert Newton Lewis (1875–1946), an American, had sought to establish a more comprehensive acid–base theory when he defined a base as a substance able to donate a pair of electrons, thereby forming a chemical bond. An acid accepted a pair of electrons. However, these definitions exceeded the interest or need of the clinician in his evaluation of a patient's acid–base status.

In *Determination of Hydrogen Ions* (1920), William Mansfield Clark (1884–1964) reviewed the status of acid–base chemistry, methodology, and applications to biology, bacteriology, and industry. The book went through three editions and helped to advance the technique and theory in the application of acid–base balance. Clark was professor of physiological chemistry at Johns Hopkins University from 1927 to 1952.

GLASS ELECTRODE

The glass electrode, which eventually replaced the hydrogen electrode, had its origins in 1906 when Max Cremer (1865–1935) discovered the electrochemical properties of certain glasses. He noticed that a thin glass membrane separating two aqueous solutions of different acid concentration could generate a potential across the membrane that varied with the difference in hydrogen ion concentrations. This phenomenon was investigated and developed more extensively by Fritz Jacob Haber (1868–1934) and his student Zygmunt Klemensiewicz (1886–1963) in 1909. For some years the glass electrode (Fig. 9.8) was known in Germany as the "Haber electrode" [46]. They discovered that not all types of glass were suitable for use in the construction of glass electrodes. Not until the 1920s did interest again turn towards the use of glass electrodes when Duncan Arthur MacInnes (1885–1965) and Malcolm Dole (b. 1903) discovered a suitable type of glass for the construction of the membrane [47], and the introduction of vacuum tube voltmeters made very accurate measurements possible. Their improved construction could be applied to measure the pH of very small volumes of solutions.

Far-reaching applications resulted from the discovery of the hydrogen ion function of glass membranes and the subsequent development of convenient, practical glass electrodes and pH meters. However, until the commercial availability of glass electrodes about 1933 and the construction of a sturdy pH meter a few years later by Arnold Orville Beckman (1900–   ), the hydrogen electrode was the only method available for

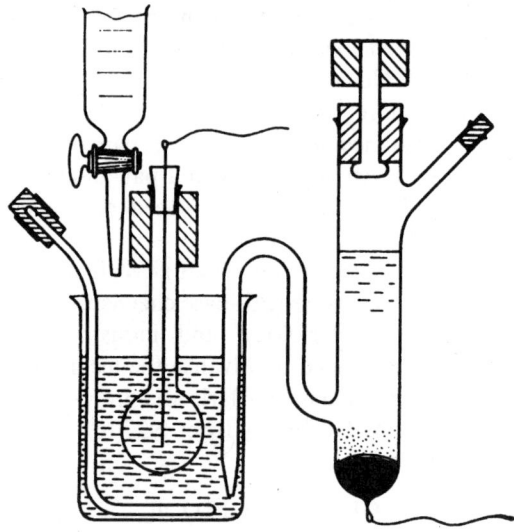

FIG. 9.8. Haber's Glass Electrode. (*Zeitschrift für Physikalische Chemie*, 1909, 77: 385–431, p. 411.)

electromotive measurement of hydrogen ion concentration. But, it had numerous drawbacks—especially for protein-containing fluids.

BECKMAN pH METER

In 1934, Glenn Joseph, a former classmate of Arnold Beckman from the University of Illinois, who was working for the California Fruit Growers Exchange, needed help on a project. He was looking for a better way to measure the acidity of lemon juice treated with sulfur dioxide that was interfering with other indicators of acidity. Joseph had been using the standard procedure of a thin (and fragile) glass electrode connected to a high-sensitivity (and delicate) galvanometer, and one or the other was always breaking. A thicker and more rugged glass electrode had to be used. This would have a higher resistance and would require a different current measuring device that would also stand up to the rough handling in the processing plant.

Beckman's experience with vacuum tubes from his post-college employment at Western Electric (later part of Bell Laboratories) suggested the solution. He realized that the electronic voltmeter would be better matched as a measuring device for the high resistance glass electrode and ultimately constructed a two-stage, direct-coupled amplifier that fed an ordinary millivoltmeter as the null-point detector for balancing the Wheatstone bridge. The greater sensitivity of the electronic amplifier had the added advantage of allowing the use of a thicker and sturdier—although with higher resistance—glass electrode [48].

Several months later, Joseph asked Beckman for a second "acidity meter" because other chemists were always borrowing the first one. Suspecting that there may be a commercial potential to his "acidimeter," Beckman visited many of the eastern apparatus supply companies that could market the meters to the scientific community. Only the Arthur H. Thomas Company in Philadelphia, acting on the evaluation of F. William Sunderman, Sr. (1898–   ), saw the possibilities of this instrument.

FIG. 9.9. Arnold O. Beckman and original model of pH meter. (Courtesy of The Beckman Heritage Center, Fullerton, California.)

Beckman, and two Caltech graduates working part-time, started to make pH meters in a rented garage in East Pasadena. They soon moved to larger facilities and in 1936, the first full year of sales, Beckman's company, named National Technical Laboratories, sold 444 pH meters. By 1939 nearly 2,000 units had been sold, and Beckman resigned from his faculty position in the chemistry department at Caltech, where he had received the Ph.D. degree in 1928, to spend full time with his company. The Beckman pH meter, the first general-purpose chemical instrument to use electronics, revolutionized a basic operation in analytical chemistry (Fig. 9.9) [49] (see also Chapter 18). The popular model G pH meter (Fig. 9.10) was introduced in 1935

FIG. 9.10. Beckman pH meter, model G. (Courtesy of The Beckman Heritage Center, Fullerton, California.)

and cost $195. More than 27,000 units were sold before production was discontinued in 1969. Accurate and reproducible to ±0.01 pH with careful technique, the model G used a null-type meter system and had a temperature compensator dial for the range 10° to 40°C.

NOTES AND REFERENCES

1. HUMPHRY DAVY, *Early Miscellaneous Papers, From 1799 To 1805. With An Introductory Lecture, And Outlines of Lectures On Chemistry, Delivered in 1802, and 1804* (London: Smith, Elder and Co. Cornhill, 1839), pp. 76–79.
2. Meyer is best known for his work on the principles of the periodic classification of the elements (1870) independently of Dimitri Ivanovich Mendeleev (1834–1907), who published a more elaborate paper on the periodic table a year earlier. Meyer and Mendeleev both attended the 1860 Karlsruhe Congress where Stanislao Cannizzaro's (1826–1910) paper on atomic weights was a major stimulus in the formation of their concepts.
3. For a historical review, see FELIX HOPPE-SEYLER, "On The Development of Physiological Chemistry and its Significance for Medicine: An Address Delivered at the Celebration of the Opening of the New Institute for Physiological Chemistry of the Imperial University of Strassburg, February 18, 1884," translated by T. Wesley Mills, *New York Medical Journal*, 1884, 40: 169–171, (concluded on pp. 197–199).
4. Researches on hemoglobin have earned a Nobel Prize in Chemistry for Hans Fischer (1881–1945) in 1930, Linus Pauling (1901–1994) in 1954, and Max Perutz (b. 1914) in 1962.
5. WILHELM PRANDTL, "Johann Wolfgang Döbereiner, Goethe's Chemical Adviser," *Journal of Chemical Education*, 1950, 27: 176–181; PAUL WALDEN, "The Gmelin Chemical Dynasty," *Journal of Chemical Education*, 1954, 31: 534–541.
6. ANDREAS-HOLGER MAEHLE, MARLIES GLASE, and ULRICH TRÖHLER, "Der Göttinger Weg von der medizinischen zur physiologischen Chemie," *Biological Chemistry Hoppe-Seyler*, 1990, 371: 447–454.
7. STANLEY JOEL REISER, *Medicine and the Reign of Technology* (Cambridge: Cambridge University Press, 1978), p. 134.
8. PROF. OPPOLZER, "The Tests for Sugar in the Urine," translated by Alfred L. Haskins, *Half-Yearly Abstract of the Medical Sciences* (London), 1868, 48: 107–108, in "Clinical Observations Upon Diabetes Mellitus," *New York Medical Journal*, 1868 (October) 8: 1–38.
9. M. L. VERSO, "The Evolution of Blood-Counting Techniques," *Medical History*, 1964, 8: 149–158.
10. W. R. GOWERS, "On the Numeration of Blood-Corpuscles," *Lancet*, 1877, 2: 797–798.
11. DR. GOWERS, "Apparatus for the Clinical Estimation of the Haemoglobin in the Blood," *Lancet*, 1878, 2: 882.
12. W. R. GOWERS, "An Apparatus for the Clinical Estimation of Haemoglobin," *Transactions of the Clinical Society of London*, 1879, 12: 64–67.
13. HERMANN SAHLI, *A Treatise on Diagnostic Methods of Examination*, 4th ed., Francis P. Kinnicutt and Nath'l Bowditch Potter, eds., translated from German (Philadelphia and London: W.B. Saunders Company, 1907), pp. 619–620.
14. *Ibidem*, pp. 620–623; see also VICTOR CARYL MYERS, *Practical Chemical Analysis of Blood. A Book Designed as a Brief Survey of This Subject for Physicians and Laboratory Workers*,

2nd ed. (St. Louis: C.V. Mosby Company, 1924), pp. 165–167; H. S. NEWCOMER, "A New Optical Instrument for the Determination of Hemoglobin," *Journal of Biological Chemistry*, 1923, 55: 569–574.
15. E. BUCHANAN BAXTER and FREDERICK WILLCOCKS, "A Contribution to Clinical Haemometry," *Lancet*, 1880, 1: 361–362.
16. MYERS (1924), p. 166. Hemoglobin is usually analyzed in the hematology division of the clinical laboratory. Yet as late as 1961, hemoglobin, prothrombin time, and sedimentation rate of red blood cells were routinely tested in the clinical chemistry laboratory of the New York University Medical Center, formerly the Post-Graduate Medical School and Hospital.
17. JOHN HALDANE and J. LORRAIN SMITH, "The Mass and Oxygen Capacity of the Blood in Man," *Journal of Physiology*, 1899, 25: 331–343.
18. JOHN HALDANE, "The Colorimetric Determination of Haemoglobin," *Journal of Physiology*, 1900, 26: 497–504.
19. JOHN P. PETERS and DONALD D. VAN SLYKE, *Quantitative Clinical Chemistry*, (2 vols.), vol. 2, *Methods* (Baltimore: The Williams & Wilkins Company, 1932), p. 662.
20. DONALD D. VAN SLYKE and WILLIAM C. STADIE, "The Determination of the Gases of the Blood," *Journal of Biological Chemistry*, 1921, 49: 1–42.
21. RUSSELL J. EILERS, "Notification of Final Adoption of an International Method and Standard Solution for Hemoglobinometry Specifications for Preparation of Standard Solution," *American Journal of Clinical Pathology*, 1967, 47: 212–214. See also F. W. SUNDERMAN, "Historical Review and Assessment of Clinical Hemoglobinometry in the United States," *Annals of Clinical and Laboratory Science*, 1985, 15: 1–12.
22. RICHARD C. CABOT, "The Historical Development and Relative Value of Laboratory and Clinical Methods of Diagnosis," *Boston Medical and Surgical Journal*, 1907, 157: 150–153, p. 152.
23. S. G. HEDIN, "Der Hämatokrit, ein neuer Apparat zur Untersuchung des Blutes," *Skandinavisches Archiv für Physiologie*, 1890, 2: 134–140; see also RUDOLF V. JAKSCH, *Clinical Diagnosis: The Bacteriological, Chemical, and Microscopical Evidence of Disease*, 4th ed., translated from German by James Cagney (London: Charles Griffin and Company, Limited, 1899), pp. 29–30 (illustration on p. 29).
24. JOHN C. DACOSTA, JR., *Clinical Hematology. A Practical Guide to the Examination of the Blood With Reference to Diagnosis*, 2nd ed. (Philadelphia: P. Blakiston's Son & Co., 1905), pp. 91–94. For a detailed description of Daland's centrifuge, see also JUDSON DALAND, "Estimation of the Corpuscular Richness of the Blood: A New Haematokrit and a New Technique," in *Cyclopaedia of the Diseases of Children, Medical and Surgical*, vol. 5, Supplement, William A. Edwards, ed. (Philadelphia: J. B. Lippincott Company, 1899), pp. 537–545.
25. THOR STENBECK, "Eine neue Methode für die mikroskopische Untersuchung der geformten Bestandt-heile des Harns und einiger anderen Secrete und Excrete," *Zeitschrift für Klinische Medizin*, 1892, 20: 457–475, p. 462 (illustration). First published in *Hygiaea* (Stockholm), 1891, pp. 40–51; see also M. LITTEN, "Die Centrifuge im Dienste der Klinischen Medicin," *Deutsche Medicinische Wochenschrift*, 1891, No. 23: 749–752; FREDERIC E. SONDERN, "The Value of The Centrifugal Apparatus for Diagnostic Purposes," *New York Medical Journal*, 1893, 57: 218–220; an illustration and explanation of Stenbeck's Sedimentator is also found in JAKSCH (1899), pp. 256–257.
26. SONDERN (1893); HENRY L. ELSNER and HIRAM B. HAWLEY, "On the Clinical Value of the Centrifuge," *New York Medical Journal*, 1894, 60: 526–531; GEORGE C. FREEBORN,

"A Method of Rapidly Collecting Deposits for Microscopical Examination," *Medical Record*, 1892 (February 27), p. 248; ARPAD G. GERSTER, "The Precipitation of Tubercle Bacilli by Centrifugal Force," *New York Medical Journal*, 1892, 56: 303–304; see also GUSTAV GAERTNER, "Die Kreisel-Centrifuge," *Wiener Klinische Wochenschrift*, 1892, 5: 365–366.
27. GEORGE DOCK, "Clinical Pathology in the Eighties and Nineties," *American Journal of Clinical Pathology*, 1946, 16: 671–680, p. 674.
28. ERNEST CHILD, *The Tools of the Chemist. Their Ancestry and American Evolution* (New York: Reinhold Publishing Corporation, 1940), p. 141.
29. DOCK (1946), p. 674.
30. L. v. BABO, "Mittheilungen aus dem chemischen Laboratorium zu Freiburg," *Annalen der Chemie und Pharmacie*, 1852, 82, 301–311.
31. POUL ASTRUP, PETER BIE, and HANS CHR. ENGELL, *Salt and Water in Culture and Medicine*, translated from Danish by Kirsten Skovbjerg and Andrew L. Cameron-Mills (Copenhagen: Munksgaard, 1993), pp. 116–117.
32. CHARLES W. PURDY, *Practical Uranalysis and Urinary Diagnosis. A Manual for the Use of Physicians, Surgeons, and Students*, 4th ed. (Philadelphia: The F.A. Davis Company, Publishers, 1898), p. 148.
33. *Ibidem*, pp. 148–151; see also footnote, p. 149 (illustration on p. 149).
34. CHILD (1940), p. 142.
35. IVAR BANG, "Über Die Verwendung Der Zentrifuge In Der Quantitativen Analyse," Festschrift Für Olof Hammarsten, *Uppsala Läkareförenings Förhändlinger*, 1905–06, 11: 1–15; see also "The Use of Centrifuge in Quantitative Analysis," *Chemical Abstracts*, 1907, 1: 197; "Über die Bestimmung des Blutzuckers," *Biochemisches Zeitschrift*, 1908, 7: 327–328.
36. ROGER G. BATES, *Determination of pH. Theory and Practice*, 2nd ed. (New York: John Wiley & Sons, 1973), pp. 279–280.
37. POUL ASTRUP and JOHN SEVERINGHAUS, *The History of Blood Gases, Acids and Bases*, translated from Danish by Patrick Graham Jørgensen (Copenhagen: Munksgaard, 1986), pp. 199, 249–250.
38. JOEL H. HILDEBRAND, "Some Applications of the Hydrogen Electrode in Analysis, Research and Teaching," *Journal of the American Chemical Society*, 1913, 35: 847–871.
39. S. P. L. SÖRENSEN, "Enzymstudien," *Biochemische Zeitschrift*, 1908, 7: 45–101; see also V. HENRIQUES and S. P. L. SÖRENSEN, "Über die quantitative Bestimmung der Aminosäuren, Polypeptide und der Hippursäure im Harne durch Formoltitration," *Hoppe-Seyler's Zeitschrift für Physiologische Chemie*, 1909, 63: 27–40.
40. S. P. L. SÖRENSEN, "Enzymstudien. II. Über die Messung und die Bedeutung der Wasserstoffionenkonzentration bei enzymatischen Prozessen," *Biochemische Zeitschrift*, 1909, 21: 131–304, pp. 134, 301. With the development of chemical thermodynamics, it became evident that pH did not actually measure hydrogen ion concentration, but depended in a complex manner on the *activity* of the electrolytes in the solution. Because the pH unit lacks precise fundamental definition, regulatory organizations have accepted the operational definition of pH. Thus, the pH is essentially the determination of a difference, and the experimental value is largely that of the value assigned to the buffer solution used as the standard for comparison. BATES (1973), preface pp. v–vi.
41. LAWRENCE J. HENDERSON, "Das Gleichgewicht zwischen Basen und Säuren im tierischen Organismus," *Ergebnisse der Physiologie*, 1909, 8: 254–325; see also "The Theory of

Neutrality Regulation in the Animal Organism," *American Journal of Physiology*, 1908, 21: 427–448.

42. K. A. HASSELBALCH, "Elektrometrische Reaktionsbestimmung kohlensäurehaltiger Flüssigkeiten," *Biochemische Zeitschrift*, 1910, 30: 317–331.
43. K. A. HASSELBALCH and CHR. LUNDSGAARD, "Elektrometrische Reaktionsbestimmung des Blutes bei Körpertemperatur," *Biochemische Zeitschrift*, 1912, 38, 77–91; CHRISTEN LUNDSGAARD, "Die Reaktion des Blutes," 1912, 41: 247–267.
44. K. A. HASSELBALCH and S. A. GAMMELTOFT, "Die Neutralitätsregulation des graviden Organismus," *Biochemische Zeitschrift*, 1915, 68: 206–264, pp. 237–245.
45. K. A. HASSELBALCH, "Die Berechnung der Wasserstoffzahl des Blutes aus der freien und gebundenen Kohlensäure desselben, und die Sauerstoffbindung des Blutes als Funktion der Wasserstoffzahl," *Biochemische Zeitschrift*, 1916, 78: 112–144, p. 115.
46. Haber succeeded in finding a method for synthesizing ammonia from its elements. This could be further processed into nitrates as a source of artificial fertilizers. During World War I, the process was critical to the German war effort not only as a source of artificial fertilizers but also for producing explosives. For "improving the standards of agriculture and the well-being of mankind," Haber was awarded the Nobel Prize for Chemistry in 1919. *Nobel Prize Winners*, Tyler Wasson, ed. (New York: The H. W. Wilson Company, 1987), pp. 402–404, p. 404.
47. DUNCAN A. MACINNES and MALCOLM DOLE, "Tests of a New Type of Glass Electrode," *Industrial and Engineering Chemistry. Analytical Edition*, 1929, 1: 57–59; "A Glass Electrode Apparatus for Measuring the pH Values of Very Small Volumes of Solution," *Journal of General Physiology*, 1929, 12: 805–811; "The Behavior of Glass Electrodes of Different Composition," *Journal of the American Chemical Society*, 1930, 52: 29–36; see also MALCOLM DOLE, "The Early History of the Development of the Glass Electrode for pH Measurements," *Journal of Chemical Education*, 1980, 57: 134.
48. D. S. TARBELL and A. T. TARBELL, "The Development of the pH Meter," *Journal of Chemical Education*, 1980, 57: 133–134.
49. ARNOLD THACKRAY and JEFFREY L. STURCHIO, "The Education of an Entrepreneur: The Early Career of Arnold Beckman," in *The Beckman Symposium on Biomedical Instrumentation* (April 12, 1985, New York City), Carol L. Moberg, ed. (New York: The Rockefeller University in association with Beckman Instruments, Inc., 1986), pp. 3–17. For a delightful and much illustrated account of the company, see HARRISON STEPHENS, *Golden Past. Golden Future: The First Fifty Years of Beckman Instruments, Inc.* (Claremont, California: Claremont University Center, 1985).

CHAPTER X

EARLY DEVELOPMENT AND USE OF THE MICROSCOPE

Magnification was referred to in the first century C.E., and the ability of a convex lens to form a magnified image was known as far back as the eleventh century. Eyeglasses to improve human vision may have been first used in Italy near the end of the thirteenth century. This certainly encouraged the study of optics and lens making, but also greatly facilitated the revival of learning. The period following their introduction coincided with the rediscovery of the Greek and Latin classics and the invention of printing. It also did away with the thick lettering of the medieval manuscripts and the heavy type of the early printed books, which were designed for elderly far-sighted readers.

Not until the late sixteenth and early seventeenth century did investigators, using combinations of magnifying lenses, study natural objects invisible to the naked eye. These lenses were incorporated into devices that later became known as the telescope and the microscope. Although the invention of the single lens microscope is surrounded by obscurity, the first useful compound microscope (one having at least two separate lenses) was constructed in the Netherlands, sometime between 1590 and 1610. However, the resolution of the early compound microscope was limited by blurring due to spherical and chromatic aberration.

Micrographia, published by Robert Hooke in 1665, was one of the earliest books devoted entirely to microscopic observations, and contains many fine plates illustrating vegetable histology. Using the word "cell" for the first time in science, he described the cellular nature of plant tissue in thin razor slices of cork, but he failed to recognize the cell's significance as a fundamental biological building block.

Also in the second half of the seventeenth century, Anton van Leeuwenhoek (1632–1723), a Dutch lens maker, mastered the difficult skill of grinding small lenses into perfectly rounded surfaces and designed many excellent single-lens instruments. He attracted considerable attention with his accounts of biological structures. However, microscopic study was dismissed on the grounds that no logical relationship existed between the macroscopic and microscopic worlds. The distortions in the visual field caused by the technical imperfections of early microscopes contributed to a general distrust of this instrument. The speculations and elaborate theories based on scant evidence or optical illusion mistaken for natural appearance, led to contradictory hypotheses and false conclusions about what was seen through the lens. In any event, many discoveries were made with these early imperfect instruments. An interest in microscopy, particularly in the compound version, was kept alive by only a few investigators, skilled instrument makers, and enthusiastic amateur scientists. These devotees from the leisured class contributed numerous accounts of the minute appearances of insects, plants, and minerals, to the eighteenth-century literature (Fig. 10.1) [1].

Microscopic examination of clinical material did not enter the mainstream of medical research and teaching until after the problem of spherical aberration in achromatic lenses had been corrected in 1827. Because spherical and chromatic blurring was a major problem with the compound microscope, scientists preferred the better-resolving but lower-power single-lens instrument.

Achromatic lenses for the microscope were developed near the end of the eighteenth century and improved in the early years of the nineteenth century, but single achromatic lenses provided low magnification. Increasing the power by addition of achromatic lenses often increased spherical aberration also. During the 1820s, practical compound objectives with increased power and corrected chromatic aberration were being built, but the resulting spherical aberration could only be kept to a minimum by selecting and positioning the achromatic lenses by trial and error, so that the spherical errors of one lens opposed those of another. The problem of accumulated spherical errors in achromatic lenses for the compound microscope was solved in 1827 by Joseph Jackson Lister (1786–1869), the father of Joseph Lister who introduced antisepsis in surgery. The elder Lister was a prosperous London wine merchant and amateur scientist who spent his leisure hours on problems of optics. By experimentation he developed a theoretical basis for selecting fully corrected achromatic lens combinations and carefully positioning them at the aplanatic foci, so that no additional spherical error was introduced. Lister's work for the first time removed the fuzziness of the image. Scientific design replaced empirical alignment and opened the way for the eventual commercial production of the compound microscope [2].

Lister helped transform the compound microscope from an unreliable device to a powerful instrument of investigation and furthered the development of its scientific uses. No longer a pastime with a toy, microscopy

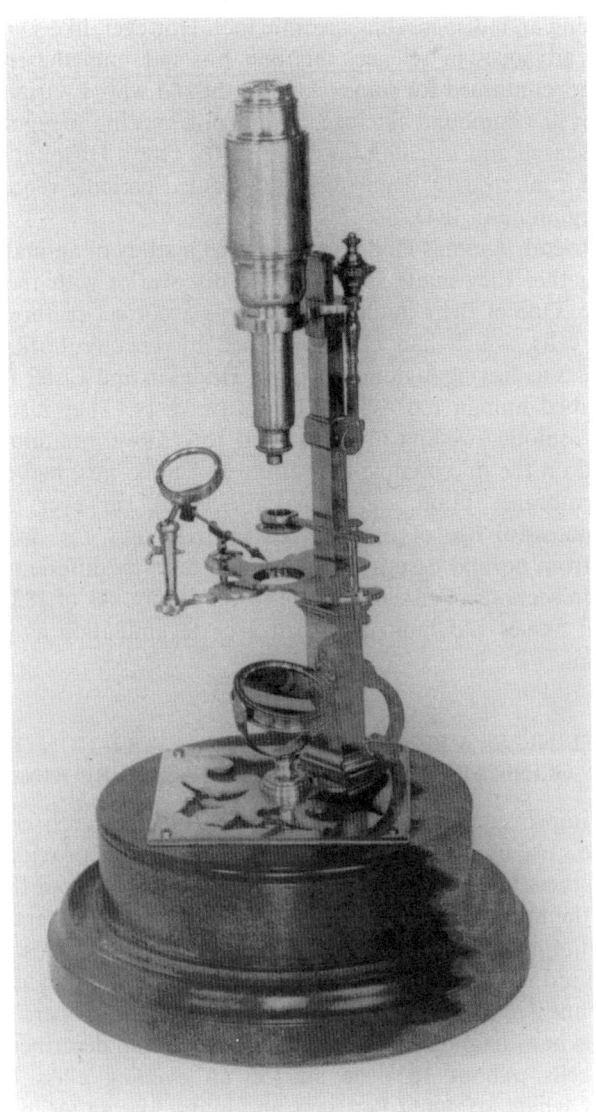

FIG. 10.1. Microscope of the type used by A. F. Fourcroy, made by C. S. Passemant, Paris. (*Roots of Clinical Chemistry*, J. Büttner and C. Habrich, eds., Darmstadt: Git Verlag Gmbh, 1987, p. 82.)

was becoming an indispensable scientific tool. However, like so many other innovative advances in science, this one was not immediately adopted. Lens-makers continued for some years, in the old way, by trial and error. The improved compound microscope did not become a regular part of medical research and teaching until the 1840s. Guy's Hospital in London established a microscopy department that issued periodic reports of their findings beginning in 1843.

Lister's friend, Thomas Hodgkin, suggested a study of animal fluids and solids. Together they made some notable discoveries with the improved microscope. One of their initial findings was a correct picture of human red blood cells, which had been described by Leeuwenhoek and other microscopists as having a globular shape. Hodgkin and Lister for the first time described human red cells as biconcave discs. They also described crenation, osmotic swelling in water, and rouleaux formation, and they commented on the microscopic appearances of milk, pus, nerves, arteries, cellular membranes, brain, and animal muscle [3]. However, the technology for successful use of microscopes in pathology for distinguishing abnormal from normal tissues on the cellular level by differential staining of ultra-thin sections of tissue was still in the future. As of 1827, study of organs and tissues was limited to their gross examination as part of the postmortem.

CLINICAL MICROSCOPY AND THE EVOLUTION OF CLINICAL LABORATORIES IN GREAT BRITAIN

With the improvements in optics in the early nineteenth century, the microscope's use in the everyday problems of diagnosis provided a strong impetus to the study and growth of clinical pathology. From 1840 to 1890 the use of the microscope in clinical medicine was quite general. Whether in observation of blood, tissues, parasites or bacteria, microscopy developed as an aid to the investigation of all illnesses of the patient. In Paris, Alfred Donné (1801–1878), one of the pioneer practitioners and advocates of this new diagnostic tool, examined blood and described an excess of white cells. About 1837 he started an evening class in medical microscopy which attracted Parisians and many foreign visitors. By 1845 he was the first to use the newly discovered process of photography to make photomicrographs of various body fluids and their cellular contents as well as of urinary deposits. At mid-century the microscope was increasingly being used for examining urinary sediments and intestinal parasites [4].

In London, the introduction of the microscope as a clinical diagnostic instrument was largely due to Lionel Smith Beale (1828–1906) who, in 1852, established a private laboratory near Kings College Hospital and gave a course of lectures on "The Microscope in Medicine." His course included practical demonstrations and instructions in microdissection and preparation of tissues for examination with the microscope to "afford some assistance to practitioners and students in medicine who employ the microscope in clinical investigation, or in physiological and pathological inquiries." He considered the application of chemical analysis to microscopical investigation as essential for the correct diagnosis and proper course of treatment [5]. Beale described and drew illustrations of the many varieties of deposits found in the urine, including bacteria; yet as late as 1872 he rejected the germ theory of disease. He also made the diagnosis of leukemia, first described by Virchow and John Hughes Bennett (1812–1875) in 1845, by noting the increase in white cells relative to red cells. Beale is best known by his books, *The Microscope in Medicine* (1st ed. 1854) and *How to Work With the Microscope* (1858), which were the outgrowth of his lecture course.

Clinical laboratories were uncommon during the eighteenth century, but became more prevalent in the nineteenth century. Initially, they were private research laboratories set aside by some doctors in a room in their homes, e.g., Henry Bence Jones. It is clear from his own writings that Browne Langrish had a considerable private laboratory in his home where he carried out research in therapeutics and toxicology as well as routine investigations of the blood and urine of his patients. This was the usual arrangement until the 1880s. Joseph Lister's houses always had a "microscope room." Lionel Beale must have had an extensive laboratory in his home because he held practical classes there [6].

At this time there were no clinical laboratories anywhere, not even in the teaching hospitals of London, and few of the largest hospitals provided any facilities for clinical pathology. Beale pointed out this deficiency in the 1878 edition of *The Microscope in Medicine* and urged that this could be remedied by expenditure of small sums of money. Complaining of the limited numbers of medical practitioners engaged in scientific medical research, Beale wanted laboratories and microscope rooms to be established at the major hospitals, for research and teaching, and that government grants be voted annually to support original investigations. He was a persistent advocate for the improvement of facilities for the scientific laboratory investigation of disease. Yet, opposition, especially to microscopic inquiries, was voiced by some of the senior and most influential members of the medical profession. Medical research was held in low regard because it had little if any money-earning value [7].

There was, he believed, "only one hospital in London in which there are efficient means for conducting scientific inquiries into the nature of disease, and I do not believe there is one, the managers of which would allow a very moderate sum, say £300, to be set apart for working expenses." Rich establishments like Guy's, St. Thomas's, and St. Bartholomew's should take the lead in this matter. "One would think that £1,000 of their large incomes might be spent very advantageously in scientific work, but I fear it will be difficult indeed to convince the authorities who have command of the purse." The persons most suitable to engage in medical research, said Beale, were the young physicians and surgeons attached to the medical schools and hospitals. As inducement, Beale suggested providing them with a place in which to work, "and an income just sufficient to provide the necessaries of existence—say, £100 a year." Would this not be time better spent by a talented physician and surgeon than for him to devote "fifteen or twenty years to seeing out-patients?" [8].

In 1890 a book reviewer commenting on clinical pathology admitted that "our London hospitals do not always provide adequate facilities for this kind of work, the arrangements for clinical research being, as a rule, far inferior to those afforded for *post-mortem* investigations. In Continental hospitals, where clinical instruction is given, we find clinical laboratories attached to the principal wards, and such an arrangement is really necessary if the work is to be properly done" [9].

Despite individual achievements, England lagged behind the continental countries in the laboratory investigation of disease. This was also the case earlier in the century when students flocked to Paris to learn the new pathology that correlated patient symptoms with findings at autopsy, and the use of the stethoscope. The main reason England trailed was that all the great hospitals in the country were established as charities for the relief of the sick poor and it was considered inappropriate to spend these funds on laboratories or research workers. Furthermore, the senior physicians in academic medicine were engaged in private practice, teaching, and clinical work in the wards—a professional routine which is not suited for purely scientific investigation [10]. Clinical medicine flourished, while laboratory studies languished, except in the cases of private laboratories. It took until the end of the century before the governors of the charity teaching hospitals realized that such laboratories were of the greatest service to the sick poor.

On the Continent, especially in Germany, the academic profession enjoyed a favored status during a period of unprecedented economic growth of the middle of the nineteenth century. Latecomers to the industrial revolution, the Germans, more than their wealthier English and French competitors, had to rely on scientific methods for the improvement of technology. As a result, physics and chemistry found a more favorable environment for development under the patronage of the multitude of petty feudal governments in Germany, than in the progressive and prosperous manufacturing centers in England and France [11].

In the pre-bacteriology days, "clinical pathology" was known as "clinical microscopy" because the microscope was the central piece of equipment in the room. With the introduction of bacteriological techniques and the discovery of cultivatable pathogenic microorganisms, the microscope room was converted into a laboratory with sterilizers, incubators, and tables and shelves for glass flasks and test tubes [12]. The diseases that mandated the establishment of laboratories were tuberculosis, cholera, typhoid fever, and diphtheria, and it was from medical health officers in particular that the call went out for facilities. The first hospital laboratory of this type in Great Britain may have been that of John Chiene, Professor of Surgery in Edinburgh. He began work in bacteriology about 1883 and the following year reported on "The desirability of establishing bacteriological laboratories in connection with hospital wards" [13]. Probably the earliest hospital bacteriology laboratory in England was the one started at St. Thomas's Hospital about 1886. Sterilizers, incubators, and the other apparatus used in bacteriological work had been presented to the Pathological Department by one of the physicians [14].

Although in general, the establishment of comprehensive clinical laboratories in English hospitals had hardly begun before the end of the nineteenth century, there was one exception. In 1828 Guy's Hospital in London had set aside two clinical wards with a small laboratory adjacent specifically for clinical purposes, where interested students would serve a three-month clerkship. In 1896, citing the best hospitals on the Continent and at other locations in Great Britain, the medical and surgical officers of London's St. Thomas's Hospital appealed to the administrators for a centralized clinical laboratory for scientific investigation of diseases of patients in the hospital. The governors responded and the following year a clinical laboratory was opened. During its first year of operation, it did mostly histological and bacteriological examinations and less than a combined total of 250 red and white cell counts, stained blood film examinations, hemoglobin estimations, and urine examinations other than for tubercles. During its first seven years, the number of all laboratory procedures averaged less than 1,500 annually [15].

To facilitate the routine and frequent microscopic examination of blood, the *British Medical Journal*, in 1896, suggested that a little rearrangement could bring the pathological laboratory in part to the bedside and that "small clinical laboratories should be fitted up, attached to or opening out of the wards," [16]. It points to the achievements of Dr. Osler's wards at the Johns Hopkins Hospital.

EMERGENCE OF CLINICAL LABORATORIES IN THE UNITED STATES

William Osler (1849–1919), physician-in-chief at Johns Hopkins from 1889 to 1905, always appreciated the role clinical pathology could play in medicine and in teaching medical students to observe patients on the wards. His interest in the use of the microscope in the diagnosis of disease originated at McGill University in Toronto where, about 1875, he initiated a course on the microscope in medicine [17]. It was at his suggestion in 1896 that a $10,000 bequest to the Johns Hopkins Hospital (opened in 1889) was employed for the construction of a clinical laboratory which provided the latest facilities for the teaching and practice of clinical pathology. This was accomplished by the addition of two extra stories to an existing building.

The instructional program of the department of medicine at Johns Hopkins Medical School (opened in 1893) emphasized that particular attention would be paid to clinical microscopy in the third year. "The students will learn the use of the instruments of precision employed in clinical research—the Stethoscope, Microscope, Ophthalmoscope, etc.—by daily routine manipulations. Particular attention will be paid during this year to Clinical Microscopy in the study of the urine, sputum, and blood" [18].

In that first rather simple clinical laboratory, a generation of Hopkins medical students learned to count the red and white cells of the blood, to recognize the parasites of malaria and of amoebic dysentery, and to carry out examinations of feces, urine, sputum, and gastric juice, which would yield important information for the diagnosis of disease. It represented a notable achievement at the time since, in the United States at least, the microscope had been slow in making its way into the clinic [19]. Microscopes and reagents were provided and each student paid a rental fee of five dollars a year for the microscope [20]. Speaking about the clinical and ward laboratory, Osler said: "They are to the physician just as the knife and scalpel are to the surgeon" [21].

The trustees of the Johns Hopkins Hospital were very grateful for the bequest and expressed their hope that "this gift will influence the construction of many similar laboratories in connection with other hospitals, and thus inaugurate a movement which in the end will prove of untold benefit to the sick and suffering throughout the United States" [22].

Laboratory facilities of one sort or another were already part of the hospital scene in the United States. In 1847 the trustees of the Massachusetts General Hospital in Boston, taking note of the powerful aid that the science of medicine "has received from the study of organic chemistry and the knowledge and use of the microscope," authorized the purchase of a microscope at a cost not to exceed $50. In 1851 they established the position of "Chemist-Microscopist"—duties included assisting at autopsies—and in 1855 they separated the position of chemist from that of microscopist for whom a "Pathological Cabinet" was provided. In 1874 the hospital built an amphitheater, morgue, and small laboratory, but the facilities soon became inadequate, and this was noted by the trustees in 1893. A new "Clinico-Pathological Laboratory" was opened in 1896 on the fiftieth anniversary of the first public demonstration of ether during surgery. Shortly after, a chemical laboratory was provided and a chemist appointed [23].

The Philadelphia General Hospital was the second oldest hospital in Philadelphia and the first city hospital in America to evolve from an almshouse infirmary. As the hospital housing large numbers of the sick poor, it was regarded as an important institution in which to enjoy visiting or attending privileges. In 1866 it created the position of microscopist. Three microscopic observations were recorded for that year and four the next. By 1871 the number of examinations had risen to twenty-one. The microscope was valued at $250 [24]. With such simple beginnings as these in Boston and Philadelphia, it is not surprising that, as of 1870, the average medical student or average practitioner in the United States had barely a nodding acquaintance with chemistry and could not use a microscope. What is now called "clinical pathology" was nothing more than simple examination of the urine [25].

Nevertheless, the importance of pathology in medicine was becoming well enough recognized so that a pathologist was appointed at the Philadelphia General Hospital in 1871 and in 1885–86 a small two-story building was erected for a laboratory and morgue. Rooms upstairs were set aside for clinical microscopy and bacteriology. Osler, Professor of

Clinical Medicine at the University of Pennsylvania (1884–1889), managed the morgue and performed the autopsies from 1885 to 1889. In the early 1890s the work in the laboratory consisted of postmortems, urinalyses, and examination of sputums. The infrequent blood counts and smears were performed by the residents in small rooms attached to many of the wards. Gas lights and reflectors were used for microscopic work, for it was not until after 1905 that the small laboratory building was wired for electricity [26].

The establishment in 1874 of the Hospital of the University of Pennsylvania (HUP)—the nation's first university-owned hospital—signaled an expansion of opportunities for elite hospital practice in Philadelphia. In 1893 the Pennsylvania state legislature's appropriation for the University Hospital stipulated that the $80,000 set aside for facilities be matched privately; $25,000 of the total raised was to be used to construct a laboratory of clinical medicine. William Pepper, Jr. (1843–1898), Professor of Medicine, personally contributed $50,000 and raised another $30,000 from friends and colleagues. Pepper, who was Provost of the University, named the laboratory in memory of his father (1810–1864), who was Professor of the Theory and Practice of Medicine (1860–1864). Half of his original gift was used for the building, the other half was set aside for an endowment to earn funds to cover some of its operating expenses.

The William Pepper Laboratory of Clinical Medicine at HUP, opened on December 4, 1895, was the first laboratory of its kind in the United States amply equipped for both routine work and research and provided with its own four-story brick building. On the first floor were laboratories for microscopic, chemical, and bacteriological investigations. The second floor contained a laboratory for anthropometry and the Director's office. The third floor had a laboratory for postgraduate students, and the fourth floor housed a library, a conference room, and another laboratory. The donor expressed the hope that a special research fund would be established to study tuberculosis or heart disease or infectious fevers—"all of which destroy so many thousands of precious lives annually" [27].

The Pepper Laboratory was conceived as a radically new type of institution. The terms of the gift specified that it was to promote the interests of the patients by clinical studies and original researches. Only postgraduate teaching was to be given. The clinical laboratory was placed under the control of the medical school's department of medicine, where it remained until 1968. William Pepper, Jr. was named the first director of the laboratory.

At the time of its opening in December 1895, the Pepper Laboratory was hailed as a key innovation in both medical education and hospital facilities. The major address on that occasion was given by William H. Welch (1850–1934), pathologist at the Johns Hopkins Hospital. Welch traced the development of laboratories from Alexandria of the early Ptolemies to the late nineteenth century. He referred to the first pathological laboratory, established by Rudolf Virchow in Berlin in 1856, and quoted him as having said: "As in the seventeenth century anatomical theatres, in the eighteenth clinics, in the first half of the nineteenth physiological institutes, so now the time has come to call into existence pathological institutes, and to make them as accessible as possible to all" [28].

Welch noted that the conception of a thoroughly equipped laboratory as an integral part of a hospital and intended for the study and investigation of disease is of recent origin. The idea may be traced to men such as Hughes Bennett and Lionel Beale in Great Britain, and to Frerichs and Traube in Germany, who applied microscopical and chemical methods in their hospital work. Germany's first clinical institute ("Klinisches Institut") was opened by Hugo von Ziemssen (1829–1902) in Munich in 1878. It contained a chemical, physical, and bacteriological department, library, lecture hall, and rooms for practical courses and the examination of out-patients. A similar laboratory facility was opened by Curschmann in Leipzig in 1892. The growing recognition of the need for such laboratories, said Welch, is the result of the great progress in scientific medicine during the last quarter of the nineteenth century. However, he added, it takes more than a microscope, some test tubes, and chemical reagents for simple tests of the urine. Examinations now included the blood, stomach contents, fluid from serous cavities, sputum and various secretions, and fragments of tissue removed for diagnosis. Alluding to the important role for the new laboratory in teaching and research, Welch said: "A hospital, and especially one connected with a medical school, should serve not only for the treatment of patients, but also for the promotion of knowledge. Where this second function is prominent, there also is the first most efficiently and intelligently carried out" [29].

However, the Pepper Laboratory could not live up to its founder's expectations. In 1909 the HUP began to supplement the Pepper endowment income in exchange for diagnostic services. In 1928 the building erected in 1895 was torn down to make room for a clinic, in which the laboratory was assigned space. An internal survey of the University of Pennsylvania medical school in 1931 reported the sharp decline in research activity of the Pepper Laboratory and proposed that the trend be allowed to continue. It was noted that the relationship of laboratory research to hospital needs had changed, and extensive laboratory procedures have become routinely necessary in the proper care of patients [30].

In 1898 Mrs. Josephine Mellen Ayer bequeathed $50,000 to the Pennsylvania Hospital, founded in 1751, to establish the Ayer Clinical Laboratory. Her son, Frederick Fanning Ayer, later added $25,000 and an additional $51,000 for the publication of bulletins of the laboratory's work [31]. The first director of the laboratory left after only one year and was succeeded by Dr. Simon Flexner (1863–1946) [32], the newly appointed professor and chairman of pathology at the University of Pennsylvania (1899–1903). While a portion of the laboratory's work would necessarily be diagnostic, it was Flexner's intention that the laboratory's primary function should be research. However, here too, the endowment was simply inadequate to support major investigative work [33].

After Flexner left Philadelphia in 1903 to become the first director of the newly established Rockefeller Institute for Medical Research in New York, the poorly-endowed Ayer laboratory soon settled into routine diagnostic and therapeutic activity. Flexner's greatest contribution to American medical science was in the organization and administration of the Rockefeller Institute as a center for fundamental and applied research (see Chapter 12).

Research in the central clinical laboratory of hospitals was secondary to the service function of routine diagnostic testing. With no control over patients and deluged by routine work, these laboratories, despite the best of research intentions, gradually became merely service organizations. This trend was helped along soon after 1910, when it became evident that the ability to charge patients for individual tests made the laboratory a source of financial gain for the hospital.

MEDICINE AND THE SCIENTIFIC METHOD

During the first half of the nineteenth century, the great increase in medical knowledge sharpened the distinctions between the art of medicine—knowing what to do, and the science of medicine—knowing why it is done. Art relied on intuition and guesswork, based on experience— there was nothing else to go on. Empiricism, not of choice but of necessity, pervaded the best medical practice of the mid-nineteenth century. In the second half of the century, laboratory research, especially in microbiology, greatly altered the conceptual foundation of medicine. Tools, such as the ophthalmoscope and laryngoscope, permitted direct examination of parts of the body previously inaccessible. Still other instrumental measurements, such as body temperature, levels of hemoglobin, and blood cell counts, provided the clinician with precise data rather than rough guesswork [34].

The rapid growth of "modern science" prompted a practicing physician and essayist, John Brown of Edinburgh, in 1882, to write:

"Let us by all means avail ourselves of the unmatched advantages of modern science, and of the discoveries which every day is multiplying with a rapidity which confounds; let us convey into, and carry in our heads as much as we safely can, of new knowledge from Chemistry, Statistics, the Microscope, the Stethoscope, and all new helps and methods; .... "

"Chemistry and Physiology have become, to all men above forty, impossible sciences; they dare not meddle with them; and they keep back from giving to the profession their own personal experience in matters of practice, from the feeling that much of their science is out of date; and the consequence is, that, even in matters of practice, the young men are in possession of the field" [35].

Technical progress in the 1880s and 1890s was rapidly having an effect on practice in America. In 1897 a prominent figure in medicine wrote: [36] "The microscopical and chemical examinations of the secretions and excretions and of the blood have become our daily duty... The time is indeed at hand in which, without the ready access to a laboratory manned by experts in all these lines, or the association with a trained laboratory assistant, no physician can do his patients, himself, or his science justice." The implication was clear. To give his patient proper care, the physician had to use proper tests to obtain precise data. "Medicine is no longer an art founded only upon empirical observation, no longer a purely deductive science; it is becoming also an inductive one."

The art of medicine and the science of medicine began to fuse into a poorly defined concept known as "scientific medicine." The precision of the new technology provided increasingly sharp discriminations in clinical diagnosis. However, precision alone was not enough, a critical attitude was essential. The old distinction between science and art was being replaced by a new distinction, that between precision and critical judgment. By 1900 the leading physicians recognized that the optimal practice of medicine, as well as advances in general clinical knowledge of medicine, would depend on tools of precision that only a laboratory could furnish. The terms "precision," "laboratory," and "science" became closely associated [37].

Other voices cautioned that there was more to science than collecting accurate data. In 1900 an editorial writer objected to the narrow view that scientific activity required a laboratory. More important was the "scientific method." "The 'scientific' physician is he who observes facts in a scientific manner, under whatever form those facts may be presented," not whether laboratory work was involved [38]. In 1901 interpretation and the pitfalls in diagnosis arising from reliance on the laboratory—since tests are subject to error—was the theme of a paper. "Therefore the first requisite for a diagnostician is not technical skill, but a judicial mind, capable of weighing the value of evidence" [39].

New blood testing instruments, i.e., hemoglobinometer, blood cell counter, hemacytometer, often made their first appearance in American hospitals and doctors' offices when physicians returned with them from visits to hospitals and clinics in Germany and France, where the teaching facilities for scientific study were more numerous and more highly developed than in America. They also brought back a new attitude of respect for scientific methods that resulted in laboratories being set up in hospitals in many locations. This growth in the number of laboratories was greatly stimulated by charitable gifts from philanthropists and private citizens.

On the Continent, every capital and large city had a great school of medicine, each with its own laboratories. Research was energized by the rivalry among the scientists, each seeking the reward of advancement to more prestigious posts in larger institutions in recognition for outstanding achievement.

THE $300 LABORATORY

In 1881 albumin (protein) and glucose were the only chemical constituents in urine of real importance. Urea and uric acid received much attention, although their physiologic mechanisms and clinical significance were far from clear. The literature gave considerable attention to crystals and efforts were made to identify them. It was still believed that crystals of the amino acids leucine and tyrosine were indicators of liver disease. Many other substances were found in urine, but they were curiosities and of no known diagnostic significance. In addition to albumin and glucose, urines were examined for acetone and diacetic acid, indican, and bile pigments. For protein Esbach's method was used. Quantitative estimation of glucose was usually made by fermentation. Later, the Einhorn tube was used routinely (Fig. 10.2). Following the introduction of the stomach pump in 1871 by Wilhelm Otto Leube (1842–1922) and the use of test meals, gastric analysis quickly became an important laboratory procedure. George Dock (1860–1951) related that following his arrival at the University Hospital in

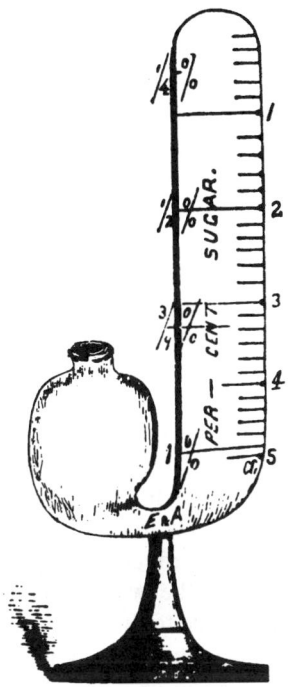

FIG. 10.2. Einhorn tube. (*Medical Record*, 1887, 31: 91–94.)

Ann Arbor in 1891 as head of medicine, almost all patients on admission had examinations of stomach contents and stools in addition to urine and blood tests. Everyone looked for evidences of cancer. In a few years, Dock "had an admirable arrangement for clinical pathology, but no room for class training." During his seventeen years at Ann Arbor, "my office was the clinical laboratory" [40].

In 1890 almost the only laboratory work entrusted to the house staff at New York Hospital was the routine analysis of the urine and an occasional hemoglobin test. Clinical bacteriology was still in its infancy and even the staining and examination of sputum for tubercle bacilli was so complicated and technical that it required the services of the hospital pathologist. Efforts at counting blood cells were only barely begun. Except for these few simple tests and the occasional use of the stomach tube and the sphygmograph, the recognition of disease rested entirely on a study of the patient's history, symptoms, and physical signs [41].

Advances in the new applied clinical sciences of chemistry, bacteriology, immunology, serology, and hematology, generated enthusiasm for laboratory analysis. By the turn of the century, increasing numbers of physicians were relying on laboratory tests. However, these "diagnostic" procedures provided essentially qualitative (yes/no) results. Writing in 1900, Camac said that the time had now come to bring the practical application of the knowledge and methods developed in the clinical research laboratories, to the bedside in the hospital or ward clinical laboratory. Although such ward laboratories had been established in the 1890s at the Royal Infirmary in Edinburgh, St. George's Hospital in London, Johns Hopkins Hospital in Baltimore, Bellevue Hospital in New York City, and soon at St. Luke's Hospital in New York City, there were objections raised by some authorities that such laboratories are scientific luxuries, because they require space, are expensive, and impose on the busy schedule of the internes. Camac's answer was that space requirements are only about 6×6 to 10×10 feet, a window and a table being the only essentials. "While running water and gas are helpful adjuncts, the ordinary water-bottle and spirit-lamp are all that is necessary. Probably no part of a hospital is more modest in its demands [than] the ward laboratory." Only the simple methods and apparatus applicable to disease detection would have a place in the hospital or ward clinical laboratory, and can be transported to the bedside [42].

Such a laboratory could be fully equipped for $300 with an annual maintenance cost of $50. It would include "two large items"—a microscope with oil-immersion lens and a cabinet in which apparatus is to be kept—costing $125 to $150. The objection that patients may drink some of the poisonous reagents, can be addressed by locking these reagents as well as the drugs in the cabinet. As for "the appropriation of apparatus by members of the interne staff," this, said Camac, can be dealt with by the hospital's "examining board, in ascertaining more correctly the character of applicants for the position of internship." Camac also recommended that clinical tests be a full-time assignment for one or more of the ward's internes, for bacteriology, clinical microscopy, and chemistry [43].

A similar word of caution was sounded in 1918. "To have ward laboratories where the internes make the examinations at their sweet pleasure is to invite slovenliness, inefficiency, and inaccuracy in all the routine examinations." Internes should do laboratory work for assigned periods "under the guidance and with the assistance of the trained workers,...." [44].

Wherever the laboratory tests were done, there was no attempt at specialization. Blood counts and hemoglobin levels were done alongside of urinalysis for albumin, sugar, and specific gravity, acid fast stains on sputum, and protein in spinal fluid. Hospital laboratory facilities, always a low priority, were assigned to small cramped quarters in the basement or on the top floor of a wing or an annex, usually not distant from the autopsy room. When one spoke of the laboratory one meant pathology whether it was analytical or anatomical.

CLINICAL LABORATORY TESTING (1900-1920)

The importance of the microscope in clinical studies emerged in the 1890s with the expansion of hematology as a new discipline. As hematology developed, the clinicians began to dominate the new hospital laboratories, for this type of examination had no real roots in classical pathology, but was primarily tied to the live patient. This was the clinician's domain. Though chemistry had been making contributions to physiology, and pathology, these were being communicated in the classroom. As of the end of the nineteenth century, the clinical applications of chemistry were limited to qualitative procedures on urine [45]. These often evolved into semi-quantitative tests.

The clinical pathology in William Osler's first edition of *Principles and Practice of Medicine* (1892) was limited to hemoglobin estimation, red and white cell counts, search for parasites of malaria, simple urinalysis, and microscopy of the sputum for tubercle bacilli. There was no blood chemistry. In the third edition (1898), the presence of aminoaciduria in liver necrosis was mentioned. In the eighth edition (1912), Osler describes an oral glucose tolerance procedure (100 grams glucose) and testing for glycosuria for suspected diabetes mellitus. In these early editions advances in bacteriology outpaced those in chemistry. The dominant laboratory position at that time was held by bacteriology (including serology) owing to the rapid advances made in this field during the second half of the nineteenth century. Consequently, the earliest professional clinical pathologists were the bacteriologists.

It was the advent of bacteriology that finally transformed the hospital morgue and museum into a clinical laboratory. Robert Koch's (1843-1910) identification of the tubercle bacillus in 1882 hastened this process by increasing an interest in discovery of other pathogens. His methodology for bacteriological research was easily adapted to the hospital and speeded the development of new techniques for culturing and staining bacteria. Fernand Widal's (1862-1929) discovery in 1896, that a substance in the blood serum of patients with typhoid fever caused the agglutination of a culture of live typhoid bacteria, introduced serological testing—a new method which did not require the visual identification of the pathogen [46].

As late as 1914, Osler noted that there still were many hospitals of 200 to 300 beds without clinical laboratories, and those he saw were usually located in a poorly equipped and badly-lighted room in the cellar [47]. On the eve of World War I, judging from P. N. Panton's popular text of *Clinical Pathology* (1913), blood chemistry in England was in a very elementary state. It consisted of spectroscopic examination of the absorption bands of hemoglobin and its normal and abnormal variants; detection of bilirubinemia by visual observation of the increased yellow color of serum and confirmation by Gmelin's test; and, demonstration of excess uric acid by Garrod's "thread" technique. An "index of alkalinity" of the blood was determined by utilizing serum and a series of standard acid solutions and a color indicator. The author acknowledges that "little information of any clinical value is to be obtained from such methods." Furthermore, he suggests "that no purely chemical examination of the blood has at the present any wide application in clinical medicine" [48].

Urine examinations showed no advance over the previous fifty years: pus, red cells, albumin, and glucose were the only useful determinations. Other blood examinations consisted of cell counts, stained films, hemoglobin, coagulation time, agglutination reactions, Wassermann test, examination for parasites such as malaria, and blood cultures. Fecal examination was little more than simple inspection and a few simple qualitative tests for fat and stercobilin. The test meal and gastric analysis, introduced about thirty years earlier, was described.

MEDICAL TECHNOLOGISTS, COMMERCIAL LABORATORIES, AND WORLD WAR I

Following the initial interest and activity in laboratory diagnosis and bedside teaching that accompanied the opening of clinical laboratories in the 1880s and 1890s, practitioners began to delegate the laboratory work to an auxiliary staff. This gave rise to a new professional specialty— medical technologist.

To cope with the growing number of chemical tests, the physician in private practice, rather than do the tests himself or hire an assistant, would usually enlist the help of chemists or physicians skilled in chemistry. This demand led to the establishment of private laboratory services. As early as 1883, a Philadelphia physician, Judson Daland, advertised his willingness to make "*thorough chemical and microscopical urinary examinations* . . . in the most careful manner, and to furnish promptly a written report of the results. A moderate fee will be charged" [49].

By the start of the twentieth century, commercial laboratories were springing up in most large cities in the United States to serve private hospitals and to meet the demand of physicians in private practice for chemical and microscopical examinations. One such laboratory was opened in New York City in 1898 by Dr. Frederick Sondern [50]. At first, clinical pathology laboratories were staffed by physicians with laboratory training who then taught others to perform the simpler tests. When World War I reduced the number of male technicians in the laboratories, assistance was rendered by women, many of whom had training in chemistry and bacteriology [51].

Medicine is the only winner in wartime. The contributions of military experience to civilian medical practice, surgery, and nursing are milestones of history. But the impetus that World War I gave to the spread of laboratory medicine is less generally recognized. Early in 1918, the American Army laboratories overseas as well as competent personnel to staff them, were few, and equipment was makeshift. Worse yet, the usefulness of the pathologists was limited by the indifference of many medical officers who were unaccustomed to the consultant services of laboratory doctors in civilian practice. Laboratory personnel were not infrequently assigned to "more important" storeroom or mess hall duties. In most base hospitals simple urine and sputum examinations were something to fill the time when things were slow [52].

Soon an efficient pathological service was organized and by early fall in 1918 there were nearly 300 laboratories in use in the American Expeditionary Forces, still understaffed, but much less so than before [53]. The percentage of autopsies in hospital deaths had risen from 25% to 92%, and a reasonably adequate review of diagnosis and treatment was being provided. And best of all, in many units the internists and surgeons were beginning to accept the laboratory doctor as a worthy partner in the medical fraternity [54].

Faced with reduction in medical staff due to America's entry into World War I, hospitals found that it was possible "to get much work done by securing at a comparatively small salary the services of a well trained laboratory helper who is not a physician,.... Women are now fitting themselves for these positions in rapidly increasing numbers" [55].

Following the war, rapid development in the number of clinical laboratories resulted in additional shortage of medically trained men. Forced to rely on nonmedical technicians to do the laboratory work, hospitals became the setting for the emergence of a new professional identity—the nonmedical technician. The negative prefix was soon dropped.

Despite the training of many technicians by the army during the war, and the laboratory courses offered by several medical schools, the demand for reliable, well-trained technicians greatly exceeded their availability [56]. Citing the army's example, Gradwohl proposed the establishment of schools for the proper training of laboratory technicians and for the organization of a "laboratory examining board" to pass on their qualifications for employment. "Technicians have come to stay;..." In order that "a better class of women be urged to enter this field," women's clubs should be informed of this new career opportunity "so that the best possible raw material may be selected and utilized in the upbuilding of this highly technical branch of medical specialism." Gradwohl also suggested "the formation of some kind of national association by the technicians themselves" to bring about recognition by the medical profession [57].

Thousands of doctors who first experienced the regular use of a diagnostic laboratory in their war service demanded similar competent facilities in their community hospitals when they returned home [58]. As methods increased in number and clinical importance, the practice of medicine witnessed an increasing tendency to refer even the simplest examination to the clinical pathologist. Many commercial laboratories of varying degrees of competence and trustworthiness sprang up to fill this need. This raised the question whether these should not be required to meet certain minimum standards of accuracy [59]. Hospitals also responded by hiring additional technicians, scientists, and physicians to run the hospital laboratories and scheduled portions of the house physician's time for laboratory work only.

The introduction of chemistry into medicine established the laboratory as an important element of diagnosis. Recognition by the medical community came when the American Society of Clinical Pathologists (ASCP) was organized in 1922. Four years later the American College of Surgeons recognized the new specialty. It required that hospitals seeking accreditation have a clinical laboratory service in the charge of a physician especially trained in clinical pathology [60]. In Great Britain, the Association of Clinical Pathologists was formed in 1927; a European Association took shape in 1947.

The Board of Registry of the ASCP was organized in 1931 to certify the qualifications of clinical laboratory personnel. By 1993 over 150,000 technologists and technicians at various levels of expertise and experience, working in hospital, physician office, commercial, industrial, or other clinical laboratory settings, were registered in twenty-one levels and categories, e.g., chemistry, hematology, microbiology, blood bank, histology, cytology, immunology, nuclear medicine, phlebotomy, hemapheresis, and laboratory management.

LABORATORY SUPPLIES AND SUPPLIERS

Laboratory apparatus after the turn of the nineteenth century no longer included furnaces, pneumatic troughs, and blowpipes. The usual equipment now consisted of physical measuring devices such as the spectroscope and colorimeter. In use were the polariscope, microscope, and analytical balance. Soon to come—the introduction and gradually increasing use of the centrifuge.

Coal gas had come into use as a local heat source in the 1850s. Hotter than candles and spirit lamps, and more flexible and controllable than the furnace, the gas was piped directly to the benches. New laboratories had massive teak-topped benches custom-made by carpenters, with sinks built into them for water and drainage. With the passage of time, additional services including electricity, compressed air, steam, demineralized water, and suction, became available from a service manifold at the bench. Sometimes the controls were brought to the front of the bench to avoid having to reach through the equipment on the bench top to operate them. The laboratory was planned around the benches. They had become part of the building structure. Following the rapid growth of science after World War II, production of laboratory furniture became a commercial enterprise. In 1946 Fisher Scientific (Pittsburgh, Pennsylvania) began marketing a line of prefabricated steel units to be assembled in the laboratory.

In America, until the 1850s, there was hardly any domestic production of chemical apparatus or reagents. Most of these items were imported from Germany, France, and Great Britain and sold through dealers in scientific instruments, opticians, or pharmacies. A few firms manufactured scales, weights, thermometers, and hydrometers. American chemists bought their burners, tripods, and related hardware from plumbers; water baths and drying ovens from tinsmiths or coppersmiths; and beakers, flasks, bottles, and tubing from glass factories that specialized in druggists's glassware. Other items were made by shops that did similar work in other lines. A wood turner might make wood supports for burettes, condensers, and funnels [61].

In 1847, as a result of meeting with American chemist Eben Norton Horsford (1818-1893) studying at Giessen, Liebig's assistant, Bernard

G. Amend (1821-1911), decided to come to America. His expectation was to join Horsford, appointed Rumford Professor of chemistry at the newly established Lawrence Scientific School of Harvard University. On arriving in New York, Amend found a job as a chemist in a small pharmacy owned by a physician. When the owner retired in 1851, Amend came into control of the business, located at the corner of 18th Street and 3rd Avenue. It was the address the company occupied for its entire existence.

Amend was joined later by a school friend, Carl Eimer. The firm of Eimer & Amend soon became the country's leading importer of drugs and chemicals. In 1874 the company added chemical laboratory supplies and introduced the first line of reagent chemicals that displayed the actual lot analyses on each label instead of merely stating maximum impurities. From the glassblowing shops established by Eimer & Amend and by Emil Greiner (also in New York) during the early 1880s came many expert glassblowers and precision calibrators. Greiner later expanded his business to include chemical laboratory supplies [62].

Eimer & Amend was bought by Fisher Scientific Company in 1940 and was integrated into the company in 1950, its 100th year. Fisher Scientific had been founded as the Scientific Materials Company in 1902 by Chester G. Fisher (1881-1965) with the purchase of the laboratory apparatus section (stockroom) of the Pittsburgh Testing Laboratories. At the time, the latter was western Pennsylvania's only local source of laboratory supplies, most of which were purchased from Eimer & Amend who imported them from Europe. Although Fisher was an engineer and not a chemist, he sensed the growth of the iron and steel industries in the Pittsburgh region and the opportunity it presented to a supplier of laboratory materials. The company—name changed to Fisher Scientific in 1925—eventually became a leading supplier and manufacturer of laboratory equipment in the United States, with branches and offices overseas. The Fisher burner was the most important improvement in burners since the Bunsen-Desaga design of 1855 for use wherever coal gas was available [63] (see Chapter 14).

Until the latter part of the nineteenth century, there was no American company totally into the production of chemical laboratory apparatus. The study of chemistry was taken up by comparatively few, and therefore, the demand for apparatus was never great enough to justify large scale production. Various events during that period stimulated the growth of an American laboratory supply industry. The founding of the American Chemical Society in 1876 and later, the campaign by the Department of Agriculture for legislation for pure foods and drugs, made the American public more conscious of the role that chemistry played in practically all industries [64].

ORIGINS OF THE CHEMICAL GLASSWARE INDUSTRY IN AMERICA

In 1890, a simple test for estimating the quantity of butterfat in milk, developed by Stephen Babcock at the University of Wisconsin, revolutionized the dairy industry. With dairymen able to make rapid determination of the most valuable constituent of milk—the fat—not only was there a drop in the adulteration of milk, but now, pricing could be based on quality. The test produced an immediate need and demand for the hand-operated centrifuge, graduated test bottles, pipettes, and small graduated cylinders for measuring the sulfuric acid used in the test. All these items had to be manufactured at a price affordable to dairy farmers. The great demand for Babcock bottles led to the first real quantity production of graduated glassware in America [65].

The earliest reference to the manufacture of chemical glassware in America, may be an advertisement in the New York *Gazette* in 1754. During the early part of the nineteenth century, the limited needs for chemical glassware were supplied by the few lamp glassblowers who had come to America from England, Italy, and France. Glass-stoppered reagent bottles, glass dishes, and a few heavy molded glass articles, i.e., with raised lettering to identify the contents, were imported from France. Later on, several factories were making limited amounts of glass tubing and druggist's glassware. Not until 1873, when a demand for chemical glassware became evident, was there an organized attempt in this country to make molded glassware for the laboratory and the druggist's shop. Until then, for the most part, only glassware imported from Bohemia (Austria) and Germany was sold by American dealers [66].

The pioneer in the American chemical glassware industry was Whitall Tatum & Company, Millville, New Jersey. In 1876 this company first produced reagent bottles with molded chemical names. It also started to produce laboratory ware from a potash glass very similar to that used by the Bohemian factories. Output was limited because quantity users, viz., educational institutions and government laboratories, could import laboratory glassware duty-free. With American manufacturers unable to compete because of the price differential, the market for American-made flasks, beakers, and test tubes was limited almost entirely to the small needs of industrial and private laboratories, and did not justify increased domestic production. Some graduated glassware was produced in small glassblowing shops mostly from tubing imported from Germany. Chemists who had studied abroad preferred the apparatus they had used in their training. For years the notion persisted, with some justification, that quality apparatus came from abroad. Europe had the skilled artisans, the competitive advantage of early development, and the lower wage scale [67].

Growth of the American chemical glassware industry began shortly after the beginning of the new century. With the passage of the Pure Food and Drug Act of 1906, government laboratories were established throughout the country to enforce the Act. Canners and other food processors and drug manufacturers were compelled to hire chemically trained personnel and to provide laboratory facilities to maintain accurate methods of chemical analysis and control of purity. This encouraged scientific investigation and the expansion of chemical education, beginning in high school. There was a corresponding increase in demand for apparatus, glassware, chemicals, and laboratory accessories. A major supplier responding to this need was the Kimble Glass Company [68].

American-born Evan Ewan Kimble (1868–1956) began his career in the glass industry at twelve years of age, by tending a wood-burning oven at the Whitall Tatum glass factory. His experience in glass manufacture led to a job as a manager of the glass vial department for the Thomas Shelton Glass Works in Gas City, Indiana. Convinced of the market opportunities for glass manufacture, Kimble formed his company in Chicago in 1901, with vials as the principal product. In 1905 the company began to produce graduated glassware by the manufacture of Babcock bottles for measuring butterfat in milk. Recognizing the shortcomings of total manual operation, Kimble moved toward mechanization. In 1911 Kimble began production in Vineland, New Jersey [69].

With the onset of World War I, imports from Germany and Austria had come to a halt. The sharp and immediate increase in demand for chemical glassware was a stimulus to American glass manufacturers. Kimble expanded production and invested in the development of automated machines for production of tubing and rod, and for measuring, dividing, and graduating. With increased production came reduced cost. The uniformity in the quality of calibrated and graduated glassware was greatly improved by automation and machine production of the glass tubing. The Kimble Glass Company led the way in mass production of graduated glassware. In 1932 the company introduced its Exax blue line calibration marking. In 1946 the company became a unit of Owens-Illinois Glass Co.; it is now part of Gerresheimer Glass AG, Dusseldorf, Germany.

Another glassware name familiar to all current laboratory personnel is the "Pyrex" brand, manufactured by the Corning Company, founded more than a century ago. In 1864, Amory Houghton, Sr. (1813–1882), who operated the Union Glass Company at Somerville, Massachusetts, closed down the works because of labor disputes. He purchased an interest in the South Ferry Glass Works, Brooklyn, New York, and changed the firm's name to Brooklyn Flint Glass Works. Houghton operated the plant until 1868, when he moved the equipment to Corning, New York, where natural gas was available at the time, and established the Corning Flint Glass Works. In 1875 the name was changed to Corning Glass Works. Signal lights and other railroad glassware, and later on, electric lamp bulbs and electrical insulators were the main products. The company ventured into the scientific field in 1896, when Arthur A. Houghton, grandson of the founder, invented the method of drawing thermometer tubing vertically. This revolutionary development contributed to the manufacture of more accurate thermometers, with none of the twisting defects that came from the old horizontal drawing method [70].

The expanded demand for laboratory glassware in 1914 prompted Corning to enter the market. After much experimentation, they improved their glass-blowing technology for the very viscous borosilicate glass which was already available in pressed baking ware as *Pyrex*. More resistant to chemical action, its extremely low thermal expansion allowed Corning, in 1916, to increase wall thickness in order to produce stronger tubing, flasks, and beakers. With the end of the war, the new industries faced renewed foreign competition. It was obvious that tariff protection was necessary to insure the survival of the new glassware and laboratory supply industry and to encourage American manufacturers to invest in research and development to achieve economical quantity and quality production. Consequently, the duty-free provision enjoyed by educational institutions was omitted from the Tariff Act of 1922, and an increase in tariff was placed on some goods to offer greater protection [71].

Meanwhile, Congress had recognised the growing need for greater accuracy in analysis and established the National Bureau of Standards in 1901. The Bureau is equipped for the testing of all kinds of apparatus and instruments for accuracy, and for assisting manufacturers in the standardization of their products, and in the calibration of volumetric glassware. In 1988 the Bureau merged with other services to become the National Institute of Standards and Technology (NIST).

BAUSCH & LOMB OPTICAL CO.

Another name familiar to the clinical chemistry laboratory is Bausch & Lomb, manufacturer of a popular version of the Duboscq-type visual colorimeter (1920–1960) and other optical instruments. The company traces its origin to 1853, when John Jacob Bausch (1830–1926), a German immigrant, set up a tiny optical goods shop in Rochester, New York. He sold eyeglasses, magnifiers, microscopes, and other products imported from Europe. The business grew and Bausch formed a partnership with his good friend Henry Lomb (1828–1908). Early in the company's history, Bausch discovered that a hard rubber, called Vulcanite, could be molded into eyeglass frames, which he then fit with lenses imported from Europe and ground by hand. These frames were more durable and less costly than the gold-filled metal or horn-rimmed frames being sold at the time. This product was the basis of the company's eventual financial security [72].

Bausch & Lomb then diversified into a variety of optical products, specializing in those that required a high degree of manufacturing precision. Production of microscopes was begun in 1878 and continued until 1987 when that line—along with its photogrammetry and ophthalmic instruments—was sold to the Cambridge Instrument Company.

After setting up an experimental glass laboratory in 1912, Bausch & Lomb soon became the first American producer of optical quality glass. By the end of 1917, the company was producing more than two-thirds of the government's wartime needs for glass for binoculars, riflescopes, telescopes, and searchlights. During World War II, the company again supplied a variety of precision optical products to the government, including optical glass for aerial camera lenses and periscopes.

In the 1950s the company bought several smaller companies that made precision instruments for the scientific community. Soon, spectrophotometers, spectrometers, and other instruments for analysis were added to the B&L product line. In 1960 the company dropped the word "optical" from its name in recognition of the wide diversity of its products. During the 1960s the company continued to expand by acquiring small companies producing scientific instruments. B&L also entered the market with soft contact lenses and associated accessory products. In 1984 the company discontinued production of spectrometers, spectrophotometers, and digital recorders and plotters, because of anticipated volatility in the market. Today, B&L is expanding into the field of ophthalmic pharmaceuticals and focusing on the personal health care products which it distributes through drug stores and other similar retail channels. Personal health care products now generate more than half of the company's annual sales.

NOTES AND REFERENCES

1. For early developments of the microscope and applications to the detection of disease during the second half of the nineteenth century, see STANLEY JOEL REISER, *Medicine and the Reign of Technology* (New York: Cambridge University Press, 1978), pp. 69–76, 77–90. See also S. BRADBURY, *The Evolution of the Microscope* (Oxford: Pergamon Press, 1967), pp. 183–187, 190–197, 200–203; see pp. 200–255 for a discussion of the microscope in Victorian times.
2. JOSEPH JACKSON LISTER, "On the Improvement of Achromatic Compound Microscopes," *Philosophical Transactions of the Royal Society of London*, 1830, 120: 187–200. In recognition of this report, he was elected a Fellow of the Royal Society in 1832. Also see *Dictionary of Scientific Biography* (*DSB*), Charles Coulston Gillispie, ed. (New York: Charles Scribner's Sons, 1973), 8: 413–415.
3. THOMAS HODGKIN and JOSEPH JACKSON LISTER, "Notice of Some Microscopic Observations of the Blood and Animal Tissues," *Philosophical Magazine*, 1827, new series, 2 (8): 130–138, pp. 131–132.
4. W. D. FOSTER, *A Short History of Clinical Pathology* (Edinburgh and London: E. & S. Livingstone Ltd., 1961), pp. 24–25.
5. LIONEL S. BEALE, *The Microscope in its Application to Practical Medicine*, 3rd ed. (Philadelphia: Lindsay and Blakiston, 1867), pp. ix, 89; see also FOSTER (1961), pp. 27–29.
6. FOSTER (1961), pp. 111–112.
7. LIONEL S. BEALE, *The Microscope in Medicine*, 4th ed. (London: J. and A. Churchill, 1878), pp. 13–15, 17. For a biographical sketch, see also W. D. FOSTER, "Lionel Smith Beale (1828–1906) and the Beginnings of Clinical Pathology," *Medical History*, 1958, 2: 269–273.
8. BEALE (1878), pp. 12, 14, 15; also see FOSTER (1961), pp. 112–113.
9. Book Review. *British Medical Journal*, 1890, 2: 846–847.
10. BEALE (1878), pp. 10, 14, 15.
11. *DSB* (1973), 7: 379.
12. "The Leicester Bacteriological Institute," *Lancet*, 1896, 1: 569–570.
13. *British Medical Journal*, 1884, 2: 653.
14. "Bacteriology at St. Thomas's Hospital," *British Medical Journal*, 1886, 1: 34.
15. H. C. CAMERON, *Mr. Guy's Hospital 1726–1948* (London: Longmans, Green and Co, 1954), p. 183; FOSTER (1961), pp. 113, 117–120.
16. "A Desideratum—Clinical Laboratories," *British Medical Journal*, 1896, 1: 1467.
17. WM. OSLER, "Introductory Remarks to, and Synopsis of, Practical Course on Institutes of Medicine," *Canada Medical and Surgical Journal*, 1876, 4: 202–207. For a review of Osler's contributions to clinical pathology, see HAROLD M. MALKIN, "The Influence of William Osler on the Development of Clinical Laboratory Medicine in North America," *Annals of Clinical and Laboratory Science*, 1977, 7: 281–297.
18. ALAN M. CHESNEY, *The Johns Hopkins Hospital and The Johns Hopkins University School of Medicine. A Chronicle*, (3 vols.), vol. 2, *1893–1905* (Baltimore: The Johns Hopkins Press, 1958), pp. 78–79. Initially, laboratories were affiliated with the Department of Medicine. With the emergence of bacteriology, serology, and chemistry, these services

became affiliated with anatomic pathology and were grouped under "clinical pathology." Both "clinic" and "clinical" (Greek: *kleenay*, bed) convey the idea of a sick-bed. The word "clinician" signifies an individual whose medical knowledge is derived from bed-side study and observation of patients, as distinct from one whose medical knowledge is acquired from books or from theoretical considerations only. CHESNEY (1958), pp. 93–94.
19. *Ibidem*, pp. 117–118.
20. WILLIAM OSLER, "Clinical Microscopy at Johns Hopkins Medical School, Baltimore, United States of America," *British Medical Journal*, 1899, 1: 69–70.
21. WILLIAM OSLER, "Discussion," *Journal of the American Medical Association*, 1900, 35: 230.
22. CHESNEY (1958), pp. 115–117, p. 116; see also HENRY M. HURD, "Laboratories and Hospital Work," *Bulletin of the American Academy of Medicine*, 1896, 2: 483–495, pp. 486–487, 490.
23. FREDERIC A. WASHBURN, *The Massachusetts General Hospital. Its Development, 1900–1935* (Boston: Houghton Mifflin Company, 1939), pp. 105–107, 110–111.
24. JEFFERSON H. CLARK, "The Development of a Pathological Laboratory at Blockley," *Medical Life*, 1933, 40: 237–252, p. 244.
25. LESTER S. KING, "XII. Clinical Laboratories Become Important, 1870–1900," *Journal of the American Medical Association*, 1983a, 249: 3025–3029, p. 3025.
26. CLARK (1933), pp. 245–249.
27. Editorial, "The William Pepper Laboratory of Clinical Medicine," *Boston Medical and Surgical Journal*, 1895, 133: 603–604; for a history of the laboratory, see DONALD S. YOUNG, MARY CREGAR BERWICK, and LEONARD JARETT, "Evolution of the William Pepper Laboratory," *Clinical Chemistry*, 1997, 43: 174–179. See reference #40.
28. WILLIAM H. WELCH, "The Evolution of Modern Scientific Laboratories. An Address Delivered at the Opening of the William Pepper Laboratory of Clinical Medicine, Philadelphia, December 4, 1894 [sic]," *Johns Hopkins Hospital Bulletin*, 1896, 7: 19–24, p. 21 (reprinted in *Archives of Pathology and Laboratory Medicine*, 1993, 117: 658–663). The correct date of the opening is 1895.
29. *Ibidem*, p. 23; see also JOHANNA BLEKER, "Medical Students—to the Bed-side or to the Laboratory? The Emergence of Laboratory-training in German Medical Education 1870–1900," *Clio Medica*, 1987, 21: 35–46, pp. 40–41 ff.
30. EDWARD T. MORMAN, "Clinical Pathology in America, 1865–1915: Philadelphia as a Test Case," *Bulletin of the History of Medicine*, 1984, 58: 198–214, pp. 207–208.
31. FRANCIS R. PACKARD, *Some Account of the Pennsylvania Hospital From Its First Rise to the Beginning of the Year 1938* (Philadelphia: Engle Press, 1938), p. 108. The Pennsylvania Hospital is not to be confused with the Philadelphia General Hospital (named in 1902) which developed from the Almshouse and House of Employment founded in 1731 (pp. 3–5).
32. Flexner had no college degree, but received the M.D. degree from the Louisville Medical School in 1889. At that time, Louisville conferred the degree automatically after eight months of lectures and with no required laboratory or practical instruction. Determined to become a pathologist, Flexner completed William Welch's course at Johns Hopkins Hospital for graduates in medicine.
33. MORMAN (1984), pp. 204–205.
34. LESTER S. KING, "XI. Medicine Seeks to be 'Scientific'," *Journal of the American Medical Association*, 1983b, 249: 2475–2479, pp. 2475, 2477, 2478.

35. JOHN BROWN, *Horae Subsecivae* (Locke and Sydenham), 1st series, 4th ed. (Edinburgh: David Douglas, 1882), pp. 68–71 (pp. 69, 71).
36. J. M. DA COSTA, "Tendencies in Medicine," *Transactions of the Association of American Physicians*, 1897, 12: 1–8, pp. 2, 8.
37. KING (1983b), p. 2479.
38. Editorial. "The Scientific Physician," *Boston Medical and Surgical Journal*, 1900, 142: 701–702, p. 702.
39. THEODORE C. JANEWAY, "Some Sources of Error in Laboratory Clinical Diagnosis," *Medical News*, 1901, 78: 700–706, p. 700.
40. GEORGE DOCK, "Clinical Pathology in the Eighties and Nineties," *American Journal of Clinical Pathology*, 1946, 16: 671–680. One writer credits Dock for the establishment in 1893 of the first clinical laboratory of medicine in the United States. RAYMOND WALLACE, "Laboratory Diagnosis—Its Relation to the General Practitioner," *Interstate Medical Journal*, 1903, 10: 148–150; MAX EINHORN, "Fermentation as a Practical Qualitative and Quantitative Test for Sugar in the Urine," *Medical Record*, 1887, 31: 91–94, p. 94. See also LOUIS ROSENFELD, "Gastric tubes, meals, acid, and analysis: rise and decline," *Clinical Chemistry*, 1997, 43: 837–842.
41. LEWIS A. CONNER, "Relation of Laboratory Aids to the Practice of Medicine and Surgery," *Journal of the American Medical Association*, 1923, 81: 871–873, p. 871.
42. C. N. B. CAMAC, "Hospital and Ward Clinical Laboratories," *Journal of the American Medical Association*, 1900, 35: 219–227, p. 223.
43. *Ibidem*, p. 224.
44. MAX KAHN, "The Department of Laboratories," *The Modern Hospital*, 1918, 11: 271–274, p. 273.
45. KING (1983), p. 3029.
46. FOSTER (1961), pp. 121, 122; MORMAN (1984), p. 202.
47. FOSTER (1961), p. 119; HARVEY CUSHING, *The Life of Sir William Osler*, (2 vols.), vol. 2 (Oxford: Oxford University Press, 1925), p. 367.
48. FOSTER (1961), pp. 50, 121–122; PHILIP NOEL PANTON, *Clinical Pathology* (London: J. & A. Churchill, 1913), pp. 92–99 (pp. 97, 99).
49. JUDSON DALAND, "Urinary Examination," *Medical and Surgical Reporter*, 1883, 48: 701.
50. JAMES B. HERRICK, "The Relation of the Clinical Laboratory to the Practitioner of Medicine," *Boston Medical and Surgical Journal*, 1907, 156: 763–768, pp. 766–767; LEWELLYS F. BARKER, "Medical Laboratories: Their Relations to Medical Practise [sic] and to Medical Discovery," *Science*, 1908, 27: 601–611, p. 608; WILLIAM MCKEE GERMAN, *Doctors Anonymous. The Story of Laboratory Medicine* (New York: Duell, Sloan and Pearce, 1941), pp. 49–52 (p. 50).
51. DOCK (1946), p. 680.
52. HELEN CLAPESATTLE, *The Doctors Mayo* (Minneapolis: The University of Minnesota Press, 1941), p. 573.
53. For a description of field laboratory organization see LUCIUS A. FRITZE, "The Laboratory Service of Divisional Laboratories," *Journal of Iowa State Medical Society*, 1919, 9: 378–382.
54. CLAPESATTLE (1941), p. 574.
55. FRANCIS CARTER WOOD, "The Hospital Laboratory," *Bulletin of the American College of Surgeons*, 1917, 3: 20–24, p. 21; see also CLARENCE F. GRAHAM, "The Clinical Laboratory of Albany Hospital," *The Modern Hospital*, 1917, 9: 158–162, p. 162; DOCK (1946), p. 680.

56. O. J. WALKER, "Organizing a Modern Hospital Laboratory," *The Modern Hospital*, 1921, 16: 502–506, p. 504.
57. R. B. H. GRADWOHL, "The Proper Recognition of the Laboratory Technician," *Journal of the American Medical Association*, 1921, 76: 127; see also "The Training and Proper Recognition of the Laboratory Technician," *Journal of Laboratory and Clinical Medicine*, 1920–21, 6: 644–647. R. B. H. Gradwohl (1877–1959) was director of the Gradwohl Laboratories and of the Gradwohl School of Laboratory Technique, St. Louis, Missouri. The first edition of his manual, *Clinical Laboratory Methods and Diagnosis. A Textbook on Laboratory Procedures With Their Interpretation* (1935), devoted only eighty-one pages to blood chemistry and sixty-four pages to urine analysis out of a total of 1,028 pages. The book grew to two volumes and five editions during his lifetime, and an eighth and final edition in 1980.
58. CLAPESATTLE (1941), p. 574; GERMAN (1941), pp. 49–50; WALKER (1921), p. 502.
59. CONNER (1923), p. 873.
60. GERMAN (1941), pp. 49–50; CLAPESATTLE (1941), p 574.
61. F. KRAISSL, SR., "A History of the Chemical Apparatus Industry," *Journal of Chemical Education*, 1933, 10: 519–523, p. 519; WILLIAM H. BROCK, *The Norton History of Chemistry* (New York and London: W. W. Norton & Company, 1993), pp. 192–193.
62. ERNEST CHILD, *The Tools of the Chemist. Their Ancestry and American Evolution* (New York: Reinhold Publishing Corporation, 1940), pp. 192, 194–195, 107; BROCK (1993), p. 193.
63. CHILD (1940), pp. 198–199; Fisher Scientific (company publication).
64. KRAISSL (1933), p. 520.
65. *Idem*. Additional demand for the bottles came from the American petroleum industry's need to determine how much water, sand, and mud was mixed in with the oil so that the refiner could pay for it accordingly. KRAISSL (1933), p. 521.
66. CHILD (1940), pp. 95–97.
67. *Ibidem*, pp. 71, 96–97; KRAISSL (1933), p. 523.
68. KRAISSL (1933), p. 521.
69. KIMBLE GLASS, "Kimble Klippings" (memorial publication for Col. Kimble). Brock (1993, p. 193) mistakenly credits Kraissl with setting up the Kimble Glass Company in New Jersey in 1887. Kraissl's paper could be read with that misinterpretation.
70. CHILD (1940), pp. 96, 111; Corning Glass (company publication).
71. CHILD (1940), pp. 72, 111; KRAISSL (1933), pp. 521, 523.
72. Bausch & Lomb, "A History of Innovation" (company publication).

CHAPTER XI

INTRODUCTION TO PHYSIOLOGICAL CHEMISTRY

As late as the last quarter of the nineteenth century, physiological chemistry had not the status in the United States or in England to warrant a laboratory for instruction or research. Nor was there any practical instruction in physiology in either country. Physiology was frequently taught as part of anatomy by the same instructor, or included with general pathology and the elements of clinical medicine. Instruction in "medical sciences" was not even thought of and physiology in particular had no standing at all outside the medical curriculum [1].

At this date, the role of chemistry in expanding knowledge of the functions of the animal body was not appreciated and consequently, "physiological chemistry" was only a name with uncertain significance. For many years it fell on the border between chemistry and physiology, uncertain as to its proper location. Even in Germany, where physiology and physiological chemistry had been greatly advanced, there was some ambiguity in the position held by physiological chemistry. German publications used the term "medical chemistry" or "animal chemistry." Yet as early as 1826, there was a publication in Leipzig by Friedrich Ludwig Hünefeld of Breslau titled *Physiologische Chemie des Menschlichen Organismus zur Beforderung der Physiologie und Medicin*. Mulder's *Allgemeine Physiologische Chemie* appeared from 1844 to 1851, and in 1844, Marchand of the University of Halle issued his *Lehrbuch der Physiologischen Chemie*. The subject matter was simply treated as a division of chemistry dealing with the identity of the many components of animal tissues and fluids [2].

Notwithstanding the occasional use of the terminology, in most German universities physiological chemistry played only a supporting role to organic chemistry and physiology. One explanation is that most organic chemists interested in the reactions of the human organism were trained as physicians, and in their investigations they stressed physiology rather than chemistry. Another factor is that, with no distinctive journal of its own, papers on physiological chemistry appeared in chemical, physiological, and medical journals and were often lost in these surroundings [3].

Some of the more conspicuous publications were: *Archiv für Physiologie, Zeitschrift für Biologie*, Liebig's *Annalen der Chemie und Pharmacie, Centralblatt für die Medicinische Wissenschaften, Journal für Praktische Chemie,* and *Archiv für die gesammte Physiologie.* In 1873, marking a milestone in the history of physiological chemistry, there appeared the first volume of the *Jahresbericht über die Fortschritte der Thierchemie* in which was collected brief summaries of all the work published in physiological chemistry for the year 1871. It was an even more significant event when Hoppe-Seyler's *Zeitschrift für Physiologische Chemie* appeared in 1877 and established a claim for this branch of chemistry as a distinct discipline [4].

Journal activity was also taking place on the American side of the Atlantic. The *American Chemical Journal,* established in 1879 by Ira Remsen (1846–1927) [5], initiated a series of *Reports on Progress in Physiological Chemistry* in its second volume. For many years it was more important than the *Journal of the American Chemical Society,* which absorbed it in 1912.

Because Germany was so advanced in developing scientific disciplines, its influence was inevitably a major force in the development of the biomedical sciences in other European countries and the United States. In some German laboratories there were more foreign than German students. Except for the Russians, the Americans were probably the main consumers of German science [6].

The ability of German universities to develop new disciplines was bound up with the system of German society, viz., the uniform and high quality secondary school system of *Gymnasia*; the importance of learning or *Wissenschaft* in the German national identity; the firm hold of bureaucratic, paternalistic governments over university budgets and planning; and the structured and elitist character of German society, with its weak entrepreneurial middle class and strong university-trained elites [7].

Physiological chemistry in the United States developed along different lines than in Germany because of distinctively American cultural values and national character. American colleges and medical schools evolved in a society without a national system of high schools, and where higher education was not the goal of its diverse, democratic, business-oriented middle class. In America the federal government was constantly under challenge by states's rights, politics, and dislike of the bureaucratic professional civil service. The distinctive features of American universities, e.g., graduate school, professional school, organization of university research, and professionalization of research were created when German standards were introduced into a utilitarian and egalitarian cultural system,

CHITTENDEN AND THE SHEFFIELD SCIENTIFIC SCHOOL

Russell H. Chittenden (1856–1943) (Fig. 11.1) represents the beginning of the teaching of physiological chemistry in the United States. He had graduated from the Sheffield Scientific School of Yale University in 1875 at the age of nineteen. At the start of his senior year (1874–75) he was placed in charge of what was the first laboratory curriculum of physiological chemistry established in the United States. As he himself has stated: "... the movement was an experiment and the authorities obviously felt it unwise to risk much in a venture that might prove unsuccessful." The new course accommodating eight students was designed primarily for the instruction of students who were planning to study medicine, yet this pioneer course originated not in a medical school, but in a university department of chemistry. The creation of this laboratory evolved from recognition of the principle, not then generally understood, that the study of medicine should rest upon a foundation of the biological sciences [9].

At this date, physiological chemistry was considered simply as a branch of chemistry, dealing with the composition and reactions of substances having physiological significance and of practical value to the physicians. This view persisted for many years. In the medical schools of Germany and other European countries, physiological chemistry was developed extensively by physicians in departments of physiology—not in separate departments. As late as 1896–97, the Harvard University catalog entry for physiological chemistry stated that "instruction in physiological chemistry is given by lectures, recitations and exercises in the laboratory where each student will be taught the chemistry of the carbohydrates, proteids and fats, the chemistry of digestion, the chemistry and microscopy of the urine and the tests for the important poisons." A part-time instructor in physiological chemistry was first provided at Harvard in 1898. In 1905 two full-time instructors—Lawrence J. Henderson and Carl L. Alsberg (1877–1940)—were assigned to the subject [10], and kept alive the instruction of medical students.

Trained in chemistry, Chittenden naturally emphasized the more purely chemical aspects of physiology. In 1874 when he began to teach the subject with laboratory work, the students were given a small text, *A Manual of Chemical Physiology, Including Its Points of Contact with Pathology* (1872) by Johann Ludwig Wilhelm Thudichum (1829–1901), an M.D. graduate

FIG. 11.1. Russell Henry Chittenden. (Reproduced from the collection of the Archives of the National Academy of Sciences, Washington, D.C.)

from the University of Giessen. As a pupil of Liebig, Thudichum's interest turned to physiological chemistry [11]. The manual was one of several books on the same general topic by different authors at the time, but it was the first text on physiological chemistry to be used in the United States. The book has detailed descriptions of experiments using volumetric methods illustrating the chemistry of the tissues and substances of the animal body. Many of the experiments are of the type that continued to appear in laboratory exercises in medical school courses well into the 1960s. Thudichum also founded the short-lived *Annals of Chemical Medicine* (1879, 1881).

Chittenden continued this instruction from Thudichum's manual after his graduation and was appointed instructor in 1877. At that time American medical men saw no physiological importance for physiological chemistry. The training and experience in experimental methods needed for research in physiology were not available in the United States. During the 1870s, Claude Bernard of Paris and Carl Ludwig of Leipzig were the major players in advancing experimental physiology to the level of other experimental sciences. Prior to their work, animal experimentation was virtually ignored and considered as useless. Germany became the magnet toward which American and English students eager to pursue experimental physiology, physiological chemistry or other biological sciences, gravitated [12].

Consequently, in the fall of 1878, three years after his graduation, and wishing to learn the applications of chemistry to physiology, Chittenden was granted a leave of absence to continue study in Germany. There, physiological chemistry was much further advanced than in the United States or England. In Germany the physiologists had come to appreciate the importance of applying chemical techniques to the solution of physiological problems. The subject was regarded as a branch of physiology and was taught by chemically trained personnel in the department of physiology [13].

Initially, Chittenden had made arrangements to work in the laboratory of Hoppe-Seyler in Strassbourg. However, when he saw the crowded facilities he realized that he would get little personal attention and he decided that this laboratory was not suited for him. Although the laboratory was a bee-hive of activity, it seemed to lack the atmosphere essential for the proper development of someone that needed guidance and encouragement. Rather than risk failure because of the unfavorable conditions, he decided to look elsewhere and went to Heidelberg, where he applied to

Wilhelm Friedrich Kühne* (1837–1900). A new set of credentials had to be sent from the United States; but Kühne remembered and had been impressed by an article written by Chittenden—his graduation thesis—originally published in the *American Journal of Science and Arts* and also submitted in German translation to Liebig's *Annalen der Chemie* where Kühne had seen it. Double publication was not uncommon at the time in view of the slow methods of communication. It was the start of a close friendship between Chittenden and Kühne of more than twenty years. Kühne was one of a distinguished group of physiologists who helped to make Germany the center of physiological activity during the last quarter of the nineteenth century [14] (see Chapter 16).

Chittenden returned to New Haven near the end of 1879, and the following spring with Kühne's approval, submitted a thesis based on his work at Heidelberg, and was awarded the Ph.D. degree. It was the first year the degree was granted by Yale. In 1882 Chittenden was appointed professor of physiological chemistry in the Sheffield School, the first such nonmedical academic position in the United States. From this modest beginning in 1874, Chittenden and his first student and colleague, Lafayette Benedict Mendel (1872–1935), pioneered in the training of American graduate students and investigators in physiological chemistry and chemical physiology. These students went on to fill positions in many of the medical schools in the United States, where they trained others. In 1898 Chittenden became Director of the Sheffield Scientific School and was requested by Columbia University to organize a department of physiological chemistry at its College of Physicians and Surgeons.

After Chittenden returned to America, he and Kühne continued a transatlantic collaboration on proteolytic digestion by pepsin and trypsin. They published a series of joint papers on the large number of relatively low molecular weight protein products resulting from digestion by gastric and intestinal enzymes. They noted that the products formed by the action of the two enzymes differed. They named the protein cleavage products peptones and gave them specific names to indicate their original state, i.e., albumoses, globuloses, myosinoses, etc. These products were too indefinite in composition to reveal protein structure, however, their work showed that digestion is a gradual process and that the end product of protein hydrolysis was often an amino acid [15].

---

*Willy Kühne changed his name to Wilhelm Friedrich Kühne. (H. Gutfreund, *Febs Letters*, 1976, 62 (Supplement): E1.)

SOCIETIES, JOURNALS, AND THE DIVERSITY OF BIOCHEMISTRY

The American Physiological Society was organized on December 30, 1887. Chittenden, a charter member, was elected president of the society in 1896 and reelected for eight successive years. The *American Journal of Physiology* was established in 1898. Of the thirty-two papers in the first volume, seven were from the Sheffield Laboratory of Physiological Chemistry, indicating the tendency to consider physiological chemistry as a part of physiology [16]. Eventually, the increasing number of workers in physiological chemistry made it necessary to form a society of their own because, owing to specialization, those working entirely in chemical physiology could not get time at the annual meetings of the physiologists.

The American Society of Biological Chemists was organized in New York on December 26, 1906. As the leading biochemist in the country, Chittenden was selected as the new society's first president. The *Journal of Biological Chemistry* had been founded the year before by Christian A. Herter of New York. In 1913 the American Chemical Society authorized creation of a Division of Biological Chemistry.

The term biochemistry had been in use during the nineteenth century, e.g., the book by V. Kletzinsky, *Compendium der Biochemie* (Vienna, 1858). Other terms in use were "chimie biologique" in France and "physiologische Chemie" in Germany. The name took hold during the first few years of the new century, in the titles of journals in Germany, England, and the United States. The first "biochemical" journal appeared in 1902—*Biochemisches Centralblatt*—an abstract journal. The *Biochemical Journal* (Great Britain) and the *Biochemische Zeitschrift* (Germany) commenced publication in 1906. In France the *Bulletin de la Société de Chimie Biologique* began publication in 1914.

Biochemists were also becoming occupationally more diverse. Most nineteenth-century biochemists worked in an academic setting. Between 1905 and 1915, however, nonacademic institutions began to create new roles for professional biochemists. Hospitals, medical research institutes, industrial research and development laboratories, government regulatory agencies, and expert commissions, all provided new opportunities for biochemists to broaden their professional base. The sudden popularity of the name "biochemistry" and "biological chemistry" applied to journals and societies here and abroad between 1900 and 1910 was not a fad of nomenclature, but a strategy for consolidation of biochemists operating in other contexts in this highly diverse discipline. It heralded a break from the traditional subservience of "physiological chemistry" to the dual constituency of physiology and medicine [17] (see Chapter 12).

The policy of the *Journal of Biological Chemistry* was to publish papers in all the biomedical sciences. When the American Society of Biological Chemists was organized, the goal of the founders was to expand their territory of interests, as well as gain recognition for biochemistry in medical faculties. They sought membership for all who are interested in the biological sciences from the chemical point of view, but have little contact with one another [18].

The variety of names affiliated with biochemistry was indicative of the different contexts in which biochemistry was perceived and practiced. William J. Gies, whom Chittenden had sponsored to head Columbia's department of physiological chemistry at the College of Physicians and Surgeons, described Philip B. Hawk, Walter Jones, A. P. Mathews, and L. B. Mendel as "physiological chemists"; Otto Folin and Philip Shaffer as "biological chemists"; Victor Myers as a "pathological chemist"; and Stanley Benedict as a "chemist" [19].

The opinion has been expressed by some that physiological chemistry referred to a static chemistry that was primarily concerned with the separation and analysis of animal material, whereas biochemistry was dynamic and focused on the chemical changes of cellular metabolism in living systems [20]. Fruton rejects this distinction and states that biochemical problems have not been approached within a defined scientific discipline, but as a complex interplay of chemistry and biology, where molecules were considered to be units of physical motion or as units of chemical reaction. The increasing use of "biochemistry" or "biological chemistry" during the early years of this century may have indicated the wish to emphasize its separation from medical physiology [21].

With the growing recognition near the close of the nineteenth century, of the importance of physiological chemistry in the science-based education of medical students, we have come to the point where clinical chemistry makes its final—and permanent—entrance onto the stage of medical practice. But first, the new analytical technique that made it possible, viz., colorimetry, had to develop and join gravimetry and titrimetry as a method of chemical analysis. Gasometry would soon follow. The manometric procedures of Donald D. Van Slyke and colleagues would determine the blood gases and other blood constituents capable of reactions producing a gas end product for analysis.

ANALYSIS BY COLOR COMPARISON

The need to match and measure the intensity of colored solutions is probably as old as the dyeing industry and wine manufacture. Furthermore, the advantage of using some chemical reagent to alter, develop, or enhance a color of an investigated substance has long been recognized. The earliest recorded report of such a color reaction was by Pliny the Elder about 60 C.E. He used gallnuts to detect the presence of iron in verdigris (blue hydrated basic copper acetate). According to Pliny, this was apparently a very profitable means of adulteration. A strip of papyrus soaked in an extract of gallnuts was then treated with the solution being examined. If iron sulfate was present the papyrus turned black. In later centuries, filter paper, cotton rag, or wood splinters were used. The reaction is used today to detect iron in vinegar [22].

Nineteen centuries after Pliny, color-forming reactions are at the heart of most analytical procedures in clinical chemistry laboratories as well as playing a major role in industry. One of the earliest documented colorimetric methods was that of Wilhelm August Lampadius (1772–1842) who, in 1838, estimated the amount of iron and nickel in a cobalt ore by comparing the colors of a sample solution with those of standard reference solutions in glass cylinders [23].

Another application of colorimetry at this time was made by Anselme Payen (see Chapter 16) for the sugar industry. Caramelization during the refining process of sugar beets imparted a brownish discoloration to the final product. Heating with activated animal charcoal removed the impurities, but the activity of the charcoal preparations varied. Color comparison was utilized to measure the decolorizing activity of the carbon [24].

In 1845 Carl Heine determined the bromine content of mineral waters by comparison with a series of solutions of known amounts of bromine in ether. Augustin Jacquelain (1804–1885) in 1846 devised a method to estimate copper based on the color of the copper–ammonim complexes. In any event, development of colorimetric methodology was slow, and there was little of this color-based chemistry prior to 1852 [25].

In that year, Thornton John Herapath (1830–1858) proposed thiocyanate as a reagent for iron. Nessler's reagent for ammonia appeared in 1856. Whereas many other reagents were found for iron, Nessler's reagent, for all its faults, has had no real competition, and the terms nesslerize and nesslerization have entered the laboratory language of the chemist. In subsequent years, numerous reagents, organic and inorganic, were discovered that gave a unique color for detecting specific organic or inorganic substances. Typical colorimetric reactions used in the late nineteenth century were phenoldisulfonic acid for nitrate (1864); methylene blue for hydrogen sulfide (1883); and picric acid for creatinine (1886) [26].

Most of these colorimetric procedures were carried out by the most direct method, viz., comparison with a series of standards of varying concentration, some of which may be artificial—but permanent—secondary standards, usually solutions of dyes or colored inorganic salts, gels or glass, as substitutes for unstable or expensive standards. The most useful visual comparison method is color matching of unknowns with a single standard in a depth-balancing procedure (see section on Duboscq Colorimeter). Two other techniques, duplication of color, and dilution to color match, are tedious procedures and have been described elsewhere [27].

The first color "measuring device" may have been the simple light-transmitting box designed by Jacques Julien Houton de Labillardière (1755–1834) in 1827 for the determination of indigo in solution. The box held two calibrated tubes, one for the standard and the other for the unknown solution, which was diluted until the colors matched by visual observation. In 1838, F. Collardeau, a scientific instrument manufacturer, produced a twin-tube colorimeter in which a standard solution could be compared with the one to be tested. The depth of the solution viewed in each tube could be varied by pushing a calibrated tube with a glass window in or out of the sample tube (or the standard), as in a telescope, until the two images matched. It was not very convenient to use. However, this device and that of Houton de Labillardière were the usual means of matching colors in the dyeing and sugar industries for thirty years [28].

The "Complementär Colorimeter" was designed in 1853 by Alexander Müller (1828–1906), a professor of agricultural chemistry in Chemnitz. Color balance was determined by reflecting sunlight with an adjustable mirror vertically through a glass disc of a color complementary to the color of the test solution and up through the changing depth of solution to the observer. If the inner viewing tube is raised, the color of the solution would predominate, while if it is lowered, the color of the disc would be seen. At some intermediate point the two would balance and an approximation to white light would be observed at the point of color transition. This is noted on the millimeter scale. A similar observation is made with a solution of known concentration. The amount of colored substance in the unknown is calculated by inverse proportion. However, since unknown and standard could only be viewed separately, there was no control over the constancy of the light source [29].

THE DUBOSCQ COLORIMETER AND ITS EARLY USES

The inherent inaccuracy of separate viewing was overcome by Duboscq's depth-balancing comparator. This instrument, originally designed and manufactured in 1854 by Jules Duboscq (1817–1886), utilized two sets of cylinders. It made possible the simultaneous side-by-side color matching of unknown and standard solutions when viewed through a monocular telescope. Little was heard of it until an improved version was described in 1868 when Charles Mène, an industrial chemist and editor, made a report to the Académie des Sciences and published a full account of it in his own journal [30]. This information and an illustration with a schematic diagram appeared in the *Chemical News* (London) of January 21, 1870 and in *Zeitschrift für Analytische Chemie* in a report of advances in analytical chemistry (Fig. 11.2) [31]. One of the early uses for this instrument was to measure the amount of caramel in syrup.

The first use of the Duboscq colorimeter in biological chemistry may have been in Paris by Jolyet and Laffont (1877) to determine the respiratory capacity of the blood in dogs by means of hemoglobin measurements. However, the procedure required a relatively large sample size and there were problems with standardization. Clinical hematologists preferred existing methods in which the diluted sample was compared visually with a fixed standard and required only the small volumes of blood obtained by capillary sampling. Venipuncture for blood collection was not yet a routine procedure. Other reported uses of the Duboscq colorimeter were in 1890 by Lapique, to measure iron as ferric sulphocyanide, and by Sidersky, to measure sugar by a copper reduction method; and in 1895 by Lévy, for ammonia in the air, using the Nessler method [32]. This instrument made its debut as a basic scientific tool for clinical chemistry in 1904, in Otto Folin's classic paper on the measurement of creatinine and creatine in urine, based on the Jaffe reaction with alkaline picrate solution (see Chapter 12). The report appeared in *Hoppe-Seyler's Zeitschrift für Physiologische Chemie*, a widely read journal [33]. Folin subsequently developed a whole system of urine (and later blood) analysis based primarily on colorimetric methods using the Duboscq colorimeter for urea, ammonia, uric acid, creatinine, and creatine.

Ph. Pellin, successor to Duboscq, modified the colorimeter by fitting longer plungers and adding a detachable cover for the compartment to exclude lateral stray light from the plungers and cups. During World War I it was impossible to obtain the French-made colorimeter or other foreign makes (Krüse, Hellige), and so Kober and Graves designed a combination

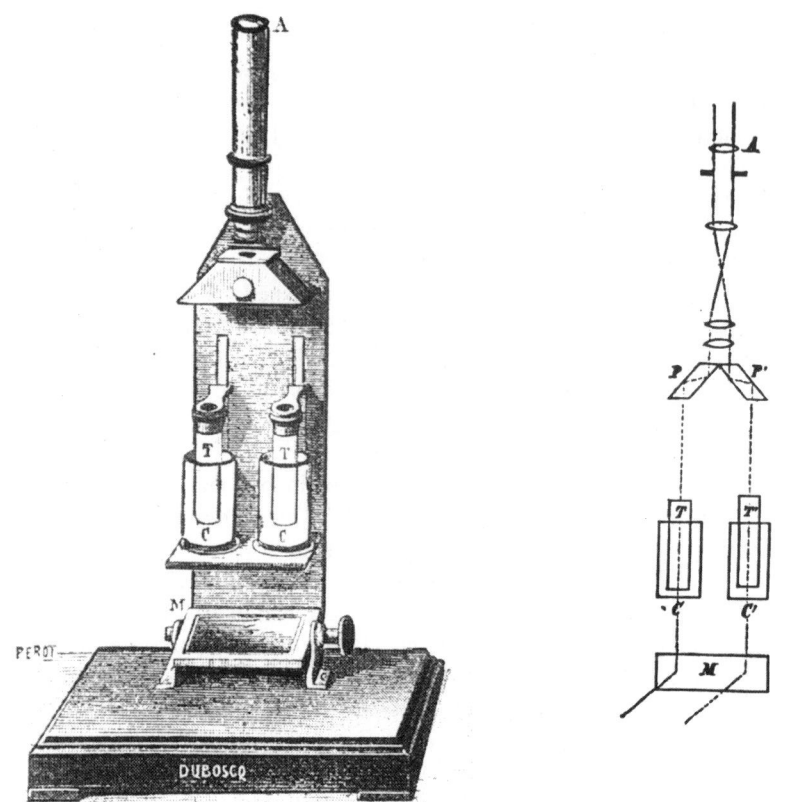

FIG. 11.2. Colorimeter of Duboscq and schematic of light path. (*Chemical News*, London, January 21, 1870, p. 31.) A, telescope viewer; M, mirror; CC', stationary glass cups; TT', movable hollow glass tubes; PP', Fresnel rhombs; rack and pinion and graduated scale with vernier are not shown.

colorimeter-nephelometer (Greek: *nephele*, cloud or mist) similar to the Duboscq model [34]. This new instrument was modified and improved in 1917 [35] and again in 1921 [36] and produced by the Klett Manufacturing Co. (New York, N.Y.) (Fig. 11.3). In 1920 Bausch & Lomb (Rochester, N.Y.) produced a Duboscq-type colorimeter that incorporated significant modifications. Other American companies—Eimer & Amend, Leitz, Spencer—also entered into the manufacture of color comparators. Prices ranged from about $40 to $165 [37]. Patents for modifications and improvements were being issued to various manufacturers up to the mid-1940s [38].

FIG. 11.3. Kober–Klett Nephelometer–Colorimeter. (*Journal of Biological Chemistry*, 1921, 47: 19–25.)

The Duboscq-type colorimeter remained in active use for routine colorimetric analysis in clinical chemistry laboratories until past 1960 when those units that survived the introduction of photoelectric colorimeters were rapidly replaced by the AutoAnalyzer of the Technicon Corporation (Tarrytown, N.Y.).

MODERN COLORIMETER

In the modern Duboscq colorimeter, the calibrated racking device raises and lowers the cups until a satisfactory color match of the standard and unknown solutions is observed. In the original design by Duboscq, the cups were stationary and the "plungers,"—two hollow glass tubes of smaller diameter, closed at the lower end by a disc of glass—were independently raised or lowered by a rack and pinion arrangement.

258 FOUR CENTURIES OF CLINICAL CHEMISTRY

Unfiltered white light from a single uniform source of intensity is reflected from below by means of a mirror and passes upward simultaneously through the clear plate glass bottoms of the two cups, the glass plungers with plane parallel bottom surfaces, double Fresnel rhomboid prisms, and an assembly of lenses (Figs. 11.4, 11.5). In later modified versions the plungers were solid glass. Illumination may be diffused daylight, reflected white artificial lamplight or, as in one of the Klett designs, from a lamp in the base of the instrument.

FIG. 11.4. Duboscq-type visual colorimeter (front view) by Bausch & Lomb, Rochester, N.Y. (*circa* 1950). Note reflecting mirror, light shield, fixed plungers, and glass bottom of sample cup.

FIG. 11.5. Duboscq-type visual colorimeter (rear view) by Bausch & Lomb, Rochester, N.Y. (*circa* 1950). Note location of graduated scale.

The two beams of light undergo double reflection in the Fresnel rhombs. On emerging, they are brought together to the eyepiece of the monocular viewer where the observer sees an illuminated split-circle image of adjacent colored halves. The standard solution cup was usually set midway on the scale, and the position of the test unknown solution cup was adjusted until both halves of the illuminated field were judged to be of equal color density—a subjective evaluation by eyesight. Several readings on the test unknown side were averaged. The attached graduated scale and vernier on each side allow readings to within 0.1 mm. Since there is an inverse proportionality between concentration and the scale reading, i.e., the optical path* (depth of illuminated solution through which the color is viewed), the product of these two values for the standard—at color match—is equal to the corresponding product for the unknown. Since three of the terms are known, the equation is easily solved for the concentration of the unknown. The standard solution is prepared for reading by treating it with the same chromogenic reagents as for the unknown.

Special attachments for the Duboscq and Kober colorimeters allow their conversion into a nephelometer with illumination by horizontally directed light. While seeking to improve nephelometry, Denis obtained reliable results with a turbidimetric technique for protein in urine and cerebrospinal fluid precipitated by sulfosalicylic acid [39]. Denis used the colorimeter as a turbidimeter, and measured light transmission of cloudy or turbid suspensions in other analytical procedures of biochemical interest. The turbidity is assumed to be a function of concentration and analogous to a colored solution. It is primarily the decrease of light energy that has been dispersed by refraction and reflection that is observed visually. However, turbidimetric methods are not very accurate and exhibit a considerable lack of precision, because the size, shape, and concentration of the suspended particles are affected by many more empirical factors than are substances in solution.

THE OPTICAL LAWS OF BEER–LAMBERT–BOUGUER

This balancing method of color comparison is based on the law of light absorption. Long before it was stated as a law, it was obvious from numerous everyday observations that there was a correlation between the color of a solution and the concentration of the colored substance in the solution. The earliest contribution to the absorption law was made by Pierre Bouguer (1698–1758) [40], French mathematician and astronomer and Royal Professor of Hydrography at Le Harve. In his *Essai d'optique sur la gradation de la lumière* (Paris, 1729), Bouguer expresses the relationship between absorption of radiant energy (diminution of light) and thickness of an absorbing medium of uniform transparency, as the light passes through successive thicknesses of that medium. Often called Bouguer's law, it states that each layer of equal thickness absorbs, not equal amounts, but the same fraction of the radiant energy that reaches it

---

*The graduated scale reading gives directly the distance between the inner side of the bottom of the cup and the outer side of the bottom of the tube (plunger).

from the previous layer of equal thickness and passes through it to the next layer. Stated another way, since all of the equal-thickness layers absorb an equal fraction, the intensity of the transmitted light energy in a beam decreases in a geometric progression, i.e., exponentially, as the thickness of the absorbing medium (length of its light path) increases arithmetically. The relationship may also be expressed in logarithmic form as follows: the absorptive capacity varies directly as the logarithm of the thickness of the absorbing medium.

This law was restated in more exact mathematical terms by Johann Heinrich Lambert (1728–1777) [41], a German physicist, mathematician, and astronomer, in *Photometria sive de mensura et gradibus luminis, colorum et umbrae* (Augsburg, 1760). Perhaps because of the rarity of copies of Bouguer's *Essai*, the law has been referred to as Lambert's law.

Nearly a century later (1852), August Beer (1825–1863), a mathematics professor at Bonn, showed that the logarithmic relation of Lambert could be extended to solutions of absorbing substances in nonabsorbing solvents. In this case, the concentration is mathematically analogous to the thickness of a pure absorbing material. According to this generalization, the absorbance, i.e., the reduction in intensity of transmitted light, is directly proportional to the concentration of the solute. Beer predated by only a few months the publication by a Frenchman, Félix Bernard, who, in the same year, arrived independently at a similar mathematical relationship between absorptive capacity and concentration [42].

However, Beer's paper makes no reference to an absorption law, as we know it, in which concentration appears as an explicit variable; nor is there any evidence in the paper indicating that he believed that he discovered a new absorption law. Beer thought primarily in terms of the amount or mass of absorbing material. The paper deals with the absorption of red light by colored solutions of varying dilutions and varying lightpaths, and demonstrates their equivalence. The first reference to Beer's law was by B. Walter in 1889 in the *Annalen der Physik*, in an article dealing not with absorption but with fluorescence. Walter refers to "the well-known Absorption Law of Lambert," and also to "Beer's Absorption Law." He characterized it as proving the equivalence of lightpath and concentration [43]. Strictly speaking, the relationship applies only to a single wavelength (monochromatic) light source. However, this is not a factor in visual comparisons; and with "white light"—real or artificial daylight—Beer's law usually holds within the limits of visual acuity.

In practical terms, the contributions of Bouguer, Lambert, and Beer, often referred to as the Beer–Lambert law or more commonly, but purely arbitrarily, as Beer's law, state that the color density of a solution is directly proportional to the concentration of colored substance, molecules, or ions, and to the depth of solution through which the light passes. For example: when matching two colored solutions of unequal solute concentration in the Duboscq colorimeter, a 1% solution of colored solute viewed through a solution depth of 20 mm would have the same color intensity as a 2% solution viewed through a depth of solution of 10 mm, because there are the same number of colored molecules or ions in the light path in each instance. Most colorimetric analyses were more rapid than gravimetric or volumetric methods, but less accurate. On the other hand, volumetric or gravimetric methods were usually unable to measure the small quantities customarily determined colorimetrically.

DUBOSCQ'S OPTICAL INSTRUMENTS

Generations of clinical chemists and laboratory technicians remember Jules Duboscq for his invention of the visual balancing colorimeter. However, during his lifetime, he was known primarily for other inventions and optical devices, especially for instruments he made to illustrate all the phenomena of light. Duboscq was a member of the firm of Jean-Baptiste-François Soleil (1798–1878), which was noted for its optical instruments. Starting as an apprentice in 1830, Duboscq became Soleil's son-in-law and when Soleil retired in 1849, Duboscq (also known as Duboscq-Soleil) took over the business.

By 1878 the company employed between forty and sixty workers and had developed a world market, including the United States, and exported two-thirds of its production. The company played an important role in the development of optics in France. Leading physicists used its facilities to test their theories and improve their instruments at a time when university laboratories could not afford the latest models. But, by the end of the century, the French scientific instrument makers were finding it increasingly difficult to compete in world markets. One of the main reasons was their insistence on producing instruments which were works of art, and consequently, expensive, at a time when the Germans and the British were offering cheaper and more robust products better adapted to a mass market [44].

Duboscq was very successful in making and marketing the stereoscope. This was an instrument presenting two slightly dissimilar pictures that could be recombined by the viewer's eyes and mind into a single three dimensional illusion. It was introduced in 1838 by the English physicist Charles Wheatstone (1802–1875) (see also Chapter 14) to illustrate his theory of binocular vision, and required slightly different pairs of pictures that had to be drawn by hand. Outline figures, or even shaped perspective drawings of simple objects, presented little difficulty. It was impossible, however, for an artist to draw the two projections necessary to form the stereoscopic view of objects as they exist in nature [45].

Then, less than a year later in 1839, the photographic process invented by Louis Jacques Daguerre (1789–1851) was announced. His pictures (daguerreotypes) together with a smaller more convenient version of the stereoscope, perfected by David Brewster (1781–1868) in 1849, made the stereoscope a phenomenal success. Now, a twin-lens camera, simulating human binocular vision, could produce the necessary pairs of pictures. Attracting no interest in England, Brewster took his invention to Paris in 1850, and to Soleil and Duboscq. Sensing the commercial possibilities, Duboscq quickly began production of the Brewster stereoscope. Duboscq also began producing binocular daguerreotypes of people, statues, flowers, and other objects of natural history. A few of his daguerreotypes have survived including the enlargement made from one of a stereoscopic pair of self-portraits taken around 1850 (Fig. 11.6) [46].

Duboscq exhibited the stereoscope in the "Physical Optics" section at the Great Exhibition of 1851 in London, as well as other instruments for which he received a Council medal—the highest award. One of the other instruments exhibited was Soleil's improved model of Biot's saccharimeter, which enabled white light to be used in polarimetry as the source of illumination. As an optician, Duboscq specialized in perfecting, simplifying, and popularizing optical instruments. In collaboration with the physicist Léon Foucault (1819–1868), he produced an improved electromagnetic regulator for the electric arc lamp to the point where it could be successfully introduced as a projector light source. This eventually permitted electric lighting in the theater [47].

Duboscq also described an overhead projector in 1876. Following the invention of photography, Duboscq applied his expertise in optics to improving various photographic processes and devising new ones. Duboscq's 1864 catalog lists the phenakistoscope with transparent photographs (instead of drawings) representing the sequential positions of a moving subject for projection, thus creating the illusion of moving pictures. This device was invented in the 1830s by the Belgian scientist Joseph Antoine Plateau (1801–1883) to demonstrate the phenomenon of persistence of vision, which he had discovered [48].

FIG. 11.6. Jules Duboscq. (Courtesy George Eastman House, Rochester, N.Y.)

NOTES AND REFERENCES

1. RUSSELL H. CHITTENDEN, *The Development of Physiological Chemistry in the United States* (New York: The Chemical Catalog Company, Inc., 1930), pp. 15–16.
2. *Ibidem*, pp. 17–19.
3. CHITTENDEN (1930), p. 25. For discipline building and the interaction of physiology, organic chemistry, and physiological chemistry in Germany and England from the mid-nineteenth to the early twentieth centuries, see ROBERT E. KOHLER, *From Medical Chemistry to Biochemistry. The Making of a Biomedical Discipline* (New York: Cambridge University Press, 1982), pp. 1–92 ff.

4. CHITTENDEN (1930), pp. 25–26.
5. REMSEN, an organic chemist who had trained with Wöhler at Göttingen, was President of Johns Hopkins University from 1901 to 1912.
6. KOHLER (1982), p. 93.
7. *Ibidem*, p. 94.
8. *Ibidem*, pp. 94–95; see also JOSEPH BEN-DAVID, *The Scientist's Role in Society. A Comparative Study* (Englewood Cliffs, New Jersey: Prentice-Hall, Inc., 1971), pp. 139–168.
9. CHITTENDEN (1930), pp. 33–35.
10. *Ibidem*, pp. 35–36; WILIIAM C. ROSE, "Recollections of Personalities Involved in the Early History of American Biochemistry," *Journal of Chemical Education*, 1969, 46: 759–763, p. 759.
11. HENRY MCILWAIN, "Thudichum and the Medical Chemistry of the 1860s to 1880s," *Proceedings of the Royal Society of Medicine*, 1958, 51: 127–132. Thudichum was born in Germany and migrated to London in 1853. In 1865, he offered his services gratis to St. Thomas's Hospital and was appointed the first lecturer on pathological chemistry and director of the newly formed Laboratory for Chemistry and Pathology (1865–71), becoming thereby perhaps the first pathological chemist on record. When he left, he was not replaced and the department, such as it may have been, was closed. Thudichum's first book was *A Treatise on the Pathology of the Urine, Including a Complete Guide to its Analysis* (1858). His major work was *A Treatise on The Chemical Constitution of The Brain* (1884); facsimile edition by Archon Books, Hamden, Connecticut, 1962. He isolated cephalin, sphingomyelin, sphingosin, phrenosin, and kerasin, and foresaw that various forms of insanity may have a chemical basis. Thudichum suggested that insanity may consist of "external manifestations of the effects upon the brain-substance of poisons fermented within the body, just as the mental aberrations accompanying chronic alcoholic intoxication are the accumulated effects of a relatively simple poison fermented out of the body" (p. xiii). He believed it was possible that chemistry will aid in precise treatment of many derangements of brain and mind. Thudichum also wrote books on... *Varieties of Wine*.... (1894), and the *Spirit of Cookery*.... (1895). See also KENDAL DIXON (pp. 60–124) in *The Chemistry of Life. Eight Lectures on the History of Biochemistry*, Joseph Needham, ed. (Cambridge: University Press, 1970), p. 91.
12. CHITTENDEN (1930), pp. 20–23.
13. *Ibidem*, pp. 27–28; ROSE (1969), p. 760.
14. CHITTENDEN (1930), pp. 28–32.
15. HENRY M. LEICESTER, *Development of Biochemical Concepts From Ancient to Modern Times* (Cambridge, Massachusetts: Harvard University Press, 1974), pp. 184–185.
16. CHITTENDEN (1930), pp. 44–46.
17. KOHLER (1982), pp. 196–198; "The Enzyme Theory and the Origin of Biochemistry," *Isis*, 1973, 64: 181–196, pp. 182–183.
18. RUSSELL H. CHITTENDEN, *The First Twenty-five Years of the American Society of Biological Chemists* (Baltimore, Maryland: Waverly Press, 1945), p. 3.
19. KOHLER (1982), pp. 366–367 (note 11).
20. KOHLER (1973), pp. 183–184.
21. JOSEPH S. FRUTON, "The Emergence of Biochemistry," *Science*, 1976, 192: 327–334, p. 332 (note 8).
22. M. NIERENSTEIN, "The Early History of the First Chemical Reagent," *Isis*, 1931, 16: 439–446; "The First Chemical Reagent," *Analyst*, 1943, 68: 212–213; FERENC

SZABADVÁRY, *History of Analytical Chemistry*, translated from Hungarian by Gyula Svehla (Langhorne, Pennsylvania: Gordon and Breach Science Publishers, 1992), pp. 7–8. Gallnut (nutgall) is a nutlike swelling produced on oak trees by insect eggs and larvae embedded in the plant tissues. It is a source of gallic and tannic acids; also called *Aleppo gall* and *Smyrna gall*.

23. *A History of Analytical Chemistry*, Herbert A. Laitinen and Galen W. Ewing, eds. (American Chemical Society, 1977), p. 134; SZABADVÁRY (1992), p. 337.
24. M. G. RINSLER, "Spectroscopy, Colorimetry, and Biological Chemistry in the Nineteenth Century," *Journal of Clinical Pathology*, 1981, 34: 287–291, p. 289.
25. LAITINEN and EWING (1977), p. 134; SZABADVÁRY (1992), pp. 337–338.
26. M. G. MELLON, "A Century of Colorimetry," *Analytical Chemistry*, 1952, 24: 924–931, pp. 925, 924; SZABADVÁRY (1992), p. 338.
27. The four basic methods of color comparison are described by PHILIP B. HAWK, BERNARD L. OSER, and WILLIAM H. SUMMERSON, *Practical Physiological Chemistry*, 13th ed. (New York: McGraw-Hill Book Company, Inc., 1954), pp. 503–509. For illustrations of various twentieth-century color comparators, see also W. B. FORTUNE, "Color Comparimeters," in *Analytical Absorption Spectroscopy. Absorptimetry and Colorimetry*, M. G. Mellon, ed. (New York: John Wiley & Sons, Inc., 1950), pp. 137 ff.
28. RINSLER (1981), p. 289.
29. ALEX MÜLLER, "Neues Colorimeter," *Journal für Praktische Chemie*, 1853, 60: 474–476; SZABADVÁRY (1992), pp. 338–340; see also LAITINEN and EWING (1977), pp. 135–136.
30. CH. MÈNE, "Le décolorimètre. Nouvel instrument inventé par M. Duboscq (opticien rue de l'Odéon 21), pour l' analyse des noirs de sucreries," *Revue Hebdomadaire de Chimie Scientifique et Industrielle*, 1868, 1: 99–101; J. DUBOSCQ and CH. MÈNE, "Nouveau colorimètre pour l'analyse des matieres tinctoriales au point de vue commercial," *Comptes Rendus Hebdomadaires des Séances de l'Académie des Sciences*, 1868, 67: 1330–1331; see also RINSLER (1981), p. 289.
31. PROFESSOR MORTON, "Duboscq's New Colorimeter," *Chemical News* (London), 1870 (January 21), 21: 31–32; see also SZABADVÁRY (1992), pp. 339–340; W. CASSELMANN, "Allgemeine analytische Methoden, analytische Operationen, Apparate und Reagentien," *Zeitschrift für Analytische Chemie*, 1870, 9: 471–481, pp. 473–475.
32. RINSLER (1981), p. 290.
33. OTTO FOLIN, "Beitrag zur Chemie des Kreatinins und Kreatins im Harne," *Hoppe-Seyler's Zeitschrift für Physiologische Chemie*, 1904, 41: 223–242, p. 224; "Some Metabolism Studies. With Special Reference to Mental Disorders," *American Journal of Insanity*, 1904, 60: 699–732. The Duboscq instrument was obtained from Eimer and Amend of New York "on approval." (p. 710).
34. PHILIP ADOLPH KOBER and SARA STOWELL GRAVES, "Nephelometry (Photometric Analysis). I—History of Method and Development of Instruments," *Journal of Industrial and Engineering Chemistry*, 1915, 7: 843–847.
35. PHILIP ADOLPH KOBER, "An Improved Nephelometer-Colorimeter," *Journal of Biological Chemistry*, 1917, 29: 155–168.
36. PHILIP ADOLPH KOBER and ROBERT E. KLETT, "Further Improvements in the Nephelometer-Colorimeter," *Journal of Biological Chemistry*, 1921, 47: 19–25.
37. VICTOR CARYL MYERS, *Practical Chemical Analysis of Blood. A Book Designed as a Brief Survey of This Subject for Physicians and Laboratory Workers*, 2nd ed. (St. Louis: C. V. Mosby Company, 1924), p. 209.

38. FOSTER DEE SNELL and CORNELIA T. SNELL, *Colorimetric Methods of Analysis, Including Some Turbidimetric and Nephelometric Methods*, 3rd ed., vol. 1 (*Theory-Instruments-pH*) (New York: D. Van Nostrand Company, Inc., 1948), pp. 44–52, footnotes on pp. 46, 48–51. This volume contains many illustrations of visual and photoelectric instruments for color comparison, nephelometry, etc., by American and European manufacturers.
39. OTTO FOLIN and W. DENIS, "The Quantitative Determination of Albumin in Urine," *Journal of Biological Chemistry*, 1914, 18: 273–276; W. DENIS and JAMES B. AYER, "A Method for the Quantitative Determination of Protein in Cerebrospinal Fluid," *Archives of Internal Medicine*, 1920, 26: 436–442.
40. *Dictionary of Scientific Biography (DSB)*, Charles Coulston Gillispie, ed. (New York: Charles Scribner's Sons, 1970), 2: 343–344.
41. *DSB* (1973), 7: 595–600.
42. AUGUST BEER, "Bestimmung der Absorption des rothen Lichts in farbigen Flüssigkeiten," *Annalen der Physik und Chemie* (Poggendorf), 1852 [2], 86: 78–88; FÉLIX BERNARD, "Thèse Sur L'Absorption De La Lumière Par Les Milieux Non Cristallisés," *Annales de Chimie et de Physique*, 1852 [3], 35: 385–438. See also DOROTHY R. MALININ and JOHN H. YOE, "Development of the Laws of Colorimetry. A historical sketch," *Journal of Chemical Education*, 1961, 38: 129–131.
43. HEINZ G. PFEIFFER and HERMAN A. LIEBHAFSKY, "The Origins of Beer's Law," *Journal of Chemical Education*, 1951, 28: 123–125; GABOR B. LEVY, "A Literature Search," *American Laboratory*, 1992, 10: 10. The first use of "Beer's Law" or "Beer's Absorption Law" may have been in two successive articles by B. WALTER, "Die Aenderungen des Fluorescenzvermögens mit der Concentration," *Annalen der Physik*, 1889, 36: 502–518, pp. 505, 512; "Ueber den Nachweis des Zerfalles von Moleculargruppen in Losungen durch Fluorescenz- und Absorptionserscheinungen," 518–532, pp. 520, 521.
44. PAOLO BRENNI, "The Soleil–Duboscq–Pellin Dynasty," in "Report on the 11th International Scientific Instrument Symposium, Bologna, September, 1991," *Bulletin of the Scientific Instrument Company*, 1991, No. 31, p. 16.
45. JANET E. BUERGER, *French Daguerreotypes* (Chicago and London: The University of Chicago Press, 1989), pp. 100–109, pp. 100–101.
46. "Stéréoscope de M. Brewster, exécuté par M. Duboscq," *Comptes Rendus Hebdomadaires des Séances de l'Académie des Sciences*, 1850, 31: 895–896; BUERGER (1989), p. 101.
47. JULES DUBOSCQ, "Note sur un régulateur électrique," *Comptes Rendus Hebdomadaires des Séances de l'Académie des Sciences*, 1850, 31: 807–809; BUERGER (1989), pp. 102, 103, 185 (notes 12, 21).
48. JOHN T. STOCK, "The Duboscq Colorimeter and its Inventor," *Journal of Chemical Education*, 1994, 71: 967–970; BUERGER (1989), p. 103.

CHAPTER XII

INTRODUCTION

With the start of the twentieth century, clinical chemistry emerged into its own space on the mosaic of medical practice. The pattern of its future growth and development took shape during the first two decades of the new century, with the United States leading the way in the decisive breakthrough. Until then, the United States had played no role in the growth or development of clinical chemistry. Afterward, the nation quickly achieved leadership which it never relinquished.

Three names dominated this period, as their papers filled the pages of the *Journal of Biological Chemistry* and other publications—Otto Knut Folin (1867–1934), Stanley Rossiter Benedict (1884–1936), and Donald Dexter Van Slyke (1883–1971). Their systematic explorations on blood and urine set the style and shaped the parameters for clinical chemistry for the remainder of the century as they developed practical and clinically applicable methods of analysis. Based on a new approach to methodology—analysis of small volumes of biological fluids—they determined normal ranges, correlated variations with pathological conditions, and elucidated metabolic pathways in health and disease. None of the three held medical degrees, yet their research and teaching of biochemistry and clinical chemistry demonstrated that chemists could make great contributions to advances in medical diagnosis and the treatment of disease.

THE FLEXNER REPORT AND MEDICAL SCHOOL REFORM

Folin, Benedict, and Van Slyke received a valuable assist in the development of clinical chemistry from an unlikely and unexpected source—the "Flexner Report." Abraham Flexner (1866–1959) (Fig. 12.1), a former Louisville, Kentucky high school teacher and educator, had published his first book, *The American College* (1908), in which he severely criticized some of the current educational practices. The book drew the attention of Dr. Henry S. Pritchett, president of the recently established Carnegie Foundation for the Advancement of Teaching, and he asked Flexner to make a study of American medical schools. Flexner asked whether he was

FIG. 12.1. Abraham Flexner (1953). (National Library of Medicine, Bethesda, Maryland.)

being mistaken for his brother Simon, director of the recently formed Rockefeller Institute for Medical Research. Pritchett replied that he wanted an educator's evaluation, not that of a medical practitioner. Published in 1910, the report exposed the disgraceful practices of the American (and Canadian) medical school systems and made specific recommendations for correcting the deficiencies. The report had a far-reaching effect on the practice of science in the laboratories of medical schools and hospitals and on the research and teaching of biochemistry. Flexner specifically referred to "clinical chemist" and "clinical chemistry." Concerning the laboratories connected with the university hospital, he wrote [1]:

> "To suffice for clinical investigation the laboratory staff must be so extended as to place, at the immediate service of the clinician, the experimental pathologist, experimental physiologist, and clinical chemist in position to bring all the resources of their several departments to bear on the solution of concrete clinical problems. Of these branches, experimental pathology and physiology have already won recognition; the next step in progress seems to lie in the field of clinical chemistry, thus far quite undeveloped in America."

His emphasis on the use of laboratory sciences in the training of medical students and in the teaching of specialties, contributed to the favorable environment for the rapid growth of clinical chemistry. Flexner's views complemented and reinforced what Folin had said two years earlier to the Harvey Society. Folin reminded the medical profession of the large variety of clinical material available for biochemical investigations in the large city hospitals. If these hospitals are to become centers for biochemical research, as they should, they must provide laboratory facilities, personnel, and independence to the trained chemist to work on the many biochemical problems seeking answers [2].

During the latter half of the nineteenth century, education in the natural sciences was slowly evolving in colleges that were mostly devoted to instruction in the classics, philosophy, and theology. University professors had few experimental facilities and little time for scientific research. Medical schools and colleges of technology carried out very little research because their faculties were committed to private practice for their livelihood. There was little time for instruction and investigation in poorly equipped laboratories. Not until 1876 was the first graduate school of significance founded as The Johns Hopkins University. American students, returning from study abroad, were motivated by the spectacular achievements of European scientists and were eager to continue their research.

What awaited them was the revelation by Abraham Flexner of the shocking weakness of most American medical colleges [3].

The celebrated Flexner Report of 1910 revealed the serious deficiencies in American medical education. Many of its recommendations had already been addressed a few years earlier in surveys made by the Association of American Medical Colleges and the American Medical Association (AMA), and corrective actions were being taken by some medical schools. Had the rate of progress been sufficiently rapid there would hardly have been a need for Flexner's special study.

In 1910 most medical schools in America were proprietary—privately owned by an individual or group, and operated for profit. The relationship of medical colleges to universities extended from the use of the name only, through affiliation, to full university control. However, not a few "university" medical departments were no more than fictitious alignments [4].

Flexner found that during the previous twenty-five years there had been "an enormous overproduction of uneducated and ill-trained medical practitioners" in the United States. The result was four or five times as many physicians in proportion to the population as in Germany. This overproduction was due mainly to the existence of very many commercial schools, sustained in many cases by advertising, which drew a mass of unprepared youth out of industrial occupations into the study of medicine. Based on the medical needs of the general population, Flexner proposed that the number of medical schools be drastically reduced. Within a decade of the report, the number of medical schools dropped from a peak of 162 in 1906, to about 80 in the 1920s [5].

Of the 155 medical schools examined (five were in Canada), 120 (77.4%) either had no endowments or received no public funds from which to meet operating expenses. Tuition fees alone paid salaries to faculty and other personnel, operated plant facilities, maintained and constructed buildings, supported library and museum, and provided laboratory support for classwork. In the proprietary schools most of the income was paid to the faculty. Other needs were neglected to the detriment of the students. Even then, the faculty was merely part time, with its major source of income from private practice. More than a third of the schools had such limited resources that they did not wash, sweep, clean, or disinfect. Laboratory sciences were not "hands on" experiences, but were didactic lectures and demonstrations or group activities, textbook study or even nonexistent. There was much ungraded teaching. Contact with patients was very limited. The final product was an oversupply of poorly trained graduates, who were a menace to society. Payment of fees virtually guaranteed a medical diploma, sometimes after only one year. The Flexner Report urged that universities take over control of medical training, so that laboratories could be provided, and be properly equipped and staffed for the first two years of the curriculum [6].

The reform movement that transformed the old-time proprietary medical college into the university medical school was driven by the new requirement of one or two years of college for admission to medical school. In 1898 almost no medical college required even a high school diploma. Only Johns Hopkins (1893) required a bachelor's degree, and in 1901 Harvard made this an admission requirement. But, both schools did admit students with three years of college. By 1915 most medical schools required two years of college, including courses in physics, biology, and chemistry. The resulting reduction in the number of qualified students created a financial crisis for proprietary medical colleges that depended on student fees. It forced them into an affiliation with the universities whose social and community connections could raise the large sums necessary to expand departments, build laboratories, and support teaching hospitals. These colleges also had to adjust to new academic values, professional goals, research standards, and university criteria for hiring and promotion [7].

The cycle of higher academic standards, reduced enrollment, financial crisis, and closer financial relationship with universities did not end until a new system of medical education had emerged. In 1910 15.3% of new medical school graduates had a B.A. degree; in 1920 it was 43%; by 1930 it was 70%. The ultimate aim of the medical reform movement was to improve clinical practice by producing qualified and competent physicians. In the process, it was the biomedical sciences that benefited most from reform. As the least developed area of medicine in the old proprietary schools, they were the obvious target for reform [8].

Between 1900 and 1910, pressure for higher admission standards came from medical colleges and universities, and not from state licensing boards or the AMA. The AMA's Council on Medical Education did not aggressively pursue higher entrance requirements. In 1905 it set a high school education "or its equivalent" as the preliminary requirement. The ideal program would be a five-year medical course, the first year of which should be devoted to chemistry, physics, and biology, and could be taken in a school of liberal arts or in the medical school. A college education was recognized as desirable for some but "it is not and never will be desirable to make such college education a requirement to the study of medicine," as it would delay graduation to age 27 or 28 years, "too old a period at which the young medical man should begin his life's work" [9].

In 1910 medical licensing boards in only seven states required one or more years in a liberal arts college, and only twenty-three states required a four-year high school education. The others had minimal or no specific requirements. Also in 1910, the AMA enlarged its minimum requirement for admission to medical school to include at least one year's college work each in physics, chemistry, and biology, and a reading knowledge of at least one modern language, preferably German or French, "as soon as conditions warrant" [10].

MEDICAL CHEMISTRY, PHYSIOLOGICAL CHEMISTRY, AND BIOLOGICAL CHEMISTRY

Prior to the revision of college course requirements for admission to medical school, medical chemistry was taught by chemists or physicians, most of whom had no interest in physiological chemistry as such. They viewed medical chemistry as an applied science analogous to agricultural or pharmaceutical chemistry. Most courses in medical chemistry consisted almost entirely of elementary inorganic chemistry, with a bit of organic and physiological chemistry thrown in toward the end, along with urinalysis and toxicology. The physician-chemists who taught these courses were out of place in the new reform-minded academic medical schools [11].

As long as medical colleges had to provide students with general chemistry, teaching opportunities for biochemists were limited and not very appealing as a career. When medical schools, in an effort to reform medical education, made general chemistry an admission requirement, this course quickly disappeared from the curriculum and was replaced by higher level courses in biological chemistry. Although some medical chemists saw the changes as an opportunity to improve their own academic status, most of them did not inherit the new departments of biological chemistry. Most medical chemists did not possess the appropriate education or experience to qualify as teachers and researchers in biological chemistry or other basic sciences. The chemists with medical training switched to other clinical specialties or accepted lesser service roles in the new departments or were reassigned to the veterinary or dental school. Some went to less prestigious medical schools, where clinicians continued to favor the older style of medical chemistry—a pattern which persisted as late as 1950 in some medical schools, where the department of chemistry was staffed almost exclusively with medical doctors [12].

Within a brief ten-year period, shortly after the start of the twentieth century, departments of medical chemistry in leading medical schools had been reorganized as departments of biological chemistry, staffed by Ph.D.–biochemists. Chittenden commented in 1908 that only a few years earlier, "it was a rarity to find a laboratory of physiological chemistry attached to a university. Now, such laboratories are to be seen on all sides...." [13]. It was in this brief period of rapid change that the American pattern diverged from the European pattern of dependence on organic chemistry or physiology. Biochemistry in the United States did not evolve gradually out of physiology as in Britain, nor was it shared between organic chemistry and physiology, as in Germany. In the United States biological chemistry emerged from medical chemistry in the early 1900s. Ironically, it was the backwardness of the biomedical sciences in the American medical schools that made rapid success possible. Progress was helped along by medical school reform. In addition, a recurrent theme in American universities was the role played by death or resignation in providing the opportunity for structural reorganization, followed then by the establishment of physiological chemistry in the medical school [14].

The medical schools in the United States could not develop biochemists because they not only lacked biochemistry departments, but also lacked graduate schools to train teachers of medical students. In many places, nonphysician doctoral scientists in biochemistry trained by Folin, Van Slyke, Benedict, and Chittenden, became the professors and chairs of biochemistry departments in American medical schools in the interval between the two World Wars.

Some medical schools made a departmental name change from physiological to biological chemistry. Physiological chemistry emphasized too much the aspect of nutrition and medicine. For many biochemists, the term "physiological chemistry" was a disagreeable reminder of their former subordinate role to physiology and medicine. Adoption of the name biological chemistry or biochemistry symbolized a separate identity.

At Harvard, where nearly all the admitted students had a B.A. degree, innovation was less complicated. Politically, however, the establishment of a department of biological chemistry in the reorganized medical school was complicated by professional rivalry between the departments of physiology and chemistry. The result was a division of chemistry into physiological chemistry in the first year and clinical chemistry in the second. General chemistry was eliminated. The plan called for appointment of a professor of biological chemistry to develop the new freshman course. It led to the appointment in 1907 of Otto Folin as associate professor of biological chemistry [15].

OTTO FOLIN

Otto Folin was 15 years old when, at his mother's urging, he left home in Sweden in 1882 to join his older brother in Minnesota. What followed is the familiar American success story. Immigrant boy arrives penniless and in debt, works as unskilled laborer, learns English and the new customs, acquires an education, and becomes a professor at Harvard [16].

Folin graduated from the University of Minnesota and entered the University of Chicago for graduate study in 1892, the year it opened. At Chicago he completed his doctoral work on urethanes under Julius Stieglitz (1867–1937) in 1896. Then, acting on the advice of the eminent physiologist, Jacques Loeb (1859–1924), Folin decided to take additional training in Europe in the newly emerging field of physiological chemistry, which hardly existed in the United States. Six months of the first year were spent in the laboratory of Olof Hammarsten (1841–1932) at the University of Uppsala, not far from Folin's childhood home. Here he examined the properties and composition of a hydrolysis product of a glycoprotein, mucin, from submaxillary glands. A short paper on this subject was published in *Hoppe-Seyler's Zeitschrift für Physiologische Chemie* in 1897. It was Folin's first contribution to biochemistry (Fig. 12.2).

During the summer of 1897 he worked in the Berlin laboratory of Ernst Leopold Salkowski (1844–1923). There, his first contact with hospital patients led to an interest in the urinary end products of nitrogen metabolism. He improved a quantitative method for urinary uric acid and published it in 1897 as sole author. The analysis of uric acid remained a lifelong interest. Six months of the next year (1897–98) were spent with Albrecht Kossel (1853–1927) in Marburg, where he applied his knowledge of organic chemistry to biological problems and where his interests in the intermediary stages of protein metabolism had their beginning. Two more papers appeared in *Hoppe-Seyler's Zeitschrift*. It was while in Germany, possibly at Marburg, that he discovered the new technique of colorimetry used in the brewing industry and the color comparator invented by Jules Duboscq. This instrument was to become the basis of his major contributions to developing simple, reliable, and convenient colorimetric methods for clinical chemistry. On his return to Chicago in mid-1898, he was awarded the Ph.D.

There were no academic positions available in physiological chemistry in the U.S.A. In the few universities and medical schools where this subject was being taught, it was assigned to instructors in physiology, pharmacology, or medical chemistry. Only Yale had a department of physiological chemistry, established in 1882 by Chittenden. Consequently, Folin accepted a job as

FIG. 12.2. Otto Folin at Uppsala (1897). (Courtesy of S. Meites.)

chemist in a commercial laboratory in Chicago, specializing in analysis of water, food, and medical products. A teaching opportunity came in the summer of 1899. He accepted a position as assistant professor of chemistry at West Virginia University, where he offered courses in quantitative analysis and elementary physiological chemistry.

METABOLIC STUDIES AT MCLEAN HOSPITAL

In 1900 Folin received an offer from the McLean Hospital in Waverley, Massachusetts, a suburb of Boston. Edward Cowles (1837–1919), the medical superintendent of this private psychiatric hospital, had established a laboratory for physiological chemistry in 1889—one of the first of its kind in the United States to support research. His objective was to advance the understanding of mental diseases by searching for a connection between abnormal mental states and urinary excretion, especially of urea and uric acid. Cowles believed, as did others, in a correlation between insanity and chemical toxins produced by faulty metabolism and poor nutrition. He expected to find evidence of this in the patient's urine. Research conducted by resident physicians was begun in 1891–92. Blood changes in hemoglobin, red and white blood cell counts, differential count of white cells, and specific gravity were also studied. When larger laboratory facilities were built in 1895, Cowles planned a special research department to be run by a professional biochemist. Folin was asked to plan, equip, and develop his own program of research toward achieving Cowles's objective, i.e., uncovering an association between mental status and urinary excretion [17].

Nineteenth-century clinical chemistry involved chiefly the examination of the urine. This was understandable. Its collection offered no technical difficulties or risks, and the quantities of fluid available allowed utilization of the gross methods of gravimetric and volumetric analysis already in use and requiring large volumes of specimen. The objective at that time was to isolate the particular substance in pure form, then weigh it or titrate it.

Finding no evidence of toxicity in urine of insane patients, as was claimed by some French writers, and no qualitative differences, Folin turned to study the protein metabolism of normal versus mentally disturbed individuals by measuring as accurately and completely as possible all of the known nitrogenous and other products excreted in the urine of patients fed a standard diet. He would thereby establish the normal range of the nitrogenous fractions and then consider whether differences were due to an abnormal metabolism in mental disease. Normal patterns were then unknown. To establish norms would by itself be an important undertaking; but first, he had to devise additional and improved quantitative methods before any survey could be started. This was to lead to his lifelong interest in quantitative methods for nitrogenous end products in urine.

When Folin began his detailed studies of nitrogen metabolites in urine, there was no commercial source of purified chemicals, water, standards, calibrated glassware, or instruments designed for use in the clinical chemistry laboratory. Procedures for testing of urine—mostly qualitative, some quantitative—filled major portions of books on clinical diagnosis by laboratory methods. Practical quantitative chemical analysis of blood was virtually nonexistent or was described only briefly. The development of blood chemistry was hampered by a shortage of blood for experimental and diagnostic purposes, as a result of the gradual abandonment of bloodletting as a therapy late in the nineteenth century. Large volumes of blood were required for chemical analysis and there was no well-developed or convenient technique for drawing the large amounts needed. Furthermore, the plasma proteins (and red cells) interfered markedly with the methods; consequently, blood was rarely tested. Hematological procedures, on the other hand, viz., blood counts, hemoglobin, and white cell differential, were readily supplied by finger stick.

The first years at McLean were spent mainly in devising and testing methods for the determination of nitrogenous constituents in urine, most of which were known qualitatively, e.g., urea, ammonia, uric acid, creatinine, and creatine. Previously, quantitative methods for these constituents were frequently laborious or complicated and, as in the case of urea, were nonspecific; or in the case of uric acid, they required relatively large amounts of specimen.

Folin's first colorimetric method was developed for urine creatinine in a reaction with picrate ion in alkaline medium at room temperature to form a red color. Color comparison was made with an artificial standard—N/2 potassium bichromate—after establishing a correlation with pure creatinine. Although other color reactions had been used long before this to estimate biological products, e.g., Nessler's reagent for ammonia in water analysis, Folin's use of the Duboscq colorimeter for color comparison in the quantitative analysis of creatinine in urine in 1904 ushered in the modern era of clinical chemistry [18].

Creatinine was discovered in urine by Pettenkofer in 1844 (see Chapter 6 note 44 and Chapter 13). It was the name given by Liebig in 1847 to the substance he obtained by heating creatine with mineral acids, and was synthesized by Horbaczewski in 1855. The color reaction, discovered by

Max Jaffe* (1841–1911) in 1886 [19], is the longest continuously in use colorimetric procedure for blood or urine analysis; and until 1936, when the reaction with 3,5-dinitrobenzoic acid was described [20], it was the only method for creatinine.

Until Folin's work, there was no reliable information regarding creatinine, second only to urea in quantity in urine. In 1905 his classic papers reported on the composition of urine in human subjects maintained on diets of varying nitrogen content. Folin helped clarify the significance of the nitrogenous waste products eliminated by the kidneys. He explained the changes in the nitrogen composition of the urine in terms of two kinds of protein metabolism—a variable exogenous metabolism and a constant endogenous metabolism. This was shown in the contrast between excretion of urea and creatinine. In the exogenous metabolism, the daily output of urea varied with protein intake, but remained constant in the absence of dietary protein. With endogenous metabolism, urinary excretion of creatinine (on a meat-free diet) was independent of protein intake and nearly constant for each individual. Folin believed that the "endogenous" metabolism was a measure of the body's true metabolic state and had eluded previous researchers because the much larger "exogenous" metabolism fluctuated with diet. Uric acid excretion was intermediate—about half endogenous and half exogenous on an ordinary diet [21].

It followed that dietary protein in amount greater than the small needs of the animal for replacement of nitrogen loss due to wear and tear on the body tissues is essentially surplus. This surplus is excreted as urea or utilized for energy without participating in the nitrogen metabolism of the tissues. In nitrogen equilibrium, the bulk of urinary nitrogen was of dietary origin. Only a small amount of protein is needed for endogenous metabolism—not the larger amounts included in the standard diets of the time.

Folin's theory of exogenous and endogenous protein metabolism was taught in medical schools and dominated clinical investigation for about thirty years. Its influence on quantitative clinical chemistry was evident from the prominence given by laboratory manuals to the analysis of nonprotein nitrogen and its major components—urea, uric acid, creatinine, ammonia, and creatine. But, the theory was incorrect as was shown by the remarkable labeling studies with stable isotopes deuterium and ^{15}N by Rudolf Schoenheimer (1898–1941), David Rittenberg (1906–1970), and

---

*Jaffe's name is often mistakenly shown with an é. See Michael M. Lubran, "To é or Not to é?," *Clinical Chemistry*, 1995, 41: 1204–1205.

others at Columbia University's College of Physicians and Surgeons during the 1930s. By chemical means, they introduced the stable isotope into selected chemically stable locations of the molecule, e.g., fatty acids, cholesterol, or amino acids. This work made it possible to follow the fate of a particular metabolite in the living animal and to detect its participation in a multiplicity of hitherto unrecognized reactions.

Mice were fed deuterium-containing fat. The animals were on a diet insufficient to maintain weight, and the expectation was that almost all of the ingested fat would be burned and relatively little would be deposited. They found instead, that the largest part of the dietary fat, even when present in small quantities, is deposited in the fat tissues before it is utilized, indicating that the fat which was burned was not oxidized directly after absorption, but had been taken from the existing fat depots [22].

Schoenheimer and Rittenberg used ^{15}N, the stable isotope of nitrogen, placed in the alpha amino position, to study the intermediary metabolism of amino acids. They revealed details in the intermediary metabolism of proteins indicating that these molecules are not static, but are constantly undergoing breakdown and resynthesis. The isotope was found not only in the tyrosine that was fed, but also in other amino acids in the tissues of the test animal. Up to that time, the components of body tissues were regarded as being structural in function and metabolically inert.

Schoenheimer's *The Dynamic State Of Body Constituents* (1942) described how dietary nitrogen-containing compounds become indistinguishable from similar molecules already in the metabolic pool. Folin's "wear and tear" theory was replaced by the concept of a dynamic metabolism. According to the modern view, a metabolic reservoir is maintained in a steady state of numerous reactions in equilibrium. By constant controlled change, the body achieves the needed freedom from its external environment.

Folin's studies at the McLean Hospital revealed no metabolic evidence related to mental disease, but in the course of his work he had developed methods for biochemical research that promised to deliver significant results of a more general physiological interest and importance. While at McLean, a personal misfortune struck Folin. In the spring of 1903, a benign tumor was removed from his left parotid gland. During the surgery the facial nerve was cut. The procedure permanently altered Folin's appearance.

COLORIMETRIC METHODS FOR BLOOD ANALYSIS

Folin's simple colorimetric method for the quantitative estimation of urinary creatinine in 1904 was the breakthrough that opened up the possibilities of this rapid, simple, and inexpensive technique for analysis. It gave great impetus to the development of additional methods for quantitative analysis of other nonprotein nitrogen products in urine. The increased sensitivity of colorimetric procedures allowed use of smaller samples and resulted in greater accuracy than was previously possible with the older gravimetric and volumetric (titrimetric) methods. What followed in laboratories in the United States and abroad was the design of research protocols using Folin's small-sample methods to study the composition of urine from normal individuals and patients with various disorders. However, the analysis of urine had limited clinical use. It gave information primarily about the excretion of abnormal amounts of urine constituents.

Folin then turned to improving and refining analytical methods to make them applicable to the same constituents in blood, but in much smaller samples than were required by other methods. He recognized that, since blood plasma reflected the condition of the extracellular fluids as a whole, blood analysis was a better guide to metabolic reactions and clinical evaluation of nephritis than was urinalysis. It was much more important to know what metabolic products the kidneys failed to excrete and accumulated to harmful concentrations in the blood, than it was to know what and how much was excreted in the urine [23].

In 1914, ten years after introducing the alkaline picrate colorimetric reaction for creatinine in urine, Folin described the first satisfactory method for determining creatinine in blood and published the first extensive data with this reaction for normal individuals and in various pathological states [24]. Folin followed with colorimetric methods for urea, uric acid, creatine, ammonia, and nonprotein nitrogen in blood. These provided a tool for quick and reliable assessment of the retention of the ordinary nitrogenous waste products caused by failing kidney function. Their practical value as an aid in diagnosis and determination of operative risk represented an important advance for medicine and surgery. Other investigators in America and Europe followed Folin's lead and modified his procedures or developed their own practical colorimetric methods.

NONPROTEIN NITROGEN (NPN)

Many components of nonprotein nitrogen are present in blood in small amounts; but the term NPN usually refers to the nitrogen-containing substances that remain in the filtrate after the proteins have been precipitated, viz., urea, ammonia, uric acid, creatine, creatinine, and amino acids.

These constitute nearly the entire nonprotein nitrogen of plasma. Since much of the NPN in blood is in the red cells, and includes substances of unknown composition, this determination, as well as that of almost all other blood constituents, should be made on plasma [25] or serum.

All the methods for the quantitative determination of the NPN involve the same two principles: removal of the protein by some suitable precipitant, and analysis of the filtrate for nitrogen by some modification of the Kjeldahl method. Numerous protein precipitants have been used and most of them have been discarded because of unfavorable properties or other inconveniences [26]. Acetone-free methyl alcohol was first suggested by Folin and Denis [27] as the protein precipitant, then m-phosphoric acid [28], and later, in Folin's widely used and highly successful system of blood analysis, tungstic acid (see next page).

The protein-free filtrate is digested in a mixture of sulfuric acid, phosphoric acid, and copper sulfate solution. Nitrogen is converted to ammonium ions which combine with the sulfuric acid to form ammonium sulfate. This can be determined colorimetrically by addition of Nessler's reagent, or by making the reaction mixture alkaline, distilling the liberated ammonia into an acid receiver, and back-titrating with standard alkaline solution.

FOLIN JOINS HARVARD MEDICAL SCHOOL

The publication of several papers on a new theory of intermediary metabolism of ingested protein, along with the growing popularity of his methods of chemical analysis, brought Folin to the attention of the biochemical profession. It led, no doubt, to his appointment in 1907 as associate professor of biological chemistry, and in 1909, as Hamilton Kuhn Professor of biological chemistry and head of the department at Harvard Medical School, the first nonphysician on the faculty. He remained at Harvard until his death in 1934, teaching biochemistry to first-year medical students and building his department into a center of graduate study and research with a strong emphasis on analytical methods and clinical applications.

Two of his graduate students later won Nobel prizes: James B. Sumner (1887–1955) in 1926 for the first crystallization of an enzyme (urease), and Edward A. Doisy (1893–1986) (with Henrik Dam) in 1943 for the isolation and synthesis of vitamin K. A third member of his department, George H. Hitchings, a teaching fellow for the 1929–30 academic year, would share a Nobel Prize in Medicine or Physiology in 1988 for his part in the discovery of important principles for creating a rational method of designing new compounds that selectively operate against various disease states. Other students were Walter Ray Bloor (1877–1966), who developed methods for the determination of cholesterol and other blood lipids, Philip A. Shaffer (1881–1960), and Cyrus H. Fiske (1890–1978) who, with Yellapragada SubbaRow (1895–1948), discovered phosphocreatine in 1927 and developed a popular method for serum phosphorus.

Willey Glover Denis (1879–1929), a Ph.D. graduate of the University of Chicago, was assistant chemist at the Massachusetts General Hospital and the first woman elected to its staff. She collaborated with Folin on many methods for nitrogen compounds in the blood. In 1913 Folin was appointed chemist at the hospital, and acted in a consulting and supervisory capacity from his base at Harvard Medical School. Folin's best-known collaborator was Hsien Wu (1893–1959) with whom a number of different analytical procedures were combined into a widely used, simplified and compact system of blood analysis on a protein-free filtrate. It was a welcome response to the rapidly increasing number of chemical blood analyses being performed in hospitals. Wu was a doctoral candidate in biochemistry in 1917; two years later, "A System of Blood Analysis" was published by Folin and Wu. It was a landmark development in clinical chemistry [29].

THE FOLIN–WU PROTEIN-FREE FILTRATE AND OTHER PROTEIN PRECIPITANTS

In most analyses of biological fluids, the proteins interfere with the added chemical reagents and must be removed beforehand. When Folin and Wu discovered tungstic acid, a new protein precipitant, they introduced their new system of blood analysis as follows:

> "The main purpose of the research recorded in this paper has been to combine a number of different analytical procedures into a compact system of blood analysis, the starting point for which should be a protein-free blood filtrate suitable for the largest possible number of different determinations. It need scarcely be pointed out what a convenience and advantage it would be if one could take the whole of a given sample of blood and at once prepare from it a protein-free blood filtrate suitable for the determination of all or nearly all the water-soluble constituents, non-protein nitrogen, urea, creatinine, creatine, uric acid, and sugar.
>
> In connection with our work on the problem we have also had in mind the desirability of reducing as far as practicable the amount of blood filtrate to be used for each determination, for by means of such reduction the total usefulness of the filtrate is correspondingly increased" [30].

Tungstic acid was better adapted than other reagents to completely precipitate all the blood proteins without loss of the nonprotein constituents by adsorption on the protein coagulum or by interfering in the different analyses. The approximate neutrality of the tungstic acid filtrates and their freedom from heavy metals made them very suitable for analyses utilizing enzymes, as in the determination of urea and fermentable carbohydrates. The tungstic acid filtrate was formed by the addition of seven volumes water to one volume blood, followed by one volume of 10% sodium tungstate and one volume of 2/3 N sulfuric acid. Their protocol, which called for 10 cc blood, resulted in sufficient filtrate for colorimetric analysis of the following constituents: sugar, NPN, urea, uric acid, creatinine, and creatine. Analyses for amino acid nitrogen, chloride, and phosphorus, were later adapted to the protein-free filtrate as were improved assays for sugar and uric acid. These colorimetric methods, together with titrimetric assays for chloride and calcium, and gasometric procedures for carbon dioxide combining power and oxygen, soon became the core of the systematic application of chemical analysis in the clinical laboratories of hospitals.

Haden [31] simplified the Folin–Wu reagent by combining acid and water in a single addition of eight volumes of N/12 sulfuric acid to avoid adding the 2/3 N acid very slowly, or in two fractions 20–30 minutes apart, as suggested later by Folin [32], to prevent loss of some of the uric acid with the protein precipitate. Van Slyke and Hawkins [33] simplified the procedure even further by mixing the three reagents in advance and adding them as one reagent (stable for two weeks). The tungstic acid protein-free filtrate was adopted worldwide and remained in use until the introduction of "automated" analysis in the late 1950s when deproteinization was accomplished by dialysis; and in subsequent years, by other innovative technologies and methods that did not require prior removal of proteins.

Trichloroacetic acid, introduced by Greenwald [34], yields a highly acid filtrate which holds calcium and phosphorus in solution. In the absence of free acid, these inorganic constituents might be removed with the protein coagulum. Trichloroacetic acid yields a smaller amount of precipitate and a greater volume of filtrate which filters more rapidly than tungstic acid filtrates. Both acid reagents produce very similar results for nonprotein nitrogen [35].

Of the many protein precipitants that have been employed—chiefly acids, salts of heavy metals, alcohols, and other organic compounds—only a few have survived to the mid-twentieth century because of their general usefulness or adaptability for a specific purpose. Tungstic acid and trichloroacetic acid are best suited for general purpose precipitation of protein. Zinc hydroxide (zinc sulfate with sodium hydroxide or barium hydroxide) is useful for a specific purpose [36]. It removes not only protein, but also some nonprotein nitrogen constituents whose reducing capability contributes a false elevation of about 10 mg per 100 mL in the determination of blood glucose.

Sulfosalicylic acid had been used as a protein precipitant for urine by Roch in 1889 and independently by MacWilliam in 1891 [37]. Folin and Denis [38], using Kober's nephelometric method for the determination of protein in milk [39] and in digestion mixtures [40], introduced the turbidimetric determination of protein in urine with sulfosalicylic acid. The turbidity was compared on a Duboscq colorimeter with that formed with a standard protein solution. Folin regarded sulfosalicylic acid as the best reagent for the quantitative precipitation of albumin and thought it applicable also to exudates, transudates, and cerebrospinal fluid. He cautioned against testing urines deeply colored with blood or bile pigments. Folin also introduced benzoic acid and formaldehyde as preservatives for standard solutions.

Modern chemical analysis of small quantities of blood would have been impossible without the colorimetric methods developed by Folin and others. These procedures entailed some sacrifice of accuracy since the product to be analyzed was not isolated and the reaction occurred in a complex chemical milieu—even if the proteins were removed before the analysis. The methods were far from specific and the results obtained were often falsely high because of nonspecific components which gave the same color reaction as the substance analyzed. However, in spite of the shortcomings of these early methods, the analyses were of distinct clinical usefulness in the diagnosis of diabetes, uremia, gout, and other diseases.

Folin helped found the American Society of Biological Chemists in 1906 and served as its third president in 1909. After the establishment of the *Journal of Biological Chemistry* in 1905, he submitted most of his papers there. He joined the editorial committee in 1919 and was its chairman for many years.

In 1908 Folin proposed that American hospitals employ clinical chemists to advance "our ability to differentiate between the physiologic and the pathologic." He cautioned that although hospitals should become involved in biochemical research, clinicians can neither do nor direct chemical work. Systematic biochemical research requires the "ingenuity, resourcefulness and critical judgment of the trained chemist" [41].

Much of the early work in methods and applications in clinical chemistry was published in the *Journal of Biological Chemistry*. To a large extent, during the first quarter of the century, biochemistry *was* clinical chemistry. The *Journal of Laboratory and Clinical Medicine*, founded in 1915, served as another major outlet for publication. After the clinical chemists formed the American Association of Clinical Chemists in 1948 (see Chapter 19), it began to publish *Clinical Chemistry* in 1955. It had been preceded in 1949 by publication of the *Scandinavian Journal of Clinical & Laboratory Investigation* and followed in 1956 by an international journal, *Clinica Chimica Acta*, based in the Netherlands.

Folin's new methods, based on visual colorimetry and small volumes of specimen, were a stimulus to the growth of clinical chemistry. This activity coincided with the beginning of the institutional reform of biochemistry during the first two decades of the twentieth century. The professional prestige of biochemists was largely advanced by their success in developing diagnostic tests for the practicing physician.

STANLEY BENEDICT

Numerous "competitors" followed in the wake of Folin's publications, attempting to devise newer or improved methods for estimation of the constituents of blood and urine. Stanley Rossiter Benedict (Fig. 12.3) [42] was one of the most persistent and prolific. Benedict graduated from the University of Cincinnati with a B.A. in 1906. He went to Yale University for post-graduate study in physiological chemistry at the Sheffield Scientific School under Mendel and received the Ph.D. in 1908. Until Folin and Benedict established their reputations as physiological chemists, it was Mendel who had the greatest influence upon the medical profession.

After a brief period at Syracuse University and Columbia University, Benedict joined the staff of the Cornell University Medical College in 1910, where he remained until his death in 1936. He was on the editorial board and one of the directors of the *Journal of Biological Chemistry* and president of the American Society of Biological Chemists in 1919 and 1920.

Benedict's research paralleled the scientific work of Otto Folin. Their research objectives were identical and their years of creative work nearly coincided. They successfully devised and refined analytical procedures for the determination of minute amounts of the main nonprotein constituents of blood and urine. For the first time, chemical analysis became a useful and reliable technique for revealing the chemical processes of normal metabolism and of pathological processes of the human body and in

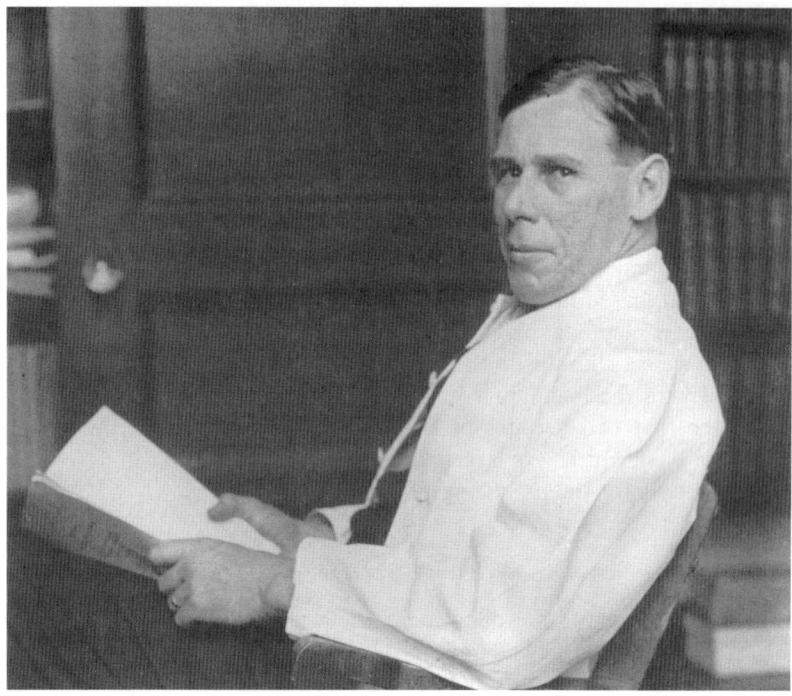

FIG. 12.3. Stanley R. Benedict. (Reproduced from the collection of the Archives of the National Academy of Sciences, Washington, D.C.)

experimental studies. Their work opened a new era in the application of quantitative clinical chemistry to the solution of problems in biochemistry, physiology, and medicine.

Benedict was especially interested in formulating reagents for the determination of glucose in blood and urine that would not be affected by non-glucose substances [43]. Uric acid was another constituent of blood and urine that absorbed and challenged the talents of Folin and Benedict. Benedict's studies revealed that the end product of purine metabolism in man, other primates, and the Dalmatian dog, is uric acid. In all other mammals, the breakdown goes one step further to allantoin through the action of the enzyme uricase.

Although Folin's methods for sugar, uric acid, creatine, creatinine, etc., when first published, were the best available, Benedict consistently subjected them to analytical review, and made modifications and revisions that improved them in some important detail, or replaced them with his own procedures. Despite the regular publication of criticisms of Folin's methods by Benedict, and criticisms of Benedict's revisions by Folin, they still managed to find opportunities to express favorable comments about one another's work. First person commentary was then the style in scientific publication. In any event, Folin and Benedict always remained on good terms.

Both Folin and Benedict were on a committee to reorganize the biochemical laboratory of the Metropolitan Life Insurance Company. Asked to become director of the laboratory, Folin preferred to remain at Harvard (Fig. 12.4). However, he and Benedict shared responsibility for the laboratory activities until a director was appointed in 1928. Folin and Benedict continued as consulting biochemists.

THE ROCKEFELLER INSTITUTE FOR MEDICAL RESEARCH

A great impetus to scientific medicine was provided by the Rockefeller Institute. Its founding provided a driving force to scientific medicine in the still barren period of American science. In well-equipped laboratories of the new and unique institute, eager scientists found freedom from long hours of formal teaching; adequate salaries enabled physicians to devote their lives to research unimpeded by the distractions of private practice. Hundreds of young scientists came to the institute to learn the methods of research from eminent scholars who had been recruited from many countries. From the institute, scores of future professors returned to universities, where they built schools of medicine and departments of graduate study; and, where they played major roles in the spectacular scientific progress of the first half of the twentieth century [44].

The idea came originally from Frederick T. Gates, advisor in philanthropy to John D. Rockefeller. Gates, after reading Osler's *Principles of Medicine* in 1897, realized how neglected the "scientific" study of medicine had been in the United States. He believed that an institute of medical research should be established in this country, comparable to the Koch Institute in Berlin and the Pasteur Institute in Paris. His urging had the desired effect, and The Rockefeller Institute for Medical Research was incorporated on June 14, 1901. At first, research in existing laboratories was promoted through annual grants-in-aid ranging from $300 to $1,200. In 1902, William H. Welch, who played a leading role in the formation of the institute, urged the establishment of an independent research facility. Temporary quarters were opened in 1904, and a permanent laboratory was dedicated in 1906. Simon Flexner (Fig. 12.5), who had trained under Welch

FIG. 12.4. Otto Folin. Portrait by Emil Pollak–Ottendorf (1934). (National Library of Medicine, Bethesda, Maryland.) Folin died one month before this portrait was to be presented to the University in his honor.

at Johns Hopkins and then went to the University of Pennsylvania, was chosen as director and to plan the organization of the institute. As its outlet for publication, the institute took over the *Journal of Experimental Medicine* in 1905.

FIG. 12.5. Simon Flexner. (Courtesy of the Rockefeller Univeristy Archives, New York, N.Y.)

Rockefeller authorized the creation of a hospital early in 1907. In 1909, Rufus Cole (1872–1966), previously the head of the biology laboratory at the Johns Hopkins Hospital, was chosen as the new director of the hospital. All staff were salaried and full-time. The Hospital of The Rockefeller Institute was formally opened on October 17, 1910. Here, patients would be studied with an unprecedented degree of thoroughness, and any new principles of treatment that resulted would be available to other patient populations. The careful study of individual patients required well-equipped laboratories and competent personnel skilled in research. This was the setting into which Donald Van Slyke was soon to enter.

DONALD DEXTER VAN SLYKE AND THE ROCKEFELLER HOSPITAL

Donald D. Van Slyke (Fig. 12.6) received his Ph.D. from the University of Michigan in 1907 in organic chemistry under Moses Gomberg (1866–1947), the discoverer of organic free radicals. Van Slyke had expected to follow his father's career as an agricultural chemist and had actually been offered a job with the Bureau of Chemistry in Washington. The elder Van Slyke (Lucius L.) was a chief chemist at the Geneva Agricultural Experiment Station in New York. However, a chance encounter between Van Slyke's father and Phoebus Aaron Theodor Levene (1869–1940) (Fig. 12.7) at an American Chemical Society meeting in 1907, led to a job offer from Levene at the newly formed Rockefeller Institute for Medical Research [45].

Levene, a Russian-born and educated physician, had come to the United States in 1892. He was selected by Simon Flexner in 1905 to be in charge of the chemistry laboratory. As a visiting scientist at Marburg, Levene had worked with Albrecht Kossel and became interested in nucleoproteins and their component nucleic acids. Kossel, who succeeded Willy Kühne as professor of physiology at Heidelberg in 1901, advanced the chemical characterization of the nucleoproteins—the major constituent of cell nuclei—much further than had Johann Friedrich II Miescher (1844–1895), who had discovered them. Kossel and his students discovered adenine, thymine, cytosine, and uracil as breakdown products of nucleic acids. Kossel also discovered a new amino acid, histidine. For his work on proteins and nuclear material in the cell, Kossel received the Nobel Prize in Physiology or Medicine in 1910. Levene continued his investigations on nucleoproteins at the Rockefeller Institute and made many contributions to the structure of the breakdown products of nucleic acids. Levene and

FIG. 12.6. Donald D. Van Slyke. (National Library of Medicine, Bethesda, Maryland.)

associates identified the specific purine bases and also identified d-ribose and deoxyribose as components of two important nucleic acids.

From 1910 to 1913, the chemistry laboratory at the hospital of the Rockefeller Institute was led by Francis H. McCrudden who had a B.A.

FIG. 12.7. Phoebus Aaron Levene. (Courtesy of the Rockefeller University Archives, New York, N.Y.)

in chemistry from the Massachusetts Institute of Technology and an M.D. from Harvard. McCrudden resigned in 1913 to become professor of applied therapeutics at Tufts Medical School. Meanwhile, Flexner and Cole recognized that internal medicine was moving rapidly ahead along chemical lines.

To guide this advance at the institute, they believed that the hospital should now have an experienced chemist in a senior position to conduct his own research while serving as a general advisor to physicians on chemical problems. The chemist would have to develop an interest in medical problems and be temperamentally able to cooperate with physicians, for whom the patients came first.

There were few biochemists in America qualified for such a post. In 1913, Flexner offered the position with full membership in the institute to Franz Knoop (1875–1946), an internationally known biochemist at the University of Freiburg, Germany. When Knoop declined the offer, the position remained vacant until Flexner realized that the ideal candidate was already at the institute. He turned to Van Slyke who was working with Phoebus Levene. After studying the chemistry of proteins and amino acids and their analysis at the institute for seven years, Van Slyke was selected in 1914 to develop a department of chemistry in the hospital, related to clinical chemistry. Cole had spent a year in Levene's laboratory while the hospital was being built and was impressed with the young biochemist (Fig. 12.8) [46].

Although Van Slyke had no experience in clinical work, Flexner was impressed by his training in organic chemistry and his publications in biochemistry. Van Slyke had spent the year 1911 in Berlin working with Emil Fischer and Emil Abderhalden (1877–1950) and published a paper with each. Levene, who had worked with Fischer before, arranged it. Uncertain as to whether he could do the clinical work, Van Slyke agreed to try it for one year, provided he could return to Levene's laboratory if he didn't like it. Van Slyke found that the young doctors in the hospital were all just about his age, and they welcomed him and his assistant, Glenn Cullen (1890–1940), "into their group, and almost by force imposed their enthusiasm and their problems upon us." He began to pick up medicine pretty fast, found it fascinating, and stayed—for the rest of the time he was at Rockefeller. There he applied chemistry to the solution of clinical problems related to diseases under investigation at the hospital [47].

Because the Rockefeller Hospital was a research hospital, the research worker served as the chief and the physician or surgeon as the assistant. As a service chief, Van Slyke had free access to blood and urine specimens. He taught himself kidney physiology and disease and soon found himself in charge of a ward of patients with Bright's disease [48]. He went on hospital rounds and instructed the resident staff on special diets or other preparations required for the patients under study. He also instructed them on what to look for in the patients and then to report from

FIG. 12.8. Chemistry Laboratory at the Rockefeller Institute for Medical Research (1908). (l to r) Walter A. Jacobs, Donald D. Van Slyke, Gustave M. Meyer. (Courtesy of the Rockefeller University Archives, New York, N.Y.)

a clinical point of view. This arrangement was an unusual privilege. In most institutions at that time, clinicians dominated the partnership with biochemists and infrequently acknowledged any help from them in the study of disease [49].

Biochemistry was in its infancy. Accurate methods for blood constituents in small specimens were just beginning to be available. The concentration and distribution of many of the inorganic constituents of the body were not known. Proteins were not yet regarded as chemical entities; enzymes had not yet been isolated and chemically characterized; the existence of hormones and vitamins was suspected, but they had not yet been clearly identified. Although Van Slyke began his career as an organic chemist, his interest in physiological function in health and disease resulted in an acceleration of new knowledge and the development of quantitative clinical chemistry. Van Slyke's design of accurate analytical methods for measuring gas and electrolyte equilibria in blood and the transport of blood gases furthered the understanding of respiratory physiology.

Institutional context was crucial to Van Slyke's success in integrating chemistry and clinical medicine, because the Rockefeller Hospital encouraged a cooperative attack on a problem from all sides—chemical, physiological, and clinical [50]. Clinical problems provided opportunities to develop, extend, and improve analytical procedures; new techniques led to discoveries in the physiology of disease. Other laboratories at the hospital carried out research in various other medical sciences, e.g., microbiology, pathology, immunology, organ culture, nutrition, etc.

Van Slyke's first clinical problem was one in diabetic acidosis. Severely ill diabetic patients, under the most efficient treatment (low-calorie diet) that was available in the pre-insulin days, sometimes developed acidosis which, by the time it had become clinically noticeable, progressed at a very rapid rate to a fatal coma. What was needed was a method for detecting earlier stages of the acidosis. Van Slyke began by defining acidosis in chemical terms instead of descriptive medical language. He devised an instrument and developed a simple and reliable gasometric method for measuring the carbon dioxide content of plasma (serum). This rapid and relatively simple test for quantifying blood acidity made it possible to anticipate and prevent the fatal acidosis. The instrument and method were soon in use in hospital laboratories everywhere (see Chapter 8). Van Slyke's technique for studying acidosis by using quantitative biochemical analysis dramatically increased understanding of the disease processes and provided a basis for rational treatment before the discovery of insulin.

Van Slyke made an important contribution to enzyme kinetics in the study of urease, then widely used to determine urea in biological fluids. He made fundamental contributions to the understanding of buffer action, acidosis, and carbon dioxide transport by hemoglobin and oxyhemoglobin. Other areas of his research interests included kidney function, acid–base balance, fluid and electrolyte equilibrium, and their relation to disease states. Van Slyke developed a method for determining clearance of urea from the blood (see Chapter 13) and a rapid procedure for the determination of red cell, hemoglobin, and plasma protein concentration under battlefield conditions by measurement of specific gravity. Results were interpolated from an empirically prepared nomogram (see Chapter 17). During the 1920s, he described a new amino acid present in gelatin, which he later showed to be hydroxylysine.

Van Slyke's investigations into protein metabolism helped to resolve an unsettled question surrounding protein digestion and absorption.

Nineteenth-century medical physiologists were interested in the chemical conversion of food into the constituents of animal tissues—what they termed nutrition, and included what later came to be called metabolism. The dominant theory of animal metabolism about 1850, at least in Germany and England, was the one presented by Liebig during the previous decade. He accepted the idea that respiratory oxygen is entirely converted to carbon dioxide and water in the animal body, and that this combustion is the cause of animal heat. For him, the non-nitrogenous materials of the diet, such as the fats, starch, and sugar, are oxidized in the blood. The fats are deposited only when there is insufficient oxygen for their complete combustion. He believed that the food proteins are converted to the proteins of the blood and tissues without extensive chemical change, because elementary analysis had shown little difference between the two kinds of protein. Similar ideas were offered in France by Dumas who asserted that food proteins are transferred to the tissues without any change whatever, and that animals, in contrast to plants, cannot perform synthetic reactions [51].

By the 1850s, it was clear that an essential chemical step in protein metabolism is the solubilization of food proteins to form what were called "peptones." It had been widely believed for many years that the albumin and fibrin of the blood are regenerated in the intestinal wall from dietary proteins solubilized in the digestive organs. Although it was known that amino acids are released from ingested protein in the gastrointestinal tract, leading physiological chemists during the second half of the nineteenth century, did not accept the idea that there was extensive breakdown to amino acids [52]. However, by 1910 it was clear that proteins are nearly completely digested in the intestine to amino acids; that these amino acids can replace intact dietary proteins; and that attention had to be paid to the "quality" of a dietary protein, not merely to its nitrogen content, in considering its nutritional value to the organism.

Since ordinary methods failed to reveal any considerable concentrations of amino acids in the blood, it was believed that the amino acids were resynthesized to protein during absorption in the intestinal mucosa. Emil Abderhalden made this proposal in 1912, but it was quickly put to rest by Van Slyke with a new method and glass apparatus he devised for measuring nitrogen produced by the action of nitrous acid on amino acids and peptides. With this, the first of his many ingenious gasometric methods, Van Slyke and his colleague Gustave M. Meyer (1875–1945) studied the fate of protein digestion products in the body [53].

They consistently found 3 to 5 mg amino acid nitrogen per 100 cc in the blood of dogs after twenty-four hours fasting. After a meal of meat, the level rose to 10 to 11 mg in the same animals. Their data showed that: "Ingested proteins are hydrolyzed in the digestive tract setting free most, if not all, of their amino-acids. These are absorbed into the blood stream, from which they rapidly disappear as the blood circulates through the tissues." The "extra" amino acids are merely absorbed from the blood by the tissues, without undergoing any immediate chemical change. The response was similar for intravenously injected amino acids. They concluded that the amino acids of the blood appear to be *in equilibrium* with those of the tissues [54].

Van Slyke and Meyer also showed that the liver is the primary site of amino acid deamination and the subsequent rapid conversion into urea of any resulting ammonia [55]. The rapid disappearance of amino acids (absorbed by the liver) was accompanied by an increase in the urea of the blood. This disproved the idea that urea arises by direct oxidation of tissue protein. There could be no doubt that the concept of protein resynthesis in the intestinal wall was superfluous, and that free amino acids in the blood are intermediates between food proteins and urea. The formation of urea was explained in 1932 when Hans Krebs (1900–1981) and Kurt Henseleit (b. 1907) demonstrated the workings in liver slices of a metabolic cycle involving three amino acids, ornithine, citrulline, and arginine—which eliminated carbon dioxide and ammonia from the body as urea—without being used up. It was the first time that an explanation of a biochemical synthesis was presented in terms of chemical reactions in a specific biological system and not merely inferred from the known chemical behavior of the presumed reactants.

In 1942 Van Slyke was awarded the Kober Medal by the Association of American Physicians for his contributions to clinical medicine [56]. Van Slyke stood out among the clinical chemists of his time for his achievements in research. Unlike many of his colleagues, not only did he excel in analytical chemistry but he had mastered physiology and many aspects of clinical medicine as well. His elegant yet precise and accurate analytical methods produced quantitative data that clarified the physiological and pathological states of humans.

Van Slyke was as influential as a teacher in the hospital laboratory as Folin was in the medical school. Although Folin trained more professional biochemists, Van Slyke's wide ranging investigations of disease states helped bridge the gap between biochemistry and internal medicine. Most of the many individuals who passed through Van Slyke's laboratory went on to professorships of internal medicine or biochemistry or to other important posts in the United States and abroad [57]. Some of these were Vincent P. Dole, Julius Sendroy, Alma E. Hiller, Frederick C. McLean, Walter W. Palmer, Irvine Page, Glenn Cullen, Christen Lundsgaard, John P. Peters, Michael Heidelberger, and A. Baird Hastings, who succeeded to Folin's chair of biochemistry at Harvard.

Van Slyke spent several months in 1922–23 as a Visiting Professor of Biochemistry at the Peking Union Medical College in China—a Rockefeller-endowed institution—in collaboration with Hsien Wu on important studies of gas and electrolyte equilibria in the blood. Wu spent the spring of 1925 with Van Slyke at the Rockefeller Hospital.

In 1914 the directors of the *Journal of Biological Chemistry* asked the Rockefeller Institute to take over its publication, with Van Slyke (age 31) to join the editorial board as managing editor. The *Journal* was nine years old and the only publication in the United States devoted solely to biochemistry. As such, the standards it set for the publication of experimental data determined the direction and quality of biochemical research in this country. Toward the end of 1925, the ownership of the *Journal* and its management were transferred from the Rockefeller Institute to the American Society of Biological Chemists; and the editorial office was moved to Cornell University Medical College in New York City. Stanley Benedict was appointed managing editor. Van Slyke, who remained on the editorial board until 1950, had also served as president of the society in 1921 and 1922.

Shortly after his retirement from the Rockefeller Institute in 1948, Van Slyke joined the newly formed Brookhaven National Laboratory of the Atomic Energy Commission in Upton, New York, as Assistant Director for Biology and Medicine. He also was Chief of the Chemical Division of the Medical Department with laboratory facilities for his research. During this second productive career of research in radioisotopes and metabolism of amino acids, he devised a more accurate and sensitive gasometric procedure for the measurement of Carbon-14. He remained at Brookhaven until his death at age 88.

Clinical chemists are especially indebted to Van Slyke for his collaboration with John P. Peters of Yale University in the writing of the two-volume classic *Quantitative Clinical Chemistry* which for more than thirty years was the authoritative source for clinical chemistry and even today remains a valuable resource for the history of methods (see Chapter 18). Van Slyke, a member of the National Academy of

Sciences, was awarded the National Medal of Honor by President Lyndon B. Johnson in 1965.

IVAR BANG AND MICROMETHODS

At about the same time that Folin, Benedict, and Van Slyke were developing methods, Norwegian-born Ivar Christian Bang (1869–1918) (Fig. 12.9) [58], an Oslo M.D. and professor of medical and physiological chemistry at the University of Lund in Sweden, was developing a new system of microanalysis of the blood. His micromethods for the estimation of sugar, chloride, nitrogen, proteins, fats, and cholesterol, were applied to the limited amounts of blood available from the finger tip of humans or the ear vein of rabbits. The technique was based on approximately 100 mg of blood, i.e., one or two small drops (100 to 150 microliters) absorbed onto prewashed and dried filter paper and quickly weighed on a torsion balance to determine the amount of blood applied. Proteins were fixed to the paper by a boiling solution of acid and salts and the extractable substances removed for analysis by micro titration methods [59].

Bang's procedure for blood sugar, devised in 1907, was based on reduction of copper sulfocyanate followed by iodometric titration and was the first practical micromethod for estimating blood sugar. In the initial step, Bang added alcohol to precipitate the blood proteins and removed them by centrifugation. In this procedure, he may have been the first to use the centrifuge for quantitative separation and washing of precipitates, as reported in a 1907 *Festschrift* to Olof Hammarsten [60].

In 1918 Bang introduced a new principle for determination of fat. After extraction from the weighed blood sample with organic solvents, the fat was oxidized by dichromate in the presence of concentrated sulfuric acid. The excess dichromate was then titrated iodometrically. Under his conditions, the oxidation was incomplete, and an empirical factor had to be applied in calculating the results. The method, with improvements yielding nearly quantitative oxidation, was the basis of Bloor's system of fat analysis a decade later [61]. Bang also invented a microburette and introduced steam distillation of ammonia in micro-Kjeldahl estimation of protein, nonprotein nitrogen, and urea. His productive life was cut short prematurely at age 49.

In a tribute to Ivar Bang, Van Slyke characterized him as "the complete clinical chemist. For clinical chemistry includes, not only the development of methods, but study of all the phenomena of the body's normal chemical processes, and of the alterations that they undergo in disease" [62].

FIG. 12.9. Ivar Christian Bang. (*Roots of Clinical Chemistry*, J. Büttner and C. Habrich, eds., Darmstadt: Git Verlag Gmbh, 1987, p. 105.)

Bang's analytical methods were impractical for clinical use. The micro scale of operation for specimen handling, weighing, preparation of extracts, titration and centrifugation, use of small-scale apparatus—no doubt home-made—required highly sophisticated skill, much patience, and lots of time. This technology was not readily transferable to the clinical laboratory, where turnaround time was a major concern. Although Bang's application of microtechniques was far ahead of his time, he did not adopt colorimetry—which became the key to the expansion and growth of clinical chemistry—as a basis for developing his new analytical methods. Publishing in German, much of it shortly before and during World War I, his work escaped notice in the United States at a time when American leadership in clinical biochemistry began its rise to dominance.

Although the use of less than 0.1 mL specimen is not practical for routine chemical analyses because of the time involved due to the special handling necessary—there has always existed a need for methods on a reduced scale to accommodate the limited amounts of blood that may be available in certain situations. From some subjects such as infants, children, small experimental animals or when frequent sampling is required, as for glucose tolerance tests, it is difficult or undesirable to obtain the needed amounts of blood by venipuncture and it is preferable to use capillary blood from finger tip or heel. Obtaining sufficient blood for analysis is also a problem if superficial veins are not accessible or adequate in cases where the patient is obese, edematous, a burn victim or when available veins have to be reserved for intravenous feeding or medication drip. Furthermore, even if adequate amounts of blood could be obtained by venipuncture, the frequent bleeding of any hospitalized, undernourished or otherwise weakened patient, especially the elderly, can rapidly lead to anemia.

With the increased use of laboratory testing as an aid in diagnosis and treatment, it is often necessary to analyze a single sample of blood for more than one or two constituents—sometimes as many as five or six or more. Scaled-down versions of "macro" methods of analysis (1.0 mL and greater) are not adaptable to apparatus and procedures designed for larger volumes. Reduced amounts of specimen require special techniques and smaller glassware and apparatus, e.g., pipettes, balance, centrifuge. Consequently, by the 1950s, although many analytical procedures were now being carried out on the "micro" scale (between 0.1 and 1.0 mL), testing in the "ultramicro" range (less than 0.1 mL) was gaining increasing attention because of the challenges presented by the needs of special patients [63].

VENIPUNCTURE AND BLOOD ANALYSIS

Micromethods, using finger tip capillary blood, never became widespread because the advance of modern chemical pathology depended upon the acceptance of venipuncture with a needle and syringe as a routine method to obtain adequate samples of blood. At about 1900, the introduction of blood culture as a valuable diagnostic measure, particularly in typhoid fever, showed that venipuncture was safe as well as convenient. Venipuncture led to other diagnostic applications, although initially it was by no means the routine procedure it has become today [64]. Venipuncture, as we know it, came into general use after 1910 and greatly facilitated and stimulated further chemical studies on human blood. Folin's development of methods requiring small volumes of blood hastened its use even more.

The discovery of a variety of colorimetric reactions during the last two decades of the nineteenth century; Folin's first method with the Duboscq visual colorimeter; and the convenient availability of blood by venipuncture; all came together during the opening decade of the twentieth century to make possible the replacement of urine analysis by blood analysis. As a result, between 1912 and 1922, many methods requiring small volumes of blood were published. The laboratory was chiefly occupied with determinations of urea, uric acid, creatinine, creatine, nonprotein nitrogen, carbon dioxide combining power, and glucose in the blood. Even for these constituents, the significance of blood levels in some diseases was not well understood. During the 1920s, clinical research laboratories began to show the importance of estimations of serum chloride, cholesterol, total protein, calcium, and phosphorus. Slowly, these analyses were absorbed into the laboratory repertoire, first in the large hospitals, then in the smaller ones, as gravimetric and most volumetric procedures were replaced by better colorimetric procedures. It was a remarkable period of development.

The event that brought the chemist to the patient's bedside and was a stimulus for the rapid growth of biochemical testing, was the discovery of insulin in 1922. Previously detected glycosuria now had to be thoroughly evaluated before using this new remedy which could be lethal in wrong dosage or curative when properly used. Once stabilized, each diabetic patient faced a lifetime of periodic and frequent blood sugar determinations for monitoring control of this condition.

The next thirty-five years witnessed a great deal of improvement in all chemical procedures. There was a gradual increase in the variety and range of determinations—keeping pace with advances in basic biochemistry and physiology—though, without any important change in the fundamental technology. Colorimetric methods made possible, previously impossible determinations, and made them more sensitive and accurate. This remarkable period of development provided the major methodology of the clinical chemistry laboratory.

The men who made this possible were the first generation of clinical biochemists. Unlike their successors, they were defined by employment opportunities rather than by their education. Folin and Van Slyke were trained as organic chemists. Stanley Benedict and Victor Myers (see Chapter 18) were both students in Chittenden's department of physiological chemistry and were drawn into clinical chemistry by their jobs at Cornell and the New York Post-Graduate Medical School and Hospital, respectively.

EVOLUTION OF THE HYPODERMIC SYRINGE

Unlike urine, which is readily available in large volumes, the chemical analysis of blood requires methods that are adaptable to small volumes of fluid. Development of the hypodermic syringe made this possible. However, the syringe developed as a tool for purposes unrelated to its eventual use in the administration of drugs and the withdrawal of blood. It evolved from appliances used to irrigate various bodily orifices. Such instruments have been in use for hundreds of years. Simply attach an animal skin to a metal tube or bird's quill, fill it with liquid and squeeze. After William Harvey's discovery of the circulation of the blood, it was inevitable that someone would experiment with introduction of substances into this systemic conduit of animals or humans.

The consensus in the historical literature [65], at least in the western world, credits Sir Christopher Wren (1632–1723), the English architect, experimenter, and Renaissance man, with the first injections in animals, using a syringe-like device in 1657. It consisted of a slender quill attached to a pouch or bladder filled with the drug. Gravity rather than pressure infused the liquid into the vein. The quill was soon replaced by a silver or gold tube, and the gravity-fed reservoir by a bladder which could be squeezed. These early intravenous infusions were made by first opening a vein with a sharp instrument, then inserting the "tube." An integrated system using a needle for direct penetration of the vein was not invented until well into the nineteenth century. Yet even in the early part of the twentieth century, it was usual to expose the vein before puncture. Bloodletting by venesection was an old practice and the technique persisted.

A seventeenth-century Dutch physician, Regnier de Graaf (1641–1673), constructed a device which resembles the modern instrument and used it to inject mercury to outline the arteries and veins of cadavers for anatomical study of circulation. The next development occurred in 1827, when a

German physician, A. Neuner, constructed a syringe with a glass barrel that allowed him to measure the volume of liquid injected and was resistant to the chemicals he used. He injected cadaveric lens with mercurial compounds to teach medical students about cataract removal. The syringe was made of silver or platinum. His reference to syringes of horn, bone or ivory, suggests that even his device was preceded by others.

In 1853, a French physician, Charles-Gabriel Pravaz (1791–1853), devised a metal syringe that resembles the hypodermic syringes that came later. He used it to inject iron perchloride into the veins of sheep and horses to study clotting. About the same time, and knowing of Pravaz's injections, an Edinburgh physician, Alexander Wood (1817–1884), is credited with the first subcutaneous injection as a therapeutic procedure in a human. Dr. Wood injected an elderly woman suffering from neuralgia, with muriate of morphine in an attempt to relieve her pain. He did not describe the needle. After Wood's paper appeared in 1855, a Dublin physician, Francis Rynd (1801–1861), claimed priority for the localized treatment of neuralgia by injection of morphine ten years earlier. His results were published in 1845 in an obscure Dublin journal of limited circulation. He did not describe the delivery instrument until 1861. It turned out not to be a true syringe, but rather a trochar and cannula that made a puncture, then served as a gravity-fed receptacle for an infusion of the drug, which was poured in through an aperture in the base of the cannula. Dr. Fordyce Barker, a New York City physician visiting Edinburgh in 1856, brought back a syringe of the type used by Wood and pioneered the introduction of hypodermic medication in the United States.

In 1859, Dr. Charles Hunter, an English surgeon, adopting Wood's technique, introduced the term "hypodermic" to distinguish his use of subcutaneous injection as a mode of systemic administration of drugs, from Wood's use of the same technique for supposed local analgesia action.

During the second half of the nineteenth century, syringe designs evolved from metal cylinders with leather plungers to glass barrels with silver or platinum fittings. About 1896 Hermann Wulfing Luer of Paris introduced the first all-glass hypodermic syringe. In America, syringes were first commercially produced by George Tiemann and Co. in New York. By the 1870s, the company catalog listed several dozen different varieties of syringes.

The modern syringe metamorphosed out of devices which were inextricably bound to the experiments involving the hypodermic administration of drugs. The controversy over priority publicized the new method of drug administration. By the 1880s, the hypodermic route had become a major avenue for administration of every conceivable kind of drug, extract, elixir, and medication, as well as some food nutrients.

Even after the concept of antisepsis by Joseph Lister in 1867, the notion that syringes might transmit infection was not widely held until the turn of the century. Although numerous cases of infection attributed to dirty syringes and needles appeared in the literature, general practitioners were rather casual about disinfection even after surgeons accepted the need for this precaution. When obtaining blood from finger tip (preferable in most instances) or ear lobe, DaCosta did not sterilize but merely wiped the puncture-needle with a towel wet with alcohol, unless the patient was syphilitic or septic. Then it was "safer to pass the blade through an alcohol flame after having used it." Blood was also obtained with a small steel trocar blade, a lancet, spear-pointed surgical needle or new sharp-pointed steel pen from which one nib had been twisted off [66]. "Wet-cupping" and "blistering" were used to obtain sufficient blood for analysis of glucose or uric acid [67].

Wet-cupping is a procedure whereby small glass cups are rinsed with a flammable material, such as rubbing alcohol or grease, ignited, and the open end immediately pressed firmly onto freshly scarified skin. The suction of the partial vacuum created in the cup by the rapid consumption of the oxygen during the brief burn, draws blood through breaks in the skin [68].

Initially, the major use of the syringe was for the hypodermic administration of medication. Although its value for withdrawing blood by venipuncture for analysis was readily apparent, this use was slow in being adopted in the laboratory. Scientists, long accustomed to find their own solutions to problems, resorted to other techniques of their own making.

In 1912 Folin used neither cannulae nor syringes to draw blood, but simply hypodermic needles and pipettes. The sterilized needles were immersed in a dilute solution of vaseline in ether and then allowed to drain and dry on a clean paper for a few minutes before being used. The needle is attached to the tip of a two or five cc pipette by means of a short piece of narrow pure gum tubing. A small pinch of powdered potassium oxalate is entered into the upper end of the dry and clean pipette and is allowed to run down into the tip and needle. The other end of the pipette is connected with a rubber tube, which in turn connects with a mouth piece consisting of a short tapering glass tube. Close to the pipette the rubber tube carries a pinchclamp. Drawing blood was a two-person operation. One person inserts the needle into the vein or artery, and the other regulates the flow of the blood by means of the pinchclamp and by suction. The exact quantity of blood desired is thus obtained without any waste and without clotting [69].

A different design for drawing blood by venipuncture is shown in Fig. 12.10. A bottle of 35 to 60 cc capacity is fitted with a two-hole rubber stopper through which two glass tubes pass. One of these is made of heavy-walled glass tubing, is slightly bent and is ground to receive a needle. To the other is attached a short piece of heavy, small-bored rubber tubing, to which slight suction may be applied, if needed. The needle is attached to the short rigid glass tube, which makes for easy manipulation of the needle. Prior to drawing blood, 2 to 4 drops of a 20% solution of potassium oxalate is added to the bottle to prevent clotting. But first, the bottle is placed in a hot air oven to dry the solution and disperse a uniform layer of oxalate. After the bottle is cooled, the blood can be drawn directly into it and immediately mixed with the oxalate by a gentle rotary motion. When a glass syringe is used, the barrel may be moistened with potassium oxalate solution before withdrawal of blood [70]. To obtain plasma, the oxalated blood is centrifuged. As late as 1943, a popular laboratory manual gave the following directions for drawing blood: "Plunge the needle into the vein, and allow the blood to run into the bottle containing the oxalate" [71].

An early application of a vacuum tube containing fluoride-thymol as anticoagulant and preservative for collecting, preserving, and mailing

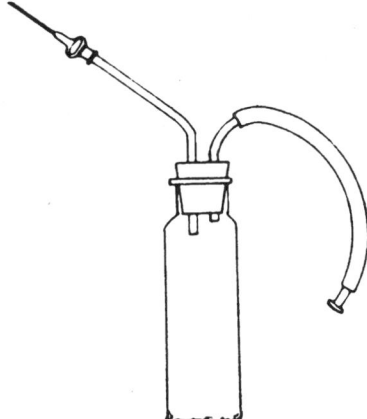

FIG. 12.10. Blood drawing outfit. (V. C. Myers, *Practical Chemical Analysis of Blood*, 2nd ed., St. Louis: C. V. Mosby Co., 1924, p. 30.)

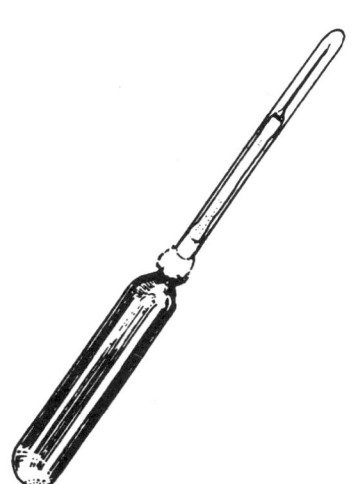 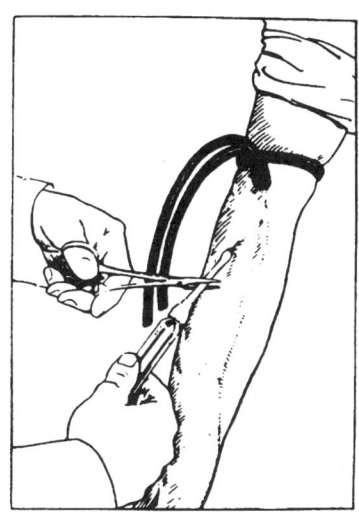

FIG. 12.11. Vacuum assisted venipuncture. (*Archives of Pathology and Laboratory Medicine*, 1926, 1: 227–236. "Copyright 1926, American Medical Association.")

blood samples, was described in 1926. The glass tube which protects the sterile needle and the stylet is removed, and the needle is inserted into the vein. The end of the sealed tube is crushed with a hemostat—as shown in Fig. 12.11—allowing the suction of the vacuum to draw blood into the tube. When a sufficient amount has been obtained, the tourniquet is released, the needle is withdrawn from the vein, the stylet is reinserted, and the glass tube placed over the needle. The specimen was then ready to be mailed to the laboratory. The tube was manufactured by Hynson, Westcott & Dunning, Baltimore, Maryland [72].

BECTON DICKINSON AND COMPANY (1897–1997) AND THE VACUTAINER TUBE

The story of the hypodermic syringe is deeply written into the corporate history of Becton Dickinson and Company (BD). The hypodermic syringe was the first product sold by the newly formed company back in 1897 and from the earliest days was a major contributor to sales and profits. Almost every BD employee has, at one time or another, been involved in the manufacture, sale, or distribution of hypodermic-related products [73].

The company opened for business at 45 Vesey Street in lower Manhattan, in September 1897, not far from New York's City Hall. The syringes that were offered for sale at the time were all manufactured in Europe. The Luer syringe, produced in France, was the first all-glass hypodermic syringe. It was a marked improvement over earlier types made of hard rubber or metal and glass, which wore out quickly or eroded in sterilization. Maxwell W. Becton (1868–1951) and Fairleigh S. Dickinson, Sr. (1866–1948) soon learned about the problems of relying on foreign imports, viz., delayed deliveries, damaged and defective products. Deciding to manufacture on their own, the partners purchased the American patent rights to the Luer glass syringe in 1899. By 1906 the company had outgrown its rented facilities in Plainfield, New Jersey and moved to a new two-story plant in East Rutherford, New Jersey.

World War I propelled the company into increased production of glass syringes, which were easier to sterilize than those made of metal with leather plungers. The medical kits ordered by the U.S. Army for field use specified glass syringes. During the war, the company became the major manufacturer of domestically produced glass syringes. President Woodrow Wilson named Fairleigh Dickinson to be chairman of an industry advisory board to coordinate the manufacture of medical and surgical equipment for the armed forces.

In the years up to 1924, improvements in the original all-glass Luer syringe provided a firmer grip that enabled the physician to inject as well as aspirate with one hand. The newer syringes were made of alkali-free hard glass such as Pyrex, which resisted erosion caused by sterilization and medications. However, these improvements did not address the demand for a simple, secure method of removing and attaching the needle to the syringe. This problem was solved in 1925 with the introduction of the Yale Luer-Lok syringe designed and patented by Dickinson. This new hypodermic syringe had a metal Luer-Lok tip with an inside thread that securely locked the needle to the syringe by a one-half turn. This virtually eliminated the danger of a needle slipping off the syringe while in use and reduced breakage of syringe tips. The needle could be easily removed by a reverse one-half turn. The new syringe allowed free manipulation of both needle and syringe—easy and fast to attach and detach. Meanwhile, the search went on for a syringe with completely interchangeable parts.

The first American-made interchangeable syringe was developed by Joseph J. Kleiner (1897–1974). Kleiner had entered the surgical supply business in 1920, establishing the Goodman-Kleiner company to develop and market a line of diagnostic instruments. One product in the line was a Japanese syringe with interchangeable parts. However, this syringe failed to maintain a vacuum in the barrel. A tight fit between barrel and plunger is necessary to prevent seepage past the plunger and to insure full delivery of medication. The fault had nothing to do with interchangeability, but with the fact that these syringes were made of lime glass, which deteriorated after repeated sterilization.

Kleiner decided to have the interchangeable syringes made of borosilicate glass—much harder than lime glass—and according to his specifications. These required more time and labor to grind to an interchangeable fit and could only be produced in small lots of a dozen matched barrels and plungers. In 1935 Kleiner and a group of associates founded Multifit, Inc. to import these syringes from Japan and package them for sale. Despite the additional work and cost, the selling price in the U.S. could still be competitive because of the low wage scale in Japan. America's entry into World War II ended the Japanese source of these interchangeable syringes.

In 1943 when Kleiner joined Becton Dickinson and Company, he brought with him the rights to the Multifit syringe with interchangeable parts. Becton Dickinson—determined to develop its own interchangeable syringe—kept the name and built up the line. The Multifit syringe represented a major technological advance, making it possible to produce syringes of a quality equal in performance and durability to the older, custom-ground plunger and companion barrel, individually matched to be a perfect fit, and numbered—but when one of the two pieces broke, both had to be discarded. With the new syringes, unbroken parts could be refitted into any intact matching part.

In 1952 the company introduced Multifit syringes engineered so precisely that every Multifit plunger fits every Multifit barrel that was manufactured and distributed in each of three geographic zones of the United States. Only a few millionths of an inch in diameter separated the syringes in one group from those in another. Within each one of the zones all syringes had completely interchangeable parts.

However, making crafted glass syringes reusable did not diminish the laborious process of washing and sterilizing. While Kleiner's Multifit syringe had helped to deal with the problem of glass breakage, the instrument was too expensive not to reuse. Responsibility to clean and sterilize after each use belonged elsewhere—to someone in the doctor's office or in the hospital. After use, the syringe is taken apart, immersed and flushed in a cleaning solution. Sterilization could then be achieved by one of several techniques: boiling in water, baking in hot air at 320°F, treatment with chemicals, or autoclaving at 15 to 17 pounds per square inch of pressurized steam at 250°F. Once the syringe parts were dried, they could be recombined and again sterilized in ovens. All this handling and potential for breakage could be made irrelevant by development of a hypodermic syringe and needle that were manufactured and packaged sterile and restricted to one-time use on only one patient and then discarded. Having finally succeeded in producing hypodermic syringes with completely interchangeable parts, the company made a major commitment to the development of the disposable syringe and needle. Today, Becton Dickinson and Company is the largest single manufacturer of disposable syringes in the world.

Years earlier, Kleiner had approached the company about purchasing his Evacutainer blood collection tube. This device consisted of an evacuated test tube sealed with a rubber stopper, a double-pointed needle, and a holder for fitting the two together. When in use, the short-end needle is pressed against the stopper as the other end makes the venipuncture. The vacuum in the tube is then broken by pushing the tube so that the needle pierces the stopper. This allows the blood pressure to push, and the vacuum to pull, the blood into the tube. The tube served as both a syringe for drawing blood from the vein and as a sealed container for transporting the sample. The system offered another advantage—many blood samples could be drawn by leaving the needle in the vein and using vacuum tubes containing different anticoagulants to obtain plasma for specific laboratory tests. In Kleiner's early design, the stopper sometimes leaked, and over a period of time, the tubes often lost their vacuum and wouldn't work. Because of these quality issues, and because the company was already developing a similar product it felt was superior, it declined Kleiner's offer.

Kleiner got his idea for the Evacutainer from observing the problems associated with blood drawing in medical offices and seeing how difficult it was to obtain a small sample of blood. To secure a proper quantity was just as difficult. The procedure was slow, and spillage could occur during transfer of the blood from syringe to test tube. Sometimes more than one venipuncture was needed for additional samples for different blood tests.

After joining BD, Kleiner continued to refine and improve his Evacutainer tube, despite management's reluctance to promote this product. Eventually the company agreed to introduce it rather than risk having a competitor produce something similar. One of the early steps taken to maintain the vacuum in the collection tube was to package the evacuated tubes in a vacuum can similar to vacuum coffee cans. Kleiner took the lead in marketing the product by visiting hospitals and blood banks and demonstrating it at public health shows. Within a very short time, the product, renamed the Vacutainer, became the company's largest dollar sales item. Key to the success of the Vacutainer tube was the growing importance of the clinical laboratory in medical diagnosis. Blood was the medium for much of the diagnosis and the Vacutainer tube made its collection and handling easy and reliable.

A patent for the Vacutainer blood collecting apparatus was granted on February 1, 1949. In 1975 Becton Dickinson added an inert polyester gel to the empty Vacutainer tubes to separate and prevent transfer of the blood's constituents between cells and serum (plasma) after centrifugation. A large variety of collection tubes are now available for different procedures. The company also manufactures and sells a broad range of medical supplies and devices, diagnostic systems, and needles for their syringes.

In the early years, most of the needles sold by BD were imported from Europe. After introduction of new and improved machinery, they were able to produce a good quality of hypodermic needle at small cost. By the start of World War I, BD offered 53 styles of needles in 230 variations of size and length for a wide variety of uses. The war halted imports from Europe. Forced to increase its productivity, the company embarked on a program to mechanize production. For many decades afterwards, BD was the only firm in the industry with fully automated needle manufacturing facilities.

NOTES AND REFERENCES

1. ABRAHAM FLEXNER, *Medical Education in the United States and Canada. A Report to the Carnegie Foundation for the Advancement of Teaching. Bulletin Number Four* (Boston: Merrymount Press, 1910), pp. 101–102. Flexner also surveyed *Medical Education in Europe* (1912) for the Carnegie Foundation. His reports brought him to the attention of philanthropists who sought his advice on the creation of new medical schools, the improvement of clinical teaching in existing institutions, as well as making endowments to colleges and universities. He was also instrumental in raising the money for the creation of the Institute for Advanced Study at Princeton University, which he directed until 1939. See also *Abraham Flexner: An Autobiography* (New York: Simon & Schuster, 1960); SAMUEL MEITES, "Abraham Flexner's Legacy: A Magnificent Beneficence to American Medical Education and Clinical Chemistry," *Clinical Chemistry*, 1995, 41: 627–632.
2. OTTO FOLIN, "Chemical Problems in Hospital Practice," *The Harvey Lectures*, 1907–08, 3: 187–198; also in *Journal of the American Medical Association*, 1908, 50: 1391–1394.
3. DETLEV W. BRONK (Foreward), in George W. Corner, *A History of the Rockefeller Institute, 1901–1953. Origins and Growth* (New York: The Rockefeller Institute Press, 1964), p. v.
4. FLEXNER (1910), p. 12.
5. HENRY S. PRITCHETT (Introduction), in Flexner (1910), p. x; ROBERT E. KOHLER, *From Medical Chemistry to Biochemistry. The Making of a Biomedical Discipline* (New York: Cambridge University Press, 1982), pp. 121–122; see also FLEXNER (1910), pp. 14–19.

6. MEITES (1995), pp. 628–629; FLEXNER (1910), pp. 7–10, 140.
7. FLEXNER (1910), pp. 12, 28–29; KOHLER (1982), pp. 122–124.
8. KOHLER (1982), pp. 122, 124, 125.
9. "Standards of Medical Education Adopted by the American Medical Association, July 1905," *Journal of the American Medical Association*, 1906, 47: 627; KOHLER (1982), p. 126.
10. "Report of the Council on Medical Education," *Journal of the American Medical Association*, 1910, 54: 1974–1975.
11. KOHLER (1982), pp. 160–162.
12. *Ibidem*, pp. 164–167.
13. RUSSELL H. CHITTENDEN, "Some of the Present-Day Problems of Biological Chemistry," *Science*, 1908, 27: 241–254, p. 253.
14. KOHLER (1982), pp. 167–168, 172, 174. For the background to the development of biochemistry in the United States, and the priorities and styles of various departments, schools, personalities, and groups in medical colleges and universities promoting reform, and those opposed to it, see KOHLER (1982), pp. 127–157 ff.
15. *Ibidem*, pp. 198, 176–179.
16. SAMUEL MEITES, *Otto Folin. America's First Clinical Biochemist* (Washington, D.C.: American Association for Clinical Chemistry, Inc., 1989); see also PHILIP ANDERSON SHAFFER, "Otto Folin, 1867–1934," *Biographical Memoirs of the National Academy of Sciences*, 1952, 27: 47–82; "Obituary," *Science*, 1935, 81: 35–37; "Tribute to Professor Folin," pp. 37–38.
17. G. STANLEY HALL, "Laboratory of the McLean Hospital, Sommerville, Mass.," *American Journal of Insanity*, 1895, 51: 358–364.
18. OTTO FOLIN, "Beitrag zur Chemie des Kreatinins und Kreatins im Harne," *Zeitschrift für Physiologische Chemie*, 1904, 41: 223–242, p. 224; "Some Metabolism Studies. With Special Reference to Mental Disorders," *American Journal of Insanity*, 1904, 60: 699–732, p. 710; "Approximately Complete Analyses of Thirty 'Normal' Urines," *American Journal of Physiology*, 1905, 13: 45–65, pp. 48–49.
19. M. JAFFE, "Ueber den Niederschlag, welchen Pikrinsäure in normalem Harn erzeugt und über eine neue Reaction des Kreatinins, *Zeitschrift für Physiologische Chemie*, 1886, 10: 391–400.
20. ADOLPH BOLLIGER, "The Colorimetric Determination of Creatinine in Urine and Blood With 3,5-Dinitrobenzoic Acid," *Medical Journal of Australia*, 1936, 2: 818–821; STANLEY R. BENEDICT and JEANETTE ALLEN BEHRE, "Some Applications of a New Color Reaction for Creatinine," *Journal of Biological Chemistry*, 1936, 114: 515–532; WILSON D. LANGLEY and MARGARET EVANS, "The Determination of Creatinine with Sodium 3,5-Dinitrobenzoate." *Journal of Biological Chemistry*, 1936, 115: 333–341.
21. OTTO FOLIN, "A Theory of Protein Metabolism," *American Journal of Physiology*, 1905, 13: 117–138; see also pp. 45–65, 66–115.
22. RUDOLF SCHOENHEIMER and D. RITTENBERG, "Deuterium as an Indicator in the Study of Intermediary Metabolism. III. The Role of the Fat Tissues," *Journal of Biological Chemistry*, 1935, 111: 175–181; "Deuterium as an Indicator in the Study of Intermediary Metabolism," *Science*, 1935, 82: 156–157. Some of the early isotope work is described by Sarah Ratner, "The Dynamic State of Body Proteins," *Annals of the New York Academy of Sciences*, 1979, 325: 189–209. For an obituary of Schoenheimer and review of his work, see HANS T. CLARKE, *Science*, 1941, 94: 553–554.

23. OTTO FOLIN and W. DENIS, "Nitrogenous Waste Products in the Blood in Nephritis. Their Significance and the Methods for Their Determination," *Boston Medical and Surgical Journal*, 1913, 169: 467–468.
24. OTTO FOLIN and W. DENIS, "On the Creatinine and Creatine Content of Blood," *Journal of Biological Chemistry*, 1914, 17: 487–491; OTTO FOLIN, "On the Determination of Creatinine and Creatine in Blood, Milk and Tissues," *Journal of Biological Chemistry*, 1914, 17: 475–481.
25. HSIEN WU, "Separate Analyses of the Corpuscles and the Plasma," *Journal of Biological Chemistry*, 1921, 51: 21–31.
26. JOHN P. PETERS and DONALD D. VAN SLYKE, *Quantitative Clinical Chemistry*, vol. 2, *Methods* (Baltimore: The Williams & Wilkins Company, 1932), pp. 63–64; ALMA HILLER and DONALD D. VAN SLYKE, "A Study of Certain Protein Precipitants," *Journal of Biological Chemistry*, 1922, 53: 253–267.
27. OTTO FOLIN and W. DENIS, "New Methods for the Determination of Total Non-Protein Nitrogen, Urea and Ammonia in Blood," *Journal of Biological Chemistry*, 1912, 11: 527–536.
28. OTTO FOLIN and W. DENIS, "Nitrogen Determinations by Direct Nesslerization. II. Non-Protein Nitrogen in Blood," *Journal of Biological Chemistry*, 1916, 26: 491–496.
29. FREDERIC A. WASHBURN, *The Massachusetts General Hospital. Its Development, 1900–1935* (Boston: Houghton Mifflin Company, 1939), p. 108; SAMUEL MEITES, "Willey Glover Denis (1879–1929), Pioneer Woman of Clinical Chemistry," *Clinical Chemistry*, 1985, 31: 774–778. After seven years at Massachusetts General Hospital, Denis resigned in 1920 to accept a position as assistant professor in the department of physiology (and physiological chemistry) in the medical school of Tulane University; see also CHARLES BISHOP, "Hsien Wu (1893–1959): A Biographical Sketch," *Clinical Chemistry*, 1982, 28: 378–380.
30. OTTO FOLIN and HSIEN WU, "A System of Blood Analysis," *Journal of Biological Chemistry*, 1919, 38: 81–110, p. 81.
31. RUSSELL L. HADEN, "A Modification of the Folin–Wu Method for Making Protein-Free Blood Filtrates," *Journal of Biological Chemistry*, 1923, 56: 469–471.
32. OTTO FOLIN, "A System of Blood Analysis. Supplement IV. A Revision of the Method for Determining Uric Acid," *Journal of Biological Chemistry*, 1922, 54: 153–170.
33. DONALD D. VAN SLYKE and JAMES A. HAWKINS, "A Gasometric Method for Determination of Reducing Sugars, and its Application to Analysis of Blood and Urine," *Journal of Biological Chemistry*, 1928, 79: 739–767, p. 741.
34. ISIDOR GREENWALD, "The Estimation of Non-Protein Nitrogen in Blood," *Journal of Biological Chemistry*, 1915, 21: 61–68; "The Estimation of Non-Protein Nitrogen in Blood," *Ibidem*, 1918, 34: 97–101.
35. HILLER and VAN SLYKE (1922), pp. 257, 259, 266.
36. H. C. HAGEDORN and B. NORMAN JENSEN, "Zur Mikrobestimmung des Blutzuckers mittels Ferricyanide," *Biochemische Zeitschrift*, 1923, 135: 46–58; MICHAEL SOMOGYI, "A Method for the Preparation of Blood Filtrates for the Determination of Sugar," *Journal of Biological Chemistry*, 1930, 86: 655–663; "Determination of Blood Sugar," 1945, 160: 69–73.
37. THOMAS B. MAGATH, "Minimal Albuminuria and Tests for Albumin in the Urine," in *The Kidney in Health and Disease*, Hilding Berglund and Grace Medes, eds. (Philadelphia: Lea & Febiger, 1935), p. 444.

38. OTTO FOLIN and W. DENIS, "The Quantitative Determination of Albumin in Urine," *Journal of Biological Chemistry*, 1914, 18: 273–276.
39. PHILIP ADOLPH KOBER, "Nephelometric Determination of Proteins; Casein, Globulin and Albumin in Milk," *Journal of the American Chemical Society*, 1913, 35: 1585–1593; "Nephelometry in the Study of Proteases. II.," pp. 290–292.
40. PHILIP ADOLPH KOBER, "Nephelometry in the Study of Proteases and Nucleases," *Journal of Biological Chemistry*, 1913, 13: 485–498.
41. FOLIN (1907–08), pp. 190, 188; (1908), pp. 1392, 1391.
42. ELMER VERNER MCCOLLUM, "Stanley Rossiter Benedict, 1884–1936," *Biographical Memoirs of the National Academy of Sciences*, 1952, 27: 155–171.
43. STANLEY R. BENEDICT, "A Reagent for the Detection of Reducing Sugars," *Journal of Biological Chemistry*, 1908–09, 5: 485–487.
44. BRONK (Foreword), in CORNER (1964), pp. v–vi.
45. DONALD DEXTER VAN SLYKE, *Oral History* (Bethesda, Maryland: National Library of Medicine, 1971), pp. 6–8; see also A. BAIRD HASTINGS, "Donald Dexter Van Slyke. The 20th Century Iatro-Chemist," *Federation Proceedings*, 1964, 23: 586–591; *Biographical Memoirs of the National Academy of Sciences*, 1976, 48: 309–360 (bibliography, pp. 336–360). For other tributes and obituaries, see *Clinical Chemistry*, 1963, 9: 645–663; 1971, 17: 670–672; *Journal of Biological Chemistry*, 1972, 247: 1635–1640.
46. CORNER (1964), pp. 97–98, 269, 106, 274. Folin, while still at McLean Hospital, had been approached in 1906 to join the Rockefeller Institute, but accepted the offer from Harvard instead (p. 73).
47. *Oral History* (1971), p. 18; DONALD D. VAN SLYKE, "Acceptance of the Kober Medal Award," *Transactions of the Association of American Physicians*, 1942, 57: 42–43.
48. *Oral History* (1971), p. 19.
49. E. C. DODDS, "American Impressions," *Middlesex Hospital Journal*, 1932, 32: 49–62, pp. 54–55; see also KOHLER (1982), pp. 242–243.
50. KOHLER (1982), p. 245.
51. JOSEPH S. FRUTON, "Biochemistry and Clinical Chemistry – A Retrospect," *Journal of Clinical Chemistry and Clinical Biochemistry*, 1982, 20: 243–252, p. 244.
52. *Ibidem*, p. 245.
53. DONALD D. VAN SLYKE, "A Method for Quantitative Determination of Aliphatic Amino Groups. Applications to the Study of Proteolysis and Proteolytic Products," *Journal of Biological Chemistry*, 1911, 9: 185–204; see also "The Quantitative Determination of Aliphatic Amino Groups. II.," 1912, 12: 275–284; "The Gasometric Determination of Aliphatic Amino Nitrogen in Minute Quantities," 1913, 16: 121–124; "Manometric Determination of Primary Amino Nitrogen and Its Application to Blood Analysis," 1929, 83: 425–447.
54. DONALD D. VAN SLYKE and GUSTAVE M. MEYER, "The Fate of Protein Digestion Products in the Body. III. The Absorption of Amino-Acids From the Blood by the Tissues," *Journal of Biological Chemistry*, 1913, 16: 197–212, pp. 204, 206, 212.
55. DONALD D. VAN SLYKE and GUSTAVE M. MEYER, "The Fate of Protein Digestion Products in the Body. IV. The Locus of Chemical Transformation of Absorbed Amino-Acids," *Journal of Biological Chemistry*, 1913, 16: 213–229, pp. 219–220, 228–229.
56. VAN SLYKE (1942), pp. 42–43.
57. KOHLER (1982), p. 246.
58. VAGN SCHMIDT, "Ivar Christian Bang (1869–1918), Founder of Modern Clinical Microchemistry," *Clinical Chemistry*, 1986, 32: 213–215.

59. IVAR BANG, "Ein Verfahren zur Mikrobestimmung von Blutbestandteilen," *Biochemische Zeitschrift*, 1913, 49: 19–39.
60. IVAR BANG, "Über Die Verwendung Der Zentrifuge In Der Quantitativen Analyse," Festschrift Für Olof Hammarsten, *Upsala Läkareförenings Förhändlinger*, 1905–06, 11: 1–15; see also "The Use of Centrifuge in Quantitative Analysis," *Chemical Abstracts*, 1907, 1: 197; "Über die Bestimmung des Blutzuckers," *Biochemische Zeitschrift*, 1908, 7: 327–328.
61. W. R. BLOOR, "The Determination of Small Amounts of Lipid in Blood Plasma," *Journal of Biological Chemistry*, 1928, 77: 53–73. Accurate lipid analysis is never an easy procedure, particularly with micromethods. Exacting quantitative care must be taken at every step (p. 68).
62. D. D. VAN SLYKE, "Ivar Christian Bang," *Scandinavian Journal of Clinical & Laboratory Investigation*, 1958, Supplement 31, pp. 18–26, p. 21.
63. WENDELL T. CARAWAY and HERBERT FANGER, "Ultramicro Procedures in Clinical Chemistry," *American Journal of Clinical Pathology*, 1955, 25: 317–331; Edwin M. KNIGHTS, JR., RODERICK P. MACDONALD, and JAAN PLOOMPUU, *Ultramicro Methods for Clinical Laboratories* (New York: Grune & Stratton, 1957); SAMUEL NATELSON, *Microtechniques of Clinical Chemistry* (Springfield, Illinois: Charles C. Thomas, Publisher, 1957, 1961, 1971); E. J. KING and I. D. P. WOOTTON, *Micro-Analysis in Medical Biochemistry*, 3rd ed. (New York: Grune & Stratton, 1956); for special techniques see PAUL L. KIRK, *Quantitative Ultramicroanalysis* (New York: John Wiley & Sons, Inc., 1950). See also ALBERT E. SOBEL, "Quantitative Microchemistry in Clinical Laboratories," *Industrial and Engineering Chemistry. Analytical Edition*, 1945, 17: 242–245.
64. W. D. FOSTER, *A Short History of Clinical Pathology* (Edinburgh and London: E. & S. Livingstone Ltd., 1961), p. 49.
65. The sections on the hypodermic syringe are derived mainly from the following: JOHN S. HALLER, JR., "Hypodermic Medication. Early History," *New York State Journal of Medicine*, 1981, 81: 1671–1679; G. A. MOGEY, "Centenary of Hypodermic Injection," *British Medical Journal*, 1953, 2: 1180–1185; NORMAN HOWARD-JONES, "The Origins of Hypodermic Medication," *Scientific American*, 1971, 224 [1]: 96–102; see also "A Critical Study of the Origins and Early Development of Hypodermic Medication," *Journal of the History of Medicine and Allied Sciences*, 1947, 2: 201–249.
66. JOHN C. DACOSTA, JR., *Clinical Hematology. A Practical Guide to the Examination of the Blood With Reference to Diagnosis*, 2nd ed. (Philadelphia: P. Blakiston's Son & Co., 1905), pp. 33–34.
67. *Ibidem*, pp. 144, 145.
68. The same technique using several cups applied to the patient's chest and back *without* cutting into the skin, i.e, dry-cupping, drew blood to the surface of the body, which helped to reduce fever in patients with chest colds (bronchitis, influenza, etc.). Signs announcing "We do Cupping" were prominently displayed in barber shop windows in New York City as late as the 1940s.
69. FOLIN and DENIS (1912), pp. 527–528.
70. VICTOR CARYL MYERS, *Practical Chemical Analysis of Blood. A Book Designed as a Brief Survey of This Subject for Physicians and Laboratory Workers*, 2nd ed. (St. Louis: C. V. Mosby Company, 1924), pp. 29–30. For a similar design also see CLYDE L. CUMMER, "Device for Withdrawing Blood From Veins," *Journal of Laboratory and Clinical Medicine*, 1919, 5: 257–259.

71. R. B. H. GRADWOHL, *Clinical Laboratory Methods and Diagnosis. A Textbook on Laboratory Procedures With Their Interpretation*, 3rd ed., vol. 1 (St. Louis: C. V. Mosby Company, 1943), pp. 153–154.
72. HENRY J. JOHN, "Preservation and Transportation of Blood for Chemical Study," *Archives of Pathology and Laboratory Medicine*, 1926, 1: 227–236, pp. 234–235. An illustration of the device also appears in volume 100, January 1976, p. 2.
73. *The Echo*, 1991 (Spring) 11: 3–5; (September) 11: 5–7; 1996 (December) 16: 1 (publication by Becton Dickinson and Company, Franklin Lakes, New Jersey).

CHAPTER XIII

INTRODUCTION

By the start of the twentieth century, methods for the quantitative determination of glucose depended upon three of its properties: reduction of cations of certain heavy metals (copper, bismuth) or of nitroaromatic acids (picric acid) in alkaline solution; dextrorotation of plane polarized light; and generation of carbon dioxide during fermentation with yeast. None of these properties, however, is specific to glucose. Neither the polariscopic nor fermentation methods lend themselves to rapid, high-volume analyses. Polariscopic methods require an expensive apparatus with little other use in the laboratory and are not adapted to the determination of the small amounts of sugar found in blood. Determination of fermentable sugar requires two measurements of reducing substances—before and after reaction with yeast.

Since the content and variability of non-glucose reducing substances in blood and urine is low in comparison with the increases in blood glucose that show up in glycosuria or hyperglycemia, for most clinical purposes the reduction methods were satisfactory. These have been developed in an almost endless variety and may be classified according to the oxidizing agent (usually copper) and the chromogenic analytical procedure for measuring the amount of this agent reduced. Of course, blood could not be analyzed until protein-precipitating agents were developed—mostly by Folin.

BENEDICT'S REAGENT FOR URINE GLUCOSE

In 1911 Benedict [1] devised a new and successful formulation of a copper sulfate reagent that is ten times more sensitive for glucose in urine than Fehling's solution and less susceptible to reduction by non-glucose reducing substances (creatinine, uric acid). He retained the principle of Fehling's method, but replaced the strong alkali with sodium carbonate which is less destructive of glucose in hot solution, and replaced the tartrate with sodium or potassium citrate. The resulting single reagent, stable indefinitely, also contains potassium ferrocyanide to keep cuprous oxide in solution, and potassium sulfocyanate. By addition of the sulfocyanate, Benedict achieved what had eluded other chemists, namely, a sharp endpoint when the sugar solution is titrated directly into the boiling copper reagent. This time the cuprous copper formed by the reducing sugar is precipitated not as the red oxide but as a white cuprous sulfocyanate. At a readily observed endpoint, the blue color of the copper sulfate solution is completely replaced by the white of the cuprous sulfocyanate precipitate.

Benedict's reagents for qualitative and semi-quantitative analysis of sugar in urine were employed in clinical laboratories well past 1960. Both methods appear in *Clinical Chemistry. Principles and Technics* (1964) by Richard J. Henry; the qualitative procedure is described again in the second edition (1974); and in *Todd-Sanford Clinical Diagnosis by Laboratory Methods* (1974), 15th edition, by Israel Davidsohn and John Bernard Henry; and in *Fundamentals of Clinical Chemistry* (1976), 2nd edition, by Norbert W. Tietz.

ANALYSIS OF BLOOD GLUCOSE

The widespread clinical application of sugar determinations in blood began with the almost simultaneous publication in 1913 of a micro-copper procedure by Bang in Sweden [2], and a picrate procedure by Lewis and Benedict in the United States [3]. These methods used small volumes of blood. Bang's method was applicable to the analysis of 0.1 cc capillary blood obtained by puncture of the skin. Benedict used larger amounts of blood (2 cc) which were now available by venipuncture with hypodermic needle and syringe. The perfecting of methods requiring only small quantities of blood drawn from a vein or a few drops of blood from a finger, was hailed in 1914 as "perhaps the greatest recent advance in diagnosis by chemical means." Now that blood glucose analysis could be made daily, "a new chapter in the study of diabetes is being opened" [4].

In Bang's micromethods (see Chapter 12), one or two small drops of blood (about 100 mg) were absorbed on prewashed, dried filter paper and weighed on a torsion balance. After the proteins were chemically fixed to the paper, an eluate was boiled with a copper reagent for a specific brief period. The reduced copper was titrated iodometrically using starch solution as indicator. Comparison was made with standard solutions of glucose similarly treated. Although the method was cumbersome, it was adapted by Bang for determination of other blood constituents. Bang objected to venous section (incision) as it was then practiced and advocated skin puncture for true microanalysis. He saw little difference between 2 cc and 50 cc or more of blood for analysis because in both cases a venesection must be made [5].

Lewis and Benedict utilized a saturated solution of picric acid that served a dual function—as protein precipitant and as sugar oxidizing agent. The application of heat to the protein-free filtrate made alkaline with sodium carbonate produces a red color which can be quantitated in the visual colorimeter. The method was simple enough for clinical as well as scientific investigation. The response by chemists and clinicians was immediate and led to numerous modifications. Myers and Bailey [6] simplified the procedure by using a smaller dilution of the blood and adding sufficient dry picric acid to precipitate the proteins and make the filtrate saturated and ready for color development without the heating (evaporation) step. This procedure was used by Charles Best in his analyses of blood sugar in the pioneer study of the effectiveness of insulin administered in diabetes (see Chapter 8).

Many clinical studies of blood sugar determination appeared during the next few years. Early European workers followed Bang in application of methods which were designed for the small volumes of capillary blood obtained by skin puncture. Americans, on the other hand, for a considerable time, worked with venous blood. In view of Chauveau's early finding (1856) with Barreswil's reduction method, that sugar levels in arterial nonfasting blood resemble those in capillary blood and are higher than in venous blood—by now a frequently confirmed observation—it is not surprising that the early results of European and American investigators frequently disagreed.

The availability of a simple method for blood sugar made it possible to follow blood sugar levels at timed intervals after the administration of carbohydrate, usually glucose. No distinction had been made previously between the use of venous blood or finger blood for glucose tolerance. Foster, a former student of Folin, showed that the sugar concentration in the two samples is the same in the fasting state, but following the ingestion of glucose there is a greater and more rapid rise in finger capillary blood than in venous blood [7].

METHOD MODIFICATIONS BY FOLIN AND BENEDICT

In 1915 Folin described a test for the small amounts of glucose he recognized to be present in normal urine, a fact which, he said, was "not adequately shown or recognized by the current qualitative 'tests for sugar' in urine." [8]. In 1918 Benedict and colleagues abandoned copper reduction in favor of a modified picrate–picric acid method based on his earlier procedure (Lewis and Benedict) for sugar in blood [9]. They also recognized that sugar is continuously eliminated in all normal urine and believed that, "Progress in the study of carbohydrate metabolism will probably be more rapid if the term 'glycosuria' can be abolished." The word was created "by the inefficiency of the copper tests." Since glycosuria implies the *new* appearance of sugar in the urine, they considered the term to be misleading [10].

In 1919 Folin and Wu introduced a new copper reduction method as part of a new system of blood analysis. The picrate–picric acid method of Benedict was criticized as subject to errors that produced higher results than the new method [11]. That same year, at a meeting of the American Society of Biological Chemists, Benedict condemned Folin and Wu's new blood sugar method "on the ground of excessive, inevitable, and uncontrollable reoxidations of cuprous oxide." Folin and Wu responded by introducing their special tube with the constricted neck connecting the bottom bulb with the upper portion (Fig. 13.1). By thus greatly reducing the surface of the solution exposed to the atmosphere, this innovation lessened the reoxidation of cuprous oxide to a negligible amount. They "gladly give Benedict credit for having compelled us to reexamine our method with reference to the effect of reoxidation" [12].

Then in 1925, Benedict acknowledged that the picric acid methods yield values higher than the actual glucose content of the blood, but he doubted "whether a clinician has ever been misled in his interpretation of a diabetic case by the figures obtained for these analyses when properly carried out." He presents a new method for blood sugar determinations with a more specific alkaline copper reagent than before. Another modification is the addition of sodium bisulfite, which increases the formation of cuprous oxide. He also reminds the reader that many years earlier (1907) he had pointed out the "greater delicacy and specificity of copper reagents containing carbonate in place of hydroxide...." Instead of the constricted Folin and Wu tubes, Benedict took different precautions. He used ordinary test tubes and, prior to heating, added several drops of benzene to layer the reaction mixture, then stoppered the tubes with cotton. During the heating, the benzene (flammable and toxic!) is completely vaporized, but remains in the tube and excludes the air. Benedict's reagent, practically unaffected by the non-sugar reducing substances present in blood, yields values distinctly lower than those with the Folin and Wu copper method and presumably closer to the true glucose levels. Although the average of analyses for seven healthy individuals was 73 mg per 100 cc, he believed

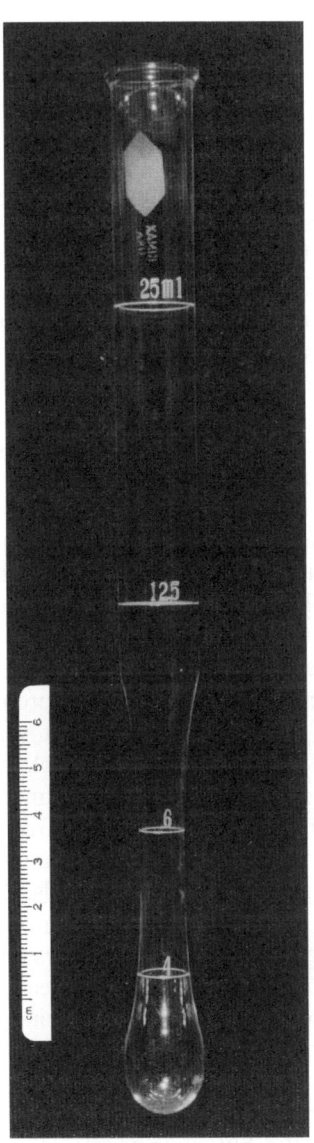

FIG. 13.1. Folin–Wu Blood Sugar Tube with additional markings at 6 and 12.5 mL.

that the actual glucose content of normal fasting human blood was even lower and does not usually exceed 60 mg per 100 cc [13].

Folin answered with another modified copper reagent for use with blood and urine and explained Benedict's dissatisfaction with the new Folin and Wu tube as due to "compensating experimental errors" [14]. Folin could not confirm lower blood glucose values with Benedict's new reagent and concluded that the bisulfite was unstable. Benedict agreed but, continuing his criticism of Folin, showed by a recovery experiment—which Folin had not carried out—that Folin's new procedure and copper reagent "is not adapted to the determination of sugar in normal urine" [15].

The search for increased specificity and sensitivity continued. Every physical and chemical parameter of a new test procedure, e.g., timing, temperature, reagents, etc., was examined and modified in a continuing quest for constant improvement. Findings were reported in great detail by Folin and Benedict, with interpretations by each keyed to the previous reports by the other. Newer methods by Folin [16] and Benedict [17] yielded results that averaged about 20 to 30 mg less per 100 cc than by other reported methods because they were less susceptible to non-glucose reducing substances in the blood. These glucose levels were closer to the "true" value obtained by yeast fermentation.

As already noted, yeast fermentation procedures require two analyses. The difference in reducing material in filtrates prepared from two samples of the same specimen of blood—one without yeast and one following fermentation with yeast—represents fermentable sugar, which is almost entirely glucose. In normal blood, this treatment leaves a residue of non-glucose reducing substances equivalent to from 10 to 30 mg of glucose per 100 cc. Fermentation of the sugar is apparently complete in twenty minutes. During prolonged fermentation periods, the continued growth of the yeast may produce its own reducing substances. Slightly higher values were repeatedly observed at the end of two hours than at the end of twenty minutes of incubation. Apparently, secondary reactions in the yeast–blood mixture result in a slow formation of reducing substances [18]. This procedure, however, does not distinguish glucose from other fermentable hexoses such as fructose, mannose or sometimes galactose. Analytical procedures such as fermentation, osazone formation, and separation by chromatography, are useful but cumbersome, slow, and difficult to quantitate and are not practical for a "routine" testing clinical laboratory.

A NEW GLUCOSE REAGENT—ALKALINE FERRICYANIDE

An alkaline potassium ferricyanide procedure for blood glucose was introduced by Hagedorn and Jensen in Denmark in 1923 [19]. Quantitative values were based on the inverse reaction of the reducing sugar and a new reagent, potassium ferricyanide. What occurred on heating, was a progressive diminution of the reagent's yellow color, with increasing concentrations of glucose. The protein-free filtrate was prepared with zinc sulfate, sodium hydroxide, and only 0.1 cc of whole blood. Their procedure, an iodometric titration of excess ferricyanide (oxidizing reagent) following reaction of filtrate and reagent, displaced Bang's method and quickly achieved almost universal use in Europe. The advantage over copper reagents was that the reduction product, ferrocyanide, remains in solution and, being more stable than cuprous oxide, is not readily oxidized by the air. The unreacted ferric salt can be easily determined. After addition of potassium iodide and sulfuric acid solutions, iodine separates out and colors the solution. This is titrated with sodium thiosulfate solution until the mixture is almost clear of the iodine color. Starch solution is added and titration is continued until the characteristic starch-iodine blue color disappears.

Because of its advantages, ferricyanide methods were widely accepted in the United States for blood glucose analysis. Even Folin [20] devised a ferricyanide procedure using 0.1 cc whole blood. In his procedure, the ferrocyanide produced by the reducing sugar is measured as Prussian blue following addition of acidified ferric sulfate. Both of his sugar methods appear in the fifth edition of Folin's *Laboratory Manual of Biological Chemistry* (1934). The diminution of the ferricyanide's yellow color in glucose analysis was the first reaction adapted to the AutoAnalyzer in 1957 (see Chapter 19).

SOMOGYI PROTEIN-FREE FILTRATE

The reducing property of glucose is exhibited by many other substances found in appreciable concentration in normal blood, e.g., creatine, creatinine, uric acid, and glutathione. This non-glucose reducing fraction tends to be rather constant. For many years, established methods challenged investigators to devise reagents that improved the specificity of the analytical procedure.

A new approach to minimizing non-glucose reduction was introduced by Somogyi [21]. Instead of investigating the oxidizing reagent, he developed a combination of zinc sulfate and barium hydroxide (forming zinc hydroxide) that would remove the non-glucose reducing substances with the proteins of whole blood, plasma or serum. Zinc sulfate worked well with sodium hydroxide for whole blood, but fails to completely deproteinize plasma. With barium hydroxide, the filtrates are almost entirely free from non-fermentable reducing substances. The high concentration of sodium sulfate in the copper reagent protects the cuprous oxide against reoxidation by atmospheric oxygen, so that neither the constricted Folin–Wu tubes nor covered tubes are necessary. The zinc–barium filtrate is heated with the alkaline copper reagent and then reacted with an arsenomolybdate color reagent. The Nelson–Somogyi modification was a popular method that was widely used during the years preceding the introduction of automated analysis [22].

TABLETS, POWDERS, AND DIPSTICKS

In the 1940s, copper and bismuth reagents were incorporated into a tablet (Clinitest) and dry powder (Galatest), respectively, for semi-quantitative estimation of glucose in urine, as determined by the intensity of the resulting color formation when reacted with urine in a prescribed procedure. The tablet copper reduction test was introduced in 1941 by the Ames division of Miles Laboratories (now Bayer), Elkhart, Indiana.

Clinitest tablets consist of anhydrous copper sulfate, anhydrous sodium hydroxide, citric acid, and sodium carbonate. The amount of copper sulfate was selected to yield maximum color with a concentration of two grams glucose per 100 mL when a specific volume of urine mixed with water is tested. Upon solution of the tablet, the sodium hydroxide provides sufficient heat for the reduction–oxidation reaction. The citrate holds the cupric ions in solution, and the carbon dioxide released by the small quantity of sodium carbonate blocks oxygen from contact with the reducing sugar. The color produced after fifteen seconds of reaction is compared against the manufacturer's color scale derived with glucose standards.

Reduction of bismuth as a test for glucose was developed by E. Nylander in 1884. A dry powder test, patented by A. Galat (1942) and marketed as Galatest by the Denver Chemical Company, New York, well into the 1960s, consisted of a mixture of bismuth oxychloride (BiOCl), water-free magnesium silicate, and caustic alkali. The alkali furnished the heat when the powder was mixed with an aqueous solution. The resulting coloration of the black precipitate of free bismuth varied from light gray to dark black and represented a glucose range from less than 0.2 gram to greater than 1.0 gram per 100 mL of test solution.

Clinitest (still in use) was followed in 1956 by Clinistix (for glucose), the first color-change dip-and-read test that depends on the action of an enzyme impregnated into a strip of stiff paper. These were followed by multiple-analyte dipsticks introduced in 1959 by Ames/Miles/Bayer. In 1975 the Boehringer Mannheim Corporation, Indianapolis, Indiana, marketed its own multi-test dipsticks. In the 1980s, these companies manufactured instruments that read and interpreted the dipstick color changes and produced a print-out of the results.

ANOTHER NEW GLUCOSE REAGENT—ORTHO-TOLUIDINE

Various aromatic primary amines react with glucose in hot acetic acid solution to produce colored derivatives. Dubowski [23] developed this reaction using ortho-toluidine (6%) in glacial acetic acid solution on protein-free filtrates of serum, at 100°C for ten minutes. The chemical reaction, specific for aldohexoses, involves the condensation of a primary aromatic amine with the aldehyde portion of the glucose molecule. Additional rearrangements and reactions take place after the initial condensation to produce a stable green chromogen whose intensity is proportional to the glucose concentration. Although this reaction does not distinguish galactose or mannose from glucose, the method is reliably selective since the concentration of those aldohexoses in human blood is usually insignificant. This simple and rapid procedure, readily carried out without deproteinization, yields almost identical results as the ferricyanide method. It experienced a period of popularity in a manual and automated mode in the 1960s and early 1970s prior to the introduction of glucose-specific enzymatic techniques which eventually replaced all other methods for glucose.

ENZYMATIC METHODS FOR "TRUE" GLUCOSE

The determination of true glucose became a practical possibility with the discovery of a specific enzyme for glucose, glucose oxidase (notatin), by Müller [24], and its preparation in highly purified form in 1945. The pronounced specificity of the enzyme for beta-glucose was established by Keilin and Hartree [25]. The introduction in 1956 by Froesch and Renold [26] of an enzymatic method for the determination of glucose in biological fluids gave clinical biochemists the first specific method for its determination. In the presence of glucose oxidase, glucose is oxidized to gluconic acid and hydrogen peroxide. In their method, reducing substances were determined by the Nelson–Somogyi method using alkaline copper and arsenomolybdate reagents, before and after the enzymatic treatment.

Later that same year, Keston [27] carried the principle of enzymatic oxidation of glucose a step further. He suggested the use of a coupled enzyme system in which the hydrogen peroxide formed in the enzymatic oxidation of glucose can, in the presence of added peroxidase, oxidize a suitable substrate (oxygen acceptor) o-tolidine, which changes color on reaction with hydrogen peroxide. The solution of combined reagents was applied to filter paper which, when dried in air and protected from light, can be employed as a sensitive reagent paper to detect glucose, when dipped in and out of urine and the color change compared with colors produced by standard glucose solutions. A few months later, Teller [28] applied the principle of the coupled enzyme system, with o-dianisidine as the chromogenic oxygen acceptor, to the direct photometric measurement of glucose in serum or plasma, omitting deproteinization. The chromogen's color intensity is proportional to the amount of hydrogen peroxide formed and hence to the amount of glucose originally present. Color comparison is made to standards similarly treated. Shortly afterward, Huggett and Nixon [29] in England, and Saifer and Gerstenfeld [30] in America, introduced methods of glucose estimation in plasma and protein-free blood filtrate based on this principle, using o-dianisidine as oxygen acceptor.

Before long, two commercial filter paper strip tests incorporating these enzyme-mediated reactions for the semi-quantitative specific determination of glucose in urine became available—Clinistix (Miles Laboratories) [31] and Tes-Tape (Eli Lilly) [32]—covering the range from negative to 2% glucose. Both used o-tolidine as the chromogenic reagent. Combination test strips followed in 1959 with reagent-impregnated segments on the strip that estimated glucose, protein, pH, ketone, blood (hemoglobin), bilirubin, urobilinogen, and phenylpyruvic acid, in a variety of combinations and from several manufacturers of diagnostic products. At present, strips may contain segments that also estimate specific gravity, and levels of nitrite and leukocytes. Reagents specific for the particular analyte, layered into separate test portions of the strip, produce a color change for a positive reaction over a range of levels and are interpreted according to the manufacturer's color chart. Although the semi-quantitative estimation of glucose is approximate, the dipsticks are useful as a convenient rapid screening procedure and to help distinguish diabetic coma from insulin shock or other coma—glycosuria in the former, not in the latter.

CREATINE, CREATININE, AND CREATININE CLEARANCE

In a report presented in 1832 to the French Academy of Sciences—and published in 1835—Michel Eugene Chevreul (1786–1889) described a new nitrogen-containing organic constituent of meat extract, which he isolated in crystalline form and named creatine (Greek: *kreas*, meat). Evaporation with hydrochloric acid converted it into another new substance, distinguishable by its different crystalline form. However, he did not determine its composition. Chevreul is best known for his work in determining the chemical nature of animal fats, which established him as one of the major figures in the early development of organic chemistry.

Liebig was the first to point out the general distribution of creatine and to submit it to quantitative elementary analysis. In 1847 he established its empirical formula. He also discovered that creatine would lose water on heating with mineral acids and was converted into another base which he named creatinine. While this name was first used by Liebig, the substance itself had actually been encountered before. It must have been the hydrochloride of creatinine that Chevreul produced when he evaporated creatine with hydrochloric acid. In 1844, while working in Liebig's laboratory in Giessen, Pettenkofer had reported the discovery in human urine of a new nitrogenous organic substance, precipitable by zinc chloride as a crystalline double salt compound (see Chapter 6, reference 44). Liebig proved that the substance uniting with zinc chloride was creatinine.

In the early part of the nineteenth century, there had been considerable doubt about the presence of glucose and urea in normal human blood. A century later, creatinine experienced a similar challenge. The identity of the "chromogenic substance" was long a subject of controversy. Numerous modifications in method were unable to prove, by isolation or identification, that creatinine or creatine really existed in blood. Doubts were raised by Hunter and Campbell [33] in 1917, when they found that whole blood filtrates treated with alkaline picrate solution continued to increase in color intensity for a longer period than did pure creatinine. They suggested that a considerable part of the color was probably not due to creatinine.

In 1922 Behre and Benedict warned against hasty determination on the basis of a single, nonspecific color reaction of "substances whose existence in the tissues or fluids analyzed has never been proved... The modern color reactions are very attractive playthings, but the facility with which they can be employed should not lead to neglect of the more fundamental work of seeking definitely to prove exactly what these color reactions may signify" [34]. They reported striking differences in behavior between pure creatinine and the chromogenic substance in blood filtrates. The evidence suggested that only a small part, if any, of the color ordinarily formed with alkaline picrate in blood filtrates could be attributed to creatinine, even when the chromogenic substance was greatly increased after impairment of renal function. Creatinine added to blood was readily detected in the filtrate.

They concluded: "Our finding that creatinine does not exist in blood in detectable quantities need not, of course, raise any question as to the value of the determination of the chromogenic substance for clinical or other purposes." How then, does one explain the presence of creatinine in the urine? Behre and Benedict suggested that the kidneys may be able to concentrate the creatinine from extremely dilute solutions or produce the creatinine from some precursor substance in the blood [35]. Myers supposed that the chromogenic substance in blood is closely related to creatinine and readily converted to creatinine in the kidney. It was difficult to explain the increase in chromogenic substance in nephritic bloods as due to anything other than impaired excretion of creatinine. The creatinine of blood was significantly increased only after considerable retention of urea had taken place and the nephritis was far advanced [36]. At this stage, creatinine (endogenous in origin) is a more reliable indicator of diminishing kidney function than is urea (largely exogenous). Influenced by diet, urea levels are subject to greater fluctuations.

Despite complete ignorance of the nature of the substances that cause the Jaffe reaction in blood filtrates, blood "creatinine" determinations by the Folin–Wu method continued to be carried out in nearly all clinics where renal function was studied. Clinical observation had shown that in nephritis an increase in the amount of chromogenic substance in the blood is of grave prognostic significance. Van Slyke did not believe that the chromogenic substance could all be due to creatinine. "It is regrettable only that this unknown substance or mixture of substances continues to be called 'creatinine' in laboratory reports, and probably to be considered as creatinine by most physicians to whom such reports are rendered" [37].

The controversy over the actual existence of creatinine in normal blood was decided in 1937 by the isolation of bacterial enzymes capable of decomposing creatinine [38]. The difference in "creatinine" before and after enzymatic action represented the true creatinine. However, the routine use of enzyme methods had to wait for the commercial availability of pure enzyme several decades later.

Owing to the relatively large amounts of nonspecific Jaffe-positive material present in erythrocytes (50–70%)—serum or plasma, with less than 20% non-creatinine chromogenic material, is more suitable for analysis than whole blood. In urine nearly 100% of the picrate-reacting material is creatinine. Despite its shortcomings, the Jaffe reaction's low cost and easy adaptation to several automated technologies has kept it in the lead of creatinine methodology [39].

In 1926 when Rehberg [40] found that the ratio of the creatinine concentration in urine to its concentration in serum was larger than for any other biologically excreted substances, he suggested that creatinine was filtered primarily through the glomeruli and was not reabsorbed. Although a small amount of creatinine is also secreted by the renal tubule, for clinical purposes, creatinine clearance provides a reasonable approximation of glomerular filtration rate.

The creatinine clearance test that Rehberg developed is a calculated value that estimates the glomerular filtration rate from the creatinine levels in plasma or serum and in a timed urine volume—preferably a 24-hour collection, to minimize within-day variations. The procedure is considered the best measure of glomerular filtration rate and the most sensitive clinical indicator of early renal disease. Since a timed urine specimen is not always available, there is an increasing reliance on serial serum creatinine levels to monitor the patient once an initial clearance has been determined. It is the rate of change in renal function rather than any single value that reflects the course of renal disease. Creatinine clearance may then be measured periodically or if there is a special indication for it [41].

There has been one other useful application of the creatinine analysis in urine—though not without controversy. Folin had reported in 1905 that creatinine output from day-to-day was independent of urine volume and total nitrogen and showed only small variations as compared with all other nitrogenous constituents. He suggested that the 24-hour urine creatinine value may be used to verify the completeness of collection when the data of several days of collection point to an expected average value [42]. Numerous reports have appeared from time to time indicating an average variation of about 10% for the creatinine output in the same individual; therefore, creatinine excretion is not sufficiently constant to be a reliable index of completion of a 24-hour collection. However, it does appear to be a reproducible index in some individuals although not in others [43]. Because the amount formed and excreted probably is the result of various physiological factors, there is a need to determine the stability of a given individual's output before using this index to verify completeness of future collections. Of course, complete urine collection in the newborn and with infants is much more difficult than with adults.

URIC ACID

Folin's interest in uric acid originated during post-graduate work in Germany and continued to challenge his research for many years. Uric acid, thought to be a product of protein catabolism, was first identified by A. B. Garrod in 1848 in trace amounts in normal blood serum and in increased amounts in gout and some cases of nephritis [44]. In his gravimetric method, 1000 grains of serum (65 cc) was evaporated to dryness, the residue washed with boiling alcohol, dried, then extracted several times with boiling water. Large volumes of blood were available at that time since blood-letting was still being practiced.

After evaporation for some hours, the concentrated aqueous extracts deposited out crystals of sodium urate. The crystals were washed with alcohol, dried and weighed. In his later clinical work, Garrod gauged the concentration of uric acid with his famous "uric acid thread experiment" [45]. Serum was diluted ten-fold with acetic acid of "moderate strength" in a flattened glass dish, and a delicate linen thread was suspended from the mixture. After 18 to 48 hours of drying at room temperature, uric acid at levels of 0.025 parts per 1000 of serum (2.5 mg/100 cc and greater) will crystallize and be attracted to the thread and adhere to it. The crystals are easily seen with a magnifying glass, the crystal size varying with the speed of drying. This finding, according to Garrod, was indicative of an abnormal amount in the serum. Lack of suitable methods delayed for a long time the study of the presence of uric acid in normal human blood. In fact, until the early years of the twentieth century, there was considerable doubt whether normal blood ever contained uric acid.

Before Folin's introduction of colorimetric micromethods, determinations of uric acid were limited to the urine. The amounts present in blood were too small to be analyzed in the limited volumes and by the means available. From 1875 to 1892, the accepted procedure for urine was the method of Salkowski and Ludwig (1870), in which the uric acid was precipitated with ammoniacal silver nitrate and magnesium chloride, the precipitate was decomposed with hydrogen sulfide, and the free uric acid was crystallized from hydrochloric acid and weighed.

In 1892 Hopkins showed that uric acid could be quantitatively precipitated by ammonium salts, particularly by saturation with ammonium chloride, and could be titrated in the redissolved precipitate by means of potassium permanganate [46]. This method, modified by various investigators including Folin, remained for many years the accepted volumetric analysis for uric acid in urine, until colorimetric estimations were introduced.

In 1894 Offer [47] reported that uric acid in alkaline solution with phosphomolybdic acid produces a blue color. Frabot [48] noted in 1904 that uric acid added to alkaline tungstate solution developed an intense blue color. There was no noteworthy advance in uric acid analysis until 1912 when Folin and Macallum [49] made a preliminary study of the conditions of these reactions and then Folin and Denis [50] applied them to blood analysis. In the course of these studies they discovered two reactions yielding a blue color. In one, urates were reacted with phosphotungstic acid; in the other, phenolic compounds were reacted with phosphotungstic–phosphomolybdic compound [51]. The color reaction, especially with aromatic compounds containing a hydroxyl group in the para position, e.g., tyrosine, was later developed into a quantitative method for plasma proteins—in conjunction with a fractionation step—on the basis of the tyrosine content of the particular protein, by Wu [52] and improved by Greenberg [53]. The method is suitable for the assay of single proteins of known and constant tyrosine content, such as fibrin, but it is not generally recommended for protein mixtures such as serum or plasma. For other color reactions of proteins, see Chapter 17.

The highlights of Folin's method, and those of others, for analysis of uric acid, are briefly described here, not only because uric acid was a lifelong interest of Folin, but because the stepwise changes and improvements which the analysis experienced are typical—more or less—for every other constituent of blood and urine. Only the details are different. Some blood components, like glucose and cholesterol, have had an even more varied adventure in the unending quest for increased specificity, sensitivity, accuracy, and precision—while at the same time utilizing as small a specimen as practical, with as rapid and simple a procedure as possible, and as low a cost as reasonable. This is what clinical chemists did!

In the Folin and Denis (1912–13) method for uric acid, the blood proteins were removed by coagulation with boiling weak (0.01 N) acetic acid and the filtrate evaporated to a very small volume. This was treated with silver lactate, magnesia mixture, and ammonia, in a small centrifuge tube, to dissolve the silver chloride that formed and to precipitate the uric acid as the silver salt. The precipitate was redissolved and decomposed with saturated hydrogen sulphide water; the excess sulphide was removed by heating on a boiling water bath; and the color was developed with phosphotungstic acid reagent and sodium carbonate. The blood sample (15 to 20 cc) was a weighed specimen and results were reported per 100 grams of blood.

As originally performed the method was not very accurate. Numerous difficulties and sources of error were encountered, including turbidity, difficulty in preparing stable standards, difference in rate of color development for blood filtrate and standard solution, faintness of color produced, and interfering blood substances which produce either positive or negative errors.

In 1915 Benedict [54] improved the accuracy, sensitivity, and simplicity of this method for uric acid in urine and blood by combining the reagents. Then, instead of decomposing the precipitated uric acid as is done with hydrogen sulphide water, he dissolved it with two drops of a 5% solution of potassium cyanide. Addition of cyanide greatly increases the intensity of the color reaction and markedly slows the rate of fading. According to Folin [55], probably less than ten percent of the color obtained results from direct reduction of the uric acid reagent by the urate, while the other ninety percent represents indirect or induced reduction due to the presence of the cyanide. Benedict also added a second precipitation step (by means of colloidal iron) because not all the protein was removed in the preliminary coagulation by heat and acetic acid as in the method of Folin and Denis. Benedict used 20 cc of blood for this determination.

In 1919 Folin and Wu estimated uric acid directly in the protein-free filtrate prepared by their new system of blood analysis utilizing tungstic acid precipitation of proteins [56]. This technique greatly simplified the removal of proteins and, by eliminating the heat and acid treatment, avoided loss of uric acid in the protein coagula. Then, they precipitated the uric acid from the tungstic acid filtrate with a lactic acid–silver lactate reagent, prior to developing the color reaction with sodium cyanide, sodium carbonate, and the uric acid reagent of Folin and Denis.

Benedict improved the analysis still further in 1922 when he bypassed the precipitation of uric acid and developed the color directly on the Folin–Wu tungstic acid blood filtrate, as well as directly on urine. Benedict used an arsenophosphotungstic acid reagent and cyanide solution (but no carbonate) to intensify the color and improve the accuracy of the visual colorimeter. Heating developed the color which was nearly seven times as great as with a previously used reagent. Benedict's accurate method developed the color on 5 cc of filtrate, the equivalent of 0.5 cc of blood. But the time constraints of the final heating step—in order to provide uniformity in handling and to finish before any turbidity may develop—limited a series of analyses to five tubes including the standard. An "ultra-micro" modification required only 0.2 cc of blood [57].

Folin [58] adopted some of Benedict's modifications, but continued to use the original Folin and Denis uric acid reagent because he believed that the improvement of Benedict's new uric acid reagent was in the use of the cyanide as the only alkali. Uric acid posed a constant challenge to Folin and in 1930 he published another "improved method" [59]. Urea was added to stabilize the cyanide solution and inhibit its tendency to produce turbidity with blood filtrates. He also made slight changes in the preparation of the uric acid standard to increase its stability.

Van Slyke was "doubtful,... that any method now available for uric acid determination in whole blood is accurate" [60]. Most of the early work on uric acid was with whole blood. Normally however, the uric acid level in plasma is approximately two times as high as in the red cells. Consequently, since whole blood levels are influenced by the hematocrit, which varies widely in health and disease, and since most of the nonspecific reducing substances are in the red cells—specificity and accuracy are increased by analysis of serum or plasma instead of whole blood.

Uric acid held the attention of Folin throughout his career as he sought to prepare a stable uric acid standard and a uric acid reagent free of molybdate which interferes by forming molybdenum blue with phenolic compounds such as tyrosine in the filtrate. Molybdenum is always present in tungsten ores from which sodium tungstate is derived [61].

The search for greater sensitivity, specificity, and accuracy was joined by Benedict as their friendly rivalry on uric acid played itself out on the pages of the *Journal of Biological Chemistry*. Revisions in method and modifications in reagents were frequently reported. Folin characterized the colorimetric method for the determination of uric acid as "probably the most complex reaction that we have in the whole field of practical colorimetry" [62]. Every time Folin thought he had finally prepared a molybdate-free uric acid reagent, he discovered later with different chemicals that the problem had not been solved. When this happened again after his 1933 paper, he described a new process for the preparation of sodium tungstate which, according to spectrum analysis, was completely free from molybdenum [63]. Although this pure sodium tungstate was commercially available, Folin still advised that it be checked with the potassium xanthate test for molybdate [64].

In what proved to be his last publication, he concluded by saying: "This is presumably my last paper on the preparation of the uric acid reagent and I hope that the method for the determination of uric acid described last year is also final." The paper appeared in the August 1934 issue of the *Journal of Biological Chemistry*. Folin died on October 25, 1934 [65].

Neither Folin's death in 1934 nor Benedict's death in 1936 ended the search for an improved uric acid reagent and procedure of analysis. Others picked up the challenge of improving the proportion between color intensity and uric acid content, recovery of uric acid added to tungstic acid filtrates, and specificity of the reaction, as well as lower blanks and the use of different additives to prevent or delay formation of turbidity.

The introduction of cyanide ion into the reaction medium greatly increased the sensitivity of the method. Offsetting this advantage is the poisonous nature of cyanide, along with the instability of cyanide solutions, and the development of considerable color in blank controls. The replacement of visual methods by the more sensitive photometric measurements of color made it possible to return to less sensitive procedures in an attempt to avoid the use of cyanide. In the original Folin and Denis method for uric acid—a lengthy procedure requiring 15 to 20 cc blood—color was developed by employing only phosphotungstic acid and sodium carbonate. Caraway [66] added these reagents directly to a tungstic acid filtrate (or centrifugate) prepared from 1.0 mL serum, to develop color whose absorbance may be determined with a spectrophotometer or photoelectric colorimeter.

The blue color formed by the reduction of phosphotungstic acid by alkaline solutions of uric acid formed the basis of all uric acid methods in general use for blood, plasma or serum analysis until the introduction of uricase-catalyzed coupled enzymatic methods in 1967 [67]. In the enzymatic procedures, uric acid is oxidized by uricase in the presence of peroxidase. Hydrogen peroxide is produced and quantitated by reaction with a chromogenic oxygen acceptor, e.g., o-dianisidine. Earlier versions utilizing uricase measured the formation of "tungsten blue" before and after the enzyme action [68]. Another method with enzymatic breakdown of uric acid, determined the decrease in ultraviolet absorbance at 293 nm, which is proportional to the disappearance of uric acid originally present. However, the ultraviolet method required an expensive spectrophotometer and quartz cells, which were not generally available in the routine clinical chemistry laboratory at that time [69].

UREA AND UREASE

Urea can be decomposed into two molecules of ammonia and one of carbon dioxide when heated to 150° to 200°C in acid or alkaline solution.

This probably was the first method used for its quantitative determination. Bunsen (1848) heated urea with barium hydroxide in sealed tubes for four hours at 220° to 300°C and determined the carbon dioxide formed from the amount of barium carbonate precipitated. This method was used for analysis of blood urea in 1882 by von Schröder. Later, heating methods used acid solutions and a temperature of 150°C in an autoclave [70].

Determination of urine urea by titration with a standard mercuric nitrate solution was introduced by Liebig (1853). A complex salt is formed which is insoluble in alkaline solution. While urea remains in excess, a drop of the mixture added to sodium carbonate solution in a test tube produces a white precipitate of the mercury–urea complex. At endpoint, an excess of mercury is present and a brown coloration or precipitate of mercuric oxide is formed when a drop of reaction mixture is tested with sodium carbonate solution. This indicates that all the urea is complexed. However, the reaction of the mercuric salts with urinary nitrogen compounds is so general that the titration measured practically the total nitrogen [71].

A gasometric hypobromite method introduced by W. Knop (1870) and G. Hüfner (1871) measures the nitrogen gas formed by the reaction of urea with alkaline hypobromite:

$$CO(NH_2)_2 + 8NaOH + 3Br_2 = 6NaBr + Na_2CO_3 + 6H_2O + N_2$$

Evolution of the nitrogen gas is completed in one or two minutes. The method requires no standard solutions; but for more accurate results, preformed ammonia in urine must be removed by shaking with *permutit*, an insoluble sodium aluminum silicate (zeolite) discovered by Gans (1905). The sodium is easily replaced by ammonia which can then be released in strongly alkaline solution—and nesslerized for a separate colorimetric analysis of ammonia. Although the yield of nitrogen gas in the hypobromite method is less than 100%, the negative error is more or less balanced out by positive errors from non-urea substances [72]. The results obtained with these less than specific reactions of heat, mercuric nitrate, and hypobromite, were sufficiently close approximations for most purposes and were in general use. They were eventually abandoned when the more convenient and highly specific urease method was developed in 1913.

As for blood analysis, Folin's method at that time began by precipitation of the proteins with methyl alcohol and an alcoholic solution of zinc chloride. After removal of the alcohol by gentle heat and an air current produced by the vacuum suction of a water pump, the urea in the protein-free filtrate was decomposed by heating with 25% acetic acid and dry potassium acetate, with a melting temperature indicator in the mixture to detect the target range of 153° to 158°C. Excess alkali is added and the liberated ammonia is driven into an acid receiver by an air current, followed by nesslerization [73].

Urea is quantitatively decomposed by the enzyme urease into ammonium carbonate and is usually estimated by determining the ammonia released when the reaction mixture is made alkaline. Urease was discovered in soy beans by T. Takeuchi in 1909, but was later found to be about fifteen times more abundant in jack beans [74]. The enzyme's extreme specificity allowed its use for quantitative estimation of urea directly in a mixture as complex as blood. Marshall developed the first methods for urea in urine [75] and blood [76], which required no preliminary treatment for removal of protein. The procedure called for 10 cc of serum, but satisfactory results could be obtained with as little as 3 cc. The reaction with urease was allowed to proceed overnight at room temperature. Ammonia is released by addition of sodium carbonate powder and transferred by a rapid air current, as suggested by Folin, into an acid receiver. "With a good suction pump, one hour suffices." The excess acid is back-titrated with dilute alkali.

Van Slyke and Cullen improved the method by preparing the enzyme in the form of a soluble and very active powder that can be accurately standardized and maintains its activity indefinitely. Without complicating the technique, they were able to reduce the time required for an analysis from several hours to a few minutes. They described a simple apparatus (Fig. 13.2) that became a familiar sight in clinical laboratories until the introduction of the AutoAnalyzer in the late 1950s [77] (see also Fig. 13.3).

A few years later, in their "system of blood analysis," Folin and Wu abandoned *direct* nesslerization of the urease-treated protein-free filtrate for the determination of blood urea because of interferences by amino acids or peptones. In the new protocol, hydrolysis of urea was accomplished by incubation of the filtrate (with jack bean urease, 10–15 minutes, not exceeding 55°C) or by heating in the autoclave (10 minutes, 150°C). Each of these procedures may be followed by aeration or distillation of the resulting ammonia into a second tube containing hydrochloric acid as receiver. These choices made four possible combinations available for analysis by nesslerization. Starting with a tungstic acid protein-free filtrate, they ended with a nesslerized solution for color comparison against a standard of ammonium sulfate nesslerized at the same time [78].

Van Slyke developed rapid and accurate micromethods on 0.2 cc of blood, plasma or Folin–Wu filtrate, for the carbon dioxide also released by

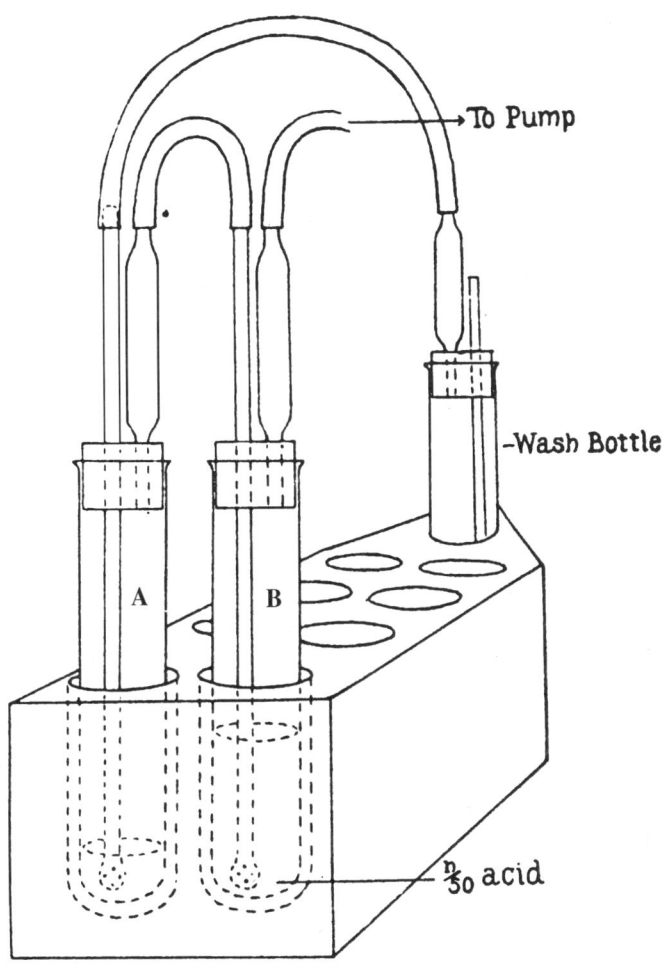

FIG. 13.2. Van Slyke's aeration apparatus for determining urea nitrogen by means of urease. (*Journal of Biological Chemistry*, 1914, 19: 211–228, p. 217.)

the urease reaction. This procedure, performed on Van Slyke's manometric blood gas apparatus, never became popular [79]. It was easier to measure ammonia after aeration or distillation.

In 1939 Fearon [80] showed that urea and many substituted ureas react with diacetyl monoxime in the presence of strong acid and heat to produce

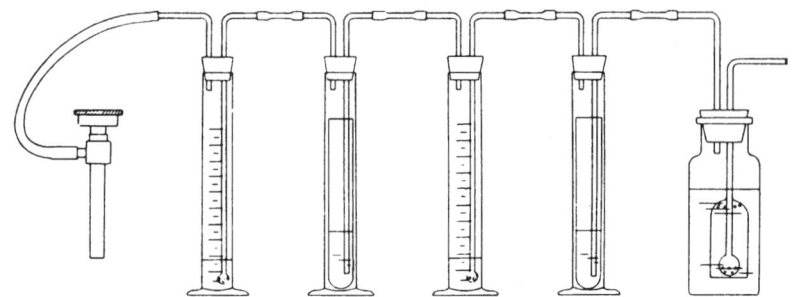

FIG. 13.3. Myers's aeration apparatus for estimations of nonprotein nitrogen, urea, and ammonia nitrogen. (V. C. Myers, *Practical Chemical Analysis of Blood*, 2nd ed., St. Louis: C. V. Mosby Co., 1924, p. 44. Originally in *The Post-Graduate*, 1914, 29: 505–512.)

a colored compound, yellow in the case of urea. The color deepens to yellow-orange on oxidation with potassium persulfate. In 1942 Ormsby [81] applied this reaction to the determination of urea in diluted urine and in protein-free filtrates of blood. By not determining ammonium ions, he did away with the need for a distillation/aeration apparatus and the problems of nesslerization. However, the color produced with diacetyl monoxime is not linear with urea concentration and calibration is necessary; and the unpleasant odor, irritant fumes, and corrosive nature of the reagent, make it advisable to work in a fume hood. The procedure was later adapted to the AutoAnalyzer for serum and urine and as late as 1993 was used by less than 0.4% of the more than 6400 laboratories reporting to the College of American Pathologists (CAP) quality assurance surveys.

The main attraction of urease methods is the extreme specificity of this enzyme for hydrolysis of urea. Consequently, the method of choice for 85% of CAP survey participants in 1993 (on twenty-seven different instruments from twelve different manufacturers) was the coupled urease/glutamate dehydrogenase (GLDH) enzymatic system employing an NAD/NADH indicator reaction which is monitored in the ultraviolet range at 340 nm, either kinetically or as an endpoint reaction [82]. A conductivity method was used by 14% of participating laboratories. In this technique, urease converts urea to ammonium carbonate which dissociates to ammonium and carbonate ions. The increased ion concentration raises the conductivity of the reaction mixture. The rate of change is measured [83].

UREA CLEARANCE

In renal insufficiency, diminished permeability of the kidneys becomes evident by a retention in the blood of the three major components of nonprotein nitrogen: uric acid, urea, and creatinine. The value of urea levels as a reliable measure of kidney function and disorder was recognized as far back as 1904 by Widal and Javal. In 1910 Ambard attempted to estimate the functional ability of the kidneys by a study of the ratio between the urea of the blood and urine. This led to an attempt to express the rate of urea excretion in the form of mathematical formulae.

Experience with Ambard's method was disappointing to American investigators. It led to development of a formula by Van Slyke and associates, which they believed to more correctly express the influence of the various factors governing excretion. As a measure of the functional ability of the kidneys, they determined "urea clearance"—the volume of plasma cleared of urea in a given period of time [84]. The concept stemmed from their studies that compared the concentration of urea in plasma and urine and the volume of urine excreted in two successive one-hour periods. Clearance is calculated from one of two formulas (or a nomogram) depending on whether the volume per minute is less than or greater than two cc per minute. For small individuals, a correction is applied to volume per minute, based on the body's surface area [85]. The test was gradually discontinued by the mid-1970s—hastened by the availability of reliable automated creatinine analysis for determination of creatinine clearance, a better indicator of glomerular filtration rate than the urea clearance.

PHENOLSULPHONEPHTHALEIN EXCRETION

In 1909, while investigating the pharmacological action of some phthalein derivatives, Abel and Rowntree [86] discovered that phenolsulphonephthalein (PSP, phenol red) injected subcutaneously, intramuscularly or intravenously, was excreted rapidly and almost entirely by the proximal tubules of the kidneys and to a small extent by the liver into the bile. This made it possible to measure excretory function of the kidneys. The strategy of challenging the patient's body to reveal a pathologic state had first been used in the form of a test meal for the analysis of gastric acidity. Now, a clinical test to determine the percentage of dye excreted in two hours was developed by Rowntree and Geraghty [87] to monitor the course of tubular cell damage as observed in nephritis. In individuals with normal kidney function, from 60% to 85% of the 1.0 ml (6 mg) injected dye solution is excreted in two hours [88]. By the early 1970s, the dye was no longer commercially available and the test was discontinued—replaced by the creatinine clearance, a more reliable and less invasive procedure.

In that same study, Abel and Rowntree also found that phenoltetrachlorphthalein given subcutaneously was excreted almost entirely in the bile. This observation was applied to the study of liver function in 1913 by Rowntree and associates who determined the amount of injected dye that was excreted in the stools. However, the procedure was not practical or accurate. In 1922 Rosenthal [89] showed that if the dye was injected intravenously, most of it remained in the circulation until it was removed by the liver for excretion. Rather than measure the dye excreted in the feces, it was simpler and more accurate to determine the amount of dye remaining in the serum at given intervals after the intravenous injection. Although the test provided a quantitative measurement of the extent of impaired liver function, there were occasional side effects.

BROMSULPHALEIN RETENTION

Rosenthal and White [90] later showed that a different but structurally related dye, phenoltetrabromphthalein (bromsulphalein, BSP), remained in the circulation until it was excreted by the liver more rapidly and completely than the phenoltetrachlorphthalein and was more suitable for colorimetric determination. Originally, the test specified intravenous injection of two mg per kg body weight and analysis of blood samples for the dye remaining in the circulation at five and thirty minutes after injection. The serum was quantitated by comparison with a series of standards prepared from the dye. Normally, in the absence of liver disease, the dye is completely removed from the blood in thirty minutes [91].

In 1939 Macdonald [92] increased the dose to five mg per kg body weight to improve the sensitivity of the test to liver impairment. The procedure eventually was standardized with a single dose of 5 mg/kg and a single serum sample tested at 45 minutes. Normally, there is less than 6% dye retention at the end of this period. The procedure was used as a "liver function" test for many years because there were no other reliable quantitative blood tests of liver function. The procedure was discontinued in the early 1970s due to its potential toxic side effects from extravasation of the dye or an anaphylactoid reaction and a reported fatality [93]. By then, other non-invasive and more specific liver function blood enzyme tests were available for the diagnosis of liver disease.

BLOOD VOLUME

Another application of injected dye was made as early as 1915 [94], in the determination of blood volume. Vital red, a non-toxic, slowly absorbable dye, is injected intravenously into the circulation and remains in the plasma long enough to mix thoroughly. Its concentration in the plasma is determined colorimetrically by comparison with a suitable standard mixture of dye and the patient's plasma before injection. The blood volume can be calculated from the plasma volume and hematocrit values. Indirect methods have included dilution with colloids, other dyes, e.g., Evans Blue, and inhalation of carbon monoxide followed by determination of this gas in the patient's blood.

Estimation of circulating blood volume by the dilution principle (using radioisotopes) is currently the method of choice and is performed in the hospital's department of nuclear medicine. The preferred procedure is to determine the red cell volume by labeling a specimen of the patient's red blood cells with ^{51}Cr. The red cells are returned to the circulation and thirty minutes later a new specimen of blood is drawn and its radioactivity is measured. Knowing the level of radioactivity of the injected red cells, the resulting decrease due to dilution allows one to calculate the red cell volume. From this value and the venous hematocrit, the total blood volume can be calculated. The earliest known attempt to determine blood volume was made in 1850. It consisted of bleeding two criminals to death.

NOTES AND REFERENCES

1. STANLEY R. BENEDICT, "The Detection and Estimation of Glucose in Urine," *Journal of the American Medical Association*, 1911, 57: 1193–1195; see also "A Reagent for the Detection of Reducing Sugars," *Journal of Biological Chemistry*, 1909, 5: 485–487; "The Detection and Estimation of Reducing Sugars," *Ibidem*, 1907, 3: 101–117.
2. IVAR BANG, "Zur Methodik der Zuckerbestimmung. II.," *Biochemische Zeitschrift*, 1913, 49: 1–18.
3. ROBERT C. LEWIS and STANLEY R. BENEDICT, "A Method for the Estimation of Sugar in Small Quantities of Blood," *Proceedings of the Society for Experimental Biology and Medicine*, 1913–14, 11: 57–58; *Idem, Journal of Biological Chemistry*, 1915, 20: 61–72. This report has a lengthy review of the literature on blood sugar determination.
4. ALFRED E. GARROD, "Medicine from the Chemical Standpoint," *British Medical Journal*, 1914, 2: 228–235, p. 233.
5. VAGN SCHMIDT, "Ivar Christian Bang (1869–1918), Founder of Modern Clinical Microchemistry," *Clinical Chemistry*, 1986, 32: 213–215.
6. VICTOR C. MYERS and CAMERON V. BAILEY, "The Lewis and Benedict Method for the Estimation of Blood Sugar, With Some Observations Obtained in Disease," *Journal of Biological Chemistry*, 1916, 24: 147–161.

7. G. L. FOSTER, "Studies on Carbohydrate Metabolism. I. Some Comparisons of Blood Sugar Concentrations in Venous Blood and in Finger Blood," *Journal of Biological Chemistry*, 1923, 55: 291–301.
8. OTTO FOLIN, "A Qualitative (Reduction) Test for Sugar in Normal Human Urine," *Journal of Biological Chemistry*, 1915, 22: 327–329, p. 327.
9. STANLEY R. BENEDICT and EMIL OSTERBERG, "A Method for the Determination of Sugar in Normal Urine," *Journal of Biological Chemistry*, 1918, 34: 195–201; STANLEY R. BENEDICT, "A Modification of the Lewis–Benedict Method for the Determination of Sugar in the Blood," *Ibidem*, pp. 203–207.
10. STANLEY R. BENEDICT, EMIL OSTERBERG, and ISAAC NEUWIRTH, "Studies in Carbohydrate Metabolism. II. A Study of the Urinary Sugar Excretion in Two Normal Men," *Journal of Biological Chemistry*, 1918, 34: 217–262, p. 258.
11. OTTO FOLIN and HSIEN WU, "A System of Blood Analysis," *Journal of Biological Chemistry*, 1919, 38: 81–108, pp. 106–107.
12. OTTO FOLIN and HSIEN WU, "A System of Blood Analysis. Supplement I. A Simplified and Improved Method for Determination of Sugar," *Journal of Biological Chemistry*, 1920, 41: 367–374, pp. 369, 370.
13. STANLEY R. BENEDICT, "The Determination of Blood Sugar," *Journal of Biological Chemistry*, 1925, 64: 207–213, pp. 207, 210, 208, 211, 213.
14. OTTO FOLIN, "The Determination of Sugar in Blood and in Normal Urine," *Journal of Biological Chemistry*, 1926, 67: 357–370, p. 363.
15. STANLEY R. BENEDICT, "The Estimation of Sugar in Blood and Normal Urine," *Journal of Biological Chemistry*, 1926, 68: 759–767, p. 766.
16. OTTO FOLIN and HAQVIN MALMROS, "Blood Sugar and Fermentable Blood Sugar as Determined by Different Methods," *Journal of Biological Chemistry*, 1929, 83: 121–127; OTTO FOLIN, "Two Revised Copper Methods for Blood Sugar Determination," *Ibidem*, 1929, 82: 83–93; "A New Blood Sugar Method," *Ibidem*, 1928, 77: 421–430.
17. STANLEY R. BENEDICT, "The Analysis of Whole Blood. II. The Determination of Sugar and of Saccharoids (Non-Fermentable Copper-Reducing Substances)," *Journal of Biological Chemistry*, 1931, 92: 141–159; "The Determination of Blood Sugar. II.," *Ibidem*, 1928, 76: 457–470.
18. ALMA HILLER, G. C. LINDER, and D. D. VAN SLYKE, "The Reducing Substances of the Blood," *Journal of Biological Chemistry*, 1925, 64: 625–638.
19. H. C. HAGEDORN and B. NORMAN JENSEN, "Zur Mikrobestimmung des Blutzuckers mittels Ferricyanid," *Biochemische Zeitschrift*, 1923, 135: 46–58.
20. FOLIN (1928); Supplementary Note on the New Ferricyanide Method for Blood Sugar," *Journal of Biological Chemistry*, 1929, 81: 231–236; *Folin and Malmros* (1929).
21. MICHAEL SOMOGYI, "Determination of Blood Sugar," *Journal of Biological Chemistry*, 1945, 160: 69–73; see also "A Method for the Preparation of Blood Filtrates for the Determination of Sugar," *Ibidem*, 1930, 86: 655–663.
22. NORTON NELSON, "A Photometric Adaptation of the Somogyi Method for the Determination of Glucose," *Journal of Biological Chemistry*, 1944, 153: 375–380; see also MICHAEL SOMOGYI, "A New Reagent for the Determination of Sugars," *Ibidem*, 1945, 160: 61–68, p. 63.
23. KURT M. DUBOWSKI, "An o-Toluidine Method for Body-Fluid Glucose Determination," *Clinical Chemistry*, 1962, 8: 215–235.
24. D. MÜLLER, "Studien über ein neues Enzym Glykoseoxydase. I.," *Biochemische Zeitschrift*, 1928, 199: 136–170.

25. D. KEILIN and E. F. HARTREE, "Properties of Glucose Oxidase (Notatin)," *Biochemical Journal*, 1948, 42: 221–229; "Specificity of Glucose Oxidase (Notatin)," *Ibidem*, 1952, 50: 331–341; "The Use of Glucose Oxidase (Notatin) for the Determination of Glucose in Biological Material and for the Study of Glucose-Producing Systems by Manometric Methods," *Ibidem*, 1948, 42: 230–238.
26. E. RUDOLF FROESCH and ALBERT E. RENOLD, "Specific Enzymatic Determination of Glucose in Blood and Urine Using Glucose Oxidase," *Diabetes*, 1956, 5: 1–6.
27. ALBERT S. KESTON, "Specific Colorimetric Enzymatic Analytical Reagents for Glucose (Abstract), 129th Meeting, American Chemical Society, Division of Biological Chemistry, Dallas, Texas, April 1956, pp. 31C–32C.
28. JOSEPH D. TELLER, "Direct, Quantitative, Colorimetric Determination of Serum or Plasma Glucose (Abstract), 130th Meeting, American Chemical Society, Division of Biological Chemistry, Atlantic City, New Jersey, September 1956, p. 69C.
29. A. ST. G. HUGGETT and D. A. NIXON, "Use of Glucose Oxidase, Peroxidase, and O-Dianisidine in Determination of Blood and Urinary Glucose," *Lancet*, 1957, 2: 368–370.
30. ABRAHAM SAIFER and SHIRLEY GERSTENFELD, "The Photometric Microdetermination of Blood Glucose with Glucose Oxidase," *Journal of Laboratory and Clinical Medicine*, 1958, 51: 448–460.
31. ALFRED H. FREE, ERNEST C. ADAMS, MARY LOU KERCHER, HELEN M. FREE, and MARION H. COOK, "Simple Specific Test for Urine Glucose," *Clinical Chemistry*, 1957, 3: 163–168.
32. J. P. COMER, "Semiquantitative Specific Test Paper for Glucose in Urine," *Analytical Chemistry*, 1956, 28: 1748–1750.
33. ANDREW HUNTER and WALTER R. CAMPBELL, "The Probable Accuracy, in Whole Blood and Plasma, of Colorimetric Determinations of Creatinine and Creatine," *Journal of Biological Chemistry*, 1917, 32: 195–231.
34. JEANETTE ALLEN BEHRE and STANLEY R. BENEDICT, "Studies in Creatine and Creatinine Metabolism. IV. On the Question of the Occurrence of Creatinine and Creatine in Blood, *Journal of Biological Chemistry*, 1922, 52: 11–33, p. 11.
35. *Ibidem*, pp. 29, 31.
36. VICTOR CARYL MYERS, *Practical Chemical Analysis of Blood. A Book Designed as a Brief Survey of This Subject for Physicians and Laboratory Workers*, 2nd ed. (St. Louis: C. V. Mosby Company, 1924), pp. 67, 69.
37. JOHN P. PETERS and DONALD D. VAN SLYKE, *Quantitative Clinical Chemistry*, vol. 2, *Methods* (Baltimore: The Williams & Wilkins Company, 1932), p. 599.
38. RENÉ DUBOS and BENJAMIN F. MILLER, "The Production of Bacterial Enzymes Capable of Decomposing Creatinine," *Journal of Biological Chemistry*, 1937, 121: 429–445; BENJAMIN F. MILLER and RENÉ DUBOS, "Studies on the Presence of Creatinine in Human Blood," *Journal of Biological Chemistry*, 1937, 121: 447–456.
39. BENJAMIN F. MILLER and RENÉ DUBOS, "Determination By a Specific, Enzymatic Method of the Creatinine Content of Blood and Urine From Normal and Nephritic Individuals," *Journal of Biological Chemistry*, 1937, 121: 457–464. The Jaffe reaction remains today the most widely used procedure for the quantitative determination of creatinine in plasma, serum, and urine. About 15% of the more than 6400 laboratories reporting to the College of American Pathologists in 1992 used a direct-reading alkaline picrate reaction; 23% reported an enzyme procedure adapted to a dry film slide technology (Kodak); 62% used a kinetic alkaline picrate reaction on twenty-two different automated instruments.

40. POUL BRANDT REHBERG, "Studies on Kidney Function. I. The Rate of Filtration and Reabsorption in the Human Kidney," *Biochemical Journal*, 1926, 20: 447–460.
41. P. D. DOOLAN, E. L. ALPEN, and G. B. THEIL, "A Clinical Appraisal of the Plasma Concentration and Endogenous Clearance of Creatinine," *American Journal of Medicine*, 1962, 32: 65–79.
42. OTTO FOLIN, "Laws Governing the Chemical Composition of Urine," *American Journal of Physiology*, 1905, 13: 66–115.
43. P. J. SCOTT and P. J. HURLEY, "Demonstration of Individual Variation in Constancy of 24-Hour Urinary Creatinine Excretion," *Clinica Chimica Acta*, 1968, 21: 411–414; GEORGE CURTIS and MAX FOGEL, "Creatinine Excretion: Diurnal Variation and Variability of Whole and Part-Day Measures. A Methodologic Issue in Psychoendocrine Research," *Psychosomatic Medicine*, 1970, 32: 337–350; see also DOOLAN et al. (1962), pp. 67–68.
44. ALFRED B. GARROD, "Observations on Certain Pathological Conditions of the Blood and Urine in Gout, Rheumatism and Bright's Disease," *Medico-Chirurgical Transactions*, 1848, 31: 83–97.
45. ALFRED B. GARROD, "Second Communication. On the Blood and Effused Fluids in Gout, Rheumatism, and Bright's Disease," *Medico-Chirurgical Transactions*, 1854, 37: 49–59, pp. 50–53.
46. F. GOWLAND HOPKINS, "On the Estimation of Uric Acid in Urine: a New Process by Means of Saturation with Ammonium Chloride," *Proceedings of the Royal Society of London*, 1892–93, 52: 93–99. A preliminary report appeared in "On the Volumetric Determination of Uric Acid in Urine," *Guy's Hospital Reports*, 1891, 48: 299–306. For a review of early methods see PETERS and VAN SLYKE (1932), p. 586 ff. For his research in nutrition, Frederick Gowland Hopkins (1861–1947) was knighted in 1925 and in 1929 shared the Nobel Prize in Physiology or Medicine with Christiaan Eijkman (1858–1930) "for their discovery of the growth-stimulating vitamins." See also LOUIS ROSENFELD, "Vitamine–vitamin. The early years of discovery," *Clinical Chemistry*, 1997, 43: 680–685.
47. TH. R. OFFER, "Phosphormolybdänsäure als Reagens auf Harnsäure," *Centralblatt für Physiologie*, 1895, 8: 801–802.
48. C. FRABOT, "Colour Reaction for Tungsten," *Journal of the Chemical Society*, 1904, 86: ii, 844 (Abstract); original in *Annales de Chimie Analytique*, 1904, 9: 371–372.
49. OTTO FOLIN and A. B. MACALLUM, "On the Blue Color Reaction of Phosphotungstic Acid (?) With Uric Acid and Other Substances," *Journal of Biological Chemistry*, 1912, 11: 265–266.
50. OTTO FOLIN and W. DENIS, "A New (Colorimetric) Method for the Determination of Uric Acid in Blood," *Journal of Biological Chemistry*, 1912–13, 13: 469–475.
51. OTTO FOLIN and W. DENIS, "On Phosphotungstic–Phosphomolybdic Compounds as Color Reagents," *Journal of Biological Chemistry*, 1912, 12: 239–243.
52. HSIEN WU, "A New Colorimetric Method for the Determination of Plasma Proteins," *Journal of Biological Chemistry*, 1922, 51: 33–39.
53. DAVID M. GREENBERG, "The Colorimetric Determination of the Serum Proteins," *Journal of Biological Chemistry*, 1929, 82: 545–550. See also OTTO FOLIN and VINTILA CIOCALTEU, "On Tyrosine and Tryptophane Determinations in Proteins," *Journal of Biological Chemistry*, 1927, 73: 627–650.
54. STANLEY R. BENEDICT, "On the Colorimetric Determination of Uric Acid in Blood," *Journal of Biological Chemistry*, 1915, 20: 629–631; STANLEY R. BENEDICT and ETHEL H. HITCHCOCK, "On the Colorimetric Estimation of Uric Acid in Urine," *Journal of Biological Chemistry*, 1915, 20: 619–627.

55. OTTO FOLIN, "The Preparation of Sodium Tungstate Free from Molybdate, Together With a Simplified Process for the Preparation of a Correct Uric Acid Reagent (And Some Comments)," *Journal of Biological Chemistry*, 1934, 106: 311–314.
56. FOLIN and WU (1919).
57. STANLEY R. BENEDICT, "The Determination of Uric Acid in Blood," *Journal of Biological Chemistry*, 1922, 51: 187–207, pp. 194, 206; STANLEY R. BENEDICT and ELIZABETH FRANKE, "A Method for the Direct Determination of Uric Acid in Urine," *Journal of Biological Chemistry*, 1922, 52: 387–391.
58. OTTO FOLIN, "A System of Blood Analysis. Supplement IV. A Revision of the Method for Determining Uric Acid," *Journal of Biological Chemistry*, 1922, 54: 153–170.
59. OTTO FOLIN, "An Improved Method for the Determination of Uric Acid in Blood," *Journal of Biological Chemistry*, 1930, 86: 179–187.
60. PETERS and VAN SLYKE (1932), p. 587.
61. Sodium Tungstate (Folin), molybdenum low (0.001%) is commercially available from leading reagent suppliers.
62. FOLIN (1934), p. 314.
63. *Ibidem* (1934), p. 312.
64. OTTO FOLIN and HARRY TRIMBLE, "A System of Blood Analysis. Supplement V. Improvements in the Quality and Method of Preparing the Uric Acid Reagent," *Journal of Biological Chemistry*, 1924, 60: 473–479.
65. FOLIN (1934), p. 313.
66. WENDELL T. CARAWAY, "Determination of Uric Acid in Serum by a Carbonate Method," *American Journal of Clinical Pathology*, 1955, 25: 840–845; *Standard Methods of Clinical Chemistry*, vol. 4, David Seligson, ed. (New York: Academic Press, 1963), pp. 239–247.
67. KLAUS LORENTZ and WINFRIED BERNDT, "Enzymic Determination of Uric Acid by a Colorimetric Method," *Analytical Biochemistry*, 1967, 18: 58–63; G. F. DOMAGK and H. H. SCHLICKE, "A Colorimetric Method Using Uricase and Peroxidase for the Determination of Uric Acid," *Ibidem*, 1968, 22: 219–224. In 1992 only 1% of nearly 5700 laboratories reporting in the College of American Pathologists survey were using a phosphotungstate procedure—down from 14% in 1985—the others were using uricase methods in over twenty-eight different automated instrument modalities.
68. MARY BRANNOCK BLAUCH and F. C. KOCH, "A New Method for the Determination of Uric Acid in Blood, With Uricase," *Journal of Biological Chemistry*, 1939, 130: 443–454.
69. HERMAN M. KALCKAR, "Differential Spectrophotometry of Purine Compounds by Means of Specific Enzymes. I. Determination of Hydroxypurine Compounds," *Journal of Biological Chemistry*, 1947, 167: 429–443; E. PRAETORIUS, "An Enzymatic Method for the Determination of Uric Acid by Ultraviolet Spectrophotometry," *Scandinavian Journal of Clinical & Laboratory Investigation*, 1949, 1: 222–230; THOMAS V. FEICHTMEIR and HAROLD T. WRENN, "Direct Determination of Uric Acid Using Uricase," *American Journal of Clinical Pathology*, 1955, 25: 833–839.
70. PETERS and VAN SLYKE (1932), pp. 542–543.
71. *Ibidem*, p. 544.
72. *Ibidem*, pp. 543–544; see also OTTO FOLIN and RICHARD D. BELL, "Applications of a New Reagent for the Separation of Ammonia. I. The Colorimetric Determination of Ammonia in Urine," *Journal of Biological Chemistry*, 1917, 29: 329–335.
73. OTTO FOLIN and W. DENIS, "New Methods for the Determination of Total Non-Protein Nitrogen, Urea and Ammonia in Blood," *Journal of Biological Chemistry*, 1912, 11:

527–536. Vacuum distillation by use of air current suction was introduced and popularized by Folin. See OTTO FOLIN, "Eine neue Methode zur Bestimmung des Ammoniaks im Harne und anderen thierischen Flüssigkeiten," *Hoppe-Seyler's Zeitschrift für Physiologische Chemie*, 1902, 37: 161–176. For a review of early aeration technics, see also PHILIP SHAFFER, "On the Quantitative Determination of Ammonia in Urine," *American Journal of Physiology*, 1903, 8: 330–354, pp. 344, 348.

74. J. G. MATEER and E. K. MARSHALL, JR., "The Urease Content of Certain Beans, With Special Reference to the Jack Bean," *Journal of Biological Chemistry*, 1916, 25: 297–305.

75. E. K. MARSHALL, JR., "A Rapid Clinical Method for the Estimation of Urea in Urine," *Journal of Biological Chemistry*, 1913, 14: 283–290.

76. E. K. MARSHALL, JR., "A New Method for the Determination of Urea in Blood," *Journal of Biological Chemistry*, 1913, 15: 487–494, p. 488.

77. DONALD D. VAN SLYKE and GLENN E. CULLEN, "A Permanent Preparation of Urease and its Use in the Determination of Urea," *Journal of Biological Chemistry*, 1914, 19: 211–228, p. 217.

78. FOLIN and WU (1919), pp. 91–98.

79. DONALD D. VAN SLYKE, "Determination of Urea by Gasometric Measurement of the Carbon Dioxide Formed by the Action of Urease," *Journal of Biological Chemistry*, 1927, 73: 695–723.

80. WILLIAM ROBERT FEARON, "The Carbamido Diacetyl Reaction: A Test for Citrulline," *Biochemical Journal*, 1939, 33: 902–907.

81. ANDREW A. ORMSBY, "A Direct Colorimetric Method for the Determination of Urea in Blood and Urine," *Journal of Biological Chemistry*, 1942, 146: 595–604.

82. T. O. TIFFANY, J. M. JANSEN, C. A. BURTIS, J. B. OVERTON, and C. D. SCOTT, "Enzymatic Kinetic Rate and End-Point Analyses of Substrate, by Use of a GeMSAEC Fast Analyzer," *Clinical Chemistry*, 1972, 18: 829–840.

83. WEI-TSUNG CHIN and WYBE KROONTJE, "Conductivity Method for Determination of Urea," *Analytical Chemistry*, 1961, 33: 1757–1760.

84. J. HAROLD AUSTIN, EDGAR STILLMAN, and DONALD D. VAN SLYKE, "Factors Governing the Excretion Rate of Urea," *Journal of Biological Chemistry*, 1921, 46: 91–112; EGGERT MÖLLER, J. F. MCINTOSH and D. D. Van Slyke, "Studies of Urea Excretion. II. Relationship Between Urine Volume and the Rate of Urea Excretion by Normal Adults," *Journal of Clinical Investigation*, 1928, 6: 427–465.

85. JOHN F. MCINTOSH, EGGERT MÖLLER, and DONALD D. VAN SLYKE, "Studies of Urea Excretion. III. The Influence of Body Size on Urea Output," *Journal of Clinical Investigation*, 1928, 6: 467–483; PETERS and VAN SLYKE (1932), pp. 564–570.

86. JOHN J. ABEL and L. G. ROWNTREE, "On the Pharmacological Action of Some Phthaleins and Their Derivatives, With Especial Reference to Their Behavior as Purgatives," I. *Journal of Pharmacology and Experimental Therapeutics*, 1909, 1: 231–264.

87. L. G. ROWNTREE and J. T. GERAGHTY, "The Phthalein Test. An Experimental and Clinical Study of Phenolsulphonephthalein in Relation to Renal Function in Health and Disease," *Archives of Internal Medicine*, 1912, 9: 284–338.

88. J. T. GERAGHTY and L. G. ROWNTREE, "The Phenolsulphonephthalein Test for Estimating Renal Function," *Journal of the American Medical Association*, 1911, 57: 811–816.

89. S. M. ROSENTHAL, "An Improved Method for Using Phenoltetrachlorphthalein as a Liver Function Test," *Journal of Pharmacology and Experimental Therapeutics*, 1922, 19: 385–391.

90. Sanford M. Rosenthal and Edwin C. White, "Studies in Hepatic Function. VI. A. The Pharmacological Behavior of Certain Phthalein Dyes. B. The Value of Selected Phthalein Compounds in the Estimation of Hepatic Function," *Journal of Pharmacology and Experimental Therapeutics*, 1924, 24: 265–288.
91. Sanford M. Rosenthal and Edwin C. White, "Clinical Application of the Bromsulphalein Test for Hepatic Function," *Journal of the American Medical Association*, 1925, 84: 1112–1114.
92. Dean Macdonald, "A Practical and Clinical Test for Liver Reserve," *Surgery, Gynecology and Obstetrics*, 1939, 69: 70–82.
93. J. B. Brierley, J. H. Adams, D. I. Graham, and J. A. Simpsom, "Neocortical Death After Cardiac Arrest. A Clinical, Neurophysiological, and Neuropathological Report of Two Cases," *Lancet*, 1971, 2: 560–565; Erik Juhl, Mogens Hilden, and Erik Dauv-Pedersen, "Atypical Reactions to Bromsulphthalein (B.S.P.)," *Lancet*, 1970, 2: 424–425.
94. N. M. Keith, L. G. Rowntree, and J. T. Geraghty, "A Method for the Determination of Plasma and Blood Volume," *Archives of Internal Medicine*, 1915, 16: 547–576; see also Peters and Van Slyke (1932), pp. 701–716; Richard J. Henry, *Clinical Chemistry. Principles and Technics* (New York: Harper & Row, 1964), pp. 896–902.

CHAPTER XIV

CHLORIDE

The alchemists and Robert Boyle used common salt to precipitate silver. This suggested to Friedrich Hoffman (1660–1742) of Halle that silver salts can be used to identify common salt, and in 1689 he established the presence of common salt in urine [1]. The simplicity of the reaction with silver nitrate made chloride the first inorganic component to be determined in urine.

The classical gravimetric procedure for chloride—precipitation and weighing as silver chloride—was abandoned when titration procedures were developed that were quicker and required much less material. The oldest and simplest of the titration methods is that of Mohr (1856). With a few drops of potassium or sodium chromate as indicator, the chloride is precipitated as silver chloride in neutral or slightly acid solution by addition of standard silver nitrate solution. Silver chloride and silver chromate are both insoluble, but silver chloride is much more so and no permanent precipitate of silver chromate is formed until all the chloride has been precipitated. Then, the first drop of excess silver produces a permanent brown cloud of silver chromate precipitate. The chief drawback for biochemical work was that any organic material in the urine and blood filtrates is also precipitated by silver unless an excess of free nitric acid is present. However, at the resulting low pH, the silver chromate precipitate indicating the endpoint, does not form. Consequently, the Mohr method is seldom used with biological material. It was succeeded in 1878 by the method of Jacob Volhard (1834–1910), a would-be historian, persuaded by his father to study with Liebig. Volhard was editor of Liebig's *Annalen* for thirty-nine years.

Volhard used a different endpoint for the silver reaction with chloride. Applications of his method were those most frequently used for chloride in biological analyses in the period leading to the introduction of automated methods. The chloride is first precipitated with an excess of standard silver nitrate solution. With a ferric salt as indicator, the unreacted silver nitrate is back-titrated with standard sulfocyanate (thiocyanate). A white precipitate of silver thiocyanate is formed until thiocyanate ion, in excess of the silver ion, reacts with the indicator to form ferric thiocyanate whose deep red color signals the endpoint. However, the endpoint fades because ferric thiocyanate is soluble and dissociates to favor formation of silver thiocyanate which is more insoluble than silver chloride. To remove the source of silver and avoid fading of the endpoint, the silver chloride precipitate should be removed by filtration before the back-titration with thiocyanate. However, a new error is introduced because silver ions are lost by adsorption on the precipitate.

Rappleye [2] precipitated protein and chloride with a standard solution of silver nitrate, then titrated the unreacted silver nitrate in the filtrate according to Volhard's procedure. Van Slyke and Donleavy [3] preferred to titrate the excess silver in the filtrate with standard iodide solution in an iodometric titration to the starch endpoint with free iodine. When the tungstic acid method of preparing protein-free filtrate proved so successful in the determination of several blood constituents, it was used for the determination of chloride using Volhard's principle. In this application, Whitehorn [4]—working in Folin's laboratory—avoided removal of the silver chloride precipitate by stabilizing the endpoint color with optimum concentrations of nitric acid and ferric salt indicator. Whitehorn's variation became the preferred procedure for serum (plasma) chloride until it was replaced by the simpler method of Schales and Schales.

In the Schales and Schales [5] procedure, the chloride ions in plasma, serum, whole blood or protein-free filtrate are titrated with standard mercuric nitrate solution in the presence of diphenylcarbazone. This indicator for mercuric ions was introduced by Dubský and Trtílek [6] and recommended for determination of chloride in blood filtrates by Lang [7]. As it does with silver ions, chloride also combines with mercuric ions to form soluble, but virtually undissociated mercuric chloride. At endpoint, mercuric ions in solution form a purple complex with the diphenylcarbazone indicator. Prior removal of protein from the fluid to be titrated is not necessary, but the results are one to two mEq per Liter lower, probably because of chloride loss by adsorption on the protein precipitate. The endpoint color is more intense in the absence of protein.

Liebig [8] was the first to use mercuric nitrate solution for the determination of chloride (1853). He noted that mercuric ions produce a white precipitate with urea in neutral solution, but not in the presence of chloride ions. Consequently, with urea as indicator, he titrated chloride with mercuric nitrate solution until the mixture showed an opalescence due to the mercury–urea interaction.

In 1958 a team led by Ernest Cotlove [9] designed an instrument and method that introduced the clinical chemistry laboratory to electrometric titration of chloride with silver ion. The technique is based on established principles of coulometric generation of silver ions (as reagent) and of amperometric indication of the endpoint. A constant direct current is passed between a pair of silver *generator* electrodes (coulometric circuit), causing release of silver ions into the titration solution at a constant rate. After all the chloride has been precipitated as silver chloride, the endpoint is signaled by the first appearance of uncombined or free silver ions. There occurs an abrupt increase in current through a pair of silver *indicator* electrodes (amperometric circuit). A relay switch is activated, stopping the titration and a timer that had been running concurrently with the generation of silver ions. Since the flow of current is steady, and the rate of generation is constant, the amount of chloride precipitated is proportional to the elapsed time. Chloride concentration in unknown specimens can be calculated by comparison of the endpoint time with the time required to titrate a standard chloride solution run concurrently. The Chloridometer instrument, capable of reading directly in millieqivalents per Liter, was available commercially from American Instrument Company and Buchler Instruments and provided an accurate, rapid, and automatic analysis of chloride in biologic samples (Fig. 14.1).

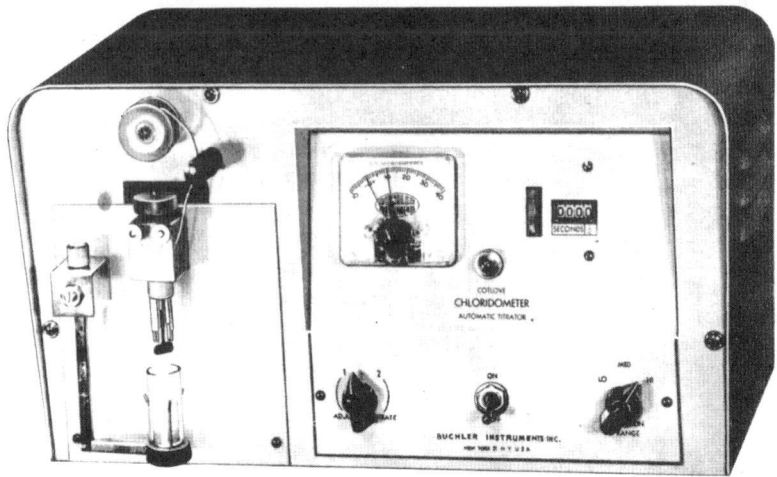

FIG. 14.1. Chloridometer.

Automated colorimetric chloride analysis in the AutoAnalyzer soon followed. Reaction of chloride with mercuric thiocyanate releases thiocyanate ion which reacts with ferric ions to yield a colored complex measured spectrophotometrically. By 1992 this reaction was used by fewer than 2% of over 6200 participants in the College of Pathologists survey. Mercuric nitrate and coulometric–amperometric methods were reported by less than 6% for each. Ion-selective electrode technology was used by over 88% of those responding to the testing survey.

For the first half of the twentieth century, chloride was routinely reported in terms of sodium chloride and in mg per 100 cc. Not until there was a better appreciation of the contribution of individual ionic species to the interpretation of osmotic pressure regulation and acid–base balance, was the reporting of chloride and carbon dioxide content (or combining power) changed from mg per 100 cc and volumes percent, respectively, to milliequivalents per Liter—units which could be added and compared with cation concentrations (see also Chapter 8). In many laboratories this conversion did not occur until the 1960s.

Early in their work on electrolyte balance, Van Slyke and Cullen [10] called attention to the shift of chloride from the red cells to plasma due to loss of bicarbonate from the plasma to the red cells. There it dissociates to form (water and) carbon dioxide which is released to the air. This was the reverse of the transfer taking place between plasma and red cells during the *in vivo* production of carbon dioxide. For precise work, where small changes in chloride content are significant, blood should be collected under oil to minimize changes in carbon dioxide tension. Otherwise, in routine work, the plasma (serum) should be quickly separated from the red cells. Because of these changes, Myers and Short [11] had once suggested that chloride analyses made on whole blood would be more trustworthy than those on plasma.

As long ago as 1850, Carl Schmidt, in his classic studies of acid–base imbalance in cholera, reported the low chloride content of whole blood and plasma. Widal and Javal in France (1905–06) noted the retention of chloride in nephritis, especially if edema was present, and were the first to restrict chlorides from the diet in their treatment of these patients.

CALCIUM

Nearly all calcium determinations, both clinical and general mineral analyses, prior to the introduction of flame photometry, atomic absorption, chelation with ethylenediaminetetraacetic acid (EDTA), and photometric determination with chromophore indicators, in the post-Word War II period, involved precipitation of calcium as the very insoluble oxalate. The classical gravimetric method was to ignite the oxalate at white heat in a platinum dish until all the carbon dioxide had been driven off, and then to weigh the resulting oxide.

In the older methods, large amounts of blood, plasma or serum were used and the proteins were destroyed by ashing. These conditions were obstacles to clinical application. In 1917 Halverson and Bergeim [12] published the first simple procedure for small amounts of blood (5 cc serum or plasma). In their method, proteins are removed by precipitation in a highly acid solution with sodium picrate and heat so that all the calcium remains in solution. Calcium in the filtrate is precipitated as oxalate, then redissolved in hot, approximately 1 N solution of sulfuric acid, and the resulting oxalic acid is titrated with 0.01 N potassium permanganate. This reagent is its own indicator, since the endpoint is signaled by the persistence of its pink color. These principles have been followed by most subsequent authors, except for those who have not removed the proteins, but chose to precipitate the calcium by addition of oxalate directly to the more or less diluted serum. The Folin–Wu tungstic acid precipitation of proteins is not sufficiently acid to prevent loss of much calcium with the protein precipitate.

Three techniques have been employed for precipitation of calcium oxalate: preliminary ashing of the sample; precipitation from a protein-free filtrate prepared with trichloroacetic acid [13] (more convenient than the original picric acid); direct precipitation from the serum sample. Fortunately, direct precipitation—which simplified the procedure—appeared to be satisfactory for the routine analysis of serum and urine. The use of ammonium oxalate for the direct precipitation of calcium from unashed serum was applied by Pribram as early as 1871. However, the technique was first introduced into routine clinical chemistry in 1921 by Kramer and Tisdall [14]. Clark and Collip [15] made some minor modifications—one wash instead of three, and decanting instead of siphoning off supernatant fluid. Their protocol was the one widely adopted for the next forty years. However, reproducibility of the results depended on the exact way in which all the steps of the procedure are carried out.

Prior to 1950, volumetric methods based on complex formation did not play a very important role in analytical chemistry. During the 1930s, it was found that certain aminopolycarboxylic acids formed stable, soluble chelate complexes with a large number of metals, notably with alkaline earths. As a result of investigations of these complexes, begun in the 1940s,

Gerold Schwarzenbach developed analytical methods for the determination of calcium and magnesium and also for the hardness of water.

The best of the large number of complexing (chelating) agents appeared to be EDTA, normally used as the disodium salt. The introduction of metallochromic indicators led next to a large number of methods involving direct titration of the metal ion solution with a standard solution of EDTA. The basis of these methods is that the metal-EDTA complex is stronger than that of the metal-indicator complex. At the endpoint, with all the metal (calcium) linked to the EDTA, there is a color change in the reaction mixture from the initial color of the metal-bound indicator to that of the free indicator. Schwarzenbach and associates [16] introduced direct "complexometric" titration of alkaline earths—especially calcium—with the first of the metallochromic indicators, murexide (ammonium purpurate). In subsequent years, similar determinations were developed for a wide range of metals by using many new metallochromic indicators and suitable adjustment of pH for selective analyses.

These procedures were very popular until they were rivaled by methods based on direct photometry of the colored reaction complex formed between calcium and certain organic chemicals. In the 1992 College of Pathologists survey, cresolphthalein was used by approximately 57% of the more than 5800 participants and arsenazo III dye by the remainder for the routine analysis of calcium in a large variety of automated discrete sample systems of analysis.

PHOSPHORUS

The phosphorus of whole blood is present chiefly as inorganic phosphates, organic acid-soluble phosphate esters, and lipid phosphorus. The distribution between cells and plasma is quite uneven, but the inorganic phosphate is practically entirely in the plasma (serum) and is the fraction usually determined for clinical purposes. Protein is precipitated with trichloroacetic acid and the phosphate-containing filtrate is treated with acid molybdate solution to form phosphomolybdic acid. On addition of suitable reducing agents—many have been used—phosphomolybdic acid is selectively reduced ($Mo^{+6} \rightarrow Mo^{+3}$) to yield a deep blue color whose intensity is a measure of the amount of phosphate present.

The method dates back to Gmelin's discovery in 1844 that phosphoric acid forms a yellow precipitate in acid solution with ammonium molybdate. There was so little phosphorus in the precipitate that he dismissed it as unessential. Sonnenschein (1851) showed that phosphoric acid was an integral part of the precipitate and suggested that it be used for the quantitative determination of phosphoric acid [17].

The reduction was at first applied to the molybdic acid precipitated as ammonium phosphomolybdate. In that method, the precipitate was redissolved and the molybdenum reduced by phenylhydrazine [18]. In a popular method by Tisdall [19], the phosphorus is precipitated as strychnine phosphomolybdate. Redissolved in alkali, the molybdenum is reduced with potassium ferrocyanide and yields a greenish solution, presumably a mixture of the blue molybdenum compound and the yellow of the ferricyanide produced by the reaction with ferrocyanide. Since proportional amounts of the two colored products are formed, the resulting combination of color is constant.

A more rapid colorimetric method of estimating the inorganic phosphorus of blood and urine without having to precipitate the phosphate as ammonium phosphomolybdate was first described by Bell and Doisy [20]. They found that certain reducing agents will reduce the Mo^{+6} of phosphomolybdic acid and at the same time have only negligible effect on the Mo^{+6} of uncombined molybdic acid in the same solution. They carried out the reduction with hydroquinone and sulfite in alkaline solution. Because the alkaline blue color fades rapidly, the procedure was modified by Briggs [21] who made the color more stable by using an acid medium. However, now the color was comparatively weak. Benedict and Theis [22] intensified the color by heating. Fiske and SubbaRow [23] suggested the use of 1,2,4-aminonaphtholsulfonic acid as reducing agent at room temperature. This method was widely used for the next forty years. Kuttner and Cohen [24] returned to the use of stannous chloride first introduced by Osmond [25] in 1887 and later by Denigès [26] in 1921. This variation was also successfully incorporated into the clinical chemistry laboratory. Of all the reducing agents used, stannous chloride produces the most blue color with phosphomolybdate, making it the most sensitive procedure. It is also more stable than aminonaphtholsulfonic acid; however, there are disadvantages. The color intensity changes continuously with time which must be carefully controlled. Deviations from Beer's law and poor day-to-day reproducibility of standard readings make daily calibration essential.

SODIUM AND POTASSIUM: INTRODUCTION

When Humphry Davy subjected aqueous solutions of the alkalies to electrolysis, only the water was decomposed. In 1807, using a powerful battery, Davy decomposed molten (fused) potash (potassium carbonate) and obtained globules of silvery matter at the negative pole. He named it potassium to imply a metallic status. He then isolated sodium from fused soda (sodium carbonate). The following year he decomposed the oxides of the alkaline earths, which previously had been considered to be elements, and isolated magnesium from magnesia, strontium from strontia, barium from baryta, and calcium from lime. Davy also showed that a certain greenish gas that Scheele had discovered and thought to be an oxide, was actually an element. He named it chlorine, from the Greek word for "green." Davy never accepted the atomic theory of John Dalton (1766–1844) because he considered it to be speculative. He believed that the simplicity and harmony of nature demanded that there be very few ultimately distinct forms of matter. It is ironic that one who held such views should have discovered six new elements. But apparently, he was not convinced. Although Lavoisier included nitrogen in his list of elements, Davy doubted its elementary nature as late as 1809 and attempted to decompose it [27].

Humphry Davy (Fig. 14.2) was knighted in April 1812 by the Prince Regent. In the autumn of 1813, with England and France at war, Davy traveled to France to receive the medal established by Napoleon and awarded to him by the Institut de France for his electrical discoveries. Michael Faraday accompanied him as his laboratory assistant and valet. While in Paris, Davy performed some experiments using a little case of apparatus that always traveled with him, and determined the essential properties of iodine—recently discovered by Bernard Courtois (1777–1838)—thereby anticipating a more detailed account by Gay-Lussac [28]. Both Gay-Lussac and Davy independently established that the substance was an element. It was Gay-Lussac who named the new element "iode" (Greek: *ioeides*, violet colored). Davy named it iodine.

GRAVIMETRIC, TITRIMETRIC, AND COLORIMETRIC METHODS

Most chemical methods for sodium and potassium are tedious and slow and require three or more hours to complete. Many require preliminary ashing or preparation of a protein-free filtrate. By the time the results are reported, they have lost much of their value. For clinical usefulness, in many cases, the result must he available in less than an hour. These chemical methods require careful technique because of the considerable manipulation of sample and the general complexity of the analysis. As with other serum constituents, there have been many reports of method

FIG. 14.2. Humphry Davy. (National Library of Medicine, Bethesda, Maryland.)

modifications designed to simplify the procedure and improve quantitative recovery and reproducibility [29].

The classical methods first described for sodium and potassium were gravimetric, following ashing. They required from five to ten cc of blood or serum and were determined together as chlorides or sulfates. The chlorides were dissolved and the potassium was precipitated and weighed, either as the chloroplatinate or as the perchlorate. With redissolved sulfates, only the chloroplatinate method can be used. Sodium is calculated by subtracting the potassium content from the combined total determined for sodium and potassium.

Later, chemical methods isolated sodium or potassium as an insoluble compound, followed by analysis for some component in this compound that bears a constant ratio to sodium or potassium. The most popular method for the determination of sodium prior to flame photometry was by precipitation as the relatively insoluble triple salt, sodium uranyl zinc acetate, a gravimetric technique introduced by Kolthoff in 1928 [30]. The technique lends itself to microanalysis because of the large weight of the composite molecule. Quantitation of the precipitate has also been made by titrimetric and colorimetric procedures. The solubility and mechanical loss of precipitate during washing is the greatest source of error. Sodium may also be precipitated as the pyroantimonate [31] and quantitated by gravimetric or titrimetric measurement.

The use of chloroplatinate for potassium analysis was recommended as early as 1847 by Carl Remegius Fresenius (1818–1897). In biologic fluids, precipitation as the chloroplatinate [32], followed by gravimetric, colorimetric, or titrimetric measurement, has been generally regarded as the most accurate of the chemical methods because the precipitate has a definite constant composition. However, preliminary ashing of the serum sample is required. The colorimetric procedure was first used by Cameron and Failyer [33] for the determination of potassium in water.

The first micropotassium method applied to blood analysis was a cobaltinitrite titration procedure devised by Clausen [34] and further developed by Kramer and Tisdall [35]. Precipitation as the potassium sodium cobaltinitrite can be made without removal of proteins or ashing. The cobalt component can be measured colorimetrically and the nitrite portion can be analyzed by titration. These methods would be ideal procedures if the cobaltinitrite precipitate could be obtained with constant composition. The content of the precipitate varies with the conditions of precipitation, its solubility increasing with temperature. The variation in weight of the precipitate is partly due to variation in the water of hydration.

Ever since the cobaltinitrite method was introduced in 1900, it posed a challenge to investigators seeking conditions under which a precipitate of constant composition could be obtained. However, under carefully standardized conditions, the ratio of potassium to cobalt or to nitrite in the precipitate is sufficiently constant to yield results with satisfactory accuracy. The cobaltinitrite method is the one by which most blood potassium analyses in the literature were obtained prior to the introduction of flame photometry. Results at this time were usually reported as mg per 100 cc.

Despite the wide variety of chemical methods for sodium and potassium, serum levels of these two cations were not usually determined or requested in the clinical chemistry laboratory until the middle of the twentieth century—when instrumentation for flame photometry became commercially available and made this analysis relatively simple and quick. With rapid accumulation of clinical data, sodium and potassium were soon shown to play an important physicochemical role in osmotic pressure regulation, electrolyte and acid–base balance, and in specialized functions. The organism has such a strong tendency to preserve total base, that only slight changes are ordinarily found, even under pathological conditions. Hence, the need for precision in analysis. The unit mEq per Liter came into regular use.

But in 1924, with little practical information available, it is not surprising that Myers [36] could write: "Nothing of special importance is known regarding pathological variations in the sodium content of the blood." As for potassium, "Although many observations have been made, there does not appear to be any significant variation in the potassium content of the serum in disease."

SPECTROSCOPY

Spectroscopy had its beginning in 1675 when Isaac Newton (1642–1727) passed a beam of sunlight through a prism and observed the solar spectrum. Flame analysis can be traced back to the middle of the eighteenth century when chemists recorded incidental observations of colors imparted to alcohol and candle flames by metallic salts and used this coloration to some extent for identification of different substances. The earliest reported use of a flame test applied to analysis was by Sigismund Andreas Marggraf (1709–1782) who, in 1762, noted the difference in the color imparted to a flame by the nitrate salts of sodium and potassium when they were placed on glowing coals: yellow for sodium, bluish for potassium. However, these low-temperature and high-luminosity flames were hardly suitable for detecting weak spectral lines. Furthermore, the chemicals were not very pure. Marggraf in 1757, had also distinguished between the bases of potash and soda by forming various salts and observing differences in solubility and crystalline form after evaporation. Henri Louis Duhamel du Monceau (1700–1782) carried out similar chemical experiments in 1736 with similar results. He also concluded that the base of common salt is identical to the base of soda [37].

The development of better prisms and spectroscopes opened new opportunities for research in spectrochemistry. As early as 1802, William Hyde Wollaston observed that the spectrum of the sun is not continuous and that black lines could be seen between the colors of the spectrum. His description was not very accurate and only now can we understand what he was trying to describe. Wollaston was not very interested in this phenomenon and did not pursue the subject any further. It was not a new discovery. A century earlier, many scientists had investigated the spectrum of the sun and, observing these lines, probably attributed them to some flaw in the prism [38].

It was known that the dark D lines in the solar spectrum, so-named by Joseph Fraunhofer (1787–1826) in 1814, were located in the same position (wavelengths) as the bright doublet of yellow lines emitted by flames containing sodium. This effect could be verified by allowing sunlight to reach the spectroscope after passing through a sodium flame. If the sunlight were sufficiently dimmed, the dark Fraunhofer lines were replaced by the bright lines from the sodium flame. The correct interpretation for this relationship, viz., that a substance capable of emitting a certain spectral line has a strong absorptive power for the same line, was made nearly half a century later [39].

BUNSEN AND KIRCHHOFF APPLY SPECTRAL ANALYSIS

The foundation of spectral analysis of chemical elements by both emission and absorption spectroscopy was set by Robert Bunsen (1811–1899) and Gustave Robert Kirchhoff (1824–1887) in their paper *Chemische Analyse durch Spectralbeobachtungen* (1860). Bunsen wanted to develop analytical techniques for the identification, separation, and measurement of inorganic substances and was exploring the analysis of metals and their salts by the distinctive colors they produced in flames. At first he used colored pieces of glass or solutions to distinguish similarly colored flames. Kirchhoff, a physicist, who had joined Bunsen, a chemist, at Heidelberg, pointed out that a sharper and surer distinction could be obtained from the characteristic spectra (lines or bands of light) obtained when the colored flame is analyzed by viewing it through a prism. Working with Kirchhoff about 1860, Bunsen designed a very efficient spectroscope (Fig. 14.3) which—utilizing the new burner that he invented—demonstrated that the visible spectral lines are due to elements, not compounds. This presented an entirely new means of chemical analysis and opened up the possibility of determining the chemical composition of the sun and other stars from the study of their optical spectra. By rigorous experimentation, the two investigators put the method on a firm basis [40]. The instrument represented no new principles or devices that had not been in use for years. What Bunsen and Kirchhoff did was to bring together on one stand the necessary collimating and viewing telescopes, with the prism enclosed inside a blackened box.

The burner invented by Bunsen in 1855 made possible the use of illuminating gas for heating. It produced a sootless gas flame of very high

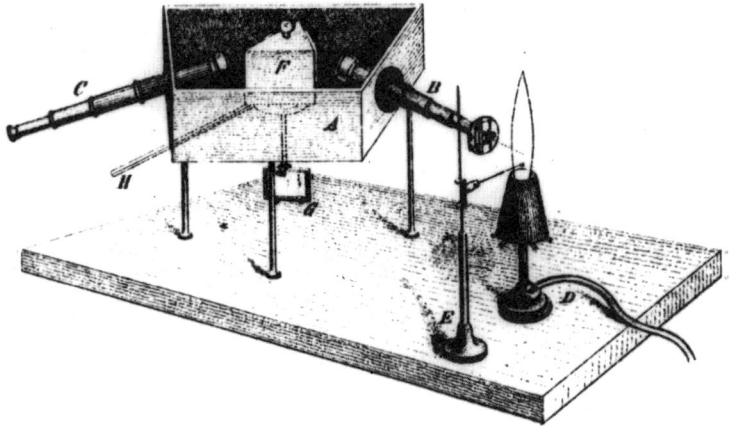

FIG. 14.3. First spectroscope of Bunsen and Kirchhoff. A, blackened box, the bottom shaped like a trapezoid, resting on three legs; B, C, telescopes; D, lamp (flame); E, support for platinum wire holding test specimen; F, hollow prism filled with carbon disulfide; G, mirror; H, handle for rotating prism and mirror. (note 40).

temperature and low luminosity that was used as the energizing source in his later studies of the characteristic emission and absorption spectra of metal salts. Bunsen was dissatisfied with charcoal burners and alcohol lamps and was seeking a means of providing better sources of heat for laboratory operations. The devices for burning gas then available all suffered from a major defect—they delivered luminous smoky flames of low heating power. The flame produced by the burner of the Swiss inventor Aimé Argand (1755–1803) was too large and difficult to regulate. Furthermore, the gas was so diluted with air that the resulting temperature was too low and the flame flickered excessively.

Bunsen overcame this problem. Instead of feeding the flame with air from the outside, he prepared the gas–air mixture before it was ignited. In this way he could produce a sootless, nonluminous flame with high heating value. He took his ideas and specifications to the university mechanic, Peter Desaga, who developed the details of construction. Soon, he was manufacturing the burner in large numbers for the Heidelberg students and all others wishing to use it. Bunsen first published his account of the burner in 1857. Neither Bunsen nor Desaga applied for patent protection on the new burner. It was not long before imitations of the burner appeared and claims of priority for the invention were made.

In Bunsen's burner, city gas enters through a narrow nozzle into a wider tube with adjustable holes at its lower end. The forward motion of the gas draws the air that is needed for combustion, through these holes, mixes it in the wider tube, and exits in support of the flame. Simple and efficient, it quickly supplanted the blowpipe flame used in the dry tests of analytical chemistry.

When analytically pure compounds of potassium, sodium, lithium, barium, calcium, and strontium were heated in the Bunsen flame, they emitted very sharp colored line spectra characteristic for each metal. Kirchhoff observed that when the colored flames of heated materials that produce bright sharp emission spectra are placed in the path of an intense light source, they absorb light of the same wavelength that they otherwise emit and produce characteristic absorption spectra.

Bunsen predicted that spectral analysis could lead to the discovery of new elements previously overlooked, either because they occurred in very small quantities or were too similar to known elements to be distinguished by conventional chemical techniques. The first systematic application led to the impressive discovery by Bunsen and Kirchhoff of two new alkaline elements in a mineral spring: cesium (Latin: *caesius*, sky-blue) in 1860, and rubidium (Latin: *rubidus*, dark red) the following year. It wasn't long before other elements were identified by spectroscopic methods: thallium (1861), indium (1863), gallium (1875), scandium (1879), and germanium (1886). Nearly a century later, the discovery by Bunsen and Kirchhoff led to the design and development of an instrument with an important application in the clinical chemistry laboratory—the quantitative determination of sodium and potassium in blood serum.

As often happens when something "new" is described, there appear claims of priority for work by others; or, old descriptions in the literature, unnoticed for decades, are rediscovered. However, what makes the difference between a discovery and just another casual report, is that earlier accounts did not lead to practical applications. Spectrum analysis did not originate from nothing. There had already been considerable progress made in spectroscopy, and many earlier workers had come close to anticipating the contribution of Bunsen and Kirchhoff. Previous investigators had commented on the analytical possibilities in the use of spectra, but they were mainly concerned with the characteristic lines of the alkali metals and their relation to the Fraunhofer lines in the sun's spectrum. Kirchhoff and Bunsen converted the examination of spectra into a usable method of "spectrum analysis" and, in the process, discovered two new elements which had been overlooked by chemical methods. There followed several attempts at quantitative analysis by various investigators, but their work was without lasting influence. Spectroscopy remained a qualitative procedure until the middle of the next century.

Their work aroused some violent opposition, mainly from English scientists. They questioned the originality of Kirchhoff's explanation of the origin of the dark lines in the sun's spectrum as due to different elements and resulting from absorption of radiation by the gases that surround the sun's glowing nucleus. The critics maintained that others discovered this phenomenon. Anders Jöns Ångström (1814–1874) was one of several others who claimed priority. Important observations about the origin of the lines, their position, and that all the elements can be identified from their spectral lines, had been made by Charles Wheatstone in 1835 in a summarized abstract of a lecture to the British Association for the Advancement of Science, but was not published until 1861 by William Crookes (1832–1919), a critic of Bunsen and Kirchhoff [41].

The newly invented spectroscope was first applied to urine analysis by Henry Bence Jones in 1865. At this time, there was no information on the process of diffusion of medicines in living tissues. Believing there was great value in knowing the rates of diffusion and lengths of time that medications might remain in the body, it occurred to Bence Jones that by means of emission spectrum analysis he could determine where diffusing substances go to; how long they take to reach the tissues; how long they remain there; and how soon they cease to appear in the excretions [42].

He administered lithium chloride orally or by subcutaneous injection to guinea pigs. The animals were killed at intervals of minutes to days. Parts of their organs were incinerated and the soluble salts extracted from the ash and examined for lithium. The element was found in all vascular tissues within fifteen minutes after oral ingestion—sooner if injected subcutaneously—and persisted in the urine up to thirty-nine days after the initial dose. With humans, lithium appeared in the urine five to ten minutes after oral ingestion of lithium carbonate on an empty stomach and continued to be detected in the urine for six to eight days [43]. This was the first study timing the distribution and fate of a drug in the body.

The description and use of what was probably the first visual spectrophotometer to attempt quantitative analysis, appeared in two volumes (1873, 1876) by Vierordt. Using a modified Kirchhoff–Bunsen spectroscope, Vierordt measured hemoglobin and its derivatives in blood, as well as other substances in blood and urine. His adaptations allowed simultaneous measurement of transmitted light by unknown and standard solution at selected wavelengths in the absorption band. Slit widths were varied until matching intensity was obtained. By comparing the amount of adjustment, the concentration of the unknown could be calculated. This type of instrument and its variations suffered from a lack of linear response relative to varying concentration of colored material in solution. Failure to obey Beer's law with biological samples was almost certainly due to the presence of substances other than that under study, which also absorbed at the wavelength chosen, and to the difference in response due to changing the slit width.

In 1890, C. Soret (d. 1931), a French physicist, began to examine the absorption patterns of materials that were colorless in solution, by using ultraviolet-rich sources of incident light. These studies extended considerably the use of spectroscopic techniques for the analysis of biological materials. By the end of the nineteenth century, the spectroscope was established as an important qualitative tool in biochemical laboratories. With the introduction of colorimetric methods in which the colored solution was the product of a chemical reaction, the limitations of the Vierordt type of visual spectrophotometer became apparent. The simpler visual colorimeter with its obvious advantages replaced the visual spectrophotometer, to more efficiently meet the quantitative needs of the newly emerging service laboratories.

ARC AND SPARK ANALYSIS

From the start of the twentieth century, until the late 1920s, spectroscopic research lay in the fields of arc and spark qualitative analysis of metals. With this technique, solid or powdered material is vaporized into the excitation region by the heat from the arc or spark. Continued interest was mainly due to progress in the technology of electricity, the continuing search for new elements, and because the complex arc and spark spectra were yielding insight into atomic structure. These developments were also speeded along by applicable advances in the optical industry.

FLAME PHOTOMETRY

The modern era of flame photometry and quantitative spectrochemical analysis began in 1929 with the work of Henrik Gunnar Lundegårdh (1888–1969), the Swedish plant physiologist and agricultural chemist. He developed a satisfactory air–acetylene flame as the energy source for investigation of trace metals involved in plant metabolism and soil–plant relationships. He was particularly interested in determining potassium in soil and tried to quantitate the intensities of color emitted by potassium and sodium in a gas flame [44].

In this technique, an aqueous solution of salts of certain metals is atomized into the flame. The metal's atoms are excited by the elevated temperature, and they emit distinct and characteristic wavelengths (lines or bands of colored light). Lundegårdh used a quartz prism for dispersion of the emitted light and a photographic plate (spectrogram) to record the emitted light energy. Quantitative evaluation of the photograph was made by microphotometer readings of the optical density of the spectral lines on the exposed plate. Calibration is made with solutions of known composition and concentration. Although it is accurate and versatile, spectrography requires a considerable amount of manipulation and equipment and was too slow and complicated for routine clinical work. Lundegårdh also tried direct photometry with a monochromator for dispersion of the light and measured the light's intensity with a photocell. However, he did most of his later work photographically.

Direct photometry was improved by German workers during the 1930s. A monochromator or narrow-band light filter was used to isolate particular spectral regions, whose light intensities were detected by a photocell. Since the photocell's electrical output is low, compared to a phototube, it had to be measured directly by a galvanometer or sensitive microammeter. A similar but simpler apparatus with photocell, galvanometer, and colored filter for the determination of potassium was developed in 1937 by Wolfgang Schuhknecht (b. 1908) and was further developed with Ferdinand Waibel (b. 1891) into a flame photometer. It was successfully used in analysis of sodium, potassium, and lithium, in soils, plant ashes, and fertilizers. Further applications were made in metallurgical products. These developments led to the first commercial production of filter flame photometers by the German companies of Siemens and Zeiss and opened the way to wider use of this technology. These filter photometers had enough optical efficiency to permit use of the simpler selenium barrier-layer photocell instead of the phototube.

In the United States, parallel development began later. Consideration of the flame by Barnes and associates led to development of a lower temperature method using air–coal gas or air–propane, which the designers called *flame photometry*. The urgent industrial need for a rapid and accurate quantitative determination of sodium and potassium in aqueous solution, in large numbers of samples, resulted in 1945 in the first practical American filter flame photometer [45]. The instrument, using filters, photocell, and galvanometer, resembled the products of Siemens and Zeiss. A year later, the American team described an improved filter photometer with a dual optical system that allowed the use of an internal standard or reference element (lithium) not normally found in the sample [46]. This principle originated with arc and spark emission spectroscopy in 1925 and was used by Lundegårdh in flame spectrography. With an internal standard, factors influencing the light intensity emitted by one element similarly affect the light intensity of the internal standard, so that the ratio of intensities is constant regardless of experimental conditions. The use of ratios rather than absolute light intensities considerably reduce the errors due to variation of the light source from fluctuations in gas or air pressure, presence of foreign ions and molecules, or viscosity differences between standards and samples due to presence of proteins.

With a low-temperature flame, only the atoms of the alkali metal and alkaline earth components in the sample are thermally excited to emit appreciable amounts of characteristic light. The fewer the constituents excited, the easier it is to isolate the light of any one component for photometry. The first commercial American flame photometer was the Perkin-Elmer model 18, developed primarily for use in industry when rapid and accurate sodium and potassium determinations on inorganic solutions are desired. It was designed in response to investigators who had used or heard of the instrument developed by Barnes and associates at the research laboratories of the American Cyanamid Company, Stamford, Connecticut [47]. This is another example of an instrumental optical device—developed largely by astronomers and physicists—being adopted by analytical chemists, first for agricultural and industrial use, and then adapted to clinical chemical analysis.

Applications of the Perkin-Elmer flame photometer (Fig. 14.4) for the determination of sodium and potassium in serum, red cells, and urine were made by several American investigators in 1947 [48]. Initially, data was reported as parts per million (ppm) and mg per 100 mL, but very soon milliequivalents per liter became the accepted mode for reporting results on biological materials. Perkin-Elmer flame photometers and those from other companies quickly found their way into clinical laboratories for the quantitative measurement of sodium and potassium in serum and urine. By the end of the 1950s, there were more than thirty manufacturers worldwide producing flame photometers. Some of the other familiar American firms were: Advanced Instruments, Baird-Atomic, Beckman, Coleman, Jarrell-Ash, Process & Instruments, Technicon, National Instrument Laboratories, and Instrumentation Laboratory.

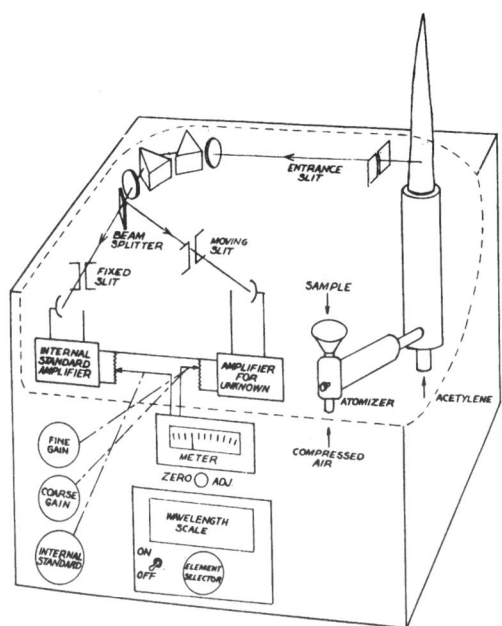

FIG. 14.4. Perkin-Elmer Flame Photometer, Model 52 A. (Courtesy Perkin-Elmer Corp., Norwalk, Connecticut.)

Advances in this technology came with improvements in each of the components of the analytical system: atomizer, burner, optical assembly and wavelength selector, photosensitive detector, and energy output recorder. The range of application of flame photometry was expanded in 1949 with the appearance of flame spectrophotometers (monochromators with flame attachment) equipped with sensitive radiation detectors, especially photomultipliers. This made possible the determination of elements whose spectral lines are relatively weak, rich in lines or located in the ultraviolet, and resulted in more elements being amenable to flame analysis. The simpler filter flame photometers benefited from advances in interference filters, radiation detectors, amplifiers, recording attachments, integrating methods, and other electronic innovations.

Although flame photometry made sodium and potassium levels in serum and urine easily available without the prior time-consuming sample preparation for chemical separation and analysis, this new procedure did not immediately enter the daily repertoire of the clinical chemistry laboratory. Instead, it was usually offered two or three times a week. However, this analysis soon renewed interest in the importance of electrolytes and acid–base balance in the explanation and identification of water and salt disorders.

The use of dangerous flammable gases in the hospital laboratory—sometimes in violation of local fire codes—drew attention to the need for alternative analytical methods for sodium and potassium which were soon elevated to the status of "available 24-hours." Eventually, a new technology of electrometric determination by ion-selective electrodes largely replaced flame photometry in the automated instruments of the clinical chemistry laboratory.

ION-SELECTIVE ELECTRODES

In 1934 Lengyel and Blum [49] reported errors in glass electrode measurement of pH. They observed that the composition of the glass caused the electrode potential to become strongly dependent on the concentration of several cations besides the hydrogen ion. In 1957 Eisenman and associates [50] showed that the specific affinity of glasses of the sodium aluminosilicate series for the alkali-metal cations is a systematic function of their composition. They concluded that practical glass electrodes for selectively measuring the sodium ion activity in biological fluids could be developed. Their general conclusions also applied to glass systems in which potassium or lithium replaces sodium.

In 1961 an electrode suitable for measurement of sodium ions in single samples of biological systems such as blood or urine, was developed by Friedman and Nakashima [51]. Unlike flame photometry, samples need not be diluted. Precision greater than that of flame photometry ($\pm 1\%$) can be obtained. A calomel electrode in the line serves as reference. Although the electrode responds to changes in ionic activity and not concentration, for convenience, calibrating solutions are made in terms of concentration to cover the range under investigation.

According to proficiency surveys by the College of American Pathologists, the use of flame analysis for sodium and potassium decreased from 48% to 26% in the period from 1983 to 1985; use of ion-selective electrodes for sodium and potassium increased from 52% to 73%. By 1992 less than 2% of the more than 6400 reporting laboratories were using flame analysis for these analytes.

ATOMIC ABSORPTION SPECTROMETRY

When Bunsen and Kirchhoff explored emission spectra in 1860, they also investigated spectra due to absorption. Atomic absorption spectrometry is basically the inverse of emission methods. The element being determined is not excited or ionized but merely dissociated from its chemical bonds in a low-temperature flame. In this ground or low energy state, the atom is capable of absorbing radiation of the same wavelength as that emitted by the element in the excited state. In modern practice, the source, a hollow cathode lamp made of the element being investigated, provides the very narrow spectral lines needed for absorption. Although the applicability of atomic absorption to analysis of metals had been recognized by Bunsen and Kirchhoff, it was virtually ignored by chemists, and left for astronomers to study the elemental composition of the stars. The explanation may be that it was a more difficult technique and was not expected to add any information to what could be obtained more easily by emission measurements [52].

In 1955 Alan Walsh, a physicist working in a government research laboratory in Melbourne, Australia, published his classic paper on the principles of atomic absorption spectroscopy and its potential for chemical analysis [53]. Experienced in emission spectrochemical analysis and in infrared absorption spectroscopy, Walsh was aware that molecules were always studied with absorption spectra, whereas atomic spectra were almost always measured by emission. According to Walsh's calculations, measurements of atomic absorption could be made with much greater sensitivity than by measurement of emission. This was an important advantage for the determination of calcium, magnesium, and trace elements in biological and agricultural material.

About the same time, two Dutch physicists, C. T. J. Alkemade and J. M. W. Milatz, independently in a letter to the editor and in a longer paper later the same year in an obscure journal, published a description of an "absorption flame photometer" that used essentially the same principles as Walsh's instrument [54]. However, they apparently did not make any detailed theoretical analysis, but were mainly interested in an alternative technique for sodium and potassium, metals that were normally determined by emission flame photometry. Walsh is credited with the innovation of atomic absorption spectroscopy because he took the time and energy to overcome resistance to this new idea. The other papers stirred very little interest in atomic absorption due to the general satisfaction with the state of emission spectroscopy.

Although atomic absorption met a need in industrial medicine for the determination of heavy metals in urine, particularly lead, at concentration levels below one mg per Liter, and in deficiency studies and in toxicology, the technique found limited, if any, acceptance in clinical chemistry laboratories, despite the rapid growth of sales of atomic absorption spectrometers in the 1960s. The Perkin-Elmer Corporation, Norwalk, Connecticut, marketed its first atomic absorption spectrometer in 1961. By 1962 eight firms had been licensed to manufacture this equipment. Perkin-Elmer's popular model 303 appeared in 1963.

Flame photometry did not do for calcium what it had done for sodium and potassium. Calcium is less easily excited in the ordinary flame; the amount of light given off does not always provide adequate sensitivity for analysis by flame emission methods. Most of the flame photometric methods required either prior removal of the protein by precipitation or ashing or separation of calcium by precipitation as oxalate. Perkin-Elmer's interest in manufacturing an atomic absorption instrument depended on being shown that the technique could determine calcium in blood in a manner competitive with existing methods [55]. Atomic absorption is not affected by interferences from hemoglobin, bilirubin or lipemia. The major drawback is the maintenance and care that the equipment requires. In 1992 this technique was not reported for calcium analysis by any participants in the CAP proficiency surveys.

NOTES AND REFERENCES

1. POUL ASTRUP, PETER BIE, and HANS CHR. ENGELL, *Salt and Water in Culture and Medicine*, translated from Danish by Kirsten Skovbjerg and Andrew L. Cameron-Mills (Copenhagen: Munksgaard, 1993), p. 43.
2. W. C. RAPPLEYE, "A Simple Application of the Volhard Principle for Blood Plasma Chlorides," *Journal of Biological Chemistry*, 1918, 35: 509–512.
3. DONALD D. VAN SLYKE and JOHN J. DONLEAVY, "A Simplification of the McLean–Van Slyke Method for Determination of Plasma Chlorides," *Journal of Biological Chemistry*, 1919, 37: 551–555.
4. J. C. WHITEHORN, "A System of Blood Analysis. Supplement II. Simplified Method for the Determination of Chlorides in Blood or Plasma," *Journal of Biological Chemistry*, 1920–21, 45: 449–460.
5. OTTO SCHALES and SELMA S. SCHALES, "A Simple and Accurate Method for the Determination of Chloride in Biological Fluids," *Journal of Biological Chemistry*, 1941, 140: 879–884.
6. J. V. DUBSKÝ and J. TRTÍLEK, "Mikromassanalyse unter benutzung von diphenylcarbazid und diphenylcarbazon als indikator (Merkurimetrie)," *Mikrochemie*, 1933, n.s. 6: 315–320.
7. KONRAD LANG, "Eine Verbesserung der Methodik der Chloridbestimmung im Blut," *Biochemische Zeitschrift*, 1937, 290: 289–290.

8. FERENC SZABADVÁRY, *History of Analytical Chemistry*, translated from Hungarian by Gyula Svehla (Langhorne, Pennsylvania: Gordon and Breach Science Publishers S.A., 1992), pp. 234–235.
9. ERNEST COTLOVE, HILLARY V. TRANTHAM, and ROBERT L. BOWMAN, "An Instrument and Method for Automatic, Rapid, Accurate, and Sensitive Titration of Chloride in Biologic Samples," *Journal of Laboratory and Clinical Medicine*, 1958, 51: 461–468; see also ERNEST COTLOVE, "Chloride," in *Standard Methods of Clinical Chemistry*, vol. 3 (New York: Academic Press, 1961), pp. 81–92.
10. DONALD D. VAN SLYKE and GLENN E. CULLEN, "Studies of Acidosis. 1. The Bicarbonate Concentration of the Blood Plasma; Its Significance, and Its Determination as a Measure of Acidosis," *Journal of Biological Chemistry*, 1917, 30: 289–346, pp. 303–305.
11. VICTOR C. MYERS and JAMES J. SHORT, "The Estimation of Chlorides in Blood," *Journal of Biological Chemistry*, 1920, 44: 47–53.
12. JOHN O. HALVERSON and OLAF BERGEIM, "The Determination of Small Amounts of Calcium, Particularly in Blood," *Journal of Biological Chemistry*, 1917, 32: 159–170.
13. BENJAMIN KRAMER and FREDERICK F. TISDALL, "The Direct Quantitative Determination of Sodium, Potassium, Calcium, and Magnesium in Small Amounts of Blood," *Journal of Biological Chemistry*, 1921, 48: 223–232.
14. BENJAMIN KRAMER and FREDRICK F. TISDALL, "A Simple Technique for the Determination of Calcium and Magnesium in Small Amounts of Serum," *Journal of Biological Chemistry*, 1921, 47: 475–481.
15. E. P. CLARK and J. B. COLLIP, "A Study of the Tisdall Method for the Determination of Blood Serum Calcium With a Suggested Modification," *Journal of Biological Chemistry*, 1925, 63: 461–464.
16. G. SCHWARZENBACH, W. BIEDERMANN and F. BANGERTER, "Komplexone VI. Neue einfache Titriermethoden zur Bestimmung der Wasserhärte," *Helvetica Chimica Acta*, 1946, 29: 811–818; see also SZABADVÁRY (1992), pp. 268–269.
17. HSEIN WU, "Contribution to the Chemistry of Phosphomolybdic Acids, Phosphotungstic Acids, and Allied Substances," *Journal of Biological Chemistry*, 1920, 43: 189–220, p. 190; EUGENE S. BAGINSKI, EMANUEL EPSTEIN, and BENNIE ZAK, "Review of Phosphate Methodologies," *Annals of Clinical and Laboratory Science*, 1975, 5: 399–416.
18. A. E. TAYLOR and C. W. MILLER, "On the Estimation of Phosphorus in Biological Material," *Journal of Biological Chemistry*, 1914, 18: 215–224.
19. FREDERICK F. TISDALL, "A Rapid Colorimetric Method for the Quantitative Determination of the Inorganic Phosphorus in Small Amounts of Serum," *Journal of Biological Chemistry*, 1922, 50: 329–337.
20. RICHARD D. BELL and EDWARD A. DOISY, "Rapid Colorimetric Methods for the Determination of Phosphorus in Urine and Blood," *Journal of Biological Chemistry*, 1920, 44: 55–67.
21. A. P. BRIGGS, "Some Applications of the Colorimetric Phosphate Method," *Journal of Biological Chemistry*, 1924, 59: 255–264; see also "A Modification of the Bell-Doisy Phosphate Method," *Ibidem*, 1922, 53: 13–16.
22. STANLEY R. BENEDICT and RUTH C. THEIS, "A Modification of the Molybdic Method for the Determination of Inorganic Phosphorus in Serum," *Journal of Biological Chemistry*, 1924, 61: 63–66.
23. CYRUS H. FISKE and YELLAPRAGADA SUBBAROW, "The Colorimetric Determination of Phosphorus," *Journal of Biological Chemistry*, 1925, 66: 375–400.

24. THEODORE KUTTNER and HARRIET R. COHEN, "Micro Colorimetric Studies. I. A Molybdic Acid, Stannous Chloride Reagent. The Micro Estimation of Phosphate and Calcium in Pus, Plasma, and Spinal Fluid," *Journal of Biological Chemistry*, 1927, 75: 517–531; see also THEODORE KUTTNER and LOUIS LICHTENSTEIN, "Micro Colorimetric Studies. II. Estimation of Phosphorus: Molybdic Acid-Stannous Chloride Reagent," *Ibidem*, 1930, 86: 671–676.
25. F. OSMOND, "Sur une réaction pouvant servir au dosage colorimétrique du phosphore dans les fontes, les aciers, etc.," *Bulletin de la Societé Chimique de Paris*, 1887, 47: 745–748.
26. G. DENIGÈS, "Détermination quantitative des plus faibles quantités de phosphates dans les produits biologiques par la méthode céruléomolybdique," *Comptes Rendus des Seances de la Societé de Biologie et de Ses Filiales*, 1921, 84: 875–877.
27. *Dictionary of Scientific Biography (DSB)*, Charles Coulston Gillispie, ed. (New York: Charles Scribner's Sons, 1971), 3: 598–604, p. 602.
28. *Ibidem*, p. 603.
29. For numerous references to the many novel analytical methods for sodium and potassium, see Richard J. Henry, *Clinical Chemistry. Principles and Technics* (New York: Harper & Row, Publishers, 1964), pp. 345–346, 350–351; John P. Peters and Donald D. Van Slyke, *Quantitative Clinical Chemistry*, vol. 2, *Methods* (Baltimore: The Williams & Wilkins Company, 1932), pp. 726 ff; Carroll F. Shukers, "Review of Estimation of Serum Sodium and Potassium," *American Journal of Clinical Pathology*, 1952, 22: 606–615, pp. 606–608.
30. H. H. BARBER and I. M. KOLTHOFF, "A Specific Reagent for the Rapid Gravimetric Determination of Sodium," *Journal of the American Chemical Society*, 1928, 50: 1625–1631.
31. BENJAMIN KRAMER and FREDERICK F. TISDALL, "A Simple Method for the Direct Quantitative Determination of Sodium in Small Amounts of Serum," *Journal of Biological Chemistry*, 1921, 46: 467–473.
32. ALFRED T. SHOHL and HELEN B. BENNETT, "A Micro Method for the Determination of Potassium as Iodoplatinate," *Journal of Biological Chemistry*, 1928, 78: 643–651.
33. F. K. CAMERON and G. H. FAILYER, "The Determination of Small Amounts of Potassium in Aqueous Solutions," *Journal of the American Chemical Society*, 1903, 25: 1063–1073.
34. S. W. CLAUSEN, "A Method for the Estimation of Potassium in Blood," *Journal of Biological Chemistry*, 1918, 36: 479–484.
35. BENJAMIN KRAMER and FREDERICK F. TISDALL, "A Clinical Method for the Quantitative Determination of Potassium in Small Amounts of Serum," *Journal of Biological Chemistry*, 1921, 46: 339–349; *Ibidem*, 48: 223–232.
36. VICTOR CARYL MYERS, *Practical Clinical Analysis of Blood. A Book Designed as a Brief Survey of This Subject for Physicians and Laboratory Workers*, 2nd ed. (St. Louis: C. V. Mosby Company, 1924), p. 176.
37. SZABADVÁRY (1992), pp. 58–59.
38. *Ibidem*, p. 319.
39. *DSB* (1973), 7: 379–383; *Ibidem* (1972), 5: 142–144.
40. The sections on the early work in spectrum analysis are derived largely from Szabadváry (1992), pp. 324–330. The sections on the Bunsen burner are from Georg Lockemann, "The Centenary of the Bunsen Burner," *Journal of Chemical Education*, 1956, 33: 20–22. Kirchhoff and Bunsen's paper on spectral analysis is in *Annalen der Physik und Chemie*, 1860, 110: 161–189, and Figures 1 to 8.
41. CHARLES WHEATSTONE, "On the Prismatic Decomposition of Electrical Light," *Reports to the British Association for the Advancement of Science*, 1835, 5: 11–12; "On the Prismatic

Decomposition of the Electric, Voltaic, and Electro-Magnetic Sparks," *Chemical News* (London), 1861, 3: 198–201 (Read August 12, 1835 at the Meeting of the British Association).
42. H. BENCE JONES, *Lectures on Some of the Applications of Chemistry and Mechanics to Pathology and Therapeutics* (London: John Churchill and Sons, 1867), p. 15.
43. HENRY BENCE JONES, "On the Rapidity of the Passage of *Crystalloid* Substances into the Vascular and Non-Vascular Textures of the Body," *Proceedings of the Royal Society of London*, 1865, 14: 63–64 (Letter); "On the Rate of Passage of Crystalloids into and out of the Vascular and Non-Vascular Textures of the Body," *Ibidem*, 1865, 14: 220–223 (Abstract). See also N. G. COLEY, "Henry Bence-Jones, M.D., F.R.S. (1813–1873)," *Notes and Records of the Royal Society of London*, 1973, 28: 31–56; LOUIS ROSENFELD, "Henry Bence Jones (1813–1873): The Best 'Chemical Doctor' in London," *Clinical Chemistry*, 1987, 33: 1687–1692.
44. The sections on flame photometry were derived from Roland Herrman and C. T. J. Alkemade, *Chemical Analysis by Flame Photometry*, 2nd ed., translated from German by Paul T. Gilbert, Jr. (New York: Interscience Publishers, 1963), pp. 1–11; see also BENNIE ZAK, ROBERT E. MOSHER, and ALBERT J. BOYLE, "A Review on Flame Analysis in the Clinical Laboratory," *American Journal of Clinical Pathology*, 1953, 23: 60–77, pp. 60–62.
45. R. BOWLING BARNES, DAVID RICHARDSON, JOHN W. BERRY, and ROBERT L. HOOD, "Flame Photometry. A Rapid Analytical Procedure," *Industrial and Engineering Chemistry. Analytical Edition*, 1945, 17: 605–611.
46. JOHN W. BERRY, DAVID G. CHAPPELL, and R. BOWLING BARNES, "Improved Method of Flame Photometry," *Industrial and Engineering Chemistry. Analytical Edition*, 1946, 18: 19–24.
47. RICHARD R. OVERMAN and A. K. DAVIS, "The Application of Flame Photometry to Sodium and Potassium Determinations in Biological Fluids," *Journal of Biological Chemistry*, 1947, 168: 641–649, footnote p. 641.
48. T. P. MARINIS, E. E. MUIRHEAD, FRANCES JONES, and J. M. HILL, "Sodium and Potassium Determinations in Health and Disease," *Journal of Laboratory and Clinical Medicine*, 1947, 32: 1208–1216; PAULINE M. HALD, "The Flame Photometer for the Measurement of Sodium and Potassium in Biological Materials," *Journal of Biological Chemistry*, 1947, 167: 499–510; OVERMAN and DAVIS (1947).
49. B. LENGYEL and E. BLUM, "The Behaviour of the Glass Electrode in Connection with its Chemical Composition," *Transactions of the Faraday Society*, 1934, 30: 461–471.
50. GEORGE EISENMAN, DONALD O. RUDIN, and JAMES U. CASBY, "Glass Electrode for Measuring Sodium Ion," *Science*, 1957, 126: 831–834.
51. SYDNEY M. FRIEDMAN and MIYOSHI NAKASHIMA, "Single Sample Analysis with the Sodium Electrode," *Anlaytical Biochemistry*, 1961, 2: 568–575.
52. A. WALSH, "Spectrochemistry since Kirchhoff and Bunsen," *Proceedings of the Royal Australian Chemical Institute*, 1975, 42: 297–303.
53. A. WALSH, "The Application of Atomic Absorption Spectra to Chemical Analysis," *Spectrochimica Acta*, 1955, 7: 108–117; *Ibidem*, see correction on p. 252.
54. C. T. J. ALKEMADE and J. M. W. MILATZ, "Double-Beam Method of Spectral Selection with Flames," *Journal of the Optical Society of America*, 1955, 45: 583–584 (Letter); "A Double-Beam Method of Spectral Selection with Flames," *Applied Scientific Research*, 1955, 4B: 289–299.
55. J. B. WILLIS, "The Birth of the Atomic Absorption Spectrometer and its Early Applications in Clinical Chemistry," *Clinical Chemistry*, 1993, 39: 155–160.

CHAPTER XV

CHOLESTEROL: INTRODUCTION

Cholesterol has become a national health obsession: you hear about it in the media, see it on food packages, and discuss it in the doctor's office. Of all the constituents of blood, it is the one best known by the most people in the closing years of the twentieth century. Mainly the result of government programs directed toward reducing the level of cholesterol in the general population, and to publicity by health awareness publications and advocacy groups, cholesterol's role in the progression of atherosclerosis and coronary heart disease has led to numerous clinical trials and studies worldwide, as well as to the inevitable clusters of nay-sayers [1]. This chapter reviews some of the early methods of cholesterol analysis and other information which has contributed to the public awareness of this interesting molecule.

Although gallstones were known by Vallisnieri as early as 1733 to be soluble in alcohol, their main constituent was first isolated as white leaflets from alcoholic solution by François Paul Lyon Poulletier de la Salle (1719–1787 or 8), sometime about 1769. The isolated material, a waxy, scaly substance, was shown by Chevreul in 1815 to be unsaponifiable, thereby differentiating it from other waxes (spermaceti) and cadaver fat. He named it cholestérine (Greek: *chole*, bile; *stereos*, solid) in 1816. Cholesterol was discovered in the blood by Prosper-Sylvain Denis (1799–1863) in 1830. In 1859 Marcellin Berthelot prepared esters of cholesterine, thus establishing the presence of a hydroxyl group and its alcoholic nature. This was followed by a name change indicated by the -ol ending—cholesterol, for both English and French.

Ever since it was first isolated from gallstones, cholesterol's complex four-ring cyclopentanoperhydrophenanthrene structure—established in 1932—and its synthesis from a simple two-carbon acetate substrate through the action of more than two dozen enzymes, has attracted the attention of organic chemists and biochemists. Indeed, 13 Nobel Prizes have been awarded to scientists for their research on cholesterol or cholesterol-related compounds—most recently in 1985, to Joseph L. Goldstein and Michael S. Brown [2] for physiology or Medicine.

COLOR REACTIONS

Modern cholesterol determinations had their beginnings in the late nineteenth century, when Salkowski (1872) described the color reaction when this constituent of gallstones, dissolved in chloroform, was treated with concentrated sulfuric acid. The chloroform layer turned deep red-purple, while the acid layer exhibited a deep green fluorescence. This procedure never caught on, perhaps because its success required low temperature and strict exclusion of moisture. The development of a practicable color reaction and the discovery that digitonin precipitates cholesterol quantitatively and relatively easily, provided a stimulus to further investigation of methods for the determination of cholesterol in human blood.

Cholesterol, as well as other steroids, reacts with many strongly acidic substances to form a variety of intensely colored products. The color response, which depends on the solvent and chromogenic reagent used, is the basis for many of the methods for its quantitative determination in whole blood, plasma, and serum. None of these reactions is specific for cholesterol, but several have been adapted as quantitative methods. The most popular and widely used method for cholesterol for the major part of this century, is based on the reaction initially described in 1885 by Carl Theodore Liebermann (1842–1914) and developed in 1890 by H. Burchard. Liebermann added sulfuric acid to a solution of cholesterol in acetic anhydride and obtained a green color. Burchard added acetic anhydride and sulfuric acid to cholesterol in chloroform and obtained a more intense green color [3].

REACTION WITH DIGITONIN

The presence of saponins along with cardiac glycosides in commercial preparations of digitals was recognized in 1875 by Schmiedeberg, who named the principal saponin component digitonin. Digitonin, as well as the other steroid saponins, possesses the specific ability to hemolyze red blood cells at high dilution. In 1901 Ransom observed that the addition of cholesterol to a saponin solution destroys the hemolytic activity. This and other observations pointing to some form of combination between cholesterol and digitonin, prompted an investigation by Windaus in 1909 that resulted in the discovery that cholesterol forms a remarkably stable and sparingly soluble 1:1 molecular complex with digitonin which is devoid of hemolytic activity. The free hydroxyl group in cholesterol is essential to the formation of the molecular compound; cholesterol esters are not precipitated by digitonin [4].

In 1910, on the basis of this reaction, Windaus [5] introduced the use of digitonin for the macrogravimetric determination of total cholesterol and its free and combined fractions in renal tissue. For the estimation of total cholesterol, saponification of esterified cholesterol was necessary since only the free form of the cholesterol is precipitated by the digitonin. His use of saponification indicated knowledge that cholesterol existed in free and ester form. Windaus's gravimetric method was the first acceptable reference procedure for the determination of cholesterol. However, the method required such large quantities of blood and was so time consuming that it was infrequently employed for clinical studies. The subsequent application of colorimetry for the determination of the cholesterol content of the digitonide precipitate led to the development of rapid and accurate micromethods of determination. Adolf Windaus (1876–1959) received the Nobel Prize for Chemistry in 1928 for his research into the constitution of the sterols and their connection with the D vitamins.

BLOOD ANALYSIS: ESTERIFIED AND FREE CHOLESTEROL

Prior to the introduction of assays using enzymes, methods for cholesterol involved two procedures: isolation of the cholesterol followed by analysis of the extract. Cholesterol was one of the first blood constituents to be determined by colorimetric assay. Almost every organic fat solvent found in the laboratory was tried, alone or in combination, to extract this water-insoluble substance, prior to color development. The Liebermann–Burchard color reaction was first applied to the quantitative determination of serum cholesterol in 1910 by A. Grigaut [6]. He treated a chloroform–acetic anhydride solution of cholesterol (free or esterified) with sulfuric acid and noted that maximum green color production occurred after 30 minutes. In 1913 the chromogenic step of Grigaut's procedure was simplified, slightly modified, and made more suitable for general clinical use by Autenrieth and Funk [7]. They added sulfuric acid and acetic anhydride to a chloroform solution of cholesterol and applied heat to speed maximum color development. Perhaps because this version was for many years a favored procedure in most laboratories, Autenrieth and Funk are often credited as the originators. This colorimetric method determines total cholesterol.

Autenrieth and Funk saponified the blood or plasma with strong potassium hydroxide before extracting it with chloroform. The process was time consuming and complicated. Then, in 1915 Bloor described a simplified adaptation of the method that could be employed in conjunction with his system for the analysis of the total fat, lecithin, and cholesterol in blood. In 1916 Bloor eliminated the preliminary saponification as unnecessary since the esterified and free cholesterol give the color reaction [8]. Blood, serum or plasma was extracted with alcohol–ether, the precipitated proteins filtered off, the extract evaporated to dryness, and the residue dissolved in chloroform, then reacted with the Liebermann–Burchard reagent. Color was compared in a visual colorimeter with a standard solution of cholesterol similarly reacted with the chromogenic reagent. However, this method produced values about 20% higher than with the method of Autenrieth and Funk because color was developed on a mixture of free and ester cholesterol, and the esters yielded more color per unit weight than free cholesterol.

REFERENCE METHODS

In 1934 Schoenheimer and Sperry [9] avoided this source of inaccuracy. They extracted the free and esterified cholesterol from serum with an alcohol–acetone solution. Saponification (hydrolysis) with potassium hydroxide converted the esters to the free form. Total cholesterol is precipitated as the digitonide and the precipitate is dissolved in acetic acid. Color is developed by the Liebermann–Burchard reagents. Free cholesterol can be determined separately before the saponification step, by precipitation from the alcohol–acetone extract as the digitonide. Total minus free equals esterified cholesterol. Time and temperature of the color development had to be controlled. Results with this procedure compared well with those obtained by the gravimetric method of Windaus and, for the next twenty years, the Schoenheimer–Sperry method was accepted as the new reference method for cholesterol.

The next significant advance in cholesterol analysis occurred in 1952 when Abell, Levy, Brodie, and Kendall [10] found that the dried petroleum ether extract of saponified serum can be treated directly for color development with a modified Liebermann–Burchard reagent (acetic anhydride–sulfuric acid–glacial acetic acid) and measured photometrically. This method yielded results identical with those by the Schoenheimer and Sperry method. It became popular and, being easier to perform, was accepted as the preferred reference method, a status it has retained to this day. In Trinder's [11] modification, the dried petroleum ether extract was dissolved in ethylene chloride and color developed by addition of acetyl chloride and sulfuric acid. Another popular procedure was introduced in 1953 by Zak and associates [12]. They applied a ferric chloride–acetic acid–sulfuric acid reagent directly to serum without any preliminary extraction of the cholesterol and obtained a purple color. This method was subsequently modified by preliminary extraction with various fat solvents or combinations, in an attempt to obtain results matching those obtained with the methods of Schoenheimer and Sperry, and Abell *et al*. In the hands of many investigators, this color reaction yielded absorption values for free and ester cholesterol within about 5% of each other.

New methods, few of which achieved general use outside the laboratory of origin, were followed by numerous modifications, as every step of every phase of the analysis was investigated and modified again and again. Throughout the twentieth century, beginning with the early work of Bloor, hundreds of papers documented the constant search for greater specificity, sensitivity, color stability, accuracy, simplicity, and speed. These efforts have been well summarized [13] by Zak, Kritchevsky, Henry, and Naito.

Until the introduction of automated analysis in the late 1950s and the opportunity it presented of adding chromogenic reagents directly to the serum without a preliminary extraction, most laboratories used some modification of the methods of Schoenheimer and Sperry, Abell, Levy, Brodie, and Kendall or Zak and associates. Direct reactions had popular appeal, despite the problem of non-specific interferences, because mechanical devices were limited in the type and number of reaction steps. The Liebermann–Burchard and Zak acid reagents for manual or automated colorimetric determination of serum cholesterol dominated the literature for the next 25 years. They were displaced following the introduction and immediate acceptance of the highly reliable enzymatic techniques that were introduced in the early 1970s. By then, owing to its marginal clinical usefulness, if any, as an indicator of liver disease, determination of serum cholesterol esters had become obsolete, hastened by the convenience and speed of automated total cholesterol analysis and other tests more sensitive for liver function.

Cholesterol is the only major lipid reported without its fatty acid component, even though a large part of the serum cholesterol—average normal value of 72%—is esterified, and both free and esterified forms are attached to different lipoproteins in different proportions for different physiological purposes.

ENZYMATIC METHODS

When it was determined that cholesterol reacts with cholesterol oxidase to form a cholestenone and hydrogen peroxide, it provided several possibilities for analysis, including colorimetry, fluorometry, and electrochemical detection of cholesterol. Most of these have been exploited to some extent. The analytical sequence begins with hydrolysis of cholesterol by cholesterol esterase followed by reaction with the oxidase. Cholesterol esterase, specific for cholesterol ester, replaces chemical saponification. Subsequent steps utilize the ability of hydrogen peroxide to oxidize any of several compounds to produce chromogens that can be measured spectrophotometrically [14]. According to the College of Pathologists Survey for 1992, all 5548 participating laboratories reported data on cholesterol with enzymatic procedures, using 29 different analyzers from 13 different manufacturers.

ATHEROSCLEROSIS, CORONARY HEART DISEASE, AND THE NATIONAL CHOLESTEROL EDUCATION PROGRAM

Cholesterol is the precursor of all the steroid hormones and is therefore essential for the life of animals and humans. As one of the most widely distributed organic compounds in the animal kingdom, cholesterol is a two-faced molecule. The very property that makes it essential in cell membranes for maintaining the barrier between cell and environment, namely, its absolute insolubility in water, also makes it potentially deadly. High concentrations of it in the blood lead to deposition in the arterial wall and eventual development of an atherosclerotic plaque.

In 1904 Marchand proposed the term atherosclerosis (Greek: *athere*, mush) to designate the type of morbid processes (arteriosclerosis) characterized by amorphous lipid accumulation in the intima. During the next ten years, investigators induced true atherosclerosis experimentally in animals and demonstrated that only cholesterol-containing foods were atherogenic. Hyperlipemia and ultimately the typical lesions of the aorta and coronary arteries were produced by feeding rabbits pure cholesterol in vegetable oil, thereby proving that cholesterol was in fact the atherogenic constituent of these foods [15]. The current practice of measuring lipids and lipoproteins in the blood to obtain useful information about cardiovascular risk had its origin in these experiments of 80 years ago.

Despite this knowledge, atherosclerosis and its clinical sequel, heart attack, were not even diagnosed at the turn of the century. Concern over hyperlipidemia and the risk of cardiovascular disease as a cause of early death in humans is a development of the second half of the twentieth century. Since atherosclerosis is a life-long process, the long interval between its silent onset in the coronary arteries and the sudden occurrence of a cardiovascular ischemic event, did not facilitate the association of hyperlipidemia with atherosclerotic vascular disease. The importance of cholesterol was evident from its widespread occurrence in the animal body, but as of 1920, it was a disputed question whether or not cholesterol is synthesized in the body. Elevated blood cholesterol levels were known to occur in arteriosclerosis, nephritis, nephrosis, diabetes, obstructive jaundice, in many cases of cholelithiasis, in certain skin diseases, and in pregnancy. Cholesterol appeared to be of diagnostic value in nephrosis. However, Denis, one of Folin's collaborators, after measuring the blood cholesterol levels in 20 normal individuals and 254 patients with assorted pathological conditions—including 12 cases of early pregnancy—and finding elevated levels only in diabetes and in a relatively small number of cases, concluded that this determination is "at present of no value in the clinical diagnosis or prognosis of disease" [16]. Her conclusion was undoubtedly due to the many variables that are now known to contribute uncertainty to a chemical analysis, viz., random error (imprecision), systematic error (inaccuracy) and, especially in the case of cholesterol, intra-individual biological and behavioral variation, interpersonal variations due to sex, age, season, and genetic factors, and preanalytical sources of variation such as sample collection and handling procedures, all of which are part of a relatively recent concept and were not recognized at that time [17].

Prior to about 1950, there was no clear perception of the many risk factors contributing to the development of coronary heart disease (CHD) and its sequel of myocardial infarction. Nor were the statistical tools available for designing and handling the large-scale demographic and epidemiologic studies that would be necessary for uncovering such a relationship. The surveys that had been conducted simply suggested a link between CHD and elevated serum cholesterol levels. Although hypercholesterolemia was frequently associated with coronary occlusion in patients of relatively young ages (below 60 years), there was a high proportion of individuals with cholesterol levels in the "normal" range who also experienced clinical arteriosclerosis or coronary thrombosis [18]. It was also noted that the degree of atherosclerosis did not correlate with the serum cholesterol level [19]. The inconsistency of such findings cast doubt on the significance of total cholesterol level in the pathogenesis of atherosclerosis. There was speculation [20] that the physicochemical configuration of cholesterol and the other lipids in blood might be more significant. Nevertheless, in their search for signs or symptoms of coronary atherosclerosis before it becomes symptomatic, investigators have focused initially on abnormally high concentrations of the total serum cholesterol.

This has culminated in a national standardization program to make cholesterol measurements by all clinical laboratories in the United States more accurate—with a maximum allowable error of 3% or better—and traceable to the reference method of the Center for Disease Control (Atlanta, Georgia) by the use of certified reference materials [21]. The need for such materials was recognized by the National Bureau of Standards—now named the National Institute of Standards and Technology (NIST)—when it issued cholesterol as a Standard Reference Material (SRM) in 1967. It was the start of a major effort to help clinical laboratories establish and improve the quality of the measurements they make [22].

In addition, the National Cholesterol Education Program (NCEP) set out to alert physicians and the general public about the dangers of high levels of blood cholesterol, by identifying cutoff points to define desirable ranges of cholesterol concentration. The objective is to reduce the incidence of high concentration of blood cholesterol in all adults in the United States and thereby contribute to reducing the morbidity and mortality of coronary heart disease [23]. At the center of this activity is the vigorous marketing of foods low in cholesterol and saturated fats and the measurement of cholesterol for screening, diagnosis, and monitoring therapy.

BILIRUBIN: INTRODUCTION

The complex chemistry and biology of the bile pigments has generated an enormous amount of literature since Gmelin's simple test for biliary pigment in urine in 1826. Bilirubin originates primarily from the breakdown of the heme moiety of hemoglobin in aging erythrocytes by the reticuloendothelial system, primarily in the spleen, liver, and bone marrow. This bilirubin is transported in plasma bound mainly to albumin, with small amounts carried by alpha globulins. When the bilirubin reaches the liver, it is conjugated with glucuronic acid and is excreted into the bile.

When there is an abnormally high level of plasma bilirubin, from whatever cause—overproduction or impaired excretion or both—diffusion of bilirubin into the skin or sclera may become noticeable and produce the clinical condition known as jaundice. In unconjugated hyperbilirubinemia, there is an increased rate of bilirubin formation that exceeds the capacity of the liver to remove it from the plasma. This occurs in hemolytic disease or impaired conjugation and uptake in the liver (Gilbert's syndrome). Because it is insoluble in water and bound to albumin, unconjugated bilirubin is not filtered at the glomerulus and does not usually reach the urine. Cholestatic jaundice (intrahepatic and posthepatic) results from destruction of liver parenchyma or obstruction to the outflow of bile after the bilirubin has been conjugated and excreted by the liver cells. The resulting back-up causes regurgitation of bile pigment into the circulation. When the blood levels are high, the bilirubin glucuronides (mono- and di-conjugates), being water-soluble and not protein-bound, are filtered at the glomerulus and excreted in the urine. In cholestatic jaundice, both conjugated and unconjugated bilirubin may be demonstrated in the serum.

ICTERUS INDEX

A useful and simple, though indirect gauge of the bilirubin level in serum was introduced by Meulengracht in 1920 [24]. Plasma or serum was diluted with physiological saline until the color intensity by visual observation matched a standard solution of 0.01% potassium dichromate. The dilution factor was reported as the "icterus index." Although theoretically unsound because of other yellow serum pigments, viz., carotenoids and hemoglobin, the method was widely used and modified for adaptation to the visual colorimeter and photometer. The icterus index was still in use in some laboratories as late as the early 1960s, despite the availability of methods for bilirubin analysis.

DIAZO, DIRECT, AND INDIRECT REACTION

The determination of bilirubin itself has more clinical significance than icterus index for the diagnosis of diseases of the liver, bile ducts, and blood-forming organs. Its estimation originated in 1883 when Paul Ehrlich (1854–1915) introduced the diazo reaction for the detection of bilirubin in urine. In 1913 van den Bergh (1869–1943) and Snapper [25] applied Ehrlich's diazo reaction to a serum filtrate after removal of the proteins by precipitation with ethanol and demonstrated the presence of bilirubin in normal serum. In this reaction, bilirubin reacts with diazotized sulfanilic acid to form an acid dye, azobilirubin. A few years later, van den Bergh and Muller [26] discovered that in cases of obstructive jaundice, color developed in aqueous solution within thirty seconds "directly" following addition of the diazo reagent to serum and without the presence of alcohol. This was called the *direct* reaction to distinguish it from the *indirect* reaction of the original procedure which required alcohol to bring the bilirubin into solution to react with the diazo reagent.

After an initial rapid development, the color continues to increase at a slower pace, so that it is difficult to know at what arbitrary point a reading should be taken to represent the direct-reacting form. Various times have been recommended. The reading at one-minute, although recognized as an arbitrary measurement, was widely accepted because of its clinical usefulness. Regardless of the direct reaction, the color resulting thirty minutes after the addition of alcohol was considered to represent the total bilirubin—the sum of direct and indirect bilirubin. Variations observed in the timing of color responses have produced other descriptive terms, e.g., "prompt direct," "delayed direct," and "biphasic." Despite the large amount of new knowledge acquired in recent years about the different forms of bilirubin, physicians remain satisfied with the broad distinction and diagnostic usefulness of the estimation of total, direct, and indirect bilirubin—analyses developed long before the underlying chemistry was well understood.

The difference between direct and indirect bilirubin was clearly established in the mid-1950s. Hydrolysis with acid and with beta glucuronidase showed that the direct-reacting bilirubin is a glucuronide conjugate, and that indirect-reacting bilirubin is not conjugated, i.e., "free" bilirubin. The direct-reacting bilirubin, which is water-soluble, is not extracted into organic solvents; but the indirect form, insoluble in aqueous medium, is extracted by chloroform and other organic solvents. Although the term "conjugated bilirubin" and "bilirubin" are more suitable than direct and indirect bilirubin, terminology long in use is difficult to dislodge.

During the 1960s, a fourth bilirubin fraction named "delta"* bilirubin, firmly bound to protein, was demonstrated in human serum. But it was not until 1981, that this was identified as an entity distinct from unconjugated and conjugated bilirubins. Delta bilirubin is covalently linked to albumin and involves a reaction between albumin and bilirubin mono- or diglucuronide. Unconjugated bilirubin does not react with albumin to form delta-bilirubin. The water-soluble delta fraction reacts like conjugated bilirubin in the direct diazo reaction and is usually present in very low (undetectable) amounts in blood. When levels of conjugated bilirubin increase, so does the concentration of delta bilirubin, resulting in an apparent increase in the conjugated fraction in most methods.

---

*The four forms of bilirubin were designated in order of elution by "high performance" liquid chromatography: unconjugated (α); monoconjugated (β); diconjugated (γ); protein-bound (δ).

In van den Bergh's original procedure, the alcohol was added before the diazo reagent, and some direct-reacting bilirubin was lost by coprecipitation with the protein. As a result, he considered the analysis an estimation rather than a quantitative determination. In 1921 Thannhauser and Andersen [27] tried to avoid this loss by carrying out the diazo reaction first and then precipitating the proteins with alcohol and ammonium sulfate. This approach was not entirely successful. Results were still too low. A few years later, van den Bergh suggested using a ten-fold dilution of serum to avoid this problem, but warned that at low levels of bilirubin the reading would not be sufficiently accurate. In 1937 Malloy and Evelyn [28] avoided precipitation of protein and loss of bilirubin by using diluted serum with an alcohol concentration of 50% methanol and adapted their procedure to a photoelectric colorimeter. The Malloy–Evelyn method has been widely used in the United States.

Alcohol and protein precipitation can be completely eliminated since many other substances will bring the unconjugated bilirubin into solution and promote its coupling with the diazo reagent in aqueous solution. A method utilizing caffeine and sodium benzoate—first used by Enriques and Sivó [29] in 1926—was improved by Jendrassik and Gróf [30]. They included a strongly alkaline buffer to transform the red acid azobilirubin into the alkaline blue form. Since they used undiluted serum, their results at low levels of bilirubin were more accurate than with the Malloy–Evelyn procedure.

Both the Malloy–Evelyn and Jendrassik–Gróf methods were adapted to the Technicon AutoAnalyzer in the early 1960s and later to other automated instruments. In a 1992 College of American Pathologists survey of laboratories reporting on thirty-three different automated systems, more than 98% of the 5,700 participating laboratories used a diazo procedure; 56% were variations of Jendrassik–Gróf; 11% used alcohol (Malloy–Evelyn). A dry chemistry slide method was used by 24% of the laboratories. Manual and other methods were also reported.

STANDARDS: ARTIFICIAL AND CERTIFIED

Originally, van den Bergh employed a "purified" bilirubin standard, but owing to its high cost and other difficulties, he proposed an artificial standard of freshly prepared solution of ferric sulfocyanate in ether. However, this was not practical because of evaporation of solvent, and an aqueous solution of cobaltous sulfate was substituted [31]. Although it was widely used, its red color does not exactly match the reddish-violet color of the azobilirubin, and it is not durable. A solution of 0.001 N potassium permanganate [32] was a frequently used artificial standard representing the azobilirubin formed by a bilirubin concentration of 10 mg per 100 mL. This also deteriorates, but is easily prepared in most clinical laboratories where dilute permanganate solution is used for other analytical purposes, e.g., calcium determination. Methyl red has also been used [33]. Artificial standards work reasonably well for visual comparison, but often fail in photometric methods unless their absorption curves show the same percentage transmission in the wavelength region used in the photoelectric determination of azobilirubin. They were routinely used until bilirubin of a certified high purity became available as a standard reference material (SRM 916) from the NIST in 1971 [34]. This could then be used to verify or correct the value of in-house preparations of purified bilirubin or commercially available, serum-based, calibrated bilirubin reference samples.

URINE ANALYSIS AND TABLET TEST

The Gmelin and iodine tests date back to the nineteenth century. These procedures form a ring at the contact zone of urine and reagent, but are not sensitive enough and are difficult to interpret when pigments other than bilirubin are present. At least sixty-eight additional or modified qualitative procedures for identification of bilirubin in urine have been proposed, but most lack sensitivity [35]. In a relatively sensitive test, Harrison treated urine with barium chloride to obtain an insoluble precipitate of barium sulfate which adsorbs bilirubin. The precipitate is filtered and Fouchet's reagent (trichloroacetic acid and ferric chloride) is added to the filter paper. A green or blue color indicates a positive response.

In 1953 Free and Free [36] introduced a tablet test for bilirubin in urine, sold as Ictotest (Ames Division, Miles Laboratories, Inc., Elkhart, Indiana, now a part of Bayer Diagnostics Division, Tarrytown, New York). The sensitivity of the test was adjusted to not detect bilirubin at the low, but normal concentration in urine of about 0.03 mg per dL. A less sensitive dipstick modification (Ictostix) is more often used, usually in a dipstick with several other test components for urine. The tablet, containing p-nitrobenzenediazonium p-toluenesulfonate, sodium bicarbonate, sulfosalicylic acid, and boric acid, is placed on an asbestos–cellulose mat previously moistened with five drops of urine. The bilirubin is adsorbed onto the mat. Two drops of water are added to the tablet. A blue-purple coloration on the mat within thirty seconds signifies a positive result. The smallest concentration detected by this method is 0.05–0.1 mg/dL. Since bilirubin may often appear in the urine before other signs of liver dysfunction (jaundice, clinical illness) are apparent, bilirubinuria is an important early diagnostic sign of liver disease, especially for industrial workers exposed to toxic agents which can cause liver damage.

KERNICTERUS IN THE NEWBORN

An important analytical application of serum bilirubin analysis is in the newborn, particularly the premature or low birth-weight infants. Kernicterus is a neurological syndrome resulting from the deposition of lipid-soluble unconjugated bilirubin in brain cells, facilitated by the increased permeability of the infant's blood–brain barrier; the conjugated form of bilirubin is not neurotoxic. Kernicterus is a potential hazard during the neonatal period of life if the unconjugated bilirubin levels rise and exceed the bilirubin-binding capacity of albumin and other plasma proteins. The precise blood level above which indirect-reacting (free) bilirubin will be toxic for an individual infant is unpredictable, but most clinicians believe that phototherapy or on rare occasions exchange transfusion, is necessary to reduce the bilirubin level and prevent possible kernicterus when the level reaches the 15 to 20 mg/dL range. The duration of exposure necessary to produce toxic effects from the bilirubin is also unknown, but the less mature the infant, the greater the risk. Low birth-weight infants can develop kernicterus at much lower total bilirubin concentration than full-term infants, and therapy should be initiated sooner.

The transfusion procedure was originally introduced as a therapeutic measure in the hyperbilirubinemia resulting from blood group incompatibility in erythroblastosis fetalis [37] and was subsequently found to be of value in jaundice of premature infants. Regardless of the cause, the goal of phototherapy with high-intensity visible spectrum light is to reduce clinical jaundice and prevent the concentration of indirect-reacting (unconjugated) bilirubin in the blood from reaching levels—depending on birth weight—at which neurotoxicity may occur. Phototherapy changes the structure of bilirubin from its most stable and least water-soluble isomeric form, to produce photoisomers that, being more water-soluble, are excreted in the bile. The therapy is supportive until the infant's underdeveloped conjugation mechanism (synthesis of glucuronyl transferase) of its liver matures. Full-term infants may also exhibit a temporary physiological jaundice. Estimation of bilirubin in amniotic fluid is an indicator of hemolytic disease in the fetus.

DIRECT SPECTROPHOTOMETRIC ASSAY

In the case of jaundiced infants, total bilirubin may be conveniently determined by direct spectrophotometric reading at the absorption maximum of 454 nm. The speed of analysis and small sample size make this method ideal for this special use. Correction for the presence of (oxy)hemoglobin which absorbs strongly at this wavelength, is made by subtracting a second absorption reading at 540 nm since hemoglobin absorbs equally at both wavelengths. A bichromatic photometer ("Bilirubinometer")—with measurement points preset at 454 nm and 540 nm [38]—designed primarily to determine bilirubin in serum from infants, particularly those affected with Rh or ABO blood group incompatibility, is manufactured by Advanced Instruments, Inc., Norwood, Massachusetts. Calibration is made with a known bilirubin reference solution. Since carotenoids, which absorb in the region of 454 nm, can cause false elevation of results, this method is not suitable for adults. The newborn have no carotene intake to interfere with the assay.

NOTES AND REFERENCES

1. MEYER TEXON, "Guest Editorial: The Cholesterol-Heart Disease Hypothesis (Critique)—Time To Change Course?," *Bulletin of the New York Academy of Medicine*, 1989, 65: 836–841.
2. MICHAEL S. BROWN and JOSEPH L. GOLDSTEIN, "A Receptor-Mediated Pathway for Cholesterol Homeostasis," *Science*, 1986, 232: 34–47.
3. C. LIEBERMANN, "Ueber das Oxychinoterpen," *Berichte der Deutschen Chemischen Gesellschaft, Heidelberg*, 1885, 18: 1803–1809; HANS BURCHARD, "Beiträge zur Kenntnis des Cholesterins," *Chemisches Zentralblatt*, 1890, 61: 25–27.
4. LOUIS F. FIESER and MARY FIESER, *Natural Products Related to Phenanthrene*, 3d ed. (New York: Reinhold Publishing Corporation, 1949), pp. 578–579, 102; A. WINDAUS, "Über die Entgiftung der Saponine durch Cholesterin," *Berichte der Deutschen Chemischen Gesellschaft, Heidelberg*, 1909, 42: 238–246.
5. A. WINDAUS, "Über die quantitative Bestimmung des Cholesterins und der Cholesterinester in einigen normalen und pathologischen Nieren," *Hoppe-Seyler's Zeitschrift für Physiologische Chemie*, 1910, 65: 110–117.
6. A. GRIGAUT, "Procédé colorimétrique de dosage de la cholestérine dans l'organisme (Note préliminaire)," *Comptes Rendus Société de Biologie*, 1910, 68: 791–793; "Dosage colorimétrique de la cholestérine dans l'organisme (Deuxième note)," *Ibidem*, 827–829.
7. W. AUTENRIETH and ALBERT FUNK, "Ueber kolorimetrische Bestimmungsmethoden: Die Bestimmung des Gesamtcholesterins im Blut und in Organen," *Muenchener Medizinische Wochenschrift*, 1913, 60: 1243–1248. This article reviews methods for cholesterol.
8. W. R. BLOOR, "Studies on Blood Fat. II. Fat Absorption and the Blood Lipoids," *Journal of Biological Chemistry*, 1915, 23: 317–326; "The Determination of Cholesterol in Blood," 1916, 24: 227–231.

9. RUDOLF SCHOENHEIMER and WARREN M. SPERRY, "A Micromethod for the Determination of Free and Combined Cholesterol," *Journal of Biological Chemistry*, 1934, 106: 745-760.
10. LIESE L. ABELL, BETTY B. LEVY, BERNARD B. BRODIE, and FORREST E. KENDALL, "A Simplified Method for the Estimation of Total Cholesterol in Serum and Demonstration of Its Specificity," *Journal of Biological Chemistry*, 1952, 195: 357-366.
11. P. TRINDER, "The Determination of Cholesterol in Serum," *Analyst*, 1952, 77: 321-325.
12. ALBERT ZLATKIS, BENNIE ZAK, and ALBERT J. BOYLE, "A New Method for the Direct Determination of Serum Cholesterol," *Journal of Laboratory and Clinical Medicine*, 1953, 41: 486-492.
13. B. ZAK and N. RESSLER, "Methodology in Determination of Cholesterol," *American Journal of Clinical Pathology*, 1955, 25: 433-446; DAVID KRITCHEVSKY, *Cholesterol* (New York: John Wiley & Sons, Inc., 1958), 1-2, 232-245; RICHARD J. HENRY, *Clinical Chemistry: Principles and Technics* (New York: Harper & Row, 1964), pp. 843-864; BENNIE ZAK, "Cholesterol Methodologies: A Review," *Clinical Chemistry*, 1977, 23: 1201-1214; HERBERT K. NAITO, "Cholesterol," in *Methods in Clinical Chemistry*, Amadeo J. Pesce and Lawrence A. Kaplan, eds. (St. Louis: The C. V. Mosby Company, 1987), pp. 1156-1178. Also see B. ZAK and J. D. ARTISS, "Some Observations on Cholesterol Measurement in the Clinical Laboratory," *Microchemical Journal*, 1990, 41: 251-270.
14. W. RICHMOND, "Preparation and Properties of a Cholesterol Oxidase from *Nocardia* sp. and Its Application to the Enzymatic Assay of Total Cholesterol in Serum," *Clinical Chemistry*, 1973, 19: 1350-1356; CHARLES C. ALLAIN, LUCY S. POON, CICELY S. G. CHAN, W. RICHMOND, and PAUL C. FU, "Enzymatic Determination of Total Serum Cholesterol," *Ibidem*, 1974, 20: 470-475.
15. LOUIS N. KATZ and JEREMIAH STAMLER, *Experimental Atherosclerosis* (Springfield, Illinois: Charles C. Thomas, 1953), pp. 3-5; N. ANITSCHKOW, "Experimental Arteriosclerosis in Animals," in *Arteriosclerosis. A Survey of the Problem*, Edmund V. Cowdry, ed. (New York: Macmillan, 1933), pp. 282-283; C. H. BAILEY, "Atheroma and Other Lesions Produced in Rabbits by Cholesterol Feeding," *Journal of Experimental Medicine*, 1916, 23: 69-84.
16. W. DENIS, "Cholesterol in Human Blood Under Pathological Conditions," *Journal of Biological Chemistry*, 1917, 29: 93-110, p. 110.
17. G. R. COOPER, G. L. MYERS, S. J. SMITH, E. J. SAMPSON, "Standardization of Lipid, Lipoprotein, and Apolipoprotein Measurements," *Clinical Chemistry*, 1988, 34: B95-B105; see also LOUIS ROSENFELD, "Atherosclerosis and the Cholesterol Connection: Evolution of a Clinical Application," *Clinical Chemistry*, 1989a, 35: 521-531, pp. 527-528.
18. LESTER M. MORRISON, LILLIAN HALL, and ALBERT L. CHANEY, "Cholesterol Metabolism: Blood Serum Cholesterol and Ester Levels in 200 Cases of Acute Coronary Thrombosis," *American Journal of Medical Sciences*, 1948, 216: 32-38; EDWIN F. HIRSCH and SIDNEY WEINHOUSE, "The Rôle of the Lipids in Atherosclerosis," *Physiological Reviews*, 1943, 23: 185-202.
19. KURT E. LANDÉ and WARREN M. SPERRY, "Human Atherosclerosis in Relation to the Cholesterol Content of the Blood Serum," *Archives of Pathology*, 1936, 22: 301-312; J. C. PATERSON, LUCY DYER, and E. C. ARMSTRONG, "Serum Cholesterol Levels in Human Atherosclerosis," *Canadian Medical Association Journal*, 1960, 82: 6-11.
20. W. C. HUEPER, "Experimental Approaches to the Problem of Arteriosclerosis," *Geriatrics*, 1947, 2: 293-296; HIRSCH and WEINHOUSE (1943), pp. 198, 193-194.

21. "Current Status of Blood Cholesterol Measurement in Clinical Laboratories in the United States: A Report From the Laboratory Standardization Panel of the National Cholesterol Education Program," *Clinical Chemistry*, 1988, 34: 193–201; HERBERT K. NAITO, "Reliability of Lipid, Lipoprotein, and Apolipoprotein Measurements," *Ibidem*, B84–B94.
22. ROBERT SCHAFFER, GEORGE N. BOWERS, JR., and ROBERT S. MELVILLE, "History of NIST's Contributions to Development of Standard Reference Materials and Reference and Definitive Methods for Clinical Chemistry," *Clinical Chemistry*, 1995, 41: 1306–1312. The NIST has no role in enforcing the adoption or abandonment of methods. These are responsibilities of the clinical and medical community and the federal and state regulatory agencies. SRM samples are sold to clinical, industrial, and governmental laboratories throughout the world.
23. "Report of the National Cholesterol Education Program Expert Panel on Detection, Evaluation, and Treatment of High Blood Cholesterol in Adults," *Archives of Internal Medicine*, 1988, 148: 36–69; see also ROSENFELD (1989a), pp. 524–525; LOUIS ROSENFELD, "Lipoprotein Analysis. Early Methods in the Diagnosis of Atherosclerosis," *Archives of Pathology and Laboratory Medicine*, 1989b, 113: 1101–1110.
24. E. MEULENGRACHT, "Die klinische Bedeutung der Untersuchung auf Gallenfarbstoff im Blutserum," *Deutsches Archiv für Klinische Medizin*, 1920, 132: 285–300.
25. A. A. HYMANS V. D. BERGH and J. SNAPPER, "Die Farbstoffe des Blutserums," *Deutsches Archiv für Klinische Medizin*, 1913, 110: 540–561.
26. A. A. HYMANS V. D. BERGH and P. MULLER, "Über eine direkte und eine indirekte Diazoreaktion auf Bilirubin," *Biochemische Zeitschrift*, 1916, 77: 90–103.
27. J. S. THANNHAUSER and E. ANDERSEN, "Methodik der quantitativen Bilirubinbestimmung im menschlichen Serum," *Deutsches Archiv für Klinische Medizin*, 1921, 137: 179–186; see also HIJMANS VAN DEN BERGH (Discussion), *British Medical Journal*, 1924, 2: 498–500, p. 499.
28. HELGA TAIT MALLOY and KENNETH A. EVELYN, "The Determination of Bilirubin With the Photoelectric Colorimeter," *Journal of Biological Chemistry*, 1937, 119: 481–490.
29. EUGEN ENRIQUES and RUDOLF SIVÓ, "Neues Verfahren zur Bestimmung des Bilirubingehaltes von Seren und Duodenalsäften," *Biochemische Zeitschrift*, 1926, 169: 152–160.
30. L. JENDRASSIK and P. GRÓF, "Vereinfachte photometrische Methoden zur Bestimmung des Blutbilirubins," *Biochemische Zeitschrift*, 1938, 297: 81–89.
31. J. W. MCNEE and CHESTER S. KEEFER, "The Clinical Value of the Van Den Bergh Reaction for Bilirubin in Blood: With Notes on Improvements in Its Technique," *British Medical Journal*, 1925, 2: 52–54, p. 53. For an improved cobalt sulphate standard see F. D. WHITE, "On Serum Bilirubin: I. The Diazo Reaction as a Quantitative Procedure," *British Journal of Experimental Pathology*, 1932, 13: 76–85, pp. 77–78.
32. B. W. RHAMY and P. H. ADAMS, "A New Standard for the Van Den Bergh Test," *Journal of Laboratory and Clinical Medicine*, 1927, 13: 87–88.
33. GEOFFREY ARTHUR DERING HASLEWOOD and EARL JUDSON KING, "The Estimation of Bilirubin in Blood Plasma," *Biochemical Journal*, 1937, 31: 920–923.
34. "Recommendation on a Uniform Bilirubin Standard," *Clinical Chemistry*, 1962, 8: 405–407; RICHARD J. HENRY, S. L. JACOBS, and NEIL CHIAMORI, "Studies on the Determination of Bile Pigments. I. Standard of Purity for Bilirubin," *Clinical Chemistry*, 1960, 6: 529–536. For a thoughtful review see also NATHAN RADIN, "What is a Standard?," *Clinical Chemistry*, 1967, 13: 55–76.
35. *The Merck Index*, 5th ed. (Rahway, New Jersey: Merck & Co., Inc., 1940); see also JOHN P. PETERS and DONALD D. VAN SLYKE, *Quantitative Clinical Chemistry*, vol. II, *Methods*

(Baltimore: The Williams & Wilkins Company, 1932), pp. 913–919. For a comprehensive review, see TORBEN K. WITH, *Bile Pigments. Chemical, Biological, and Clinical Aspects*, translated from Danish by J. P. Kennedy (New York: Acadmic Press, 1968), pp. 492–503 ff; see also HENRY (1964, 1st ed.), pp. 594–598; JAMES WINKELMAN, DONALD C. CANNON, and S. LAWRENCE JACOBS, "Liver Function Tests, Including Bile Pigments," in *Clinical Chemistry. Principles and Technics*, 2nd ed., Richard J. Henry, Donald C. Cannon, and James W. Winkelman, eds. (Hagerstown, Maryland: Harper & Row, 1974), pp. 1065–1073.

36. ALFRED H. FREE and HELEN M. FREE, "A Simple Test for Urine Bilirubin," *Gastroenterology*, 1953, 24: 414–421.
37. Erythroblastosis fetalis results from the transplacental passage of maternal antibody active against red blood cell antigens of the infant, leading to an increased rate of red cell destruction. It continues to be an important cause of anemia and jaundice in newborn infants despite the development of a method of prevention and maternal isoimmunization by Rh antigens. *Nelson Textbook of Pediatrics*, 14 ed., Richard E. Behrman, ed. (Philadelphia: W. B. Saunders Company, 1992), pp. 479–482; see also WITH (1968), pp. 479–480, 636–646.
38. SANFORD H. JACKSON, "A Direct-Reading Bilirubinometer Incorporating Hemolysis and Turbidity Correction," *Clinical Chemistry*, 1965, 11: 1051–1057; ABNER H. LEVKOFF, MILTON C. WESTPHAL, and JOHN F. FINKLEA, "Evaluation of a Direct Reading Spectrophotometer for Neonatal Bilirubinometry," *American Journal of Clinical Pathology*, 1970, 54: 562–565; S. H. JACKSON and A. H. HERNANDEZ, "A New 'Bilirubinometer' and Its Use in Estimating Total and Conjugated Bilirubin in Serum," *Clinical Chemistry*, 1970, 16: 462–465; R. T. EVANS and J. B. HOLTON, "An Assessment of a Bilirubinometer," *Annals of Clinical Biochemistry*, 1970, 7: 104–106.

CHAPTER XVI

ENZYMES: INTRODUCTION

Since prehistoric times, four chemical changes have been observed which apparently occur spontaneously. They are the fermentation of sugar with the production of alcohol; the souring of milk; the souring of urine; and the production of ammonia in urine. Each results from the growth of microorganisms that convert the substances present into other substances. All are now known to be caused by enzymes; but for a long time it was believed that these reactions were part of the life cycle of the organisms and could only occur if these forms were present and alive. The term "ferment" was used for a long time for any agent that could bring about a chemical reaction in biological material.

THE CATALYTIC FORCE

The story of enzymes in medical practice has its origins in 1833. Anselme Payen (1795–1871) and Jean-François Persoz (1805–1868) discovered that a large amount of insoluble starch granules was converted into soluble dextrin by a small amount of a water-soluble substance isolated from malt extract. They called it diastase (Greek: breaking in), now known to be a mixture of extracellular enzymes, because they believed the substance broke down an insoluble membrane on the surface of the starch granules and allowed a soluble material to escape. It was soon realized that an organic agent was at work facilitating the same reaction on starch that Gottlieb Sigismund Kirchhoff (1764–1833), a Russian chemist, had produced with dilute acids in 1811. Similar types of reactions had been reported by Humphry Davy (1817) and his brother Edmund (1820) and by others, whereby small organic molecules such as methane or alcohol, were transformed into other compounds by contact with platinum [1].

Consequently in 1836, citing the work of Payen and Persoz, Berzelius formulated the concept of a *catalytic force* to cover a wide range of chemical phenomena common to organic and inorganic nature and occurring only in the presence of some third substance which remained unchanged by the reaction. It was action by contact—a new force "such that materials may, by their mere presence, and not on account of their chemical affinities, awaken in a substance such affinities as are latent at the temperature in question." He introduced the term *catalysis* to describe the decomposition occurring through this force because it resembles *analysis*, which describes separation by the usual chemical affinity. With no theory to back it up, only intuition, it was a new name for a hidden force. Berzelius also reasoned that if there was one such example there must be "thousands of catalytic processes" taking place between the tissues and fluids in living plants and animals to produce many different chemical compounds [2].

Another example of catalysis was reported in 1836 when Theodor Ambrose Hubert Schwann (1810–1882) confirmed that something besides acidity was involved in gastric digestion and, as shown by Eberle and others, that gastric fluid is able to digest food outside the stomach. Schwann treated gastric mucosa with dilute hydrochloric acid and liberated a digestive principle which he named pepsin (Greek: *pepsis*, digestion). He believed this to occur by catalytic action since a small amount was able, on contact, to dissolve a large quantity of albumin. Liebig, who edited the journal in which Schwann's paper was published, added a footnote of caution. The terms pepsin and catalysis, said Liebig, were only representations of an idea, and should not be used unless an actual substance is shown to exist by analysis [3].

FERMENTS: ORGANIZED AND UNORGANIZED

In addition to diastase and pepsin, an albuminous substance had been isolated in 1837 by Wöhler and Liebig from an aqueous emulsion of crushed almonds. It acted like yeast on cane sugar in its ability to split amygdalin to oil of bitter almonds and hydrocyanic acid. They named it emulsin. Wöhler suspected that the decomposition was an example of Berzelius's catalysis. These examples helped generalize the concept of specific fermenting actions.

Since diastase, pepsin, and emulsin were apparently in fluids that did not require the presence of specific organs or organisms for their action and could function in a test tube, they were nonliving substances and were called "soluble" or "unorganized" ferments. On the other hand, alcoholic fermentations, one of the oldest reactions known, were thought to function only inside living yeast cells. Since yeast was able to reproduce, it was considered an "organized" being or "formed" ferment. This nomenclature was unsatisfactory and in 1876, after discovering and naming trypsin,

FERMENTS: ORGANIZED AND UNORGANIZED 397

Willy Kühne (Fig. 16.1) introduced the term "enzyme" (Greek: *in* yeast) to designate the "unformed" or "non-organized" ferments, to stress that it wasn't the yeast, but something *inside* the yeast that was responsible for the activity.

FIG. 16.1. Friedrich Wilhelm (Willy) Kühne. (National Library of Medicine, Bethesda, Maryland.)

Kühne's association of ideas about ferment and yeast probably influenced him to apply the term first to trypsin, when working with a ferment from the pancreas. In a later paper (1878) in the house journal of the Physiological Institute of the University of Heidelberg, where he worked, he specifically states that the term enzyme should be used for all *ferments* in the unorganized state and not just for those from yeast. The term gradually came to denote both forms and eventually replaced the older term "ferment" itself. In 1898 Émile P. Duclaux (1840–1904) suggested that the ending *ase* be used in the naming of enzymes to indicate their character and their origin from the name diast*ase*, the first enzyme to be isolated [4].

Using the technique of a pancreatic fistula which he learned from Claude Bernard in Paris, Kühne showed that pancreatic trypsin attacked protein fragments that had been released by the action of pepsin on the original protein substrate. He reasoned that since the stomach is not digested by its own pepsin, such ferments must have inactive protein precursors, which he called "zymogens."

CHEMICAL VS VITALIST FERMENTATION

During the 1830s, microscopic observation had identified yeast as a living organism that nourishes itself at the expense of the sugar it ferments. However, the notion of the living nature of yeast was totally rejected by Berzelius, Liebig, and other chemists. Berzelius believed in vital phenomena, however, he explained all reactions in living organisms as initiated and regulated by catalysts. He opposed biological explanations that suspended universal physical laws and believed that fermentation was merely an example of contact catalysis by a nonliving agent [5]. Berzelius considered the whole animal body as an instrument, "yet the cause of most of the phenomena within the Animal Body lies so deeply hidden from our view, that it certainly will never be found. We call this hidden cause *vital power*; and like many others, who before us have in vain directed their deluded attention to this point, we make use of a *word* to which we can affix no idea." [6]. Since in most biochemical reactions of that time only the starting and final components were known, Berzelius often resorted to a vitalistic explanation for the intermediate mechanism that he could not explain.

Liebig also believed in a vital force with a power of chemical synthesis. However, he opposed the concept of a catalytic force because, instead of helping to clear up the mechanism of chemical reactions, it introduced a new mysterious force into chemistry. According to Liebig (1839), alcoholic fermentation was a strictly chemical process closely related to putrefaction. It was due to molecular instability caused by the oxidation of albuminoid substances and resulting in molecular vibrations that stimulate the sugar molecules to decompose into alcohol and carbon dioxide. His theory conveyed the idea of a mechanical transfer of the movements of a decomposing body (ferment) to a resting body (substrate). This mechanistic theory was popular with chemists for a long time. In whatever manner it was expressed, this association with death and decay clashed with the vitalistic interpretation of the process, viz., that fermentation is an essential part of the life process of the yeast. The decisive blow to Liebig's idea was delivered by Louis Pasteur (1822–1895), the French chemist and bacteriologist, when he showed that yeast grows in the absence of albuminoid substances and that it ferments best in the absence of oxygen [7].

For Pasteur, production of beer, wine, and alcohol by fermentation was always associated with the multiplication of living, though microscopic, organisms. Furthermore, in every type of fermentation, there was a particular organism that can be isolated, cultivated, and studied. Microbial fermentation, like animal nutrition, reflects chemical activities of the living organism, and Pasteur, other microbiologists, and many scientists, considered fermentation to be a property of the living cell. With the knowledge that yeast consisted of living organisms, and from his own observations and experiments, Pasteur formulated the thesis that alcoholic fermentation was a property of the living yeast cell and not a catalytic process. Pasteur believed that fermentation required the presence of a living organism. For him, the living organism *was* the ferment [8]. Although Pasteur was mistaken, his work stimulated the rapid development of medical microbiology, as groups of physiological chemists began work on the chemical activities of bacteria and on the toxins and enzymes they produce.

ZYMASE AND THE CELL-FREE EXTRACT OF YEAST

Experimental proof that there was no need to distinguish between ferments and enzymes came near the end of the century. In 1897 Eduard Buchner (1860–1917) showed that alcoholic fermentation was a chemical and not a vital (living) process, when he prepared a cell-free extract of living yeast that converted a concentrated sugar solution to ethyl alcohol and carbon dioxide. He named the responsible agent, *zymase*. The discovery was accidental. Buchner, an organic chemist, was helping his brother Hans (1850–1902), a bacteriologist, to prepare microbial extracts that might contain antitoxins. They ground yeast with sand and diatomaceous earth and then squeezed the resulting thick paste under high pressure to obtain a cell-free fluid for therapeutic research. However, the extracts they obtained decomposed rapidly. When they added a concentrated sucrose solution to the extract as a preservative, they were surprised by the strong evolution of carbon dioxide and the formation of alcohol—in the total absence of living yeast cells [9]. Thirty years later, zymase was decoded as a mixture of twelve separate catalytic proteins that required inorganic phosphate and several organic compounds related to vitamins to produce alcoholic fermentation.

The discoveries of Pasteur in the world of microorganisms had been roadblocks to the chemical theory of enzyme action. Now, zymase revived the issues of the historic debate between Liebig and Pasteur [10] over the "chemical" and "vital" explanations of fermentation (and all of life's processes). This controversy, which began about 1857, had quieted down by 1872 with general acceptance of Pasteur's views and the distinction between "organized" and "unorganized" ferments—one vital and the other chemical. Zymase intruded on this uneasy truce and, although Liebig and Pasteur did not live to see Buchner's experiment, it reawakened the controversy of whether life processes should be explained biologically or chemically. The question—how to reconcile chemistry with life? This was the basis of the seemingly unending controversy between the mechanists and the vitalists. Meanwhile, philosophically, biochemical thinking was caught in the middle.

ENZYME THEORY OF LIFE

The distinction between reactions in the living organism and those in the laboratory, played itself out in the arena of fermentation which, in the closing years of the nineteenth century, became the bridge between chemistry and biology. Buchner's discovery that the ferment of yeast could function outside of the living cell, removed the distinction between ferments acting *in vivo* and *in vitro*. There now occurred a profound change in the way the chemical activities of living cells were explained. Aware of the importance of enzymes, investigators began to expect a specific intracellular enzyme for each vital chemical action. This was a major departure from the nineteenth-century theory that the chemical reactions of living cells were carried out by the whole cell of protoplasm. Life, it was now believed, was the self-regulating dynamic equilibrium of a system of catalytic reactions with enzymes as the key agents. This "enzyme theory of life" was the common ground of a diverse group of specialists in a wide variety of scientific fields, calling themselves "biochemists" [11]. The study of enzymes helped to transform the interface of physiology, physiological chemistry, and organic chemistry, into modern biochemistry.

After receiving the Nobel Prize in Chemistry in 1907 for his work on cell-free fermentation, Buchner was appointed to the chair of physiological chemistry at the University of Breslau in 1909 [12]. His sensational discovery provided a strong impetus to scientists to search for other enzymes and for an explanation of the chemistry of fermentation and enzymolysis. Cell-free extracts provided an easy but important new method of analysis of the constituents of living organisms and their chemical transformations in relation to their biological function. The development during the 1890s, of a new physical chemistry based on thermodynamics, electrolytic dissociation, and chemical kinetics, subsequently removed much of the mystery of catalysis, as the rules of chemical equilibrium and mass action were found to apply to biological phenomena.

PROTEIN IDENTITY OF ENZYMES

Many chemists were of the opinion that enzyme purification should yield protein-free material. It was widely believed that enzymes, like other bioactive substances known at that time, such as vitamins and the hormones epinephrine and thyroxine, are small organic molecules adsorbed onto crystalline proteins as nonspecific colloidal carriers in the enzyme preparations. Adsorption phenomena had been described qualitatively before 1900, but now, the new physical chemistry provided a consistent theory to explain them. Although Emil Fischer in 1894, had explained the specificity of the enzyme–substrate reaction as analogous to a lock accepting a key, colloid chemists held to the prevailing view that enzyme catalysis arises from physical adsorption on colloidal surfaces, rather than from chemical combination with specific groups in a defined enzyme molecule [13].

It was important to know whether enzymes are proteins, for this determined the separation strategy used for their purification. During the 1920s, not only was there uncertainty about the protein nature of enzymes, but there also was confusion about the chemical structure of proteins. Even the peptide theory of protein structure proposed in 1902 by Fischer and by Hofmeister was questioned well into the 1930s. The dominance of such negative views about proteins and enzymes explains why the claim by James B. Sumner in 1926, of isolating urease as a crystalline globulin protein, was ridiculed and dismissed as being merely the crystallization of a protein "carrier" of the nonprotein enzyme. Many leading biochemists questioned the biological significance of discrete isolated enzymes and, in keeping with then current opinions about the colloidal nature of protoplasm, considered enzymes to be artifacts [14]. As late as 1929, Joseph Needham (1900-1995) wrote [15]:

> "The fact that nobody has ever isolated an enzyme as a chemical entity in a pure state has led biochemists to abandon the hope that anyone ever will, and by all the best people enzymes are now regarded as sets of conditions (fields of force, residual valencies, etc.) associated with particular kinds of colloidal aggregates."

Sumner's claim that an enzyme could be a simple protein was soon substantiated by the isolation of crystalline proteolytic enzymes by John H. Northrop (1891-1987), beginning with pepsin in 1930, and followed by trypsin and chymotrypsin in association with Moses Kunitz (1887-1978) in the early 1930s. Sumner crystallized catalase in 1936, and Northrop crystallized diphtheria antitoxin in 1940. Sumner shared the Nobel Prize for Chemistry in 1946 with Northrop, and with Wendell M. Stanley (1904-1971) for his crystallization in 1936 of the tobacco mosaic virus, a nucleoprotein. The massive and conclusive evidence presented by Northrop that enzymes were proteins, made enzyme chemistry a branch of protein chemistry. The century-old controversy on the nature of enzymes—which began when Berzelius introduced the concept of catalysis—had come to an end.

AMYLASE

Clinical enzymology arrived on the laboratory scene in 1908 when Julius Wohlgemuth (1874-1948) in Berlin described a quantitative method for the determination of diastase (amylase) activity [16]. In this amyloclastic method, applicable to blood, urine, and duodenal fluid, a known quantity of soluble starch (2 cc of 0.1% starch solution, i.e., 2 mg) is digested in a series of test tubes containing decreasing volumes of test specimen. After a fixed period of incubation (30 minutes, 37°C), followed by immediate cooling to stop the reaction, iodine is added to determine the tube in which the smallest quantity of test solution has completely digested the starch, i.e., the last tube in the series that shows a complete absence of all blue color. Diastase units are calculated as mg starch that would be completely digested by 1.0 cc of specimen. The pH of urine specimens needs to be adjusted to the optimum for the enzyme reaction—neutral or very slightly acid. No adjustment is necessary for serum or plasma. This method in various manual and photometric modifications was still in use as late as 1970.

Saccharogenic procedures that measure the amount of reducing sugars released by the action of the enzyme have been more widely used. In 1917 Myers and Killian described a procedure for amylase in whole blood based on the picric acid method of Lewis and Benedict [17] that measures reducing sugars—in this case, resulting from the hydrolysis of starch. Somogyi's [18] method of measuring reducing sugars, and subsequent modifications, have been the most popular for the determination of serum amylase.

The introduction of a dye-coupled starch as a substrate for amylase in 1967 [19], was followed very soon by synthesis of several dye–polysaccharide substrates. Amylase methods that utilize these synthetic substrates depend on the liberation of a dye coupled to a complex insoluble polysaccharide. The amount of color released is measured after a fixed period of incubation with the test specimen. Numerous commerical products based on this chromolytic technique have been produced. The Pharmacia (Piscataway, New Jersey) Phadebas method was an early popular version. These methods often require blanks, centrifugation, and decanting and therefore, are difficult to automate. However, automated methods for amylase have been reported with iodometric and saccharogenic techniques, and with dye–polysaccharide complexes as substrate [20].

LIPASE

After observing an increase in amylase activity in urine following experimental ligature of the pancreatic duct in animals and in two patients in 1910, Wohlgemuth proposed this determination as a clinical test for pancreatic function and in cases with suspected occlusion of the pancreatic duct. However, diagnostic application proceeded slowly, despite the report of a parallel increase in serum lipase activity in acute pancreatitis by Peter Rona (1871–1949) and Leonor Michaelis (1875–1949) in 1911. In their stalagmometric method, the hydrolytic action of lipase on tributyrin reduced the surface tension of the reaction mixture—measured as drops per minute from a capillary tube. It was later shown that the hydrolysis of tributyrin was due to another enzyme, pseudocholinesterase, which is not elevated in acute pancreatitis.

In 1932 Cherry and Crandall [21] used olive oil as a substrate and measured the released fatty acids by titration with standard dilute alkali and phenolphthalein as indicator. The diagnostic value of the olive oil procedure to detect acute pancreatitis has been well established and is the preferred method. However, the analysis of serum lipase has rarely been satisfactory as a clinical laboratory procedure. In 1992, of the more than 6400 laboratories participating in the College of American Pathologists surveys, less than 1600 reported lipase values. However, lipase is determined with increasing frequency, because its sensitivity is equal to, and its specificity superior to, amylase measurements in serum for diagnosis of acute pancreatitis. Unfortunately, a wide variety of methods of varying specificity has led to discrepancies in the clinical interpretation of data [22].

ALKALINE PHOSPHATASE

The first clear concept of phosphatases as a separate group of enzymes was advanced in 1907 by Umetaro Suzuki (1872–1943) and co-workers. Continuing the experiments of earlier investigators with germinating seeds, they showed that most of the phosphorus in rice seeds was present in the form of phytin and that inorganic phosphate increased during germination. They concluded that this increase was due to the presence of an enzyme (phytase) that cleaves phytin into inositol and phosphoric acid. They predicted that this new enzyme, which could be extracted from rice and wheat bran, would be found widely distributed in nature.

In 1923 Robert Robison (1883–1941) detected phosphatase activity which splits monoesters of hexosephosphoric acid and is present to a large extent in growing bone of young rats and rabbits [23]. He also found that the optimum pH for bone phosphatase was on the alkaline side of neutral. In 1929 Herbert Davenport Kay (b. 1893), who had worked under Robison, reported an increase of phosphatase activity in plasma in osteitis deformans and other bone diseases [24], following a 48-hour incubation of unbuffered plasma at pH 7.6 with beta-glycerophosphate substrate [25]. Phosphate released during the incubation and determined colorimetrically was a measure of the enzyme activity. Bodansky [26] buffered the reaction at pH 8.6 and reduced the incubation time to one hour. An increase in alkaline phosphatase in bile duct obstruction was first described by Roberts [27].

King and Armstrong [28] introduced a popular modification for determining this enzyme by using disodium phenylphosphate substrate buffered at pH 9.3. The increased sensitivity of this method allowed a reduction of incubation time to thirty minutes. Phenol is liberated and reacts with

Folin–Ciocalteu reagent (phosphotungstic–phosphomolybdic acid) to produce a blue color measured in a colorimeter or photometer.

In 1942 Shinowara, Jones, and Reinhart [29] increased the sensitivity of the glycerophosphate procedure by raising the pH of the incubation mixture to 9.3. Procedures that determine released phosphate ions require a separate analysis of serum inorganic phosphate. Phosphatase activity is calculated as the increase in phosphate resulting from incubation of serum with substrate. Another new substrate, phenolphthalein phosphate, was described in 1945 by Huggins and Talalay [30]. The phenolphthalein released by the action of the alkaline phosphatase is measured by its absorbance at 550 nm in the presence of an alkaline buffer. Phenolphthalein phosphate was the substrate of a semi-quantitative screening test employing tablet reagents (Phosphotabs, Alkaline) sold by Warner-Chilcott, Morris Plains, New Jersey. This was also available in kit form for quantitative determination of alkaline phosphatase.

Bessey, Lowry, and Brock [31] popularized the use of p-nitrophenyl phosphate, a self-indicating substrate, for alkaline phosphatase assay. Their method was based on a study by King and Delory [32] of a wide variety of potential phosphatase substrates. Unknown to Lowry's group, the substrate had already been introduced by Ohmori [33] in 1937. The phosphatase hydrolyzes the colorless substrate to yield the yellow salt of p-nitrophenol which, at the alkaline pH of the reaction—the pH optimum with human serum was found to be 10.0 to 10.1 under the conditions described—eliminates the need for additional color-producing reagents. The modification by Bowers and McComb [34] in a kinetic or endpoint spectrophotometric analysis is widely used today because of its simplicity and sensitivity.

Alkaline phosphatase is a sensitive indicator of obstructive liver disease and may be useful for the differential diagnosis of jaundice. However, this enzyme is not specific for liver dysfunction because its concentration in serum is also elevated in bone disease. Comparison analysis of alkaline phosphatase with 5'-nucleotidase or gamma glutamyl transferase, believed to be more specific for liver disease, helps to differentiate between liver and bone as the source of the increased level of serum alkaline phosphatase.

Enzymes exist in multiple molecular forms or "isoenzymes," which have similar catalytic properties, but differ in other respects. Measuring total enzyme activity would not distinguish these forms since by definition they all catalyze the same chemical reaction. The technique of electrophoresis—migration in an electrical field (see Chapter 17)—separates the multiple molecular forms of alkaline phosphatase (or other enzymes) and helps to distinguish their organ sources and to quantitate the relative contribution of each organ to the total activity.

ACID PHOSPHATASE

After a specific acid phosphatase was found in the prostate gland [35] and in the cells of prostate carcinoma [36], Gutman and Gutman [37] developed a procedure with phenylphosphate as substrate at pH 4.9. They reported an increase of the enzyme in serum in patients with bone metastases from carcinoma of the prostate. It was the first blood test for the diagnosis of cancer. The existence of multiple forms of serum acid phosphatase was demonstrated when Herbert [38] inactivated a "prostatic" fraction by means of ethanol or by heating at 37°C for one hour.

Fishman and Lerner [39] observed inhibition of prostatic acid phosphatase by L-tartrate and determined this enzyme fraction as the difference in activity with and without the addition of L-tartrate. Some of the "non-prostatic" acid phosphatase of serum may originate in the red and white blood cells and platelets. Another approach is to inhibit the platelets's non-prostatic fraction with formaldehyde, cupric ions, or citrate [40]. The substrates used for the alkaline phosphatase procedures were also adapted for the determination of acid phosphatase levels.

In 1986 the investigation of prostatic cancer benefited from the introduction of the first commerical immunoassays for prostate specific antigen (PSA). Since then, this unique tissue-specific antigen has become established as the most useful serological marker for monitoring patients with prostate cancer [41]. The antigen was originally isolated in the 1970s from semen by three independent laboratories searching for semen-specific antigens. In 1979 Wang and associates at the Roswell Park Institute, Buffalo, New York, purified the antigen obtained from its natural source in "healthy" prostate tissue [42]. This group pioneered the use of prostate specific antigen as a tumor marker to follow patients with cancer of the prostate, proving it to be more useful than prostatic acid phosphatase in this regard [43].

Prostate specific antigen is only one of an increasing number of organ-specific "tumor-markers" that have been reported in recent years. Serum levels above their respective normal reference ranges may indicate the presence of a tumor or the reappearance of tumor after its surgical removal.

ASPARTATE AMINOTRANSFERASE AND OTHER ENZYMES

A major stimulus to diagnostic enzymology occurred with the demonstration of increased activity of serum glutamic oxaloacetic transaminase (aspartate aminotransferase) after myocardial infarction [44]. The obvious improvement in diagnosis of this illness led to intensive research into a wide variety of enzymes and their potential for diagnostic value. Of the many enzymes investigated, twelve emerged as useful to the clinician. Some of these are measured in virtually every hospital in the world. For others, there are considerable regional differences in usage. Isocitrate dehydrogenase (ICD) is of interest in English-speaking countries. Ornithine carbamoyl transferase (OCT) is popular in Scandinavian countries, while glutamate dehydrogenase (GLD) is almost exclusive with laboratories in Austria and Germany [45]. In the United States, the College of American Pathologists (CAP) regularly surveys aspartate aminotransferase (AST) (glutamate-oxaloacetate transaminase, SGOT), alanine aminotransferase (ALT) (glutamate-pyruvate transaminase, SGPT), lactate dehydrogenase (LD), and creatine kinase (CK). Methods include kinetic and spectrophotometric assays. According to a 1992 CAP survey, eleven manufacturers offered thirty-four different automated instrument-method combinations for aspartate aminotransferase.

The increasing number of different clinical enzyme reactions was accompanied by a large variety of arbitrary units usually identifying the investigator who developed the method. Reaction parameters were chosen because they were convenient rather than optimal. Assay conditions were so varied that interlaboratory comparison was difficult if not impossible. In an effort to bring order out of potential chaos, worldwide standardization was addressed in 1961 when the International Enzyme Unit was recommended by an International Sub-Commission on Clinical Enzyme Units. Wherever practicable, clinical enzyme units should be defined as micromoles of substrate transformed per minute under specified conditions and their concentrations expressed as International Units per liter or per milliliter of serum, plasma, or urine. Where this is impossible or difficult because of the complex nature of the substrate, units should be expressed in terms of the analyzable substance determined to measure the reaction [46].

Meanwhile, the importance of careful attention to test parameters for measuring enzyme activities had become apparent. Investigators had learned to define the time, temperature, pH of the reaction, substrate concentration, additives, and other test conditions, in order to achieve reproducible results from laboratory to laboratory. Even before their chemical nature was known, a mechanism for enzyme action had been worked out on the basis of the theory of intermediate compound formation. The mathematical formulation of this mechanism by Michaelis and Menten [47] in 1913 became the basis for later kinetic studies of enzyme action. As methods and spectrophotometry improved, there was a gradual shift in the clinical use of enzyme analysis from single endpoint assay to kinetic enzyme analysis.

SI UNITS

Except for the recommendation on clinical enzyme units in 1961, there were few attempts at international agreement on other units converging on the chemical landscape. As a result of the varied nomenclature for expressing concentration, the Danish Society of Clinical Chemistry and Clinical Physiology in 1963 initiated a proposal for achieving a uniform system of units. The liter (for volume) and the kilogram (for mass)—units already in use—were suggested as the most logical bases to which to refer concentrations of biological material. When the molecular weight of the dissolved substance is known, molar concentration (moles per liter) was preferred to mass concentration (kilograms per liter) [48] because the chemical processes of all living systems occur and—the proposal claimed—are most conveniently compared in molecular quantities, not in mass units. A uniform system of standardization was expected to close the existing gap between the nomenclature in clinical chemistry and that of allied fields of science.

Recommendations dealing with the basic and derived quantities were approved at the Sixth International Congress of Clinical Chemistry in Munich, 1966 and accepted by the Section on Clinical Chemistry of the International Union of Pure and Applied Chemistry (IUPAC) in Prague, 1967. These basic units were incorporated into the *Système International d'Unités* (SI units)—a system long in existence.

The use of SI units was soon adopted by many clinical chemistry laboratories and journal editors located outside the United States. American hospital laboratories, however, have generally resisted implementing what critics label as restrictions on the use of the metric system, inasmuch as mass and molar concentrations are equally acceptable in the metric system. Massive confusion was anticipated during the period of education and adjustment by the many hospitals, laboratory personnel, and physicians, in a country as large as the United States. A similar conflict of usage is already evident in the general public's lack of recognition and acceptance of the metric system with respect to commerical goods, even though both the usual and metric units are displayed side by side on packaged foods and other commodities. More serious confusion would be experienced if people who monitor their own blood glucose or cholesterol levels had to use the new units.

Whereas, biochemically related substances, such as electrolytes in metabolic studies—are best compared in molecular quantities, there is no advantage in reporting molar quantities instead of mass units for the ten or twelve chemical constituents in plasma (serum) or urine, whose reference values are well known by physicians and laboratory personnel—the principal users of this information. These analytes comprised the bulk of the selections offered by the clinical chemistry laboratory before the recommendation was made to report laboratory data in molar concentration, e.g., urea, uric acid, creatinine, creatine, glucose, cholesterol, bilirubin, protein (total and albumin).

During the early 1970s, IFCC* Committees such as the Expert Panel on Quantities and Units (EPQU), produced various documents which the AACC[†] Committee on Standards viewed as strict and unnecessary in its interpretations of the SI system. It was feared that some of the proposals would require such drastic changes of well-established laboratory nomenclature and reporting practices that the expected benefits of SI might never be realized. Most units as well as names familiar to the scientific community would become unrecognizable. The charge was made "that many of the decisions made by international agencies are arbitrary, their interpretations too rigid, and their implementation too far removed from the realities of practical scientific disciplines to be useful" [49]. Change should be considered on an individual basis and be made only when there is strong evidence that diagnosis and treatment of patients will be greatly improved and not result in chaos and confusion in the clinical laboratories. No system should be adopted unless it is greatly superior to the one it replaces. IFCC officials answered these complaints by emphasizing that their recommendations are not mandatory and that adoption of the new system has not caused any problems in other countries [50].

International organizations and their task force committees originated in nineteenth-century treaty agreements starting with the treaty of

---

*International Federation of Clinical Chemistry (IFCC).
[†]American Association for Clinical Chemistry (AACC).

1875—known as the "Convention of the Metre." This Convention and its associated regulations authorized the birth of the "International Bureau of Weights and Measures" (BIPM), a permanent scientific agency supported by the signatories of the treaty—the United States being one of them. The BIPM was placed under the authority of the "General Conference on Weights and Measures" (CGPM) and of a committee of experts—the "International Committee for Weights and Measures" (CIPM). The general purpose was to guarantee the "international unification and development of the metric system." The international treaty organizations first occupied themselves with refining the experimental definitions of centimeter–gram–second (cgs) and associated electrical units, but then moved to emphasize and finally to pronounce as official the meter–kilogram–second–ampere (MKSA) set of units as fundamental. The SI system was adopted by name by the 1960 CGPM [51].

However, the treaty committees have moved well beyond their original nineteenth-century mission of improving the accuracy, consistency, and uniformity of metric units, to a present position of strong advocacy of new units *and* strong condemnation of various existing units. Their organizational structure has taken on a bureaucratic life of its own, complete with a new self-image—no longer that of providing a service function, but now taking an active leadership role, with its own committee-oriented scientists, permanent staff, and self-serving committee-created language issuing mandates [52].

In 1989, with the debate ongoing, and many countries still uncommitted, a report was prepared on behalf of the IFCC Education Committee and EPQU to repeat the recommendations of IUPAC, the IFCC, the World Health Organization (WHO), and other organizations,* and to present some new names for familiar analytes and units of measurement, e.g., carbamide for urea; whole blood partial pressure to be reported in kilopascals (kPa) rather than millimeters of mercury (mm Hg); and for hydrogen ion concentration, nmol/L to replace pH units [53].

In the United States [54], there were objections to this limitation on the use of SI units in medicine and to the mol as base unit for the amount of substance in preference to the base unit for mass (kilogram). While the application of other units, and specifically, units in general current use is discouraged, no attempt has been made to document the benefits of the

---

*International Union of Biochemistry (IUB); International Union of Pure and Applied Physics (IUPAP); International Organization for Standardization (ISO).

proposed restrictions, by showing that, improved healthcare, cost containment or scientific advances were previously prevented or delayed by the unrestricted use of the metric system.

The user community of clinicians as well as laboratorians have not been consulted regarding either "how" or "when." In the absence of universal consensus, no separate segment of the clinical chemistry profession should attempt to impose its individual conception of a purer system by mandating a restricted use of the metric system on others who are using a slightly different version of it. There is no need to reintroduce SI since medicine has used it since its inception and traditionally expresses the quantities of mass and length in the SI units of kilogram and meter or their appropriate sub-multiples. To banish one unit (kg/L) in favor of the other (mol/L) is not only unnecessary, but often not practicable. The proper approach is to be selective. Those substances that are best measured in moles should be reported in moles and not in mass units. For example: blood levels of therapeutic drugs should not be expressed in micromoles or millimoles when the pharmacist dispenses them in grams or milligrams. Mass units are also the most appropriate and convenient for preparing a reagent. Molar units, however, are more appropriate for illustrating molecular relationships, e.g., between a nutrient and its metabolites. Rigid conformity under the cover of reform and for the sake of consistency is not practical and should be opposed.

There is no potential benefit to patient, physician or laboratory to implement the restricted use of SI units. The same units are used and understood by all physicians and laboratorians in the United States and abroad. The potential for confusion and misinterpretation will be great from the reporting and use of unfamiliar new numbers. Methods and procedure manuals will have to be rewritten. Instrument scales will have to be exchanged and computers will have to be reprogrammed. It will cause much additional and needless expense. Existing patient records and scientific literature cannot be changed. Future generations of physicians and scientists will still have to remain familiar with the full metric system of units to understand the vast amount of medical information published in the past.

There has been no demand, pressure or mandate from physicians or laboratorians for the proposed restriction of the metric system to molar units. An overwhelming majority opposes the limitations as an impediment to scientific progress, clinical education, and patient care [55]. Although there has been much public expression of support in editorials and in position papers by chairs of select committees for complete replacement of mass units by molar units, privately, there is widespread doubt about a *total* change from metric mass units to metric molar units [56]. The American Medical Association adopted SI units in 1987 for its journal (*JAMA*) and its ten specialty journals. Conventional units for some analytes were placed in parenthesis. After five years of a similar policy, the *New England Journal of Medicine* returned to conventional units in 1992. SI units were given in parenthesis. An informal survey of a dozen leading medical centers in the United States showed that none were using SI units. It seemed unwise to publish in one system, while most of the journal's readers were using the other system in their practices. The same problem existed in parts of Europe, since countries such as Germany and Italy continued to use conventional units or both systems [57].

Clinical Chemistry requests the use of SI units, but will accept conventional units. Some American hospitals that reported both units as a means of transition to SI units have returned to conventional units for most analytes. Examples are Children's Hospital of Columbus, Ohio and two small community hospitals in New Hampshire: Hitchcock Clinic in Lancaster and Upper Connecticut Valley Hospital in Colebrook.

NOTES AND REFERENCES

1. HENRY M. LEICESTER, *Development of Biochemical Concepts from Ancient to Modern Times* (Cambridge, Massachusetts: Harvard University Press, 1974), p. 166; *Comprehensive Biochemistry*, Marcel Florkin and Elmer H. Stotz, eds., vol. 30, *A History of Biochemistry* (Amsterdam: Elsevier Publishing Company, 1972), p. 265.
2. MALCOLM DIXON, "The History of Enzymes and of Biological Oxidations," in *The Chemistry of Life*, Joseph Needham, ed. (London: Cambridge University Press, 1970), pp. 18–19; MIKULAS TEICH, "On the Historical Foundations of Modern Biochemistry," *Clio Medica*, 1965, 1: 41–57, p. 49. The first catalytic reaction described in the literature is the observation by F. C. Vogel (1812) that at low temperature, oxygen and hydrogen react if charcoal be added to their mixture, FLORKIN and STOTZ (1972), p. 265.
3. TEICH (1965), pp. 48, 49; LEICESTER (1974), p. 165.
4. LEICESTER (1974), pp. 176–177; DIXON (1970), pp. 17, 21.
5. LEICESTER (1974), pp. 177–178.
6. IÖNS JACOB BERZELIUS, *A View of the Progress and Present State of Animal Chemistry of Iöns Jacob Berzelius, M.D.*, translated from Swedish by Gustavus Brunnmark (London: 1813), p. 4.
7. TEICH (1965), p. 49; JOSEPH S. FRUTON, "The Emergence of Biochemistry," *Science*, 1976, 192: 327–334, p. 330. For an extended review of Liebig's complex theorizing see FLORKIN and STOTZ (1972), pp. 141–143; see also DIXON (1970), pp. 19–20.
8. LEICESTER (1974), pp. 178, 181.
9. *Ibidem*, p. 182.

10. HAROLD FINEGOLD, "The Liebig–Pasteur Controversy," *Journal of Chemical Education*, 1954, 31: 403–406; see also TIMOTHY O. LIPMAN, "Vitalism and Reductionism in Liebig's Physiological Thought," *Isis*, 1967, 58: 167–185. In *Animal Chemistry*, Liebig believed that he had established vital force on a scientific basis. The criticism that developed was based largely upon advances in the physical sciences and placed new limits on allowable explanations. As a result, the concept of a vital force lost all validity and scientific relevance. TIMOTHY O. LIPMAN, "The Response to Liebig's Vitalism," *Bulletin of the History of Medicine*, 1966, 40: 511–524, p. 523.
11. ROBERT E. KOHLER, JR., "The Enzyme Theory and the Origin of Biochemistry," *Isis*, 1973, 64: 181–196, pp. 185, 181.
12. In 1917, after volunteering for front-line duty in World War I, Buchner was wounded by shrapnel and died two days later.
13. FRUTON (1976), p. 330.
14. *Ibidem*, pp. 330–331; for a historical review see also JOSEPH S. FRUTON, "Early Theories of Protein Structure," in *The Origins of Modern Biochemistry: A Retrospect on Proteins* (*Annals of the New York Academy of Sciences*) 1979, 325: 1–18. The 1929 edition of the *Encyclopedia Britannica* stated "enzymes were formerly thought to be proteins, but this is no longer believed." DIXON (1970), p. 22.
15. DONNA JEANNE HARAWAY, *Crystals, Fabrics, and Fields. Metaphors of Organicism in Twentieth-Century Developmental Biology* (New Haven: Yale University Press, 1976), pp. 131–132.
16. J. WOHLGEMUTH, "Über eine neue Methode zur quantitativen Bestimmung des diastatischen Ferments," *Biochemische Zeitschrift*, 1908, 9: 1–9.
17. VICTOR C. MYERS and JOHN A. KILLIAN, "Studies on Animal Diastases. I. The Increased Diastatic Activity of the Blood in Diabetes and Nephritis," *Journal of Biological Chemistry*, 1917, 29: 179–189; ROBERT C. LEWIS and STANLEY R. BENEDICT, "A Method for the Estimation of Sugar in Small Quantities of Blood," *Proceedings of the Society for Experimental Biology and Medicine*, 1913–14, 11: 57–58; *Journal of Biological Chemistry*, 1915, 20: 61–72.
18. MICHAEL SOMOGYI, "Modifications of Two Methods for the Assay of Amylase," *Clinical Chemistry*, 1960, 6: 23–35.
19. H. RINDERKNECHT, P. WILDING, and B. J. HAVERBACK, "A New Method for the Determination of α-Amylase," *Experimentia*, 1967, 23: 805.
20. RICHARD J. HENRY, DONALD C. CANNON, and JAMES W. WINKELMAN, *Clinical Chemistry. Principles and Technics*, 2nd ed. (Hagerstown, Maryland: Harper & Row, Publishers, 1974), p. 944.
21. IAN S. CHERRY and LATHAN A. CRANDALL, JR., "The Specificity of Pancreatic Lipase: Its Appearance in the Blood after Pancreatic Injury," *American Journal of Physiology*, 1932, 100: 266–273.
22. NORBERT W. TIETZ and DENISE F. SHUEY, "Lipase in Serum—the Elusive Enzyme: An Overview," *Clinical Chemistry*, 1993, 39: 746–756.
23. ROBERT ROBISON, "The Possible Significance of Hexosephosphoric Esters in Ossification," *Biochemical Journal*, 1923, 17: 286–293.
24. H. D. KAY, "Plasma Phosphatase in Osteitis Deformans and in Other Diseases of Bone," *British Journal of Experimental Pathology*, 1929, 10: 253–256; "Plasma Phosphatase. II. The Enzyme in Disease, Particularly in Bone Disease," *Journal of Biological Chemistry*, 1930, 89: 249–266.

25. H. D. KAY, "Plasma Phosphatase. I. Method of Determination. Some Properties of the Enzyme," *Journal of Biological Chemistry*, 1930, 89: 235–247.
26. AARON BODANSKY, "Phosphatase Studies. II. Determination of Serum Phosphatase. Factors Influencing the Accuracy of the Determination," *Journal of Biological Chemistry*, 1933, 101: 93–104.
27. W. MORRELL ROBERTS, "Variations in the Phosphatase Activity of the Blood in Disease," *British Journal of Experimental Pathology*, 1930, 11: 90–95.
28. EARL J. KING and A. RILEY ARMSTRONG, "A Convenient Method for Determining Serum and Bile Phosphatase Activity," *Canadian Medical Association Journal*, 1934, 31: 376–381.
29. GEORGE Y. SHINOWARA, LOIS M. JONES, and HARRY L. REINHART, "The Estimation of Serum Inorganic Phosphate and 'Acid' and 'Alkaline Phosphatase Activity'," *Journal of Biological Chemistry*, 1942, 142: 921–933.
30. CHARLES HUGGINS and PAUL TALALAY, "Sodium Phenolphthalein Phosphate as a Substrate for Phosphatase Tests," *Journal of Biological Chemistry*, 1945, 159: 399–410.
31. OTTO A. BESSEY, OLIVER H. LOWRY, and MARY JANE BROCK, "A Method for the Rapid Determination of Alkaline Phosphatase with Five Cubic Millimeters of Serum," *Journal of Biological Chemistry*, 1946, 164: 321–329.
32. EARL JUDSON KING and GEORGE EDWARD DELORY, "The Rates of Enzymic Hydrolysis of Phosphoric Esters," *Biochemcial Journal*, 1939, 33: 1185–1190.
33. YOSHIHISA OHMORI, "Über die Phosphomonoesterase," *Enzymologia*, 1937, 4: 217–231.
34. GEORGE N. BOWERS, JR. and ROBERT B. MCCOMB, "Measurement of Total Alkaline Phosphatase Activity in Human Serum," *Clinical Chemistry*, 1975, 21: 1988–1995.
35. WALDEMAR KUTSCHER and HAJO WOLBERGS, "Prostataphosphatase," *Hoppe-Seyler's Zeitschrift für Physiologische Chemie*, 1935, 236: 237–240.
36. ETHEL BENEDICT GUTMAN, EDITH E. SPROUL, and ALEXANDER B. GUTMAN, "Significance of Increased Phosphatase Activity of Bone at the Site of Osteoplastic Metastases Secondary to Carcinoma of the Prostate Gland," *American Journal of Cancer*, 1936, 28: 485–495.
37. ALEXANDER B. GUTMAN and ETHEL BENEDICT GUTMAN, "An 'Acid' Phosphatase Occurring in the Serum of Patients With Metastasizing Carcinoma of the Prostate Gland," *Journal of Clinical Investigation*, 1938, 17: 473–478; ETHEL BENEDICT GUTMAN and ALEXANDER B. GUTMAN, "Estimation of 'Acid' Phosphatase Activity of Blood Serum," *Journal of Biological Chemistry*, 1940, 136: 201–209.
38. FREDA K. HERBERT, "The Differentiation Between Prostatic Phosphatase and Other Acid Phosphatases in Pathological Human Sera," *Biochemical Journal*, 1944, 38 (Proceedings): xxiii; *Ibidem*, 1945, 39 (Proceedings): iv; "The Estimation of Prostatic Phosphatase in Serum and its Use in the Diagnosis of Prostatic Carcinoma," *Quarterly Journal of Medicine*, 1946, 15: 221–241.
39. WILLIAM H. FISHMAN and F. LERNER, "A Method for Estimating Serum Acid Phosphatase of Prostatic Origin," *Journal of Biological Chemistry*, 1953, 200: 89–97.
40. M. A. M. ABUL-FADL and E. J. KING, "Properties of the Acid Phosphatases of Erythrocytes and of the Human Prostate Gland," *Biochemical Journal*, 1949, 45: 51–60.
41. HOWARD C. B. GRAVES, "Prostate-Specific Antigen Comes of Age in Diagnosis and Management of Prostate Cancer," *Clinical Chemistry*, 1992, 38: 1930–1932.
42. M. C. WANG, L. A. VALENZUELA, G. P. MURPHY, and T. M. CHU, "Purification of a Human Prostate Specific Antigen," *Investigative Urology*, 1979–80, 17: 159–163.

43. MANABU KURIYAMA, MING C. WANG, LAWRENCE D. PAPSIDERO, CARL S. KILLIAN, et al., "Quantitation of Prostate-Specific Antigen in Serum by a Sensitive Enzyme Immunoassay," *Cancer Research*, 1980, 40: 4658–4662.
44. JOHN S. LADUE, FELIX WRÓBLEWSKI, and ARTHUR KARMEN, "Serum Glutamic Oxaloacetic Transaminase Activity in Human Acute Transmural Myocardial Infarction," *Science*, 1954, 120: 497–499; ARTHUR KARMEN, FELIX WRÓBLEWSKI, and JOHN S. LADUE, "Transaminase Activity in Human Blood," *Journal of Clinical Investigation*, 1955, 34: 126–133.
45. E. SCHMIDT and F. W. SCHMIDT, "Clinical Enzymology," *Febs Letters*, 1976, 62 (Supplement): E62–E79, p. E63.
46. E. J. KING and DIANA M. CAMPBELL, "International Enzyme Units. An Attempt at International Agreement," *Clinica Chimica Acta*, 1961, 6: 301–306; MONROE E. FREEMAN, "Clinical Enzyme Unitage," *Ibidem* p. 300.
47. L. MICHAELIS and MAUD L. MENTEN, "Die Kinetik der Invertinwirkung," *Biochemische Zeitschrift*, 1913, 49: 333–369.
48. ALAN MATHER, "Standardization of Nomenclature in Clinical Chemistry. Report of a Panel Discussion," *Clinical Chemistry*, 1965, 11: 348–353; R. DYBKAER, "Quantities and Units in Clinical Chemistry," *Clinical Chemistry*, 1968, 14: 989–992; R. DYBKAER, "International Recommendation for Nomenclature of Quantities and Units in Clinical Chemistry," *American Journal of Clinical Pathology*, 1969, 52: 637–643; see also *International System of Units (SI)*, B. N. Taylor, U.S.A. ed., NIST Special Publication #330, 1991 (62 pp.), Gaithersburg, Maryland. For sale by government printing office, Washington, D.C.
49. BASIL T. DOUMAS, "IFCC Documents and Interpretation of SI Units—A Solution Looking for a Problem," *Clinical Chemistry*, 1979, 25: 655–658.
50. R. DYBKAER, J. C. RIGG, and R. ZENDER, "IFCC Documents and Interpretation of SI Units—An Adaptable Solution," *Clinical Chemistry*, 1980, 26: 369–370.
51. ARTHUR W. ADAMSON, "SI Units? A Camel is a Camel," *Journal of Chemical Education*, 1978, 55: 634–637. For additional criticism, see also JACK L. LAMBERT, "Opinion," on p. 638.
52. *Idem*.
53. H. P. LEHMANN, H. G. J. WORTH, and O. ZINDER, "A Protocol for the Conversion of Clinical Laboratory Data," *Journal of the International Federation of Clinical Chemistry*, 1989, 1: 106–109.
54. BASIL T. DOUMAS, RONALD H. LAESSIG, and FRANK C. LARSON, "The Sanitized Système International (SI) d'Unités," *Clinica Chimica Acta*, 1987, 167: 113–116.
55. *Idem*.
56. RAYMOND GAMBINO, "Molar SI Units—An Abilene Paradox?," *Lab Report for Physicians*, 1987 (October), 9: 73–74.
57. EDWARD W. CAMPION, "A Retreat From SI Units," *New England Journal of Medicine*, 1992, 327: 49; see also letters to the editor (pp. 50–52) for supporting views.

CHAPTER XVII

PROTEINS: INTRODUCTION

Long before the chemical nature of proteins was even approximately understood, the plasma proteins were usually defined in terms of their solubility and the salt concentration employed in separating them. The process was tedious and often decidedly inaccurate. The protein was precipitated out of solution by varying concentrations of salt solutions, collected on a weighed filter paper, washed with the appropriate concentration of salt solution, coagulated with hot water, washed with water, then with alcohol, and dried to a constant weight, or its nitrogen content was determined by the Kjeldahl method. Gelatinous suspensions frequently clogged the filter paper and required much washing to remove adsorbed solution. Some precipitate was lost in the wash fluid.

None of the distinctions and refinements in salt precipitation techniques ever led to any clear-cut separations of protein fractions with sharply differing properties. There was a continuous spectrum of components with gradually decreasing solubilities and no well-defined transition. Precipitation limits were not sharply defined and tended to overlap; and there was seldom any criterion for judging purity or homogeneity. Consequently, the nomenclature was burdened by confusing designations, which made it very difficult to compare the various protein preparations.

It was very distressing to the early protein chemists that so many of their techniques, which seemed to be universally applicable, produced adverse effects when applied to protein material. Proteins were known to need careful handling and mild reagents, yet here was a class of compounds whose solubility did not improve with heating, in fact, they coagulated irreversibly even at temperatures well below the boiling of water. Exposure to acid or alkali of rather moderate concentration produced similar damage. Nevertheless, water, acid, and alkali were the first solvents to be employed, as methods for the separation and purification of proteins were developed. The task was not easy. Protein molecules are not only among the most complicated and largest which Nature produces, but they are also extremely fragile and exceedingly difficult to purify, let alone isolate from other components.

CLASSICAL SEPARATION OF ALBUMIN AND GLOBULIN

As methods evolved, a pattern to the diverse character of proteins from diverse sources began to be recognized. Albumin (Latin: *albus*, white) and fibrin were first analyzed by Liebig and Gerardus Johannis Mulder* (1802–1880), a Dutch physician, in the late 1830s. Later, other investigators noted the formation of a precipitate on dilution of slightly acidified serum with water. In 1852 Peter Ludwig Panum (1820–1885) realized that this precipitate was not albumin. He called it "serum casein" (Latin: *caseus*, cheese), probably because of a similarity to the curdy precipitate obtained from milk on acidification. In 1862 Alexander Schmidt named this insoluble protein fraction "globulin" (Latin: *globulus*, diminutive of *globus*, sphere) to describe the tiny spheres of precipitate.

The classical method for carrying out this separation of water-soluble serum proteins (albumin) from insoluble proteins (globulin) generally involved a ten-to-twenty-fold dilution of the serum with water, accompanied by the addition of acetic acid. In this manner, the solubility of globulin was so reduced that it flocculated. The precipitate obtained was insoluble in water but soluble in dilute neutral salt solutions (1–10%). These methods, however, yielded only a part of the so-called serum globulins. The whole globulin fraction was precipitated by newer methods utilizing addition of salt.

SALT PRECIPITATION

Magnesium sulfate is probably the oldest of all protein precipitants of the neutral salt class. It was first employed by Prosper-Sylvain Denis who reported that albumin alone was left in solution when blood plasma was saturated with magnesium sulfate. The systematic application of salting-out by Denis in the late 1850s provided the earliest indications of the multiplicity of the serum proteins. In 1878 Camille Méhu precipitated all the serum proteins by full saturation of serum with ammonium sulfate. In 1886 Gustave Kauder found that total globulin was precipitated by half-saturation with ammonium sulfate.

Toward the end of the nineteenth century, "albumin" was defined as the protein that remained in solution after precipitation of the "globulin" at half saturation with ammonium sulfate or full saturation with magnesium

---

*Mulder was the first to use the term "protein" (Greek: *proteios*, "primarius," of the first rank or position) at the suggestion of Berzelius in 1838.

sulfate. Precipitation with ammonium sulfate at room temperature became the classical method of protein fractionation. However, work continued on finding the salt concentration limits giving the most definitive and characteristic fractions.

When a portion of the globulins was found to be soluble in pure water, Fuld and Spiro [1] introduced the name "euglobulin" (true globulin) for the serum globulin fraction that was precipitated by dialysis (i.e., insoluble in distilled water); and "pseudoglobulin" for the fraction freely soluble in water alone. Since globulin had been originally characterized by its insolubility in (weakly acidified) distilled water, a protein soluble under these conditions could only be a "false" globulin, or *pseudo*globulin. These terms carried over to other protein–salt systems and added to the confusion of protein fractionation. Porges and Spiro [2] divided serum globulin into one euglobulin and two pseudoglobulin fractions at progressively greater concentrations of ammonium sulfate. These conclusions and the accompanying erroneous nomenclature on laboratory reports persisted well past the middle of the twentieth century.

Because ammonium ions interfere with determination of protein nitrogen with Nessler's reagent or by the Kjeldahl method, the salt must first be removed by time-consuming dialysis or by distillation after making the solution alkaline; or the protein is separated by heat-coagulation or in some other way. Removal of the ammonium salt by distillation with magnesium oxide caused extreme bumping and required constant shaking to avoid breakage of the flask [3]. Since nitrogen analysis was the only accurate means available at that time for determining protein, ammonium sulfate was replaced by magnesium sulfate for precipitation of total globulin. However, the precipitates were gelatinous and the solutions filtered slowly.

SODIUM SULFATE FRACTIONATION: EUGLOBULIN AND PSEUDOGLOBULIN I AND II

In 1921 Howe replaced ammonium sulfate with sodium sulfate which does not interfere with Kjeldahl analysis of protein nitrogen. He devised a system of separation of serum proteins into euglobulin, pseudoglobulin I, and pseudoglobulin II, by filtration at 37°C, after three hours at this temperature, in sodium sulfate solutions of progressively increasing levels of sodium sulfate [4]. The concentration of albumin remaining in solution after the globulins had been precipitated, approximated that remaining in solution at full saturation with magnesium sulfate or half-saturation with ammonium sulfate—reagents long accepted as a means of completely precipitating globulin.

Howe's analysis for serum albumin and globulin was practical because of the relatively small volumes of serum required and because sodium sulfate did not interfere with the biuret color reaction for quantitation of protein levels, whereas ammonium sulfate and magnesium sulfate did. The precipitated globulin had to be removed by filtration because centrifugation does not firmly pack the precipitate. This posed an additional source of error. The first portion of the filtrate was cloudy and had to be refiltered or discarded. Since albumin was adsorbed on the filter paper, it was necessary to analyze portions of the filtrate that came through clear—after the paper was saturated with albumin.

The albumin values obtained with Howe's method were uniformly too high because of incomplete separation of globulin from albumin, as revealed later by moving boundary electrophoresis. This gentle physical method separates serum proteins according to mobility in an electrical field, into four components, named alpha, beta, gamma globulin, and albumin (see later in this chapter). The absence of agreement was due to alpha globulin and small amounts of other globulins remaining in the albumin filtrate. Continued use of this method delayed awareness that many pathologic states (metabolic, infectious, and neoplastic) are associated with elevated alpha globulin and decreased albumin concentrations in the blood. Despite these shortcomings, the salt precipitation method was technically reproducible and clinically useful for following the variations of the two broad categories of serum proteins—albumin and globulin—and gave considerable impetus to the study of plasma proteins in disease.

Fractionation with sodium sulfate was widely used, well into the 1960s, probably because the method was a significant breakthrough in its time and had become an established procedure in many laboratory manuals and textbooks well before the heterogeneities and lack of correlation of the protein precipitates with electrophoretic analysis became generally known [5].

Salt precipitation and its imprecise classical nomenclature were eventually abandoned in favor of the new terminology and electrophoresis on solid support media, viz., filter paper, cellulose acetate membranes, and agarose gel. These methods were successors to the (free-flowing) moving boundary technique which, because of its complexity and time-consuming nature, could never be more than a research tool; the procedure eventually became obsolete and was discontinued.

IMPROVED SALT PRECIPITATION METHODS

Meanwhile, in a continuing search for improvement in salt precipitation methodology, Campbell and Hanna [6] reduced the three-hour incubation time to ten minutes at room temperature by substituting sodium sul*fite* for the sodium sul*fate* and adjusting the concentration of the sulfite to produce an albumin fraction identical to that obtained by the Howe procedure. In 1940 Kingsley [7] eliminated the lengthy incubation with sodium sulfate and avoided the adsorption of albumin on filter paper. By adding ether and shaking the precipitation mixture, he separated the globulin precipitate by centrifugation at room temperature. Without ether, complete separation by centrifugation was only possible at high speed for one hour. The ether facilitates compact packing of the finely divided globulin precipitate *above* the "undernatant" albumin–salt solution which is then easily removed and quantitated.

Other modifications using increased concentrations of sodium sulfate [8] followed, their goals being to bring the separation of globulin from albumin more in line with electrophoretic analysis. Another fractionation scheme [9] utilized only 1.0 mL of serum and separate reaction mixtures containing sodium sulfate and sodium sulfite to obtain filtrates containing, respectively, albumin and alpha globulin, and albumin alone; a saline–ammonium sulfate reagent precipitated the gamma globulin. The alpha and beta globulins were calculated by difference after analysis for total protein. A procedure using a combined sulfate–sulfite reagent [10] was the manual method of choice prior to the introduction of automated methods in the early 1960s that used albumin-binding chromogenic reagents.

ALBUMIN BINDING REAGENTS

A procedure using the anion binding of an azo dye, 2-(4'-hydroxybenzeneazo) benzoic acid (HABA) with serum albumin was introduced in 1954 [11]. It was readily applicable to the direct analysis of albumin by automated techniques. However, the method was not entirely satisfactory because of its low sensitivity and interference from salicylates and elevated levels of bilirubin that competed with the dye for binding sites on the albumin.

Bromcresol green (BCG), another anionic indicator, is also firmly bound to albumin and is the most sensitive dye-binding reagent for albumin. First introduced in 1965 [12], this sulfonphthalein dye has replaced HABA because of its greater specificity for albumin and little competitive interference from salicylate and bilirubin. BCG is not entirely specific for albumin. However, since its reaction with serum proteins proceeds in two stages, the rapid initial stage with albumin can be conveniently and accurately carried out on automated instruments such as centrifugal or kinetic analyzers, where the absorbance of the dyed protein solution can be read quickly following the mixing of serum and dye [13] (see Chapter 19).

According to the 1992 CAP survey, 69% of 5,537 responding laboratories used BCG on twenty-three different automated instruments from twelve different manufacturers. The remaining laboratories used bromcresol purple (BCP) as the binding agent.

BIURET COLOR REACTION

Before the introduction of automated analysis, the separation of globulin from albumin by salt precipitation methods was followed by the colorimetric determination of the total protein and the albumin—the difference being taken as globulin. The most frequently used procedure for this determination in the clinical laboratory has been with the biuret reagent. This reagent has also gone through several stages of development and improvement.

The biuret reaction can be traced to Ferdinand Rose [14] who, in 1833, on addition of potassium hydroxide or sodium carbonate to a mixture of copper sulfate and egg albumin, observed a violet color. A similar reaction was noted with bovine serum. The reaction is so named because biuret, a compound first obtained by Gustave Wiedemann [15] (1826–1899) in 1848 by prolonged heating of urea or urea nitrate at high temperature, also gives a reddish-violet color. Ritthausen and Pott [16] first applied the biuret reaction to the study of proteins in 1873. Hofmeister was the first to describe its use in the quantitation of proteins and peptones in 1878. In 1909 Kantor and Gies [17] dipped filter paper in biuret reagent and used it, wet or dry, as an indicator paper to detect protein.

Though long known and used as a qualitative test for proteins, the biuret reaction was first applied to a biological fluid—urine—by Riegler (1914) [18]. Autenrieth and Mink (1915) [19] used albumin dissolved in urine as a standard. In 1917 Autenrieth [20] extended the biuret reaction to total protein in serum and ascitic fluid. In 1935 Fine [21] used a diluted serum as a standard. In early applications, since the reaction is relatively insensitive at the low levels of protein in normal urine and spinal fluid, the protein was first precipitated by heat and acetic acid or by trichloroacetic acid, dissolved in sodium hydroxide, and copper sulfate added. A principal difficulty was the lack of a suitable and stable protein standard. Hiller *et al.* [22] (1927) used a commercial biuret as standard, but had to abandon it because the color produced was different from that obtained with protein solutions.

The biuret color reaction is not specific for proteins, but depends essentially on the fairly constant number of peptide bonds per unit weight of protein. Hugo Schiff (1834–1915) showed that the reaction is given by a wide variety of organic compounds whose molecules contain at least two carbamyl ($-CONH_2$) groups in a straight chain or joined through a single atom of nitrogen or carbon. The red-purple color varies with the nature of the protein and the length of the chain. Since ammonium salts react with the copper ions to form a deep-blue complex, ammonium sulfate cannot be used to separate globulin from albumin unless the albumin filtrate is dialyzed free of ammonium ions before analysis with the biuret reaction. Other salts used for fractionation, such as sodium sulfate and sodium sulfite, do not interfere. The ammonium ion concentration normally present in serum is too low to interfere. The biuret color must be standardized with protein of known concentration.

The biuret method is the most commonly used procedure for quantitation of serum protein and one of the most reproducible analyses in the clinical chemistry laboratory. A difficulty that initially kept the biuret reaction from general use for protein analysis was the need to add the sodium hydroxide and copper sulfate separately to the test specimen in order to avoid precipitation of cupric hydroxide when the two components were stored together. The precipitate had to be removed by filtration or centrifugation before measuring the color of the reaction mixture. This same problem had occurred in the formulation of the alkaline copper reagent for glucose analysis. In 1942 Kingsley [23] devised a combined biuret reagent which did not precipitate, but had limited stability due to autoreduction of the cupric ion or reduction by impurities and resulted in turbidity with serum.

In 1946 Weichselbaum [24] added sodium potassium tartrate (Rochelle salt) to the biuret reagent to keep the copper in solution, and potassium iodide to prevent autoreduction of the cupric ion and precipitation of cuprous oxide. Originally intended for visual colorimeters, a five-fold dilution was recommended for the more sensitive photoelectric colorimeters and spectrophotometers that were coming into general use after World War II.

MOVING BOUNDARY ELECTROPHORESIS

With the growing awareness of the importance of proteins as structural components of protoplasm, enzymes, hormones, transporters of oxygen and carbon dioxide, and blood clotting, the isolation, identification, and characterization of proteins posed a challenge that was met by the invention of new physical devices for their separation. Chief among these was the moving boundary electrophoresis.

The principle of electrophoresis (Greek: borne by electricity) has been known for more than a century and a half. In 1809, only ten years after Alessandro Volta (1745–1827) had built the first galvanic cell, a Russian physicist, Ferdinand Friedrich Reuss (1778–1852), reported that when electricity was passed through glass tubing containing water and clay, colloidal clay particles moved toward the positive electrode and water moved toward the negative electrode. The water rose in the tube containing the cathode and fell in the tube with the anode. In this experiment, Reuss made the first clearly recorded observation of two electrokinetic effects: electrophoresis and electroendosmosis.

Further development of electrophoresis depended on the discovery of the relationship between current and the electric field by Georg Simon Ohm (1787–1854) in 1827; the laws relating electricity and chemical change by Michael Faraday in 1834; and the development of chemistry and physics in general. After Reuss's observations, many investigations of the phenomenon were made and interesting experiments described of the migration of charged particles, inorganic colloids, and proteins, in a search for practical application of these phenomena.

In 1937 Arne Wilhelm Kaurin Tiselius (1902–1971) reported his development of a new electrophoresis instrument that overcame the difficulties of previous designs and permitted the measurement of the electric mobility of proteins in a mixture in solution. His separation of horse serum into four distinct zones, albumin, and three globulin components which he named α, β, and γ, gave proof of his instrument's value as an analytical tool [25]. The moving boundary technique showed that the proteins are not ill-defined lyophillic colloids but well-defined substances. Tiselius's apparatus became commercially available in 1945. More than forty installations were sold, mostly in the United States, by the Klett Manufacturing Company, New York City. For his researches on electrophoresis and adsorption analysis, especially for his discoveries concerning the complex nature of the serum proteins, Tiselius was awarded the 1948 Nobel Prize in Chemistry. His work had great significance for the biological sciences because it made possible the quantitative analysis of mixtures of proteins.

Electrophoresis by moving boundary produces concentration gradients in the solution as the proteins separate according to electric charge density of the individual protein molecules. This is accompanied by changes in refractive index throughout the protein solution. Through a continuous and complex photographic process, the separation of the proteins are viewed on a screen as a contour of successive peaks and valleys. Each peak represents a concentration change and denotes the position of a boundary (component) in the moving column of solution. The area under each peak is proportional to the concentration of protein in the fraction responsible for the boundary. This is determined by planimetry of an enlarged image of the contour of peaks and valleys recorded on the photographic glass negative (Fig. 17.1). The concentration of each component can be calculated from each peak's percent fraction of the total area of the contour and the total protein concentration of the test specimen. There is only partial separation and only some of the slowest and fastest components can be taken out of the electrophoresis tube.

An innovative feature of Tiselius's apparatus was the U-tube, a series of four glass sections with plane ground end-plates, each of which could be shifted to one side and separated from the rest of the tube (Fig. 17.2). This greatly facilitated filling and emptying the cell and allowed the separation and removal of a fraction with minimal disturbance of the contents of the other sections. Subsequently, Lewis G. Longsworth (1904–1981) designed a three-part cell containing a single center section, which achieved a more widespread use. By eliminating the horizontal glass plates obscuring the

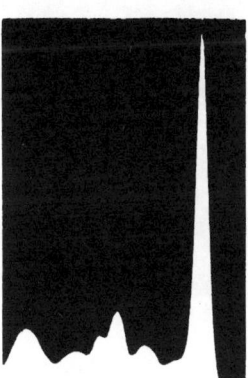

FIG. 17.1. Photographic record of moving boundary electrophoretic pattern of plasma in ascending limb of the cell. Six protein fractions are shown. The seventh peak (extreme left) is the delta anomaly due to the concentration gradients of buffer and protein components. (r to l) albumin, alpha-1 globulin, alpha-2 globulin, beta globulin, fibrinogen, gamma globulin, delta anomaly.

FIG. 17.2. Tiselius's original cell. (*Electrophoresis, Methods, and Apparatus*, Vol. 1, M. Bier, ed. Orlando, Florida: Academic Press, Inc., 1959.)

middle of the field, a more extended run was possible covering a greater distance and disclosing more detail in the pattern.

Tiselius overcame the problem of convection currents and the blurred boundaries caused by the heat-producing current during electrophoresis, by immersing the U-tube in a thermostatically controlled cooling bath several degrees below 4°C (at 1° to 2°C), the point of maximum density of water, because the temperature of the conducting solution is warmer than that of the thermostat. At this lower point there is little change of density with temperature. Heat is dissipated at a proper rate and convection can be effectively prevented.

Tiselius's moving boundary electrophoresis method was adopted by many research laboratories, but it was not suited for the routine clinical laboratory because of its inherent technical complexity. In addition to the 2-mL sample size of plasma, preliminary dialysis of 24 hours against buffer solution, and the limitation of analyzing one specimen at a time—the optical recording system required that the procedure be carried out in an air-conditioned room to prevent condensation on the lens during photography, and a dark room for photographic developing. The apparatus was 19.5 feet long and each component had to be anchored by a heavy concrete foundation to minimize vibrations (Fig. 17.3). The apparatus required a major architectural alteration and became a permanent installation. Later commercial variations were more compact and portable. Because of the need to complete a run (90 minutes), then develop the photographic record before proceeding to preparation of the cell with the next specimen, only two analyses could be done in one working day [26].

The cost of the apparatus was about $18,000. During the 1940s and 1950s, this was a major expense for an instrument. The era of government granting agencies and third-party reimbursement formulas was yet to have an impact on the contributions of industry to the instrumentation in the clinical laboratory. These external forces would soon generate an interest by many unrelated industries in the role its technology could play in the design and manufacture of instrument hardware for the clinical laboratory.

SEPARATION ON FILTER PAPER

The need for absolute separations, which was not practical with the moving boundary apparatus, gave impetus to the development of electrophoresis on solid supporting medium. The first use of filter paper as a stabilizing matrix for electrophoresis appears to be that of König (1937) [27] to determine the electric charge of a substance by its direction of

FIG. 17.3. Klett Moving Boundary Electrophoresis apparatus.

migration. His report also contained suggestions for staining, using ultraviolet light for locating the separated substances, and eluting segments for analysis in a test tube. The paper strip was suspended horizontally between the electrodes and was not covered.

The report was temporarily eclipsed by the work of Tiselius the same year and probably attracted little attention because it appeared in Portuguese in a Brazilian journal reporting on a South American chemical congress. Two years later, von Klobusitzky and König (1939) [28], writing in a German journal this time, more fully described their experiments with paper electrophoresis and the separation of a yellow pigment from snake venom—the first application of paper electrophoresis to the separation of protein mixtures.

In 1946 Consden et al. [29] separated amino acids and peptides in an electrical field (ionophoresis) on a thin slab of silica jelly in a water-cooled glass trough and first demonstrated separation into zones by staining with ninhydrin. In 1948 Wieland and Fischer [30] were the first to report the use of paper for the electrophoretic separation of amino acids and peptides. The breakthrough came in 1950, when four laboratories, in the United States, Germany, and Sweden, without knowledge of the method of König, independently and almost simultaneously, reported procedures of electrophoresis on paper for the separation of protein into components similar to those found by free electrophoresis [31].

In the United States, Emmett L. Durrum's protocol and data had actually been reported from an Army Medical Laboratory a year earlier on March 15, 1949 and had been presented to the American Chemical Society (ACS) meeting in San Francisco on March 29, 1949 [32]. It was received by the Journal of the ACS on July 28, 1949, almost three months before the brief article by Turba and Enenkel was received by *Naturwissenschaften* on October 13, 1949 and published as a "Brief Original Communication" in February 1950. Owing to the vagaries of editorial review and processing procedures, the paper by Turba and Enenkel appeared in print five months before that by Durrum. Tiselius was at the ACS meeting and was given a copy of the Army electrophoresis report by Durrum.

The simple and inexpensive paper electrophoresis method rapidly achieved popularity as a routine clinical laboratory test and led to widespread investigations of the variations of plasma proteins in disease. The advantages over the moving boundary method were speed, small sample volume, inexpensive and portable equipment, and technical ease. Durrum's one-point suspension, freely hanging, non-horizontal paper, was popular in the United States. The designs from Germany and Sweden were in a horizontal mode. An improved version of the original Durrum cell, accommodating eight paper strips, was introduced in 1954 and was soon widely accepted in the United States and abroad (Figs. 17.4, 17.5) [33]. Close to 5,000 were sold, with about two-thirds going to the U.S. and Canada, before it was discontinued in 1972. Along with a regulated power supply (Duostat) that can provide current to two cells, and a scanning densitometer (Analytrol), it gave the Spinco Division of Beckman Instruments, Fullerton, California, its very successful Model R Electrophoresis System [34].

VISUALIZATION OF DISCRETE BANDS

Absolute separation and isolation of the plasma proteins by electrophoresis on paper resulted in discrete bands which can be visualized by staining and

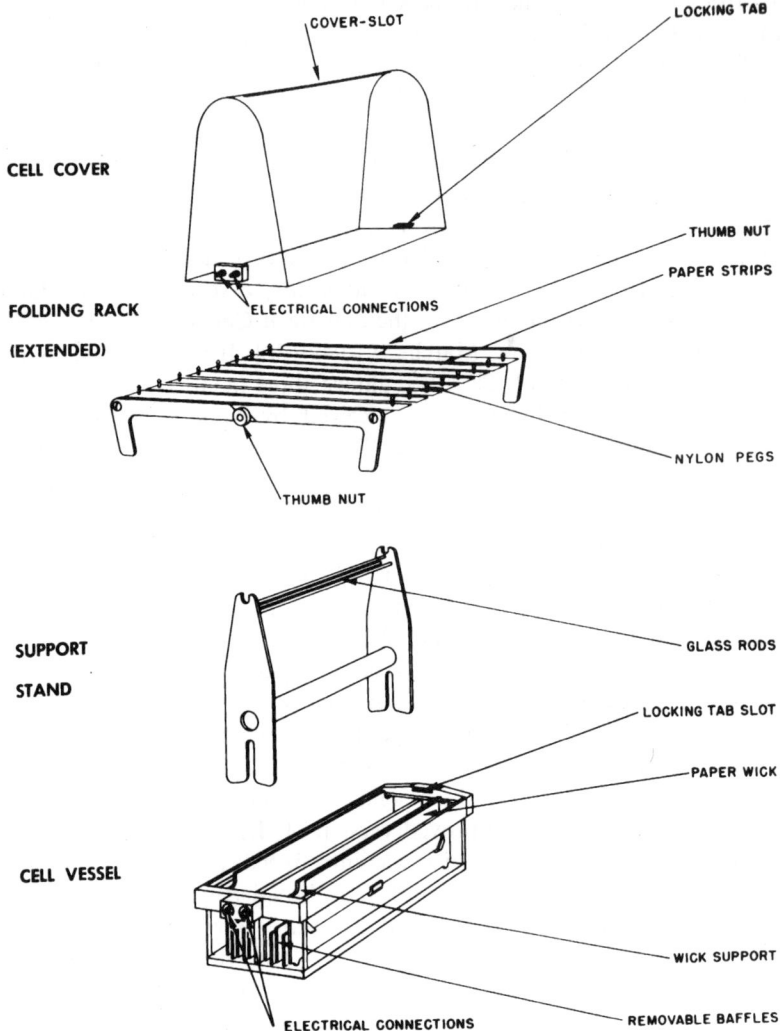

FIG. 17.4. Perspective of hanging-strip paper electrophoresis cell and rack. (Courtesy of The Beckman Heritage Center, Fullerton, California.)

quantitated by elution of the dye from each segment of paper [35]. The optical density readings of the eluates, when plotted against distances of migration, yield a pattern that is similar to that obtained with moving boundary electrophoresis. A similar pattern is obtained when the intact

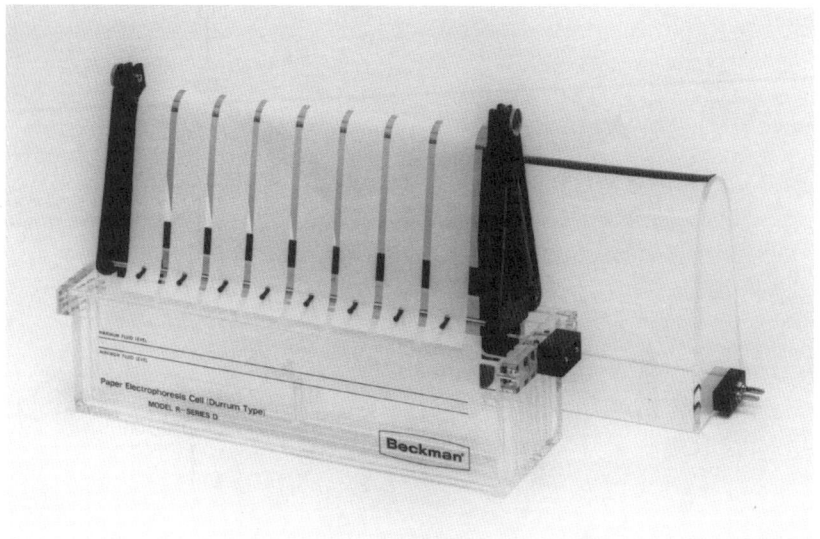

FIG. 17.5. Paper electrophoresis cell and cover. (Courtesy of The Beckman Heritage Center, Fullerton, California.)

stained strip is scanned by a densitometer that continuously and automatically integrates the area being swept out on the recording paper, as the strip moves with constant speed across a thin illuminated slit at right angles to the movement of the recording pen [36].

Paper electrophoresis is to a great degree an empirical methodology, and there is no inherent reason for values obtained on paper to agree with those of free electrophoresis or salt precipitation, for these analytical methods are all based on different properties of the protein molecule: dye binding, refractive index increments, and solubility, respectively. However, there is better agreement for normal than pathological sera.

OTHER SUPPORT MEDIA

The empirical nature of protein staining, deviation from Beer's law, and uneven structure of the paper, stimulated the search for other supports. Cellulose acetate, introduced by Kohn [37] in England in 1957, achieved better resolution of all fractions and drastically reduced the migration time from 16 hours to 20 minutes. In the United States, Grunbaum et al. [38]

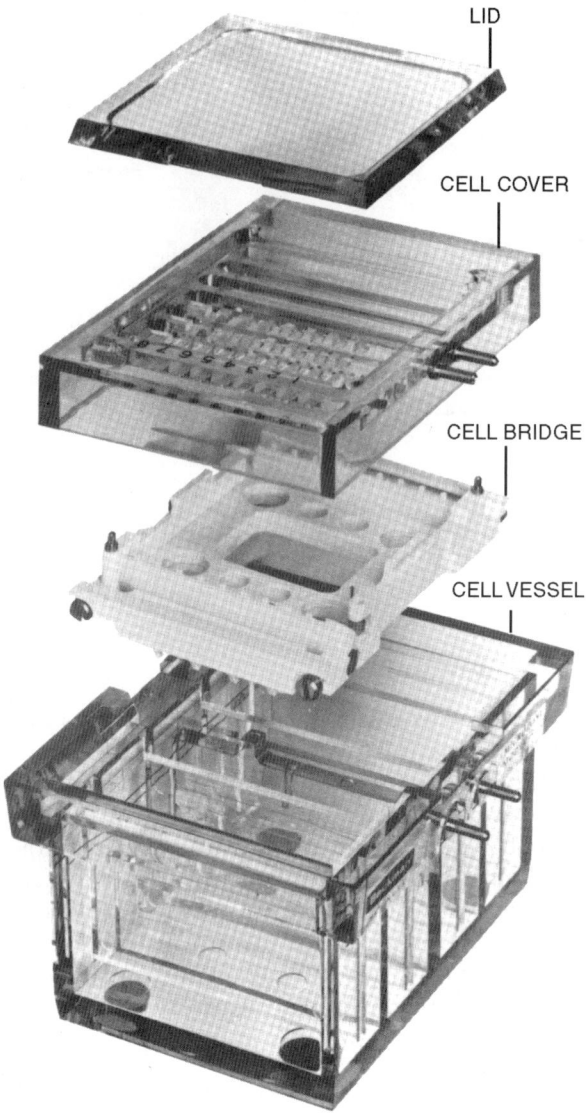

FIG. 17.6. Perspective of microzone cell for electrophoresis on cellulose acetate. (Courtesy of The Beckman Heritage Center, Fullerton, California.)

expanded on Kohn's work and developed the basic features of the Microzone system introduced by Beckman Instruments in late 1963 (Figs. 17.6, 17.7). The procedure requires as little as 0.25 μL serum compared to the 6–10 μL needed for filter paper. The entire procedure for an eight-sample membrane, including staining, clearing, drying, scanning, and calculation, can be completed within $2\frac{1}{2}$ hours.

Agarose gel was first proposed for electrophoresis by Hjertén [39] in 1961 and introduced by Elevitch in 1966. By 1985 it had almost completely replaced cellulose acetate. The scale of operation of agarose gel resembles cellulose acetate rather than paper and requires only 0.6 μL of serum. Analysis of an eight-sample gel, including migration time of 40 minutes, drying, staining, and scanning, can be completed in about 2 hours. A comparison of stained patterns for serum proteins on paper, cellulose acetate, and agarose gel, is shown in actual size in Fig. 17.8.

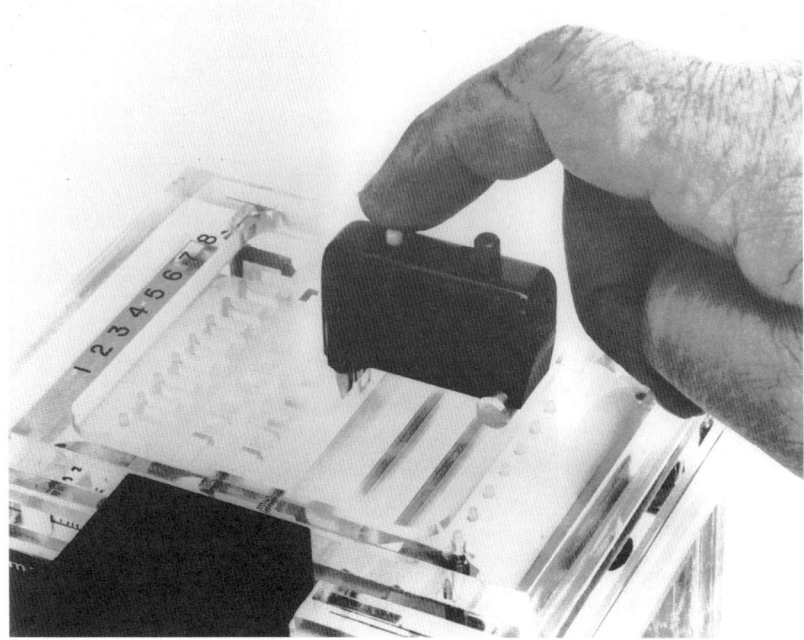

FIG. 17.7. Application of sample on cellulose acetate membrane. (Courtesy of The Beckman Heritage Center, Fullerton, California.)

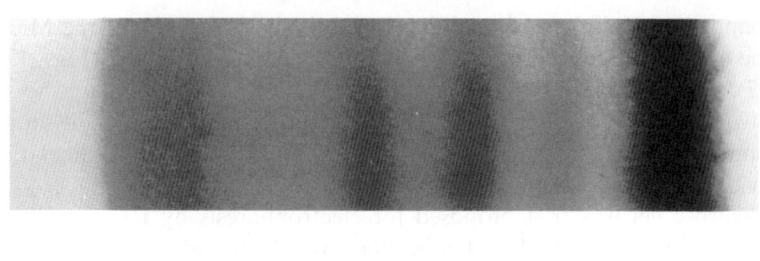
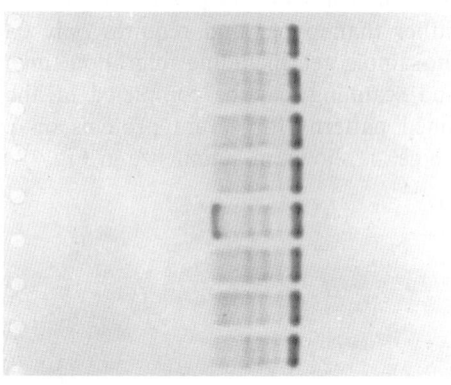
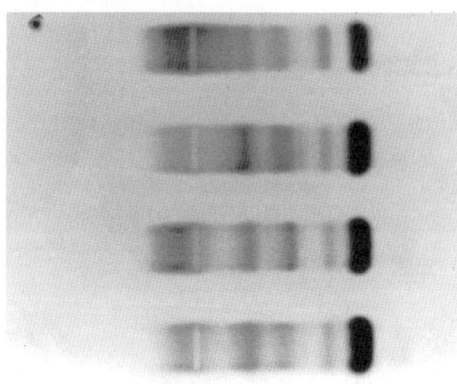

FIG. 17.8. Comparative zone electrophoresis patterns of serum proteins: (top to bottom) paper, cellulose acetate, agarose gel (actual size).

REFRACTIVE INDEX

Proteins exert relatively large effects on many physical properties as compared with the effects of simple solutes in the same solvents. Two easily measured physical properties that vary with protein concentrations and have been adapted as analytical procedures in plasma, are refractive index and specific gravity. Their simplicity, small volume requirements, and applicability directly on plasma or serum make them especially suitable as stand-by procedures for routine or emergency screening where limited facilities are available, or for verifying questionable results of other methods. The underlying assumption is that the nonprotein solutes contribute little or no effect compared with that of the protein and, if significant, can be corrected. Refractometric formulas are empirically derived and depend on a chemical reference method using biuret reagent or Kjeldahl nitrogen analysis.

The application of refractometry had its origin in the observation that a ray of light is bent as it passes from one medium into another due to the change in the velocity of light when it enters the second medium. This relationship was first expressed mathematically as a law sometime about 1621 by Willebrord Snel (1580–1626) of Leyden, long before the cause was known. The extent of bending or refraction is related to the concentration and the physical and chemical composition of the substance through which the light is transmitted. Accordingly, the refractive index, like the melting point and the boiling point, is a characteristic constant for each substance. Its determination merely requires the measurement of angles, which can be done precisely with the refractometer, an optical instrument invented by Ernst Abbe (1840–1905) in 1869.

The measurement of serum proteins by refractometry was introduced by Strubell in 1902 [40] and systematically studied by Reiss [41] and by Robertson [42]. Robertson and others determined that Reiss's experimentally derived formula yielded erroneous results in pathological sera. This led to a more time-consuming method by which Robertson [43] estimated the albumin, globulin, and nonprotein constituents separately.

The simplicity and small sample size (one drop) for the Abbe refractometer were significant advantages at a time when the alternative was a Kjeldahl analysis or a gravimetric measurement of the dried precipitated protein. The refractometric method for estimating serum proteins was used mainly in hypoproteinemias of chronic renal disease and other disturbances of water metabolism. The method fell into disuse after it was repeatedly shown that the values obtained in pathological as well as normal plasma were higher than those obtained by Kjeldahl methods [44]. Differences between the two techniques were especially pronounced in plasma from edematous patients due to the effect of the large content of lipids on the refractive index. Thus, the method suffers the grossest inaccuracies in pathological sera, where accurate protein levels are of the most interest and value. Accordingly, Peters and Van Slyke [45] made a point of not describing refractometric methods in their classic treatise. The technique was recommended by Sunderman [46] in 1944 for use in traumatic shock, as an emergency situation, where knowledge of total protein concentration is of value in determining the need for replacement therapy.

SPECIFIC GRAVITY

When Moore and Van Slyke [47] found a straight line relationship between the specific gravity of plasma and its total protein concentration, they derived an equation from which total plasma protein levels may be calculated. They determined specific gravity by measuring it with pycnometers and demonstrated that plasma protein concentration was more closely related to specific gravity than to refractive index.

Years later, this relationship led to the development of a rapid and simple flotation technique by Van Slyke and associates in a U.S. Navy Research Unit at the Hospital of the Rockefeller Institute for Medical Research [48]. The utility of blood and plasma concentration measurements is well recognized in the diagnosis and treatment of shock from wounds, burns, hemorrhage, post-operative complications, etc., and in other conditions in which plasma proteins or hemoglobin are affected, or in the dehydration of cholera, dysentery, or exposure. These measurements also help decide whether blood replacement therapy requires administration of saline solution or plasma or whole blood. During World War II, there was an urgent need for a method to determine if wounded servicemen required restoration of blood loss.

Gravity methods previously used, e.g., gravimetric, falling drop [49], gradient tube [50], require precision instruments on stable surfaces. None of these methods could well be used on shipboard. By using copper sulfate solutions of known, accurate specific gravity, the specific gravities of whole blood and plasma may be determined. Then, from these values, corrected for the specific gravity of the nonprotein constituents, and the use of line charts experimentally determined from these parameters, the plasma proteins, hemoglobin, and hematocrit may be calculated. Results are subject to large errors when the protein concentration is abnormal or the concentration of crystalloids such as urea or glucose is grossly elevated or the specimen is lipemic.

In this procedure, a small drop of test solution is allowed to fall freely by gravity from a height of about one cm into a series of standard solutions of copper sulfate of known and varying specific gravity. The solution in which the drop neither rises nor falls but remains suspended indicates the specific gravity of the test fluid. Twenty-two standards in increments of 0.001 were used to cover the range of specific gravities of plasma from 1.015 to 1.036—higher when testing whole blood. Specific gravity can be interpolated when the test drop is observed to rise and fall, respectively, in two adjacent standards. The basic contribution of this method is its ease, speed, and the use of the copper sulfate solutions, which are simple to reproduce. Calibration is unnecessary and the set of standards can be supplied ready for use from central laboratories, without restandardization or use of temperature corrections from 4° to 40°C.

CLOT FORMATION: INTRODUCTION

Although therapeutic phlebotomy had been routinely used since the dawn of medicine, the inspection of shed blood did not soon develop into a diagnostic procedure as in the case of urine. During the eighteenth century, the appearance of the blood as it flowed and then clotted in the collection bowl came to be regarded with diagnostic significance. A useful observation was the rapid sedimentation of the red cells prior to clotting and the appearance of a creamy mass of fibrin and leucocytes on the gelatinous clot. The blood was said to be "buffed" or to have an "inflammatory crust." This was a valuable diagnostic sign indicating inflammation somewhere in the body.

Near the end of the seventeenth century, Marcello Malpighi (1628–1694) examined the fibrin clot under the microscope after the red cells had been thoroughly washed out with water and described it as a white fibrous material. Malpighi recognized that the clot formed from the clear fluid (plasma) of the blood and not by some kind of aggregation of the red blood cells as was widely believed. He agreed with his contemporaries of 1686 that blood was kept liquid by its motion and he considered clot formation as analogous to the formation of cheese or gallstones or the heat precipitation of certain serum components. The concept of transformation of a precursor protein had not occurred to anyone. They all looked for a direct cause of clot formation. Aristotle had compared blood coagulation with ice formation in water and thought that the heat necessary to maintain the blood in a liquid state came from the heart. However, Malpighi was able to delay coagulation of cell-free plasma by cooling it and thereby showed that plasma clots independently of the blood cells [51].

A century later, around 1770, William Hewson (1739–1774) also showed that the clotting process was independent of the formed elements and, like Malpighi, he controlled the clotting mechanism by chilling the blood or by adding large amounts of neutral salts. When the blood cells settled out in salted blood, the cell-free "lymph" (plasma) could be decanted off and saved. Delayed clotting by addition of salt (usually magnesium sulfate) was rediscovered in the nineteenth century and remained the only way to separate unclotted plasma until decalcification with oxalate or citrate was introduced at the end of the century.

Andrew Buchanan (1798–1882), a Glasgow surgeon, was the first to realize that fibrin pre-existed in the circulation in soluble form and that some other agent caused its solidification, an effect he compared to the curdling action of rennet on milk. He did not use the terms "ferment" or "catalyst," but an editorial footnote to his 1845 paper made reference to "catalytic action." However, he did not assume that the substance which he found in blood plasma or exudates was different from fibrin.

The concept of fibrinogen is often attributed to Rudolf Virchow who, in 1847, proposed the term for a more slowly clottable substance than that which clotted normally in the same biologic liquid. Prosper-Sylvain Denis, in his *Mémoire sur le sang* (1859), was the first to recognize that plasma contained a clottable substance, not defined as a liquid fibrin, but different from fibrin, and he attempted to purify and characterize this protein. He independently proposed the name fibrinogène.

SEPARATION AND QUANTITATION OF FIBRINOGEN (FIBRIN)

Fibrinogen coagulates with heat at the lowest temperature (56°C) (first used by Frédericq in 1877) and precipitates at the lowest concentration of salts for any of the plasma proteins. Poorly soluble in water, but soluble in dilute salt solutions, it is classified as a globulin. In 1879 Olof Hammarsten precipitated and purified fibrinogen (from horse plasma) by repeated salting-out with half-saturation with sodium chloride. This has become the classical procedure for its preparation, although 25% saturation with ammonium sulfate has also been used.

Procedures for the separation and quantitation of fibrinogen, before or after conversion to fibrin by addition of thrombin or calcium chloride, are varied and numerous [52]: gravimetric, colorimetric, spectrophotometric, turbidimetric, nitrogen content, heat, salt precipitation, refractometric, viscosimetric, immunochemical, electrophoresis, and several specialized techniques that are applicable only to fibrinogen, viz., clotting time, clotting rate, and clot opacity. The earliest method may have been that of Gabriel Andral, developed in 1845. He whipped fibrin out from a measured or weighed amount of blood with a rod after it had clotted spontaneously. The fibrin was ground up and washed in a mortar with water and physiological saline until the washings no longer gave a positive biuret reaction due to the red cells and other plasma proteins. The fibrin and washings are caught on a weighed filter paper, washed free of chloride, washed with alcohol to remove the water, with ether to remove the alcohol, and dried to constant weight. The fibrin was reported as percentage of whole blood (w/v) or (w/w). To report concentration in plasma the hematocrit had to be measured.

NOTES AND REFERENCES

1. E. FULD and K. SPIRO, "Ueber die labende und labhemmende. Wirkung des Blutes," *Hoppe-Seyler's Zeitschrift für Physiologische Chemie*, 1900, 31: 132–155.
2. OTTO PORGES and K. SPIRO, "Die Globuline des Blutserums," *Beitrage zur Chemischen Physiologie und Pathologie*, 1903, 3: 277–285.
3. GLENN E. CULLEN and DONALD D. VAN SLYKE, "Determination of the Fibrin, Globulin, and Albumin Nitrogen of Blood Plasma," *Journal of Biological Chemistry*, 1920, 41: 587–597.
4. PAUL E. HOWE, "The Use of Sodium Sulfate as the Globulin Precipitant in the Determination of Proteins in Blood," *Journal of Biological Chemistry*, 1921, 49: 93–107; "The Determination of Proteins in Blood—a Micro-Method," *Ibidem*, pp. 109–113.
5. HARRY SVENSSON, "Fractionation of Serum with Ammonium Sulfate and Water Dialysis, Studied by Electrophoresis," *Journal of Biological Chemistry*, 1941, 139: 805–825; ALEXANDER B. GUTMAN, DAN H. MOORE, ETHEL B. GUTMAN, VICTOR MCCLELLAN, and ELVIN A. KABAT, "Fractionation of Serum Proteins in Hyperproteinemia, with Special Reference to Multiple Myeloma," *Journal of Clinical Investigation*, 1941, 20: 765–783; JOHN A. LUETSCHER, JR., "Electrophoretic Analysis of Plasma and Urinary Proteins," *Journal of Clinical Investigation*, 1940, 19: 313–320; VINCENT P. DOLE, "The Electrophoretic Patterns of Normal Plasma," *Journal of Clinical Investigation*, 1944, 23: 708–713; HAROLD L. TAYLOR and ANCEL KEYS, "Fractionation of Normal Serum Proteins by the Electrophoretic and Sodium Sulfate Methods," *Journal of Biological Chemistry*, 1943, 148: 379–381; MARY L. PETERMANN, NELSON F. YOUNG, and KATHERINE R. HOGNESS, "A Comparison of the Howe and the Electrophoretic Methods for the Determination of Plasma Albumin," *Journal of Biological Chemistry*, 1947, 169: 379–387.
6. WALTER R. CAMPBELL and MARION I. HANNA, "The Albumin, Globulins, and Fibrinogen of Serum and Plasma," *Journal of Biological Chemistry*, 1937, 119: 15–33.
7. GEORGE R. KINGSLEY, "A Rapid Method for the Separation of Serum Albumin and Globulin," *Journal of Biological Chemistry*, 1940, 133: 731–735.

8. C. L. H. MAJOOR, "The Possibility of Detecting Individual Proteins in Blood Serum by Differentiation of Solubility Curves in Concentrated Sodium Sulfate Solutions. II. Comparison of Solubility Curves with Results of Electrophoresis Experiments," *Journal of Biological Chemistry*, 1947, 169: 583–594; J. MILNE, "Serum Protein Fractionation: A Comparison of Sodium Sulfate Precipitation and Electrophoresis," *Ibidem*, pp. 595–599.
9. CLARENCE COHN and WILLIAM Q. WOLFSON, "Studies in Serum Proteins. II. A Rapid Clinical Method for the Accurate Determination of Albumin and Globulin in Serum or Plasma," *Journal of Laboratory and Clinical Medicine*, 1948, 33: 367–370.
10. JOHN G. REINHOLD, "Total Protein, Albumin, and Globulin," in *Standard Methods of Clinical Chemistry*, vol. 1, Miriam Reiner, ed. (New York: Academic Press Inc., 1953), pp. 88–97.
11. DAVID D. RUTSTEIN, ESTELLE F. INGENITO, and WILLIAM E. REYNOLDS, "The Determination of Albumin in Human Blood Plasma and Serum. A Method Based on the Interaction of Albumin With an Anionic Dye—2-(4'-Hydroxybenzeneazo) Benzoic Acid," *Journal of Clinical Investigation*, 1954, 33: 211–221.
12. F. LEE RODKEY, "Direct Spectrophotometric Determination of Albumin in Human Serum," *Clinical Chemistry*, 1965, 11: 478–487.
13. D. WEBSTER, "A Study of the Interaction of Bromocresol Green With Isolated Serum Globulin Fractions," *Clinica Chimica Acta*, 1974, 53: 109–115; JAN E. C. GUSTAFSSON, "Improved Specificity of Serum Albumin Determination and Estimation of 'Acute Phase Reactants' by Use of the Bromcresol Green Reaction," *Clinical Chemistry*, 1976, 22: 616–622; STEVEN W. KING, R. E. CROSS, and J. SAVORY, "Improved Specificity of the Bromcresol Green (BCG) Dye Method for Serum Albumin Using an Early (6 Seconds) Absorbance Reading," *Clinical Chemistry*, 1977, 23: 1136 (Abstract).
14. F. ROSE, "Ueber die Verbindungen des Eiweiss mit Metalloxyden," *Annalen der Physik und Chemie*, 1833, 28: 132–142.
15. G. WIEDEMANN, "Biuret. Zersetzungsproduct des Harnstoffs," *Liebig's Annalen der Chemie und Pharmacie*, 1848, 68: 323–326.
16. H. RITTHAUSEN and R. POTT, "Untersuchungen uber Verbindungen der Eiweisskörper mit Kupferoxyd," *Journal für Praktische Chemie*, 1873, 7: 361–373.
17. J. L. KANTOR and WILLIAM J. GIES, "Additional Experiments With the Biuret Reagent," *Journal of Biological Chemistry*, 1911, 9: xvii–xviii.
18. E. RIEGLER, "Eine kolorimetrische Bestimmungsmethode des Eiweisses," *Zeitschrift für Analytische Chemie*, 1914, 53: 242–245.
19. W. AUTENRIETH and FRIEDA MINK, "Ueber kolorimetrische Bestimmungsmethoden: die quantitative Bestimmung von Harneiweiss," *Muenchener Medizinische Wochenschrift*, 1915, 62: 1417–1421.
20. W. AUTENRIETH, "Ueber kolorimetrische Bestimmungsmethoden: die Bestimmung von Serum Albumin und Globulin im Harn, in der Aszitesflüssigkeit und im Blutserum," *Muenchener Medizinische Wochenschrift*, 1917, 64: 241–245.
21. JOSEPH FINE, "The Biuret Method of Estimating Albumin and Globulin in Serum and Urine," *Biochemical Journal*, 1935, 29: 799–803.
22. ALMA HILLER, J. F. MCINTOSH, and DONALD D. VAN SLYKE, "The Excretion of Albumin and Globulin in Nephritis," *Journal of Clinical Investigation*, 1927, 4: 235–251, p. 236.
23. GEORGE R. KINGSLEY, "The Direct Biuret Method for the Determination of Serum Proteins as Applied to Photoelectric and Visual Colorimetry," *Journal of Laboratory and Clinical Medicine*, 1942, 27: 840–845.

24. T. E. WEICHSELBAUM, "An Accurate and Rapid Method for the Determination of Proteins in Small Amounts of Blood Serum and Plasma," *American Journal of Clinical Pathology*, 1946, 16 (Technical Bulletin, vol. 7): 40–49.
25. ARNE TISELIUS, "A New Apparatus for Electrophoretic Analysis of Colloidal Mixtures," *Transactions of the Faraday Society*, 1937, 33: 524–531.
26. For a review of the development and operation of moving boundary electrophoresis, see LOUIS ROSENFELD, *Origins of Clinical Chemistry. The Evolution of Protein Analysis* (New York: Academic Press, 1982), pp. 162–193. See also ARNE TISELIUS, "Electrophoretic Analysis and the Constitution of Native Fluids," *Harvey Lectures*, 1939–40, 35: 37–70.
27. P. KÖNIG, "Applicação da electrophorése nos trabalhos chimicos com quantidades pequenas," in *Actas e Trabalhos do Terceiro Congresso Sul-Americano De Chimica*, Rio de Janeiro e Sâo Paulo, 1937, 2: 334–336.
28. D. VON KLOBUSITZKY and P. KÖNIG, "Biochemische Studien über die Gifte der Schlangengattung Bothrops. VI.," *Archiv für Experimentell Pathologie und Pharmakologie*, 1939, 192: 271–275.
29. R. CONSDEN, A. H. GORDON, and A. J. P. MARTIN, "Ionophoresis in Silica Jelly. A Method for the Separation of Amino-Acids and Peptides," *Biochemical Journal*, 1946, 40: 33–41.
30. THEODOR WIELAND and EDGAR FISCHER, "Über Elektrophorese auf Filtrierpapier," *Naturwissenschaften*, 1948, 35: 29–30.
31. E. L. DURRUM, "A Microelectrophoretic and Microionophoretic Technique," *Journal of the American Chemical Society*, 1950, 72: 2943–2948; F. TURBA and H. J. ENENKEL, "Elektrophorese von Proteinen in Filterpapier," *Naturwissenschaften*, 1950, 37: 93; HANS-DIEDRICH CREMER and ARNE TISELIUS, "Elektrophorese von Eiweiss in Filtrierpapier," *Biochemische Zeitschrift*, 1950, 320: 273–283; W. GRASSMANN and K. HANNIG, "Ein einfaches Verfahren zur Analyse der Serumproteine und anderer Proteingemische," *Naturwissenschaften*, 1950, 37: 496–497. The details of priority are clarified by Louis Rosenfeld (Letter), "Origins of Protein Electrophoresis," *Clinical Chemistry*, 1981, 27: 1948–1949.
32. E. L. DURRUM, "A Microelectrophoretic and Microionophoretic Technique," Report of Project No. 6-64-12-06-(18). Medical Department Field Research Laboratory, Fort Knox, Kentucky, 1949; "Microelectrophoretic and Microionophoretic Technique," (Abstract), 115th Meeting, American Chemical Society, Division of Biological Chemistry, San Francisco, California, 1949, pp. 22c–23c.
33. F. G. WILLIAMS, JR., E. G. PICKELS, and E. L. DURRUM, "Improved Hanging-Strip Paper-Electrophoresis Technique," *Science*, 1955, 121: 829–830.
34. For a review of early investigations with paper support media and its characteristics, see also ROSENFELD (1982), pp. 194–207.
35. CREMER and TISELIUS (1950).
36. For a review of the staining and visualization procedure with paper support media, see ROSENFELD (1982), pp. 208–216.
37. J. KOHN, "A New Supporting Medium for Zone Electrophoresis," *Biochemical Journal*, 1957, 65: 9p; "A Cellulose Acetate Supporting Medium for Zone Electrophoresis," *Clinica Chimica Acta*, 1957, 2: 297–303.
38. B. W. GRUNBAUM, J. ZEC, and E. L. DURRUM, "Application of an Improved Microelectrophoresis Technique and Immunoelectrophoresis of the Serum Proteins on Cellulose Acetate," *Microchemical Journal*, 1963, 7: 41–53.
39. STELLAN HJERTÉN, "Agarose as an Anticonvection Agent in Zone Electrophoresis," *Biochimica et Biophysica Acta*, 1961, 53: 514–517; FRANKLIN R. ELEVITCH, SAMUEL B.

ARONSON, THOMAS V. FEICHTMEIR, and MELINDA L. ENTERLINE, "Thin Gel Electrophoresis in Agarose," *American Journal of Clinical Pathology*, 1966, 46: 692–697; LOUIS ROSENFELD, "Serum Protein Electrophoresis. A Comparison of the Use of Thin-Layer Agarose Gel and Cellulose Acetate," *American Journal of Clinical Pathology*, 1974, 62: 702–706.

40. ALEXANDER STRUBELL, "Ueber refraktometrische Blutuntersuchungen," *Muenchener Medizinische Wochenschrift*, 1902, 49: 616–618.

41. EMIL REISS, "Der Brechungskoeffizient der Eiweisskörper des Blutserums," *Beitrage zur Chemische Physiologie und Pathologie*, 1903, 4: 150–154; "Eine neue Methode der quantitativen Eiweissbestimmung," *Archiv für Experimentell Pathologie und Pharmakologie*, 1904, 51: 18–29.

42. T. BRAILSFORD ROBERTSON, "On the Refractive Indices of Solutions of Certain Proteins. VI. The Proteins of Ox-Serum; a New Optical Method of Determining the Concentrations of the Various Proteins Contained in Blood-Sera," *Journal of Biological Chemistry*, 1912, 11: 179–200.

43. T. BRAILSFORD ROBERTSON, "A Micro-Refractometric Method of Determining the Percentages of Globulin and Albumin in Very Small Quantities of Blood Serum," *Journal of Biological Chemistry*, 1915, 22: 233–239.

44. G. C. LINDER, C. LUNDSGAARD, and D. D. VAN SLYKE, "The Concentration of the Plasma Proteins in Nephritis," *Journal of Experimental Medicine*, 1924, 39: 887–920.

45. JOHN P. PETERS and DONALD D. VAN SLYKE, *Quantitative Clinical Chemistry*, vol. 2, *Methods* (Baltimore: The Williams & Wilkins Company, 1932), p. 678.

46. F. W. SUNDERMAN, "A Rapid Method for Estimating Serum Proteins. Formula for Calculating Serum Protein Concentration From the Refractive Index of Serum," *Journal of Biological Chemistry*, 1944, 153: 139–142.

47. N. S. MOORE and DONALD D. VAN SLYKE, "The Relationships Between Plasma Specific Gravity, Plasma Protein Content and Edema in Nephritis," *Journal of Clinical Investigation*, 1930, 8: 337–355.

48. ROBERT A. PHILLIPS, DONALD D. VAN SLYKE, PAUL B. HAMILTON, VINCENT P. DOLE, KENDALL EMERSON, JR., and REGINALD M. ARCHIBALD, "Measurement of Specific Gravities of Whole Blood and Plasma by Standard Copper Sulfate Solutions," *Journal of Biological Chemistry*, 1950, 183: 305–330; DONALD D. VAN SLYKE, ALMA HILLER, ROBERT A. PHILLIPS, PAUL B. HAMILTON, VINCENT P. DOLE, REGINALD M. ARCHIBALD, and HOWARD A. EDER, "The Estimation of Plasma Protein Concentration From Plasma Specific Gravity," *Ibidem*, pp. 331–347; DONALD D. VAN SLYKE, ROBERT A. PHILLIPS, VINCENT P. DOLE, PAUL B. HAMILTON, REGINALD M. ARCHIBALD, and JOHN PLAZIN, "Calculation of Hemoglobin from Blood Specific Gravities," *Ibidem*, pp. 349–360. Preliminary reports of the method and medical applications appeared in the medical publications of the U.S. Armed Services. *Bumed News Letter* (U.S. Navy), 1943 (June), 1: No. 9, 1–16; *The Bulletin of the U.S. Army Medical Department*, 1943 (December), No. 71, 66–83.

49. H. G. BARBOUR and W. F. HAMILTON, "The Falling Drop Method for Determining Specific Gravity," *Journal of Biological Chemistry*, 1926, 69: 625–640.

50. O. H. LOWRY and T. H. HUNTER, "The Determination of Serum Protein Concentration With a Gradient Tube," *Journal of Biological Chemistry*, 1945, 159: 465–474.

51. EUGENE A. BECK, "The Discovery of Fibrinogen and Its Conversion to Fibrin," in *Fibrinogen*, Koloman Laki, ed. (New York: Marcel Dekker, Inc., 1968), pp. 25–37.

52. These and other procedures for the analysis of fibrinogen as fibrin have been reviewed by Rosenfeld (1982), pp. 298–321.

CHAPTER XVIII

VICTOR MYERS AND THE NEW YORK POST-GRADUATE MEDICAL SCHOOL AND HOSPITAL

Victor Caryl Myers (1883–1948) (Fig. 18.1) received the B.A. and M.A. degrees from Wesleyan University in Middletown, Connecticut. He then entered Yale University as a graduate student and studied under Lafayette B. Mendel and Russell Chittenden at the Sheffield Laboratory of Physiological Chemistry. After receiving his Ph.D. in 1909, he taught biochemistry and was laboratory director at the Albany Medical College in New York State. In 1911 he was invited by the New York Post-Graduate Medical School and Hospital to organize a physiological chemistry laboratory—one of the first in the nation.* He was lecturer on chemical pathology (1911–12), professor of pathological chemistry (1912–22), and professor of biochemistry (1922–24). While at the Post-Graduate Hospital he engaged in the experimental study of the metabolism of creatine and creatinine. During these early years, the American Medical Association, at its annual meeting, provided funds to pay the cost of his demonstrations of methods used in the clinical chemistry laboratory to help in the diagnosis and treatment of disease. These presentations did much to develop awareness in the medical community of the contributions made by the new profession of clinical chemists. They also helped to prepare the way for other institutions to set up their own clinical chemistry laboratories.

As the field of clinical pathology developed, he realized that individuals trained in clinical chemistry would be needed. While he was chairman of the department of biochemistry at the University of Iowa (1924–27), he initiated the first post-graduate program for training clinical chemists, primarily for hospital positions. Myers moved to Western Reserve University in Cleveland, where he was appointed the first full professor of biochemistry in 1927. He was head of the department until the creation of the new department of clinical biochemistry in 1946, when he became its director.

---

* A chemistry laboratory was in hospital service before 1899, probably the first in New York City. (Edward F. Hartung, New York State Journal of Medicine, 1961, 61: 2123–2135, p. 2132.)

FIG. 18.1. Victor Caryl Myers. (Case Western Reserve University Archives, Cleveland, Ohio.)

Myers was among the first to recognize that perception by other scientific (and medical) disciplines and by the general community are the key elements contributing to the public relations problems of clinical chemistry. Throughout his career he worked to improve the image and promote the professional identity and status of the clinical chemist. He also worked to oppose legislation that would bar clinical chemists without an M.D. degree from operating any laboratories.

In 1913, Myers and his colleague, Morris S. Fine, published *Essentials of Pathological Chemistry, Including Descriptions of the Chemical Methods Employed in Medical Diagnosis*, a series of articles reprinted from the

Post-Graduate (1912–13). The methods were designed as a guide for classes in Elementary Pathological Chemistry, although the tests described were also employed in the examination of the routine specimens from the hospital.

PRACTICAL CHEMICAL ANALYSIS OF BLOOD

In 1920–21 Myers published a series of eight articles in the *Journal of Laboratory and Clinical Medicine* (volumes 5 and 6) under the title of "Chemical Changes in the Blood in Disease." He was an associate editor of the Journal at the time. A year later, these were revised and published in book form as the first edition of *Practical Chemical Analysis of Blood*. It was one of the first books on methodology for clinical chemistry. The stated object was to present a brief discussion of the chemical blood determinations carried out in the chemistry laboratory at the New York Post-Graduate Medical School and Hospital and which had been found to be of definite value in the diagnosis and treatment of disease. Myers noted that advances since 1912 had been very rapid, and that there now were many practical tests with which the efficient physician and surgeon must be familiar. A discussion of clinical findings for the particular analyte and a brief history of the development of each method, precedes the directions for the quantitative analysis. Myers hoped that the book would be useful in indicating why and how, certain chemical blood analyses should be made.

A second edition [1] of his book (232 pages) became necessary in 1924 because of the many advances since the first edition more than two years earlier. A few quantitative micromethods of urine analysis were added as well as a discussion and illustrations of the different types of colorimeters and their retail prices and directions for the preparation of standard solutions and reagents employed in the various tests. A separate chapter on the methods of the Folin–Wu system of blood analysis was also included.

Myers correctly points out that accurate data on the chemical composition of the blood, especially of the nonprotein fraction, were of comparatively recent origin. He credits Folin, Benedict, and Van Slyke, with development of these "American methods" that gave valuable data on just those conditions on which the older methods of blood examination, cytology, bacteriology, and serology gave little information, namely, nephritis, diabetes, and gout. These methods were yielding valuable information for diagnosis, prognosis, and treatment for a wide variety of disorders. However, sounding a precaution as timely as it is today, Myers says that although the procedures are simple they are quite technical and require specially trained assistants under the direction of competent biochemists. "Obviously a physician cannot expect that his office girl or nurse can properly manipulate these methods without an appreciation of the chemical factors involved" [2].

The book has a very modern quality as it calls for accurately prepared standards and the need for recalibration of pipettes and burettes, unless specially purchased. He addresses the time for blood drawing, preservation, choice of whole blood, plasma, or serum, anticoagulants, venous, finger tip, or ear lobe as source of the specimen, and notes that the latter two sources yield arterial blood. This was the first modern book on clinical chemistry and is a remarkably informative report on the status of this new science in the mid-1920s.

For the ordinary chemical examination, Myers specified 10 to 20 cc of blood taken, if possible, before breakfast, and drawn directly into a bottle containing as anticoagulant, 2 to 3 drops of 20% potassium oxalate (previously dried in the bottle in a hot air oven) (see Chapter 12). The potassium salt was preferred to the less soluble sodium oxalate. If refrigerated, the blood may be kept suitable for some determinations for several days. Specimens sent by mail without refrigeration must be delivered and analyzed within 24 hours, or they are valueless [3].

Myers's laboratory report form lists the analytes reviewed in his book and identifies the following seven as having special practical usefulness: hemoglobin, uric acid, urea nitrogen, creatinine, sugar, chlorides (as NaCl), and CO_2-combining power. Thirteen other blood constituents were listed on the report form: total solids, total nitrogen, nonprotein nitrogen, creatine, amino acid nitrogen, diastatic activity (amylase), cholesterol, fat, acetone bodies, calcium, inorganic phosphates (as phosphorus), hydrogen ion concentration (pH), and oxygen. The book gives references for plasma proteins, sodium, potassium, magnesium, and iron, but no laboratory directions. A separate form reported "sugar tolerance test," based on an oral dose of 1.75 gm glucose per kg body weight. The clinical conditions that these chemical tests were expected to help diagnose were diabetes (renal and mellitus), gout, nephritis, mercury bichloride poisoning, urologic conditions, eclampsia, pneumonia, malignancy, intestinal obstruction, infantile tetany, rickets, diarrheal acidosis of infancy, pancreatic disease, and endocrine disorders. The two endocrine disorders specifically mentioned are hyperthyroidism and Addison's disease—identifiable by a glucose tolerance curve. A note of caution was sounded about urea nitrogen levels over 25 mg (per 100 cc) prior to surgery on kidneys, bladder, or prostate. Preliminary treatment to relieve the nitrogen retention was recommended. Urea nitrogen greater than 20 mg on the usual restricted diet of the hospital would suggest impaired kidney function [4].

ST. LUKE'S AND ST. THOMAS'S

Across town from the New York Post-Graduate Medical School and Hospital, St. Luke's Hospital published a laboratory manual in 1922 of 281 pages that covered all of the divisions that have traditionally been associated with the clinical laboratory [5]. The major portion of the book was devoted to microbiology and the preparation of culture media. In addition to microbiology with the most procedures (37), there were sections on urine (17), hematology (15), gastric and stool (12), histology (12), and chemistry (10). Only serology (9) and blood typing (3) had fewer procedures than chemistry. The ten procedures of chemistry were covered in twenty-five pages: nonprotein nitrogen, urea, uric acid, sugar, creatinine, creatine, and cholesterol—all in whole blood; bicarbonate, calcium, and chloride, in plasma.

A similar list of procedures was offered at the St. Thomas's Hospital in London, where, also in 1924, O. L. V. de Wesselow, the chemical pathologist, wrote in the preface to his book *The Chemistry of the Blood in Clinical Medicine*, that despite the claims of enthusiasts, "the clinical value of blood chemistry lies chiefly in its application to three pathological states—diabetes and glycosuria, renal disease, and those conditions in which a tendency to a change in the reaction of the blood is present" [6]. The 255-page book is mainly a discussion of the biochemistry and clinical findings in health and disease; the methods are collected in a 31-page final chapter. They are essentially American in origin and had been developed during the previous ten years. He lists procedures for the following: urea, nonprotein nitrogen, creatinine, uric acid, sugar, carbon dioxide combining power, phosphate, chlorides, and calcium. Prominent omissions from the laboratory at all three hospitals—Post-Graduate, St. Luke's, and St. Thomas's—were analyses in blood for sodium, potassium, and protein.

The 5th and final edition of Otto Folin's *Laboratory Manual of Biological Chemistry With Supplement* (1934; 1st edition, 1916) was a laboratory course manual for medical students. It presented eighteen methods for fifteen—by now, routinely tested—constituents of blood and included a lengthy titrimetric procedure for potassium, but here too, nothing for sodium or protein.

HAWK'S *PRACTICAL PHYSIOLOGICAL CHEMISTRY*

Philip B. Hawk's *Practical Physiological Chemistry* (416 pages) was first published in 1907 while he was a demonstrator (assistant professor) in physiological chemistry (1903–07) at the University of Pennsylvania. The arrangement of the book follows the plan he used during three years of classroom teaching at the university. Hawk (1874–1966) had studied physiological chemistry for two years at the Sheffield Scientific School (M.S. 1902) and was the first to receive the Ph.D. degree in physiological chemistry from Columbia University's College of Physicians and Surgeons in 1903. His book, based on Chittenden's course, developed into what was probably the most popular of the various available manuals that rapidly displaced the old textbooks on medical chemistry. Whereas Chittenden and Mendel gave verbal directions, Hawk recognized that the shorter laboratory time given to physiological chemistry at the University of Pennsylvania made brief written directions a necessity.

Hawk's book grew into a textbook on biochemistry that gave extensive coverage to the physiology of digestive enzymes, urine, and metabolism, with accompanying laboratory exercises. In early editions, the chapter on blood dealt with the formed elements and the spectroscopic examination of the various forms of hemoglobin and its derivatives and was liberally illustrated with microphotographs of crystalline forms. A separate chapter (43 pages) with several colorimetric methods for constituents of blood first appeared in the 5th edition (1916, 638 pages). The Duboscq colorimeter was used, but was illustrated in the chapter on quantitative analysis of urine (85 pages); the methods were mostly titrimetric, gasometric, and gravimetric. Purdy's electric centrifuge was illustrated for the separation of urine sediments, but not yet for the preparation of protein-free "filtrates."

This textbook-manual provided instructors with a comprehensive, high-quality, modern course. By the 14th and final edition (1965), edited by Bernard L. Oser, it had grown to 1472 pages. The biochemical preparations and analytical chemical methods described by Hawk and by Folin became the standard laboratory exercises studied by graduate and medical students for more than half a century.

Another popular resource for the clinical laboratory was *A Manual of Clinical Diagnosis* by James C. Todd (1874–1928) and Arthur H. Sanford (1882–1959). First published by Todd in 1908, it went through many editions. The 2nd edition (1912) was titled *Clinical Diagnosis. A Manual of Laboratory Methods*. After Todd's death, the 7th (1931) through 10th (1943) editions of "Todd and Sanford" were published with Sanford as sole author. George C. Stilwell collaborated on the 11th edition (1948) and Benjamin B. Wells was coauthor on the 12th edition (1953). After Sanford's death, there was a name change to *Clinical Diagnosis by Laboratory Methods* (13th edition, 1962). The new editors, Israel Davidsohn and Benjamin B. Wells, kept the Todd and Sanford names on the title page. John B. Henry was co-editor with Davidsohn on the 14th (1969) and 15th (1974) editions. The 16th edition (1979) appeared as *Clinical Diagnosis and Management by Laboratory Methods* and was edited by Henry. The heading Todd-Sanford-Davidsohn was above the title in the 16th and 17th (1984) editions. The 18th (1991) and 19th (1996, 1556 pages) editions were edited by Henry.

Two other books for the laboratory, first published in Great Britain in 1930, were *Chemical Methods in Clinical Medicine* by G. A. Harrison, and *Clinical Chemistry in Practical Medicine* by C. P. Stewart and D. M. Dunlop. Both appeared in several editions.

PETERS AND VAN SLYKE: *QUANTITATIVE CLINICAL CHEMISTRY*

The appearance of the two-volume classic by John P. Peters and Donald D. Van Slyke, *Quantitative Clinical Chemistry*, volume I, *Interpretations*, 1931, and volume II, *Methods*, 1932, was the coming of age of the laboratory science of clinical chemistry as a distinct professional discipline. It was the first time that "Clinical Chemistry" appeared in the title of a book on laboratory methods or interpretations by American authors. *Interpretations*, with 1264 pages and nearly 4000 references, dealt with the physiological and clinical significance of those substances for which quantitative methods were available. The 957 pages of *Methods* were devoted exclusively to detailed directions for a gravimetric, a titrimetric, a colorimetric, and a gasometric procedure for each body substance of importance for clinical medicine, when suitable quantitative methods were available and, when advisable, with macro and micro versions. This allowed the reader a choice best suited for his laboratory facilities and personal preferences. Methods were designed and modified to accommodate 1.0 cc of blood, plasma, or serum—considered at that time to be a small sample size.

More than thirty constituents of whole blood, plasma, serum, and urine, including sodium, potassium, and protein, are covered at some length. Colorimetric procedures utilized a Duboscq-type visual colorimeter for quantitative analyses. New methods were developed and old ones converted to yield an end product gas that could be measured with Van Slyke's gasometric apparatus for the determination of substances not already in the gaseous state, viz., amino acid nitrogen, urea, lipids, sugar, lactic acid, chloride, potassium, and calcium.

Many of the analytical methods in *Quantitative Clinical Chemistry* were developed in Van Slyke's laboratory, and most of the other procedures described were tested there before inclusion in the volume on *Methods*. Also included were discussions of the chemical principles on which the methods were based. These discussions usually covered other applicable procedures in addition to those detailed in the book. The first two chapters deal with "general chemical technique" and "special biochemical technique."

Van Slyke's collaboration with John P. Peters (1887–1955) [7] came about when the publishers, Williams & Wilkins, asked Peters to write a modest handbook for clinicians, describing useful chemical methods in a comprehensive manner and discussing their application to clinical problems. It was originally to be called *Quantitative Chemistry in Clinical Medicine*. Realizing the magnitude of the endeavor, Peters, who was professor of medicine at Yale from 1921 until his death, asked Van Slyke, with whom he had worked for a year at the Rockefeller Institute (1920–21), to join him in a collaboration between chemist and clinician. They agreed to divide the writing and also to exchange each chapter until both were satisfied with the finished product. The outcome after eight years was a work of two volumes. Apparently, the books met a need because only a year after publication there was a second printing, and a third in 1935.

Until then, physiological chemistry had concerned itself chiefly with metabolism and the analysis of those compounds for which there were practical quantitative methods. Only a few of these had been applied to clinical situations. The sources of these compounds and the processes by which they were formed were not always known. The study of intermediary metabolism was in its infancy and hormones and vitamins were still mysterious entities. Systematic knowledge about the chemistry and physiology of the essential components of food and tissues was primitive. No sooner did the book appear than the rapid advances in intermediary metabolism and the role of enzymes and hormones made a second edition mandatory. With medical research advancing at an accelerated rate of progress, responsibility for the two volumes was divided—*Interpretations* to be edited by Peters and, because of the large amount of material, to be published in two volumes; *Methods*, to be covered by Van Slyke. The association remained, only the work was divided. Only volume I of a second edition of *Interpretations* came out in 1946. Van Slyke worked on

Methods, but owing to his insistence on nothing less than proven accuracy, it was never completed.

Methods and *Interpretations* brought biochemistry and clinical medicine closer together in the understanding and treatment of disease. For many years, these two volumes were the principal sources of reliable information for teaching medical and graduate students about the chemical behavior of physiological systems. *Methods* was an especially important asset of the clinical chemistry laboratory and a stimulus to the expansion of the hospital's chemistry services during the 1930s. Both volumes remain today as useful and valuable sources of information, especially for the early literature of clinical chemistry.

Physiological Chemistry. A Text-Book and Manual for Students by Albert Prescott Mathews (1871–1957), first published in 1915, was the principal American textbook on biochemistry (with a large section on practical work and methods) for nearly three decades. The sixth (revised and final) edition appered in 1939. Written for students in medicine and science, it presented the properties of the chief groups of biochemical compounds from the viewpoint of the physical chemist. The book appeared at just the right time to inspire many students to decide on a career in the rapidly growing subject of biochemistry [8].

Other textbooks on biochemistry by American authors did not appear until the 1930s. They include: *Practical Methods in Biochemistry* by Frederick C. Koch (1934); *Biological and Clinical Chemistry* by Matthew Steel (1937); *Textbook of Biochemistry* by Benjamin Harrow (1938); *Medical Biochemistry* by Mark R. Everett (1942); *Human Biochemistry* by Israel S. Kleiner (1945). Another book that deserves notice is *The Biochemistry of Clinical Medicine* by William S. Hoffman, first published in 1954 (fourth edition, 1970, 856 pages).

PHOTOMETRY: INTRODUCTION

About the turn of the century, Duboscq's original principle of varying the light path to obtain color balance was refined, and several new designs of visual filter photometers were developed and manufactured. These achieved a weakening of the comparison light beam by other methods such as by diaphragms, polarization, or neutral wedges. The earliest of these may have been the Pulfrich instrument described in 1894. Subsequently manufactured by Carl Zeiss, it used a series of glass filters having different spectral band widths and transmittance levels. Sample and reference beams were attenuated by adjustment of diaphragms, with readings taken from micrometer drum heads. Adam Hilger, Ltd., in England, manufactured a photometer utilizing polarization attenuation. A polarization type photometer was also made by E. Leitz, Inc., in the United States about 1937. The Aminco neutral wedge photometer of the American Instrument Company, used a series of filters together with an optical wedge attenuator in the reference beam. The scale readings corresponding to wedge positions were convertible to absorbance values [9].

Photometry, though often referred to as colorimetry, differs from it in the method of measurement. Whereas in visual colorimetry, the optical part of the instrument is adjusted until equal color densities are obtained for unknown and standard, in visual photometry, monochromatic light is used mainly and the intensity of this light is diminished in several ways. The Zeiss, Leitz (and Hilger), and Aminco visual filter photometers illustrate, respectively, the application of mechanical, polarization, and neutral wedge devices. The best resources of the first half of the twentieth century's optical theory and practice went into the design of visual photometers. However, despite their versatility, constancy, and reliability, the limits of their effectiveness was set by the visual acuity of the observer. Useful in industrial settings for measuring absorption spectra of dyes and other colored substances, and in the study of keto–enol tautomerism, they found no application in the clinical chemistry laboratory. With the development of photoelectric instruments, visual photometers became obsolete [10].

THE PHOTOELECTRIC EFFECT

The photoelectric effect was discovered by Wilhelm Hallwachs in 1888 and was observed by others in subsequent years. However, it was not until 1900 that Philipp Lenard showed that electrons were ejected when metals were irradiated, and J. Elster and N. Geitel showed that the rate of emission, i.e., current, was proportional to the intensity of the light causing it. The photovoltaic effect was discovered by W. Adams and P. Day in 1876. They observed that a potential was generated when light irradiated a selenium–platinum junction. The first utilization of photoelectric detectors for absorption measurements of transmitted light may have been made by Wilhelm Berg who obtained a German patent in 1911 for a photoelectric colorimeter [11].

In 1919 a visual type of filter photometer with a photocell as receptor was built by Gibson [12] at the National Bureau of Standards. However, for various reasons, commercialization of the idea progressed slowly. As early as 1926, Reimann [13] drew a crude schematic diagram for absorptiometric methods using a photoelectric cell. It showed all the key components (Fig. 18.2). Although the photoelectric effect had been known to physicists since before the end of the nineteenth century, it was only in

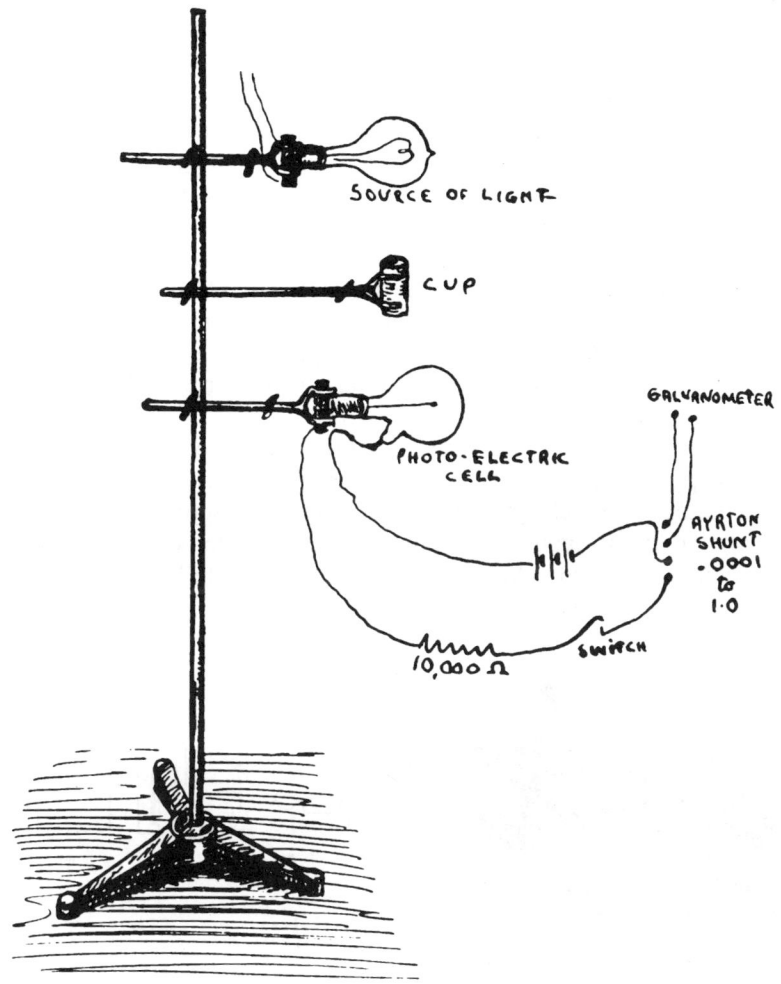

FIG. 18.2. Basic components for use of the photo-electric cell (argon filled bulb) as a colorimeter. (*Proceedings of the Society for Experimental Biology and Medicine*, 1926, 23: 520–523. Reprinted by permission of Blackwell Science, Inc.)

the early 1930s that a major breakthrough occurred when the first photometers with photoelectric cells came onto the market and were applied to chemical problems. However, for a long time, chemists in and outside of the laboratory, believed that photometers were less reliable than the long-known visual devices.

PHOTOELECTRIC COLORIMETERS (PHOTOMETERS)

In 1933 the Cenco-Sheard-Sanford Photelometer (Fig. 18.3) (Central Scientific Co.) became the first commercially available photoelectric filter photometer [14]. This single-cell instrument was connected to a constant voltage transformer to insure a steady source of white light independent of variations in the line voltage. This made possible the rapid collection of reproducible readings of colored solutions by mechanical means. It replaced the analyst's subjective color matching in a visual colorimeter with an objective instrumental measure of transmitted light intensity. From this beginning of "semi-automated" technology, analytical clinical chemistry began its slow metamorphosis into the fully automated, multi-test, computerized, data-generating systems that have had an irreversible impact on the modern practice of medicine.

Photometers measure the current produced in a photocell or phototube by light, passed first through a glass filter transmitting visible light of a selected range of wavelength, and then through a constant width of colored test solution. A galvanometer reading allows one to interpolate a result from a previously prepared calibration curve of the analyte being determined.

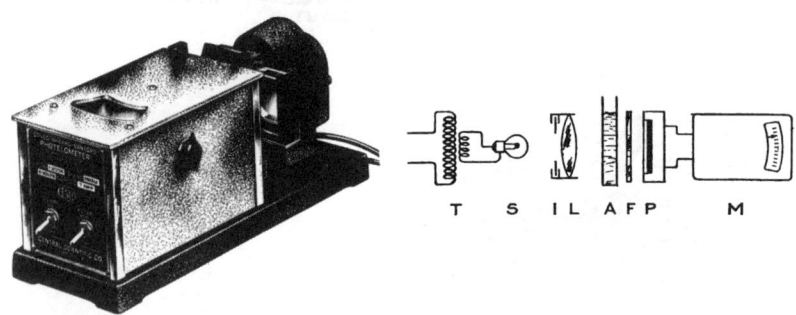

FIG. 18.3. An early design of the Cenco-Sheard-Sanford Photelometer with schematic diagram; T, constant wattage transformer; S, light source; I, iris diaphragm; L, lens; A, absorption cell; F, light filter; P, photoelectric cell; M, microammeter.

However, for more precise work, comparison against a simultaneously prepared standard is preferred. The relation between unknown and standard in the photometer is similar to that in variable depth colorimetry except that, since the depth of solution is constant, viz., test tube width, there is a direct proportionality between concentration and scale readings (optical or color density), instead of an inverse proportionality at color match between depth of solution viewed and concentration of color, as in the visual colorimeter.

During the 1930s, many types of photoelectric colorimeters became commercially available in the United States and Europe. A popular single-photocell filter photometer (Fig. 18.4) was designed in 1936 by Evelyn [15] for the Rubicon Company, Philadelphia, Pennsylvania. The light source was energized by a storage battery rather than an ordinary electrical outlet to provide stable illumination during a measurement. Another such instrument was manufactured by E. Leitz, Inc. in 1937.

To overcome disadvantages of single-photocell instruments with respect to reproducibility of measurements, filter photometers with two matched (balanced) barrier-layer photocells in a fully compensated and carefully balanced potentiometric circuit through a null-point galvanometer were developed. Among the first such instruments was the widely-used Klett-Summerson [16] photoelectric colorimeter (Fig. 18.5) introduced in 1939 and still available in 1999. It required no storage batteries or constant current regulators because the balanced circuit prevented variations in the light source—due to line voltage fluctuations—from affecting readings. The light source was operated by any ordinary electrical outlet. The Fisher Scientific Company also offered a photoelectric colorimeter with two balanced photovoltaic cells. The advent of these reliable, direct-reading instruments, sparked great interest in colorimetric analysis and in absorption spectrophotometry. Much work was done during the 1940s and 1950s to adapt the older colorimetric methods to the new photoelectric instruments and to develop the more sensitive and accurate procedures with smaller sample size that this new technology made possible. The advantages of photoelectric photometry are speed, elimination of errors from eye fatigue and the subjective variance among analysts, improved sensitivity and precision at the extremes of the visible spectrum, and ability to read very pale colors, as well as one color in the presence of other colors by careful selection of the wavelength of incident light.

When calibration standards are expensive or difficult to obtain pure or require special skill or apparatus, various "artificial" secondary standards, such as stable solutions of organic dyes, colored inorganic salts, colored

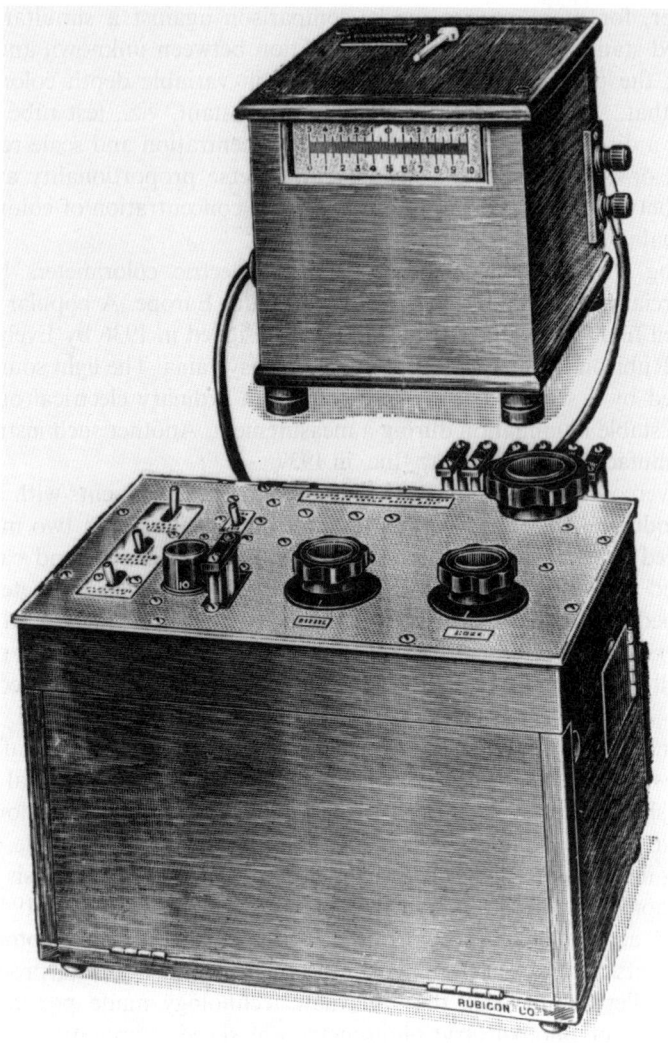

FIG. 18.4. Evelyn Photoelectric Colorimeter. The current from the photocell (forward unit) is displayed on the galvanometer (rear unit).

glass or gelatin were used, and provided approximate results as they did in visual colorimetry.

Despite their widespread use, there was a general belief that colorimetric methods were necessarily crude and approximate and likely to lack

PHOTOMETERS 457

FIG. 18.5. Klett-Summerson Photoelectric Colorimeter. A, scale knob (potentiometer dial); B, scale reading (potentiometer scale); C, pointer (galvanometer); D, galvanometer pointer adjustment; E, colorimeter tube; F, light switch; G, zero adjustment knob; H, short-circuit switch (on/off). (Bel-Art products, Pequannock, N.J.)

specificity and were resorted to reluctantly only because other methods failed. The adaptation of photoelectric photometry to colorimetry promised to revolutionize the field [17]. Yet as late as 1939, and in spite of inherent sources of error, Ralph H. Müller described the visual Duboscq colorimeter as "still the most versatile and useful instrument" for colorimetric chemical analysis [18]. Two years later, in a comprehensive review of chemical instrumentation, he added: "The assumption that any photoelectric instrument must be more accurate and reliable than a visual instrument is wholly unwarranted... However, the future does seem to lie in the direction of the photoelectric types, for there is no inherent limit in the attainable sensitivity and objectivity of the measurement" [19]. By the early 1960s, all colorimetric methods had been adapted to the photometer or the new arrival in the chemistry laboratory—the AutoAnalyzer. After briefly working alongside the photoelectric colorimeter and the single-channel AutoAnalyzer, the Duboscq-type visual colorimeter passed into history.

HOFFMAN'S *PHOTOELECTRIC CLINICAL CHEMISTRY*

In 1941 another important corner was turned with the appearance of the very successful *Photoelectric Clinical Chemistry* by William S. Hoffman of the Chicago Medical School. In addition to procedures for the, by then, established routine analytes in blood, there were methods for albumin and globulin, sodium and potassium, magnesium, lactic acid, alkaline phosphatase, ascorbic acid, sulfanilamide, ethyl alcohol, bilirubin, sulfate, and hemoglobin, for a total of twenty-three colorimetric procedures determined on the Cenco-Sheard-Sanford Photelometer. Sample calibration curves are shown for each method.

SPECTROPHOTOMETRY: VISIBLE AND ULTRAVIOLET

A significant advance in photoelectric colorimetry occurred with the development of spectrophotometers. In place of light filters to select the particular spectral region for photometric analysis, a monochromator was used to produce the complete light spectrum from which the desired wavelength of incident light is chosen. The spectrum may be produced by a quartz prism monochromator (Beckman Model DU) or by a diffraction grating (Coleman Jr.). Spectrophotometers, with their choice of wavelengths and a narrower band width than most photoelectric colorimeters, improved the sensitivity and specificity of many measurements and increased the likelihood of a colored compound obeying Beer's law. They made possible the introduction of other analyses by colorimetric methods.

The Coleman Jr. (Figs. 18.6, 18.7), a smaller version of the "Universal" (Coleman Instruments, Inc., now part of Bacharach, Inc., Pittsburgh, PA), had been designed to withstand the rugged use expected in U.S. Army hospital laboratories. It was used during World War II. The relatively low cost (about $350), ease of operation, and sturdy construction, led to its widespread introduction into clinical and research laboratories following the end of World War II. Whereas the model DU, with its ultraviolet source, was usually reserved for research studies, the visible range spectrophotometers and filter photometers soon became the workhorses in the clinical chemistry laboratory. By the late 1940s and early 1950s, these and other instruments of various designs gradually

SPECTROPHOTOMETRY 459

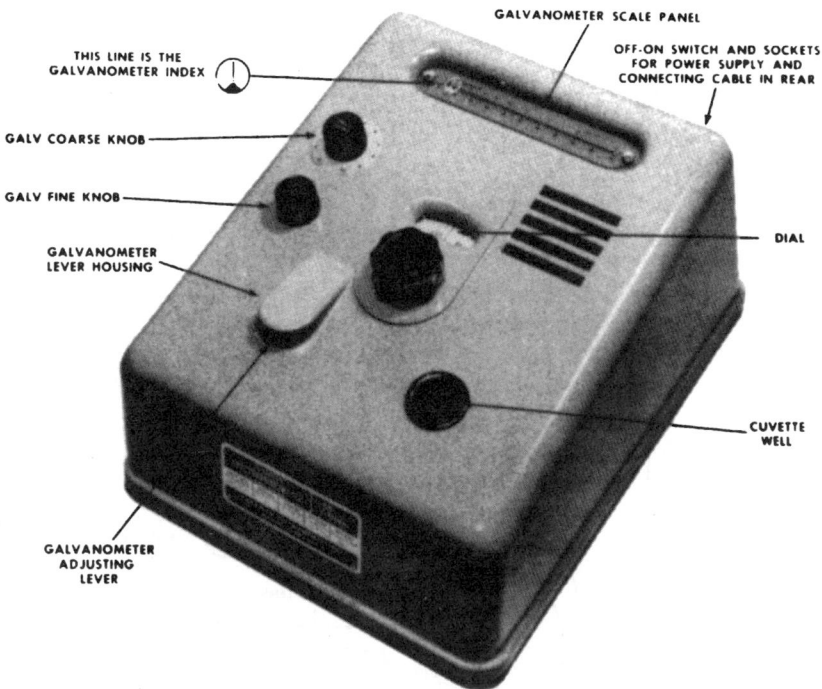

FIG. 18.6. Coleman Jr. Spectrophotometer.

began to replace the Bausch and Lomb, Klett, and other makes of visual colorimeters.

During the 1930s, there was a growing interest in vitamins, and chemical methods of analysis were sought to replace the very slow and laborious assays using animals. When it was recognized that the determination of vitamin A was readily achieved by an ultraviolet measurement in the range of 320–330 nm, at least five photometers were developed specifically for this assay. They used line emission sources that were not applicable to the majority of ultraviolet analyses. In 1940 the two most popular spectrophotometers were made by Cenco and by Coleman. They used an incandescent tungsten source that barely reached the ultraviolet region. Scientists who wanted ultraviolet photoelectric instruments had to build their own.

In early 1940, Arnold Beckman and his colleagues recognized that the dc amplifier designed for the pH meter worked well with vacuum-type

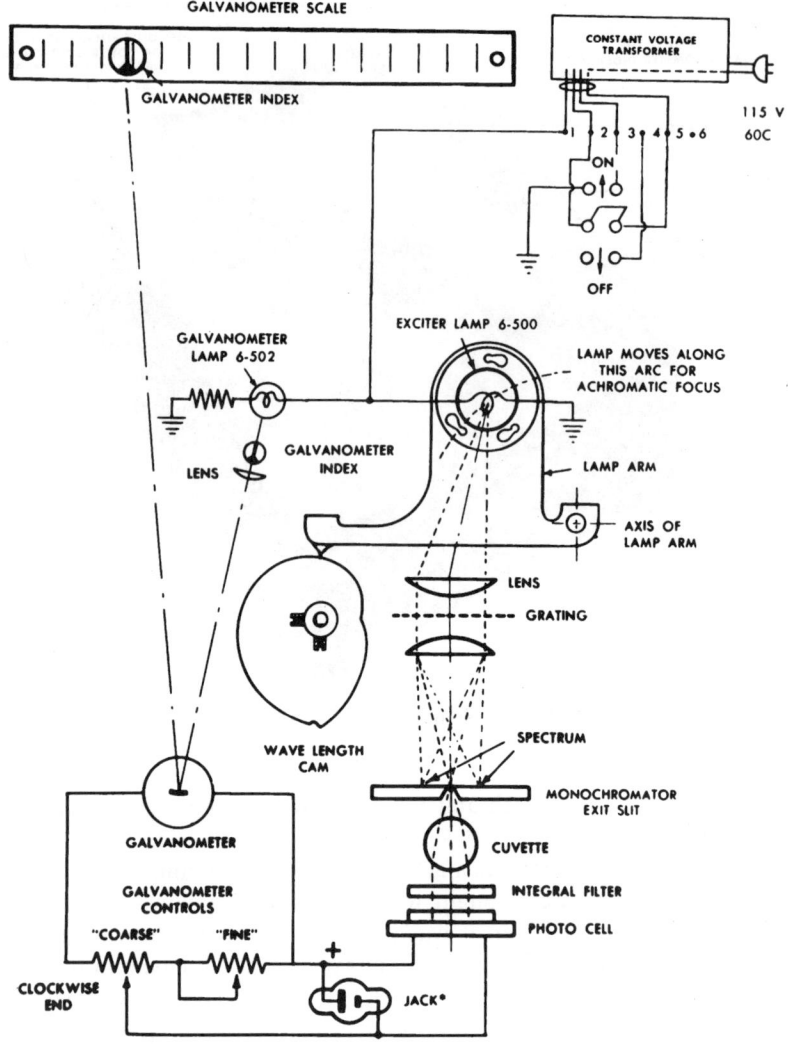

FIG. 18.7. Schematic diagram of Coleman Jr. Spectrophotometer optical and electrical systems.

photocells. The company, whose major products were pH electrodes and meters, began a spectrophotometer development program which, in about fourteen months, resulted in the model DU Quartz Photoelectric Spectrophotometer (Figs. 18.8, 18.9) [20].

VISIBLE AND ULTRAVIOLET 461

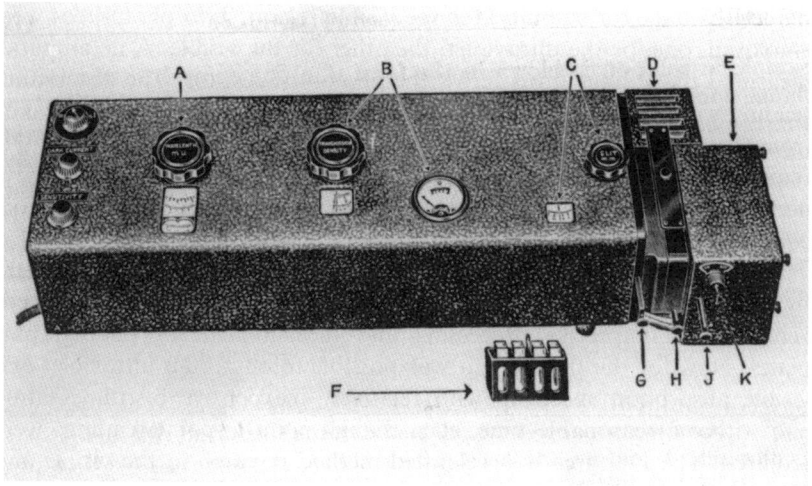

FIG. 18.8. Beckman DU Quartz Spectrophotometer. (Courtesy of The Beckman Heritage Center, Fullerton, California.) A, wavelength scale; B, built-in electronic indicating meter; C, slits with precision adjustment; D, light source; E, compartment for two phototubes; F, holder for four 10-mm absorption cells; G, filter slide; H, compartment for absorption cells; J, phototube selector; K, switch for checking dark current.

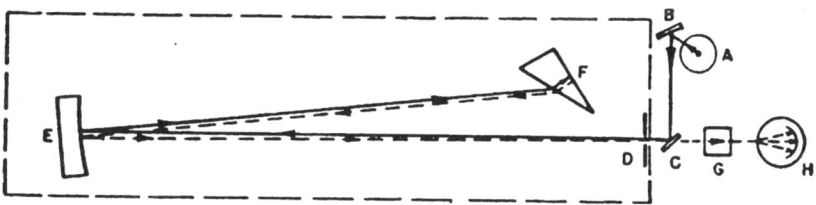

FIG. 18.9. Schematic diagram of Beckaman DU Quartz Spectrophotometer optical system. (H. H. Cary and Arnold O. Beckman, *Journal of the Optical Society of America*, 1941, 31: 682–689.) A, light source; B, C, and E, mirrors; D, entrance slit; F, prism; G, cuvette; H, phototube.

The design of the DU was carefully thought out. A prism monochromator was selected in preference to a grating to minimize stray light. The instrument featured variable slits, a hydrogen lamp source for the ultraviolet, and an incandescent automotive headlight bulb (operated at reduced voltage for stability) for the visible region. Two phototubes were employed, one for the ultraviolet, the other for the visible. A dc amplifier measured the voltage drop across the phototube resistor. The absorbance reading was obtained by null balancing of the meter, with readout taken from the calibrated dial of the balancing potentiometer. Initially, the DU required dry cells as well as a lead-acid storage battery, at that time a normal acceptable requirement. Design modification later provided a more convenient and trouble-free power supply.

The introduction of the DU in 1941 ensured the end of absorptiometry by means of the spectrograph with its dependence on the tedious, inconvenient, and imprecise processing and measurement of photographic plates. Now for the first time it was possible to obtain an ultraviolet and visible absorption spectrum with relatively inexpensive instrumentation and within a reasonable time, even though point-to-point readings were required. The DU greatly accelerated method research in the visible and ultraviolet range.

By the end of 1941, eighteen instruments had been delivered. The original selling price was $723. The DU met a need and was an immediate success. It remained unsurpassed in its field for thirty-five years. By the time the DU and its successor DU-2—introduced in 1964—were discontinued in 1976, over 30,000 of these instruments had been produced. Because the DU was more costly than the photoelectric filter photometers and spectrophotometers in use in the clinical chemistry laboratory, it was reserved for special assays and research projects requiring narrow bands of visible or ultraviolet light.

The contribution of industrial scientists to the development of clinical chemistry has been one of the characteristics of American science and may be traced to Arnold O. Beckman, the founder of Beckman Instruments, Inc., Fullerton, California. Although the two instruments for which he is best known, the Model G pH Meter and the DU Spectrophotometer, were not designed specifically for clinical chemical applications, they subsequently led to widespread use in acid–base studies and photometric measurements of many kinds.

As we approach the end of the twentieth century, it should be remembered that whatever the level of complexity and versatility to which the color measuring instrument has evolved, its basic components remain unchanged: a source of light, a container for unknown and standard, and a receptor—no longer the human eye, but an electronic device—for the light transmitted by the reaction mixture in the container.

VISIBLE AND ULTRAVIOLET 463

NOTES AND REFERENCES

1. VICTOR CARYL MYERS, *Practical Chemical Analysis of Blood. A Book Designed as a Brief Survey of This Subject for Physicians and Laboratory Workers*, 2nd ed. (St. Louis: C. V. Mosby, 1924).
2. *Ibidem*, p. 25.
3. *Ibidem*, p. 180.
4. *Ibidem*, p. 180–185; see also p. 18.
5. F. C. WOOD, KARL M. VOGEL, and L. W. FAMULENER, *Laboratory Technique. The Methods Employed at St. Luke's Hospital, New York* (East Stroudsburg, Pennsylvania: The Press Publishing Company, 1922).
6. O. L. V. DE WESSELOW, *The Chemistry of the Blood in Clinical Medicine* (London: Ernest Benn Ltd., 1924), p. 7.
7. DONALD D. VAN SLYKE, "John P. Peters," *Clinical Chemistry*, 1957, 3: 287–293.
8. E. NEWTON HARVEY, "Albert Prescott Mathews, Biochemist," *Science*, 1958, 127: 743–744.
9. *A History of Analytical Chemistry*, Herbert A. Laitinen and Galen W. Ewing, eds. (Washington, D.C.: American Chemical Society, 1977), p. 140. For a wide-ranging review with many references, see also M. G. MELLON, "A Century of Colorimetry," *Analytical Chemistry*, 1952, 24: 924–931.
10. RALPH HOLCOMBE MÜLLER, "Instrumental Methods of Chemical Analysis," *Industrial and Engineering Chemistry. Analytical Edition*, 1941, 13: 667–754, pp. 706–710. This publication has numerous illustrations of optical instruments.
11. LAITINEN and EWING (1977), p. 142.
12. K. S. GIBSON, "Photo-Electric Spectrophotometry by the Null Method," *Journal of the Optical Society of America and Review of Scientific Instruments*, 1919, 2–3: 23–26. For the general status of spectrophotometry in 1922–23 also see "Spectrophotometry. Report of O. S. A. Progress Committee for 1922–3," *Ibidem*, 1925, 10: 169–241.
13. STANLEY P. REIMANN, "The Photo-Electric Cell as a Colorimeter," *Proceedings of the Society for Experimental Biology and Medicine*, 1926, 23: 520–523.
14. ARTHUR H. SANFORD, CHARLES SHEARD, and ARNOLD E. OSTERBERG, "The Photelometer and Its Use in the Clinical Laboratory," *American Journal of Clinical Pathology*, 1933, 3: 405–420. For the development stages of this photoelectric instrument from hemoglobinometer to photelometer, see CHARLES SHEARD and ARTHUR H. SANFORD, "A Photo-Electric Hemoglobinometer. Clinical Applications of the Principles of Photo-Electric Photometry to the Measurement of Hemoglobin," *Journal of Laboratory and Clinical Medicine*, 1929, 14: 558–574; "Photo-Electrometer With One Stage of Amplification as Applied to the Determination of Hemoglobin," *Journal of the American Medical Association*, 1929, 93: 1951–1957; ARTHUR H. SANFORD and CHARLES SHEARD, "The Determination of Hemoglobin With the Photoelectrometer," *Journal of Laboratory and Clinical Medicine*, 1930, 15: 483–489.
15. KENNETH A. EVELYN, "A Stabilized Photoelectric Colorimeter With Light Filters," *Journal of Biological Chemistry*, 1936, 115: 63–75.
16. WILLIAM H. SUMMERSON, "A Simplified Test-Tube Photoelectric Colorimeter, and the Use of the Photoelectric Colorimeter in Colorimetric Analysis," *Journal of Biological Chemistry*, 1939, 130: 149–166. For illustrations of numerous photoelectric filter photometers see also FOSTER DEE SNELL and CORNELIA T. SNELL, *Colorimetric Methods of Analysis. Including Some Turbidimetric and Nephelometric Methods*, 3rd ed., vol. 1,

Theory—Instruments—pH (New York: D. Van Nostrand Company, Inc., 1948), pp. 89–100.
17. RALPH H. MÜLLER, "Photoelectric Colorimetry in Microanalysis. Photoelectric Methods in Macro- and Microanalysis," *Industrial and Engineering Chemistry. Analytical Edition*, 1935, 7: 223–226, p. 223.
18. RALPH H. MÜLLER, "Photoelectric Methods in Analytical Chemistry," *Ibidem*, 1939, 11: 1–17, p. 12.
19. MÜLLER (1941), p. 702.
20. H. H. CARY and ARNOLD O. BECKMAN, "A Quartz Photoelectric Spectrophotometer," *Journal of the Optical Society of America*, 1941, 31: 682–689; A. O. BECKMAN, W. S. GALLAWAY, W. KAYE, and W. F. ULRICH, "History of Spectrophotometry at Beckman Instruments, Inc.," *Analytical Chemistry*, 1977, 49: 280A–298A.

CHAPTER XIX

CLINICAL CHEMISTRY LABORATORY (1925–1960)

By 1925 the hospital clinical laboratory had taken on the repertoire and appearance that remained essentially unchanged for the next thirty-five to forty years, until after the debut of the Technicon AutoAnalyzer in the late 1950s. Of course, not every laboratory introduced mechanization at the same time. Modifications and improvements of existing manual methods were constantly being made, as chemists sought to impove sensitivity, specificity, and speed of analysis, while reducing sample size and cost.

Much research and ingenuity went into the development of colorimetric chemical procedures. The challenge presented by the many substances which do not lend themselves to formation of colored compounds was overcome by unique indirect methods that brought them into this analytical arena. Although the competition from photoelectric photometers was real, Müller was confident in 1941, that "the simplicity and directness of the colorimeter will continue to be attractive for some time to come." They were, for nearly twenty years, but he also knew that the future belonged to the "photoelectric types" [1].

In any event, except for a few makes of photoelectric colorimeters that were in place in many laboratories, the appearance of clinical chemistry laboratories in 1960—and in many places even later—was not very different from that of 1925. The basic technology and equipment was essentially unchanged. There was lots of glassware of different kinds— pipettes, burettes, wooden racks of test tubes, funnels, filter paper, cylinders, flasks, and beakers; visual colorimeters, centrifuge, water bath, an exhaust hood for evaporating organic solvents, a microscope for examining urine sediments, a double-pan analytical beam balance, and perhaps a pH meter. The most complicated apparatus was the Van Slyke gasometric device—manually operated. The emphasis was on classical chemical and biological techniques that did not require instrumentation. Following World War II, a few new instruments became available to the laboratory: automatic titrator, spectrophotometer, flame photometer, and paper electrophoresis. These represented the vanguard of more complex systems soon to come—and designed specifically for the clinical chemistry laboratory.

Many nonspecific methods—essentially group reactions based on the joint behavior and interrelationship of all the serum proteins—were introduced during this time period. In these reactions, albumin acted as a protective colloid to inhibit the tendency of a globulin fraction to be precipitated by the particular chemical reagents used. The protection may fail when the albumin is decreased or the gamma globulin is increased or changed qualitatively by a pathological process. Because this occurred most frequently in liver disorders, these procedures were often referred to erroneously as "liver function" tests. Positive results were manifested by varying degrees of opacity, coagulation, precipitation, flocculation, or turbidity. Some of these procedures and the year in which they were introduced, are: formol gel, 1920; Takata-Ara, 1925; Weltmann, 1930; colloidal gold, 1937; cephalin–cholesterol flocculation, 1939; thymol turbidity, 1944; thymol flocculation, 1946; zinc turbidity, 1947 [2].

Although these tests depended on the same underlying changes in the serum proteins, they responded differently in different liver and biliary tract diseases. Furthermore, there was only a fair correlation between results of closely related tests, and there was no uniform response for a given test in similar clinical situations. Consequently, these tests had limited usefulness. They were phased out, more or less in the same sequence as they were introduced, by other nonspecific procedures that developers believed would help better to distinguish between obstructive and infectious liver and biliary disease. New tests do not immediately gain acceptance and replace the old; and so, with the introduction of assays for serum alkaline phosphatase and bilirubin in the mid to late 1930s and early 1940s, and the transaminase enzyme assays and serum protein electrophoresis on paper in the mid to late 1950s—as old and new were offered side by side—the last of these nonspecific tests began a slow fade into obscurity during the early 1960s.

PROFICIENCY TESTING

The turn of the century witnessed great advances in parasitology, microbiology, and serology. Physicians began to consider microorganisms as probably the most important causative factor in disease. However, rapid advances in biochemistry soon revealed disease states that were essentially chemical disturbances and made chemical analysis indispensable for diagnosis and treatment of the patient. Out of this need, there arose the separate arena of analytical chemistry of living systems. Special methods requiring small samples soon replaced the cumbersome macrochemical procedures of previous decades. How reliable were these methods—how accurate—how reproducible from laboratory to laboratory?

Shortly after the end of World War II, directors of clinical laboratories in the Philadelphia area became alarmed over the frequency of divergent results when the same blood or serum specimen was sent to two different clinical laboratories for analysis. As a result of these discrepancies, a survey was made of the accuracy of the more common chemical measurements made in hospital laboratories throughout the state. The findings were published in 1947 by Belk and Sunderman [3]. This led in 1949 to surveys in other states and a monthly proficiency testing service that mailed ampules of sera or solutions to the participating clinical laboratories throughout the country and abroad. Statistical analyses of the values reported were subsequently returned to the laboratories with a current review of methodology and bibliography. The outcome of proficiency testing was an overall improvement in the quality of the work of the laboratories subscribing to this service and a direct benefit to the patients. The Sunderman Proficiency Testing Service continued uninterrupted for thirty-six years until it was turned over to the American Society of Clinical Pathologists in 1985.

Similar services by now had also become available from manufacturers of diagnostic laboratory instruments and from other professional laboratory organizations (College of American Pathologists, American Association for Clinical Chemistry). Although the term "quality control" (QC) is often used, it is not appropriate. QC comes from the industrial concept that every product should be tested day by day, batch by batch, to ensure that it consistently meets specifications and to prevent defective materials from reaching the consumer. In the laboratory, where the product is a numerical value, "quality assurance" (QA) is a more meaningful expression and can be best achieved by "proficiency testing." QA is a process of assuring that everything is done according to good practices by documenting all activities and reviewing records for efficiency and accuracy.

A supplement to proficiency testing was the need for continuing education for the scientific and technical staff of clinical laboratories. Workshops and seminars in the newest developments in laboratory science were initiated by the American Society of Clinical Pathologists in 1954. In subsequent years, continuing education programs have become a feature of every national and regional meeting of the many scientific associations concerned with "laboratory medicine."*

---

*The earliest use of this term may have been by Lewellys F. Barker in *Science*, 1908, 27: 601–611, p. 602.

Once the proficiency services got under way, it became recognized that standard reference materials were urgently needed for the preparation of calibration curves and for the evaluation of clinical methods. Such materials for inorganic components of biological fluids had long been available from the National Bureau of Standards. Not until 1967 did the Bureau begin to address the need for standard materials for the assay of albumin, cholesterol, bilirubin, hemoglobin, and other components in blood of interest to the clinical chemist.

In the United States, the National Committee for Clinical Laboratory Standards (NCCLS) is a nonprofit educational organization founded in 1968 to promote the development and voluntary use of national and international standards and guidelines for maintaining the high level of performance of the clinical laboratory necessary for quality patient care. NCCLS publications describe laboratory procedures, reference methods, and evaluation protocols, arrived at by a consensus process.

BIOCHEMISTRY SHEDS ITS CLINICAL CONNECTION

Prior to 1940, American biochemistry was closely connected to clinical medicine. The first generation of American biochemists to gain international prominence, viz., Folin, Benedict, and Van Slyke, were known for their achievements in clinical biochemistry. Most departments of biochemistry were in medical schools, and their success depended on the service they rendered to medical instruction and to routine testing and clinical research in the hospital laboratory. Chemists working in this cliniciandominated environment seldom questioned this arrangement. This hierarchy fell apart as medical school entrance requirements for college chemistry made room for biochemistry and a new generation of biochemists who identified with their own profession—more visible now, with a society and a journal—rather than with clinical medicine. The new relationship between biochemistry and clinical medicine was already emerging by 1910, as biochemists replaced the remaining clinician-chemists [4].

Chemistry's interface with clinical medicine, once limited to urinalysis and toxicology, branched out into new areas: digestion, respiration and blood transport, electrolyte balance, metabolism, and hormonal control. The new generation of academic clinical investigators appearing after 1900 was more receptive than their predecessors to the use of biomedical science. They brought systematic observation and scientific method to the bedside and provided biochemists with abundant experimental materials and attractive research problems. All was not smooth sailing, however.

In some of the leading schools, the politics of institutional spheres of influence tended to slow cooperation between laboratory and clinical factions. Many clinicians believed that without a medical degree a biochemist is not qualified to handle clinical material and did not cooperate with them [5].

In 1917 the AMA Council on Medical Education proclaimed that biochemists with M.D. degrees should be selected for medical school positions in preference to those with Ph.D. degrees. Since rarely did a physician choose an academic career as a biochemist, in preference to clinical research or practice, medical schools had to rely on the adaptability of biochemists to do clinical work. By the 1930s, it was common practice for clinical departments in medical schools to have Ph.D.–biochemists on their staff. Between 1918 and 1940, professorships in biochemistry in some twenty medical schools were filled by Ph.D.–clinical biochemists, many of them the students or former colleagues of Folin, Benedict, and Van Slyke. Clinical biochemists held the top positions in the discipline for nearly a generation [6].

In the 1930s, academic biochemistry and applied clinical biochemistry began to diverge. By 1940 the founding generation was nearly gone, and the new generation of American biochemists was withdrawing from the clinical connection and promoting an alternative style—a biochemistry of "general principles" relevant to all the biological sciences. It was based in the graduate division, not the medical school. As departments began to abandon the analytical interests of the founders, clinical chemistry was left to fend for itself as a separate applied science—a specialty adapted to the needs of hospital diagnostic laboratories and distinct from academic biochemistry. Progress in clinical chemistry in the United States during the first few decades of this century was the result of a close working partnership between biochemistry and clinical medicine. It prospered because the institutional infrastructure in the United States was ideally suited for it. Now, clinical chemistry was on its own [7].

In reminiscences about the development of clinical chemistry in the 1930s, Oliver Gaebler recalled that he felt "a growing sense of excommunication from biochemistry...not due to the attitude of biochemists, but to the trend of events." [8]. What happened was an abundance of interesting problems in biochemistry and nutrition as new sub-groups sprang up. The problems that had challenged Folin's generation now required new and highly specialized skills. Experiments with the total animal were replaced by newer techniques using tissue slices and extracts to study the individual chemical reactions in the breakdown and synthesis of compounds of intermediary metabolism. Enzymes participating in tissue oxidation and reduction reactions were the new frontiers of research and were more exciting and prestigious than chemical analysis. Isolation and structure of vitamins and hormones required skilled organic chemistry, whereas quantitative analysis of biological fluids, even complex procedures, had become routine and within the capability of technicians. Medical school biochemists gravitated to newer more academically prestigious lines of research. In the early years of medical reform, biochemistry in America had been nurtured by its service role in clinical chemistry, but by 1940, it could survive on its own. Development of clinical methods was abandoned to the clinical chemists who found themselves no longer at the center of the biochemical sphere. Folin, Benedict, Van Slyke, Myers, Hawk, and other clinical biochemists were a transitional generation, who resembled their predecessors, the medical chemists, as closely as they resembled their successors [9].

AMERICAN ASSOCIATION OF CLINICAL CHEMISTS

Meanwhile, the methods of the clinical chemistry laboratory continued to increase in variety. Eventually, they became sufficiently complex for a new class of specialist—clinical chemist or doctoral scientist—to emerge and take charge of research and responsibility for directing the routine work in the clinical chemistry laboratory. By the late 1940s, concerned with professional identity and the perceived low image and status of the "clinical chemist" with respect to academic biochemists, clinical chemists in New York City met to plan formation of a professional association. The broadly stated objectives were to gain understanding and recognition by government, the public, and the medical establishment.

On December 15, 1948, nine Ph.D.–clinical chemists from the major New York City private and municipal hospitals met at Mt. Sinai Hospital in New York City to organize a professional association of clinical chemists, locally at first, and then on a national scale. It marked the founding of the American Association of Clinical Chemists (AACC), whose membership by 1999 exceeded 11,000. The nine founding charter members were: Harry H. Sobotka and Miriam Reiner (Mt. Sinai Hospital); Albert E. Sobel and Samuel Natelson (Jewish Hospital of Brooklyn); Mary H. McKenna (Harlem Hospital); Max M. Friedman (Queens General Hospital); Joseph Kahn (Maimonides Hospital); Julius J. Carr (Metropolitan Hospital); Louis B. Dotti (St. Luke's Hospital). Mr. A. J. Nydick, a lawyer, was present to provide legal counsel. Aaron Bodansky of the Hospital for Joint Diseases was present, but did not become a member. Eight other hospital chemists from New York City were invited, but were unable to attend. The meeting unanimously elected Max Friedman as temporary chairman and Louis Dotti as temporary secretary–treasurer.

This group was motivated by two major concerns. One, was the need to upgrade the quality and accuracy of chemical analyses; the other, was to review, clarify, and strengthen their professional autonomous standing as Ph.D.–chemists with medical colleagues and the public. Status and authority were confused because the city denied a license as a laboratory director to anyone without an M.D. degree. They did not wish to become a pressure group for purely professional aims, but rather to promote the scientific goals of clinical chemistry. Aware of similar grievances elsewhere, it was decided to contact hospital chemists throughout the United States.

A second meeting on January 11, 1949 at Mt. Sinai Hospital was attended by 20 clinical chemists, including Joseph Benotti from Boston, the first member from outside the New York metropolitan area. The name adopted for the organization, by unanimous vote, was the American Association of Clinical Chemists. The third meeting, held on February 1, 1949 at the same place, was attended by 17 clinical chemists, including John G. Reinhold of Philadelphia. Mr. Nydick, the legal advisor, was present at both meetings. Benotti and Reinhold subsequently were active in organizing the Boston (Northeast) and Philadelphia sections, respectively. The AACC membership grew to over 150 in 1949 and to 309 by the end of 1950. At the end of 1954, there were 629 members and seven chartered local sections in the AACC, including the Southern California, Chicago, Washington–Baltimore–Richmond, and Midwest sections. Annual meetings of the AACC were held in conjunction with meetings of the American Chemical Society until 1958, when the tenth annual meeting was held independently in Iowa City, September 4–6. By then, with a membership of 748, the bank balance of the Association was $8,752.19 [10].

In 1948 the practice of medicine in general and of laboratory medicine in particular, was very different from the level of sophistication that we experience in the closing years of the twentieth century. The war against bacteria was fought mainly with sulfonamides and penicillin; mood-altering pharmacology was almost unknown; adrenal and thyroid hypofunction were treated with endocrine organ extracts; and plasma cholesterol was newly recognized as playing some sort of role in atherosclerosis and coronary artery disease. Clinical chemistry's contribution to the assessment of health and disease was secondary to that of bacteriology, hematology, and urinalysis. Thyroid status was evaluated by determination of the basal metabolic rate. The few hormones that could be measured, e.g., for determination of pregnancy, were bioassayed by using frogs, rabbits, and mice. The only enzyme assays were for amylase, lipase, and acid and alkaline phosphatase. Analytical precision was satisfied by analysis in duplicate. Medicare and the Clinical Laboratory Improvement Act of 1967 were twenty years in the future. Few states had any legislation regulating the clinical laboratory.

The new organization issued *The Clinical Chemist* (1949), a newsletter; formed regional sections; established a code of ethics (1952); initiated publication of *Clinical Chemistry* (1955), a bimonthly journal; sought licensing; and developed distinctive career patterns of training [11]. By 1995, the organization, renamed American Association *for* Clinical Chemistry, Inc. in March 1976, had more members than the American Society of Biological Chemists (renamed American Society for Biochemistry and Molecular Biology).

In January 1996, to reflect the growing list of new and nontraditional subject areas covered and the international character of *Clinical Chemistry*— 55% to 60% of its articles have come from outside the United States during the years 1992 to 1995—the subtitle *International Journal of Laboratory Medicine and Molecular Diagnostics* has been added to the new cover design.

The AACC launched a series of *Standard* (later *Selected*) *Methods of Clinical Chemistry* in 1953. The object was to provide accurate and workable methods upon which the clinical chemist can depend. Although some of the methods were difficult and not suitable for daily work, they were helpful as reference methods for evaluating those in use. Alternative methods using different principles were also included in the series.

Initially, the method was tested in the laboratory of the *Submitter*, then retested for validity and practicality in the laboratory of a referee, or *Checker*. Immunochemical methods could not be included because patent rights protected against disclosure of the details of reagent preparation; consequently, the series could not remain up to date. Eventually, the evaluation process became too slow and cumbersome, and the series was discontinued after thirty-four years, with volume eleven in 1986.

Another series, *Advances in Clinical Chemistry*, begun in 1958, aimed to provide a readable account of selected important developments, their roots in allied fundamental disciplines, and their impact upon the progress of medical science. Articles were written by experts who were working in the field that they described. Volume thirty-three appeared in 1998.

FOREIGN SOCIETIES AND THE INTERNATIONAL FEDERATION

Even prior to the organization of the American group, professional societies of clinical chemists had begun to form in Europe. The first national society of clinical chemistry was the Société Française de Biologie Clinique formed in 1942. A Swedish Association for Laboratory Doctors was organized in 1944 and included microbiologists, clinical chemists, and clinical physiologists. In Norway, the Society for Medical Physiology was formed in 1945, reorganized in 1957, and later renamed the Norweigian Society for Clinical Chemistry and Clinical Physiology. The Netherland Society for Clinical Chemistry and the Finnish Society of Clinical Chemistry were organized in 1947 and were the first to include "Clinical Chemistry" in their name. In 1953 the Association of Clinical Biochemists was founded in the United Kingdom. The present Swedish Society for Clinical Chemistry was formed in 1954.

At the same time that the AACC was being formed in the United States in response to professional needs of clinical chemists, a similar situation was developing in Great Britain. When the National Health Service (NHS) was set up in the United Kingdom in 1948, there was no mechanism for deciding the gradings and salary scales of the growing number of nonmedical employees. These were dealt with separately from those of medical staff. At this time, there were about 120 full-time biochemists employed in NHS hospitals. By 1951 it was becoming clear that there was a need for some form of professional body to deal with matters concerning the increasing number of clinical biochemists in the UK and to provide a forum for discussions on their rapidly expanding specialty.

The Biochemical Society, founded in 1911, had little interest in clinical chemistry or professional affairs. The Association of Clinical Pathologists, which included chemical pathology as one of its four specialities, limited its membership to persons with a medical qualification. Neither of the two societies nor the Royal Institute of Chemistry was able to change their organization to allow formation of a sub-section for clinical biochemists— possibilities that were explored. With little to choose from, the regional groups of clinical biochemists decided to form their own national society which would deal exclusively with their specialty. After ten months of regional committee meetings to develop working rules, the inaugural meeting was held at the Hammersmith Hospital, London, on March 28, 1953. Seventy-five people were present and signed the register. At the end of 1995, total membership was 2294.

The Association of Clinical Biochemists owes its foundation to Earl J. King (1901–1962), professor of chemical pathology at the Postgraduate Medical School in the University of London. An eminent and highly respected scientist, he was known particularly for his work in clinical enzymology and in developing colorimetric methods of analysis. King had a Ph.D. and D.Sc. degree, but no medical qualification other than an Honorary M.D., a circumstance that was encouraging to the many nonmedical clinical biochemists who were struggling locally to improve their professional status. During the months of organization and negotiation, he was able to reconcile different points of view and achieve a synthesis instead of a compromise. King was elected the Association's first Chairman—a position with much of the day to day responsibility. The office of President was primarily titular. In 1955 Donald D. Van Slyke was elected as the first Honorary Member.

In 1952 King suggested that the newly forming national societies of clinical chemistry should join into an international organization under the auspices of the International Union of Pure and Applied Chemistry (IUPAC). This was accomplished on July 24, 1952 at the Second International Congress of Biochemistry in Paris by the formation of the International Association of Clinical Biochemists. The following year in Stockholm it was proposed that the name be changed to the International Federation of Clinical Chemistry (IFCC); this was formally done at the next meeting in Brussels in 1955. In the interim, the First International Congress of Clinical Chemistry was held in Amsterdam in 1954. The initial objectives of the Federation were to "Advance knowledge and promote the interests of biochemistry in its clinical (medical) aspects." In 1967 the IFCC formally separated from IUPAC. As of 1996, societies in sixty-one nations on six continents, and thirty-one corporate groups active in clinical chemistry, have become members [12].

BIO-SCIENCE: THE ERA OF REFERRAL CHEMISTRY

Commercial laboratories for analysis of biological specimens were not uncommon at mid-century. In fact, these services were offered by enterprising physicians before the turn of the century. Following World War I, there was an increase in the number of these laboratories in response to the growing number of analytical methods with clinical relevance. At mid-century, a new kind of commercial enterprise appeared—the specialized reference laboratory—and a new name was added to the clinical chemistry vocabulary.

Bio-Science Laboratories (BSL) was started by one Navy and three Army officers who had met after World War II at Camp Detrick, Maryland, while awaiting discharge to civilian life. One was a physician, the other three had doctorates in bacteriology. Pooling their finances, they decided to enter the laboratory business by doing mainly industrial analyses backed up by a medical clinical laboratory, in southern California. The four were Richard Henry, Sam Berkman, Orville Golub, and Milton Segalove. They opened for business in February 1948 [13].

The break that gave BSL an entry into the medical community was a reproducible method for chemically measuring protein-bound iodine (PBI) of serum and, by inference, the serum thyroxine concentration for assessing thyroid status. The only other method was by measurement of the basal metabolic rate, a nonspecific and highly variable procedure. The PBI method developed by Albert Chaney at the Los Angeles County Hospital was complex and challenging and required special glassware and meticulous technique. It involved precipitation of serum proteins with its protein-bound thyroxine, washing the precipitate to remove inorganic iodine, wet digestion with chromic acid to convert thyroxine iodine to inorganic iodide, distillation of released iodine into a receiver, and measurement of that iodine by its catalytic reduction of ceric ion by arsenious acid. The diminution in the yellow color of the reaction solution was a measure of the level of iodine. Though laborious, the method measured PBI reliably. Once the test was improved by BSL, it became available to the Los Angeles medical community, and business accelerated markedly. When the alkaline dry ash method of Barker, Humphrey, and Soley [14] was published in 1951, BSL quickly adapted it to mass production. By 1966 BSL was analyzing more than 1000 specimens a day for PBI, coming from the United States and abroad.

Requests for other difficult analyses—so-called "reference" or "specialized" procedures—started to come in. Soon, BSL led the way with a wide range of specialty tests in chemistry, toxicology, microbiology, and immunology, made available to the medical community nationwide. Bio-Science's contribution to the practice of laboratory medicine was the creation (or discovery) of a place in the economy, i.e., a national market for new, specialized tests. These usually originated in specific research projects at academic institutions and were then modified for use in the specialty laboratory. The PBI method was followed by other methods for thyroid function: T3-uptake; thyroxine-binding globulin; thyroxine by column chromatography; "free" thyroxine; and thyroxine by competitive protein-binding, in the 1960s and 1970s.

In 1966 a minority interest in the ownership of the laboratory was purchased by the Dow Chemical Company. This grew to complete ownership in 1973. The Dow relationship allowed BSL to grow by acquisition, association, branch development, and foreign affiliates. For Dow, BSL provided entry into the medical and biological sciences fields, which it saw as an area of rapid growth and opportunity. Dow did not interfere in the professional operation of BSL.

Throughout the 1960s, BSL dominated the national market for specialty testing, with clients in most of the larger hospitals and clinics in the United States. However, the profits and growth possibilities of the clinical laboratory business were attracting other large companies such as W. R. Grace, Revlon, Roche, Bristol-Myers, and Upjohn Laboratories. They, like Dow, began to acquire and build chains of medical laboratories.

When the last of the founders retired in 1981, BSL had more than twenty branches and affiliates using the same analytical methods and quality control assurance systems. BSL offered several hundred different tests with annual sales greater than 85 million dollars, and employed more than 1500 people in the United States alone.

In 1982, after the founders had all retired, Dow sold BSL to the American Hospital Supply Corporation. The new owner expected BSL to blend easily into the hospital supply business. However, the partnership did not last very long. There may have been an inadequate understanding of the business difference between laboratory supplies and laboratory reports. In 1985 BSL was sold to SmithKline Beckman, an organization that knew the business. The new laboratory network was renamed "SmithKline Bio-Science Laboratories." In 1989, following the acquisition of SmithKline Beckman by the Beecham organization, the network was again renamed, "SmithKline Beecham Clinical Laboratories" (SKBCL) and the name "Bio-Science" disappeared except for references in the SKBCL directory of services.

A valuable legacy of BSL is Richard J. Henry's *Clinical Chemistry: Principles and Technics* (1964; 1128 pages). It reflected the accelerated development of clinical chemistry tests resulting from the growth in research activity in both basic and applied sciences following the end of World War II. Not since Peters and Van Slyke's classic *Methods* three decades earlier was there so comprehensive a coverage of laboratory procedures that also included a thorough discussion of the background of each analyte presented, the principles of analytical techniques and instrumentation, and a detailed introduction to the basics of contemporary quantitative analysis. Accuracy, precision, controls, statistical handling of data, normal values, interferences, and specimen collection and preservation, are also discussed. Methods are described for the more than 140 analytes determined at BSL, underscored by a thorough literature review with nearly 7000 references, many cited for historical purposes. More than ten years in the preparation, Henry personally reviewed over 15,000 publications. The book represented as significant an advance in the practice, understanding, and professional status of clinical chemistry, as was achieved by Peters and Van Slyke a generation earlier.

A second edition appeared in 1974 (1629 pages) with more than 170 methods. Due to the enormous growth of clinical chemistry, Henry was assisted by two co-editors; the book is a collaborative effort from 24 contributors—the entire doctoral and professional staff of BSL. There is much new and revised material and some that is unaltered or changed very little. The first edition included thyroxine, but no other hormones or any toxicology. To keep the size manageable and assure completion, the second edition likewise has no chapter on toxicology or hormones, and this time, not even thyroxine—the analyte that was the key to BSL's early success.

Books are a revealing gauge of the progress, growth, and expansion, of a scientific profession. Another example is the publication record of *Fundamentals of Clinical Chemistry*. Edited by Norbert W. Tietz, it first appeared in 1970 with 17 contributors, 154 methods, 983 pages. A second edition with 31 contributors followed in 1976, and a third edition with 63 contributors in 1987. A fourth edition, titled *Tietz Fundamentals of Clinical Chemistry* (1996), was edited by Carl A. Burtis and Edward R. Ashwood with 59 contributors, 881 pages. Another important work edited by Burtis and Ashwood is *Tietz Textbook of Clinical Chemistry*. The third edition (1998) has 74 contributors, 1917 pages. Other recent publications are *Methods in Clinical Chemistry* (1987), edited by Amadeo J. Pesce and Lawrence A. Kaplan, with 103 contributors, 163 methods, 1366 pages; and *Clinical Chemistry: Theory, Analysis, and Correlations* (1984, 1989, 1996).

SIGMA AND KIT METHODS

Another innovation that appeared in the 1950s was "kit" methodology from the Sigma Chemical Company of St. Louis, Missouri. All the reagents for the analysis were prepackaged and with instructions, ready for use. The user provided glassware and instrument for reading the endpoint.

The company's entry into diagnostic reagents began in 1950, the result of a chance meeting between Dan Broida (1913–1981), president of Sigma, and Oliver H. Lowry, head of pharmacology at the Washington University

Medical School in St. Louis. Lowry had developed an assay [15] for alkaline phosphatase and needed a new source for the p-nitrophenyl phosphate substrate. The chemical was no longer available from Eastman Kodak and Lowry asked Broida if Sigma would like to make this chemical for general use [16]. Sigma, whose product line consisted of only a few biochemicals, e.g., adenosine triphosphate (ATP), was willing. In 1951 the company decided to market Lowry's procedure and packaged the prepared buffer and p-nitrophenyl phosphate as a "kit" (Technical Bulletin No. 104).

In 1955 Sigma offered a second method (Technical Bulletin No. 410) for determination of glutamic oxaloacetic transaminase (SGO) by a kinetic ultraviolet technique [17]. The procedure had limited use because few laboratories had a spectrophotometer providing the near-ultraviolet range (340 nm). The following year Sigma introduced a colorimetric version of this determination (Technical Bulletin No. 505) [18]. This made the test more accessible since, by now, every clinical laboratory had a photoelectric colorimeter covering the visible range. Soon afterward, the company marketed a colorimetric procedure for determination of lactic acid dehydrogenase. In 1975 Sigma merged with the Aldrich Chemical Company, a major producer of organic chemicals.

Although the term "kit" may have been popularized by Sigma's application to enzyme analysis, the word itself has long been in the American lexicon, e.g., "the whole kit and caboodle." The concept of the self-contained total package for chemical analysis also predates its use for enzyme analysis. As early as 1919, the LaMotte Chemical Products Company of Baltimore, Maryland, began to market a series of "chemical outfits" for the commonly tested constituents of blood, serum, and urine, packaged in small portable wooden chests. The outfit contained all the reagents and apparatus needed for the complete analysis, from preparation of a protein-free filtrate, if necessary, to a comparator block for matching the test reaction's color with that of standard color tubes, also provided. Complete instructions accompanied each set.

With the introduction of immunochemical analyses in the 1970s, kit methodology became a staple of the clinical chemistry laboratory and the foundation of a multi-million dollar diagnostic reagents industry.

BERSON, YALOW, AND RADIOIMMUNOASSAY

During the mid 1950s, some unexpected findings by Solomon A. Berson (1918–1972) (Fig. 19.1) and Rosalyn S. Yalow (1921–) (Fig. 19.2)

Fig. 19.1. Solomon A. Berson. (Courtesy of Mrs. Miriam Berson.)

were destined to introduce the clinical chemistry laboratory to the marriage of physics and biology, as a new arena of analytical technique was created. The highly sensitive radioimmunoassay (RIA) technique that grew out of their work at the Bronx Veterans Administration Hospital in New

FIG. 19.2. Rosalyn S. Yalow.

York City, won a Nobel Prize in Physiology or Medicine for Yalow in 1977 [19].

An hypothesis had been advanced by I. Arthur Mirsky (1907–1974), that adult-onset diabetes might not be due to a deficiency of insulin secretion, but rather to abnormally rapid degradation of insulin by hepatic insulinase. Berson and Yalow became interested in the problem and started an investigation into whether diabetics destroyed insulin more rapidly or more slowly than normal subjects. They studied the metabolism of ^{131}I-labeled insulin following intravenous administration to nondiabetic and diabetic subjects. They were surprised to find that the radioactive insulin disappeared more slowly from the plasma of patients who had received insulin, either for the treatment of diabetes or as shock therapy for schizophrenia, than disappeared from the plasma of subjects who never received insulin. The slow rate of insulin disappearance was not due to the diabetes, but to the development of antibodies in response to the prior administration of exogenous insulin. These antibodies complexed with the labeled insulin acting as antigen in the plasma and prevented the insulin molecule from passing through the capillary walls and reaching its place of action. The adult-onset diabetics didn't degrade insulin rapidly—they made enough but failed to utilize it efficiently. Mirsky had a good theory but the wrong explanation [20].

The antibody presumed to be present could not be detected by classic immunological techniques because at such low concentration there was no precipitate formed. During these studies, Berson and Yalow developed a test tube radioisotopic method of high sensitivity for detecting soluble antigen-antibody complexes that could determine insulin concentrations with a sensitivity one thousand times greater than existing methods [21]. Radioisotopes became available from the Oak Ridge National Laboratory in Tennessee in 1946. During the next few years, gamma ray emitting isotopes were brought into general clinical use for ^{131}I uptake by the thyroid, blood volume determination, and red cell survival studies. These tests usually involved administration of isotopes to patients.

Berson and Yalow's radioimmunoassay of plasma insulin introduced a new parameter of *in vitro* laboratory measurement—gamma ray emission from radiolabeled antigens or antibodies, and a new instrument for its detection—a gamma counter. The new technology, based on competitive interaction, permitted rapid and accurate assay of biologically important hormones, polypeptides, and drugs, which previously were measured by time-consuming chemical procedures or bio-assay, if they were measured at all. Because production of antigen, antibody, and radioisotope labels for *in vitro* analysis were beyond the facilities and acceptable expense of the clinical laboratory, this new technique gave rise to a new industry, i.e., the production of packaged kits containing all the necessary components for a successful radioimmunoassay, including standards and controls for

100 to 200 tests. There also sprung up support industries feeding this new technology with accessories for the assay.

Once someone shows the way, the rest comes easy. There followed the introduction of novel configurations for the antigen–antibody reaction and the use of nonisotopic labels, such as enzymes, that can be detected colorimetrically; and molecules that fluoresce or luminesce and are sensed by specially designed photometers. This new analytical capability led to a progressive increase in the dependence by physicians on laboratory data for the diagnosis and treatment of endocrine disorders. In the study of thyroid disease, iodine analysis as protein-bound or butanol-extractable iodine (subject to the unpredictability of *in vivo* and *in vitro* contamination) was replaced by thyroxine (T4) assay in the mid-1970s and supplemented by T3-uptake, triiodothyronine (T3), and thyroid-stimulating hormone (TSH)—initially by RIA, and beginning in the late 1980s, by nonisotopic methods. This transition helped relieve the problem of isotope waste disposal.

When direct procedures for thyroxine (by column chromatography or competitive binding) replaced PBI, the analysis continued to be reported in terms of thyroxine's four atoms of iodine, hence T4-I, despite the fact that this represented only 65.3% of the thyroxine molecule. Additional ambiguity resulted from the many other new tests of thyroid function that include a number or an acronym as part of their identification. As a result, a committee of the American Thyroid Association recommended that thyroxine (T4) rather than thyroxine-iodine (T4-I) be reported. Therefore, since $T4 \times 0.653 = T4\text{-}I$, $T4\text{-}I/0.653 = T4$ [22].

RIA analysis soon replaced tedious chemical or bioassay procedures for other hormones and their metabolites in concentrates of 24-hour urine collections. The use of frogs, toads, mice, rats, and rabbits, to determine pregnancy was replaced, beginning in the late 1950s and early 1960s, with antigen–antibody reactions for chorionic gonadotropin. Previously untapped hormones succumbed to the new analytical techniques of immunoassay for the benefit of physician, patient, and laboratory. The clinical chemistry laboratory with its special expertise in quantitative analysis fell heir to these new techniques and expanded its services.

Since 1960 the increasing use of clinical chemistry's services has stimulated the commercial development of highly automated analytical instruments. New technology in reagent production and instrument design by industry-based scientists have brought the laboratory to the doctor's office and to the hospital patient's bedside and made the practice of medicine hospital-based and laboratory-centered.

TOOL—INSTRUMENT—MACHINE—AUTOMATION

Apparatus has been used since the earliest scientific investigation of nature. However, not until the beginning of modern experimental science by Leonardo da Vinci (1452–1519), Galileo Galilei (1564–1642), Robert Boyle, and Isaac Newton, was apparatus systematically applied in scientific activity. The purpose of apparatus was to extend the capabilities of man's perception, e.g., by amplification with the microscope; to manipulate experimental conditions to yield observations otherwise inaccessible to the senses, e.g., the polariscope; to transform observations of the phenomenon being investigated into information and a quantitative value [23].

The history of physical science is mainly the history of instruments. The broad generalizations and theories that have arisen from time to time have succeeded or failed on the basis of accurate measurement. Sometimes new instruments have had to be devised for the purpose. The pattern of technological evolution in the clinical laboratory was to use or adapt the instruments already developed for industrial applications—but there was a considerable time lag, and it was out of phase with the development of technology generally, i.e., tool to instrument to machine to automation. The first tools were simple utensils that extended the natural capabilities of the human hand. About 1840, at a time when the Industrial Revolution was introducing machines, the clinical laboratory experienced a transition from tool, e.g., distillation flask and condenser, burner, pipettes, burettes, etc., to instrument, e.g., microscope, analytical balance. The centrifuge, visual colorimeter (Duboscq), spectroscope, and hypodermic syringe came later, but were originally designed for other applications before finding a use in the clinical chemistry laboratory in the next century. The photoelectric effect had been known to physicists since the end of the nineteenth century, but it was only in the 1930s that the first photometers with photoelectric cells became commercially available and were adapted for analysis in the clinical laboratory [24].

Machines had long been established in general engineering and in everyday life in the home and workplace, while clinical laboratories were still largely using instruments. The first instrument designed for a specific need in the clinical laboratory was Van Slyke's volumetric gas pipette for measuring carbon dioxide combining power of serum, described in 1917.

The evolution of technology in the clinical laboratory had both a scientific and a social component. With few exceptions, scientific instruments were not made by the scientist-user, but by craftsmen with special skills in metal and later, also in glass. In the sixteenth century, watch and clock-making crafts were highly developed in Germany and France. In the seventeenth century, specialists in making thermometers, barometers, optical instruments, etc., appear, mainly in Italy and England. The choice of materials became more and more the responsibility of the instrument-maker who introduced many design improvements. In the nineteenth century, another social pattern emerges in the making of instruments for the clinical laboratory. The university mechanic, e.g., Carl Zeiss (1816–1888) of Jena in 1846, employed by a scientist at a university or institute, branches out and sets up his own business. In the second half of the century, with an increased demand for instruments for science, industry, and the home, manufacturing by craftsmen was replaced with industrial production. The craftsman became more of an engineer as instrument making was placed on a reliable scientific basis. Then, in the twentieth century, research, development, and manufacturing have all been taken over by industrial apparatus engineering companies [25].

Clinical chemical analysis is always challenged by two fundamental and inherent contradictions: (1) to use as small a sample as possible or available, without exceeding the limit of detection; and (2) to achieve speed without sacrificing precision of analysis. Physical methods and imaginative miniaturization of instruments resolved to a large degree the contradiction between quantity of specimen and limit of detection. Analytical machinery has more or less reduced the contradiction between speed and precision [26].

During the second half of the twentieth century, the clinical laboratory began to catch up with the advances made by industrial engineering. In 1957 the AutoAnalyzer became the first machine in a series of engineering marvels designed to meet specific analytical needs in the clinical chemistry laboratory. It was also the beginning of involvement by industry and commerce with the needs of the clinical chemistry laboratory beyond that of glassware, chemical reagents, and laboratory furniture.

SKEGGS AND THE AUTOANALYZER

In 1957 the first artificial orbiting satellite set off a revolution in space science with worldwide implications for the future. On earth, that same year, a quieter revolution was taking place in the clinical chemistry laboratory, with no less significant implications in a completely different realm of human endeavor. It was the introduction of the first successful automated system of blood analysis. It was the opening salvo in a laboratory revolution that shook the practice of medicine and is still going on as automation spreads from the chemistry laboratory to the other divisions of the clinical laboratory. In time, the advances in electronics, robotics, and miniaturization that evolved in the manned-space program, were utilized in the further automation and computerization of the methods and data generated by laboratory diagnostic instruments.

As noted earlier, the first tentative step toward automated analysis occurred in 1933, with the introduction of a commercially available photoelectric colorimeter. What happened then was the replacement of the subjective observation of a color match on a visual colorimeter, with the objective and impartial reading by an instrument. If we go fast forward a quarter of a century, we come to the start of the modern era of automated blood analysis.

This story begins with Leonard T. Skeggs, Jr. (1918–) (Fig. 19.3). After serving in the Navy for three years during World War II, Skeggs returned to Cleveland to complete work for his Ph.D. in biochemistry at Western Reserve University under Victor Myers. Following graduation in 1948, Skeggs accepted an appointment at the Veterans Hospital with responsibility for supervision of the clinical chemistry laboratory. Skeggs soon began to look for a way to deal with the error-prone repetitive steps in manual analysis and the increasing workload made worse by insufficient staffing in the laboratory. He wanted to build a machine that would complete a blood analysis from start to finish without any intervention by a technician. Working at home in his basement, Skeggs assembled a number of modules with specific tasks that constituted a continuous flow sequence of analysis. During the early 1950s, his prototype was rejected by four companies. Responses ranged from complete apathy to advice that he get a patent first. In January 1954, Skeggs's prototype working model #3 was demonstrated to the Technicon Corporation (Tarrytown, New York). Known primarily for the AutoTechnicon—an automatic tissue processor that moved surgical specimens through the steps of fixation, dehydration, clearing, and embedding with paraffin, in a timed sequence in preparation for staining—the company saw the possibilities of this totally new concept and went on to develop and produce the machine commercially. They brought the AutoAnalyzer on the market in 1957. The first model sold for $3,500 and was an immediate success. Fifty units were sold the first year and 4,000 by 1963 (Fig. 19.4) [27].

This analytical system was a radical departure from the usual laboratory instrument. Instead of performing a single task, it performed in automated sequence all of the analytical steps required to obtain a quantitative answer. It did this, not by mimicking the manual movements of a

Fig. 19.3. Leonard T. Skeggs.

technician, but by employing common physical and chemical techniques in a novel, heretofore unused manner, in order to perform the test.

There were three unique features of the machine, in addition to continuous flow. These were: (1) dialyzing membrane to remove proteins and

FIG. 19.4. Single channel AutoAnalyzer. (Courtesy of Bayer Corporation, Diagnostics Division, Tarrytown, New York.) (l to r) strip-chart recorder; colorimeter (forward); heating bath (rear); dialyzer with dialysis assembly submerged in water bath; proportioning pump with manifold assembly (forward); sampling tray (rear); reagent bottles.

provide a protein-free dialysate without centrifugation or filtration. Skeggs was involved in improving an artificial kidney machine and made the conceptual leap from a substitute for glomerular filtration *in vivo* to the use of dialysis *in vitro*; (2) at first polyethylene, later tygon tubing, with samples separated and segmented by air bubbles to "scrub" the tubing and prevent mixing by overlapping of successive samples en route; (3) mixing coils in which columns of liquid representing discrete samples plus reagents were propelled while being continuously inverted and mixed, as the solutions moved up and down through the coils [28].

The various volumes of sample and reagents were introduced individually in parallel into the system at appropriate points in the continuous stream by the forward propulsive pressure of a constant rate peristaltic pump. Instead of known measured volumes of reactants, the continuous flow system utilized ratios or proportionate volumes as the basis of sample-to-sample comparison. Flow rates of sample and reagents were determined by the diameter of the tubes in the pump. Heating or incubation—if necessary—was performed during continuous passage through a lengthy glass coil inside a heated bath that provided sufficient time for the reaction to take place. Photometric measurement of the exiting solution was performed by continuous monitoring of a flow-through cuvette at a given wavelength. The continuous changing response of the photocell or phototube was recorded on a moving strip-chart recorder as a series of peaks and valleys at the rate of 20 to 40 per hour. Concurrent analysis of standards or reference samples of known concentration provided data for a calibration curve from which the peaks of the unknown samples could be quantitated.

The first three serum components (and chromogenic reagents) available on the AutoAnalyzer were glucose (alkaline ferricyanide), urea nitrogen (diacetyl monoxime), and calcium (murexide). Specific configurations of tubing for additional serum constituents were soon added and were combined in multiple channel modes. In an eight-channel unit, all of the data for each sample specimen is recorded in rapid sequence on the stripchart. At a rate of 20 per hour, each sample yields a steady-state plateau instead of the familiar rounded peaks. By means of time-delay coils in the flow system, the steady-states are made to occur in sequence in each of the different analytical channels. In a single colorimeter with eight different flow cells, each can be moved successively into the light path [29]. Results in terms of concentration appear in graphic or digital format.

Later design configurations were the Sequential Multiple Analyzer (SMA) 6/60, 12/30, and 12/60 (6 or 12 analytes at 30 or 60 samples per hour). The SMAC (Sequential Multiple Analyzer with Computer) (Fig. 19.5), capable of simultaneous analysis of twenty analytes per sample at the rate of 150 samples per hour and costing $250,000, was introduced in 1972 and first shipped in 1974. The 1,000th unit was sold in 1983. The Chem-1, introduced in the late 1980s, is a selective, random access analyzer that uses continuous flow principles; and is still supported and serviced by the company.

The AutoAnalyzer also found use in quality control in numerous industries, e.g., pharmaceutical, agronomy, metallurgy, cement manufacture, textile dyeing and bleaching, and in monitoring systems for air and water pollution. The Technicon Corporation experienced a series of acquisitions and is now part of the Diagnostics Division of the Bayer Corporation, Tarrytown, New York.

FIG. 19.5. Sequential Multiple Analyzer with Computer (SMAC). (Courtesy of Bayer Corporation, Diagnostics Division, Tarrytown, New York.)

FACTORY TEACHING FACILITY

Due to the uniqueness of the AutoAnalyzer concept, and because it was felt that medical technologists, biochemists, pathologists, etc., would reject this concept unless they were throughly instructed in its proper use, it was decided to set up a teaching facility. The acceptance of both the system and the in-house teaching by the manufacturer made world acceptance of this new analytical system a complete success. As other manufacturers brought out other types of automated and semiautomated devices, they too adapted the technique of in-house teaching of the proper use of their machines. This general trend in the industry to accept the responsibility of properly instructing laboratory personnel was undertaken because it was recognized that the complexities of today's instrumentation cannot be grasped by simply reading an instruction manual. Most instruction courses are of one week in duration, although the course may be longer, depending upon the type of instrumentation purchased. The teaching program uses the latest in audiovisual materials, such as programmed instruction, and emphasizes actual laboratory participation.

Each student is normally required to work on his own system and thus obtains a true understanding of the system instead of merely learning how to push certain buttons in order to make a system work [30]. Physics and electronics are as much a part of medical technology today as were chemistry and biology 40 or 50 years ago. Bioengineering or biomedical engineering is also an important part of our medical profession today.

As we moved away from work simplification towards more complex mechanization, we freely used the word automation for describing processes which are strictly examples of mechanization. In practice, one must distinguish that which replaces man's physical efforts (mechanization) and that which replaces his intellectual control and decision-making functions (automation) [31]. Automation, as understood by industry, involves the use of machines with an element of feed-back control providing capability to self-monitor and adjust or correct. While considerable mechanization had been introduced into clinical chemistry, it was not until the late 1970s, after the development of the microprocessor—a miniature electronic device that can interpret and execute program instructions—that chemical analyzers could become self-controlling, i.e., truly automated. Experts in industrial automation pointed out that the stage of development represented by the first AutoAnalyzer was only analogous to that of industry at the time of the Industrial Revolution [32]. Although automation makes it possible to process an increased workload rapidly and reproducibly, it does not necessarily improve accuracy and cannot compensate for inherent defects in the analytical methodology.

Even before continuous flow analysis was advancing toward the limit of its exploitable potential, it was being challenged on a wide front by newly developed systems of discrete sample analysis. These have the advantage of speed and the possibility of rapid changeover, enabling the laboratory to complete small batches of different assays in rapid succession on the same instrument. With samples isolated in separate containers, the mechanical sequence of steps can closely duplicate the traditional manual procedures. Numerous discrete analyzers of novel designs began to appear by the mid-1960s in the United States and England. Companies not generally identified with hospital or healthcare products entered the marketplace with innovative automated systems and offered the laboratory a choice of instruments and applications.

ANALYZERS: DISCRETE SAMPLE AND RANDOM ACCESS

The Robot Chemist was designed to automate wet-chemical analytical procedures by mechanically duplicating the sequential steps common to most clinical chemistry procedures and performed manually by an analyst [33]. It was the first commercially available discrete analyzer, and probably the first to produce results with a digital print-out. Specimens were presented sequentially to a sampling module, aliquoted, diluted, and transferred to separate reaction tubes in a turntable, where they can be incubated at constant temperature. Further additions of reagents were made automatically and the reaction solution mixed periodically. After a selected time interval, the mixture was aspirated into a flow-through cuvette and measured spectrophotometrically. If no serum blank was required, 120 analyses per hour can be performed. Standard chemical methods similar to those in manual techniques were used [34]. There was even a separate module designed for protein precipitation, filtration, and transfer to the reaction turntable for those procedures that required it. It took about five minutes to change test assays.

The basic idea of the Robot Chemist was conceived by Hans Baruch, president of Research Specialties Co., Richmond, California, in 1957. The instrument, first marketed in 1959, was a large desk-size structure mounted on the floor. The company soon recognized that the unit was too bulky and had to be made more compact and flexible. The instrument was redesigned into small modules and released as a new bench-top unit in 1963 (Fig. 19.6). It did not meet with much success or general acceptance largely due to the complexity of its electro-mechanical and electronic components.

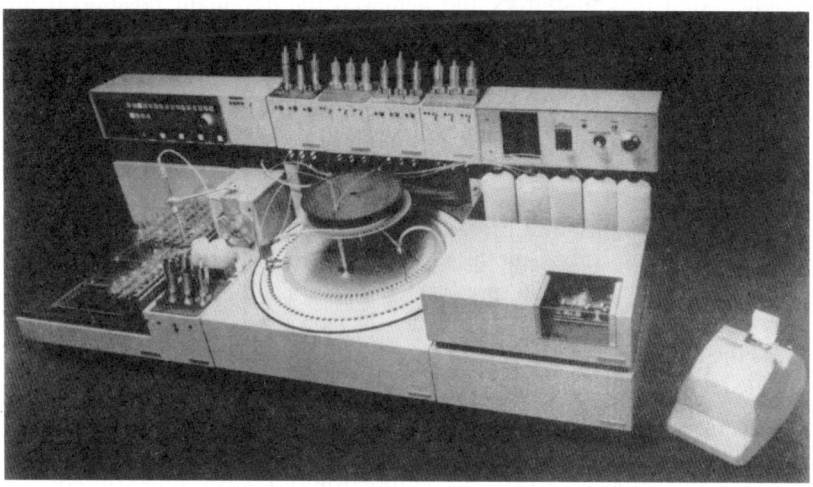

FIG. 19.6. Robot Chemist. (*Annals of the New York Academy of Sciences*, 1968–69, 153: 655–659.)

Furthermore, the company did not have adequate resources for wide distribution or for servicing problems with the instrument's numerous moving parts, nor did it provide training sessions for customers.

The company was bought in late 1964 by the Warner-Lambert Pharmaceutical Company (Morris Plains, New Jersey) and renamed Warner-Chilcott Laboratories Instrument Division (WCLID). This division included General Diagnostics. However, the new company was inexperienced in the manufacture of hardware, and the instrument did not fit into the advertising style used for the many commercial products geared to the general public. Consequently, in early 1967, Warner-Lambert bought the Analytical Instrument Division of the American Optical Corporation for their knowledge and experience in development and manufacture of scientific instruments. The Robot Chemist was taken off the market and transferred to American Optical for further development and manufacture. The intention was to broaden the machine's range of capabilities for use in the analytical and quality control laboratory. Sales were resumed in the late 1960s, but production was discontinued in 1969, hastened by competition from Technicon's SMA 12/30 and 12/60. American Optical was sold in 1982 and the General Diagnostics Division in 1985. Only about fifty units of the standard desk-size and a smaller bench-top model, designed separately and marketed later, were sold. Although supplanted by the continuous flow system of the AutoAnalyzer, it was the discrete sample concept of the Robot Chemist that eventually was adopted by instrument designers and manufacturers and achieved the dominant position in the clinical laboratory.

Another discrete sample system that made its debut in the late 1960s, the Mark 10 Multiphasic Analyzer (Fig. 19.7), also followed manual procedures. Produced by Hycel, Inc. (Houston, Texas), it experienced some success. The sample table held sixty specimens for each of which any combination of ten tests may be selected from a programming panel for simultaneous analysis, at a rate of up to forty samples per hour. Reactions take place in test tubes mounted on a conveyer system. Reagents are dispensed automatically at preselected positions for each test. The test tubes travel through heating baths as they progress towards the position where the contents are aspirated into cuvettes. All cuvettes are mounted in a single readout unit, although each cuvette is part of a discrete colorimeter with its own lamp, filter, and photocell. The cuvettes are scanned sequentially and concentrations are read from preprinted paper on a strip-chart recorder. Next, the test tubes are emptied into a waste unit, rinsed with distilled water and dried by hot air. The chemical methods employed are

ANALYZERS 493

FIG. 19.7. Hycel Mark X.

similar to those in the manufacturer's manual chemistry kits, so that back-up procedures are available. The system was limited to the tests selected by the manufacturer [35]. The Mark 10 and a subsequent model, the Mark 17, were still in use as late as 1976. Out of 2,234 laboratories reporting to a CAP survey on total protein analysis, 128 were using these Hycel instruments. It is interesting to note that 16% of the laboratories reported their protein results by manual refractometric analysis [36].

The DuPont Automatic Clinical Analyzer (aca), introduced in 1968 with a capability of eight tests, expanded its versatility with newer models and offered 85 analytes in 1995. The unique feature of this instrument is that all the reagents for each test are contained in separate compartments in a special plastic pack which also serves as the reaction chamber and cuvette for the photometric analysis. A plastic "header" at the top of the pack indicates in binary code, instructions for the test to be performed, e.g., amount of sample, diluent, wavelength of measurement, etc. Some packs include disposable chromatographic columns of resin or gel to isolate specific constituents. The aca can perform the tests in any order desired with virtually instantaneous changeover. Thus, it can handle requests in the random order received. Each specimen is followed by a choice of individual packs for each of the tests required. The results are calculated by the computer and printed out on a slip of paper which contains a facsimile photographic record of the original patient identification [37]. In 1996, the DuPont division that manufactured the aca—now with more than 90 chemistries—was purchased by Dade International, Inc. (Deerfield, Illinois).

The DSA-560, Beckman's Discrete Sample Analyzer, was an automated multi-channel modular system that prepared protein-free filtrates by precipitation and filtration instead of by dialysis. Incorporation of a filtration step allowed deproteinization and assay of whole blood. The Beckman and DuPont instruments were both introduced in 1968 at the annual meeting of the AACC [38]. Whereas the Beckman DSA-560 was not a successful entry into the field of automation, the aca and its subsequent advanced systems were very successful. The Beckman DSA's best design components were later modified for future products, especially the ASTRA. This clinical chemistry system, which automated eight of the most common chemistry blood tests, had produced one billion dollars in revenues by its fifteenth anniversary in 1994*.

Among the most expensive group of discrete sample automated analyzers entering the 1970s were the fast multi-channel machines, e.g., the Swedish AGA AutoChemist and the British Vickers MC-300. The AutoChemist was the largest clinical chemical analyzer then made. It could

---

*The Beckman Heritage Center, Fullerton, California, brochure CC94-5004.

produce twenty-four tests per specimen at the rate of 135 specimens per hour. With a price of about half a million dollars, it was suitable only for large population screening centers or commercial laboratories with very large workloads, e.g., Metpath (now, Quest Diagnostics, Inc., Teterboro, N. J.). Metpath had from six to eight units until the early 1980s when they were replaced by LKB's Prisma, a similar high-volume instrument, capable of random access, selling at approximately $250,000.

The first installation of the Vickers MC-300 in the United States was at the University of Pennsylvania Hospital. Others followed at the Mayo Clinic, the Massachusetts General Hospital, the Erie County Hospital at Buffalo, New York, Harbor General Hospital, Torrance, California, Billings Hospital in Chicago, and others. Costing $240,000, the Vickers machine processed 300 specimens per hour, with up to twenty simultaneous assays per sample. It was the fastest multi-channel analyzer developed up to that time [39]. The modular design—each test channel had its own photometer—permitted analysis of as few as one or two selected assays at the same high speed. Difficulties were experienced at first because Vickers did not appreciate the tolerances required for the chemical methodology and for reliable instrument performance. The company did not follow-up with newer designs, but provided continuous software and hardware changes and refinement of chemistry methods that led to improved performance. The unit at the Mayo Clinic was in service for about ten years, but the system required close monitoring. Less than 100 machines were sold, with about half being installed in England and some in Sweden. Numerous individual modules—capable of two tests simultaneously or one test with a blank determination—were sold. The instrument was discontinued in the late 1970s, displaced by competition from the SMAC.

Another interesting British design, the Mecolab by Joyce, Loebl & Co., Ltd., was field tested in England in 1965 and shown in the United States two years later. This is a discrete discontinuous analyzer that requires manual intervention to transfer racks of fifteen tubes from one process to another. Each preparative step, including the readout, is automatic, but the transfer from one stage to another is deliberately manual. This permits different areas of the laboratory to maintain only the modules needed for the desired group of analyses and to share the readout system, the unit's most expensive module, with other users [40].

A radically new analytical concept for clinical chemistry that utilized centrifugal force to transfer liquids was developed in the 1960s at the Oak Ridge National Laboratory in Tennessee. It seemed logical to Norman Anderson and his colleagues that they could achieve maximum benefit from the computer's capability, if they could speed up the chemical reactions to match the time scale of the computer [41]. Toward this goal, they developed an instrument in which reagent addition and mixing for a specific analyte in an entire batch of samples with standards, take place almost simultaneously. If colorimetric readings also could be made simultaneously, the reaction need only proceed for a comparatively short period. They achieved these conditions by completing the entire procedure, including colorimetry, in fifteen individual cuvettes in the rotor of a centrifuge spinning at 2000 rpm past a stationary light beam. The electronic signals from the colorimeter's photomultiplier are continuously displayed on a cathode ray tube and recorded photographically or fed to a computer to calculate concentrations.

Using the principles and most details of the new system, the CentrifiChem from Union Carbide Corporation (Tarrytown, N.Y.), GeMSAEC from Electro-Nucleonics, Inc. (Fairfield, New Jersey), and Rotochem from American Instrument Co. (Silver Spring, Maryland) entered production in the late 1960s. This innovative technology yielded many methods for the assay of enzymes, proteins, inorganic ions, metabolities, and drugs. These instruments perform rapid and uniform mixing, simultaneous starting, continuous monitoring of chemical reactions in specimens and standards run in parallel, accurate and precise temperature control, and accurate and continuous absorbance readings at precise intervals during the reaction. These capabilities make the centrifugal analyzers ideally suited for enzyme assays by measuring the kinetics of the reaction.

In 1981, after a decade of development, Eastman Kodak (Rochester, New York) introduced the first in its line of Ektachem random access analyzers for clinical chemistry analysis. Initially, protocols for twelve analytes were available; at present there are more than forty. The innovative technology, based on reflectance photometry and layered-coating of slides, evolved from Kodak's search for new products using film coating machines. This dry-phase approach to analysis offers many practical advantages over solution chemistries, e.g., waste disposal, minimal reagent storage space, no reagent preparation, no sample dilution [42]. In 1995 Kodak's Ektachem systems were bought by Johnson & Johnson (New Brunswick, New Jersey). They are now developed and sold as the Vitros chemistry systems.

Meanwhile, the focus in the clinical chemistry laboratory was shifting to improving the processing of the high volume and rapid flow of patient and quality control data resulting from the increased analytical capability of

ANALYZERS 497

the laboratory. The computer responded by providing increasingly varied services as it was fed with patient identification, test request information, and on-line and off-line signals from the laboratory instruments—to produce work-sheets, patient reports, quality control data, and warnings of malfunctions.

Marketing terms such as multiple test menu, quick turnaround, high throughput, and random access, were quickly incorporated into the language of the laboratory. Method development shifted to the instrument manufacturers. Whereas new methods could be adapted to the single and double-channel AutoAnalyzer, and existing methods modified and reagents prepared in the workplace, the new discrete sample analyzers restricted the operator to a total concept. Machine was wedded to method and prepackaged reagents in a variety of single-test reagent containers, e.g., cartridges, capsules, packs, and slides, in a closed system of analytical chemistry. This was an evolutionary path that altered the basic philosophy of clinical chemistry analysis and threatened the analytical balance—the foundation of quantitative chemical analysis—with extinction in the clinical chemistry laboratory. With increasing mechanization and computer-controlled automation, the scientific role of clinical laboratories faced being reduced almost to the point of elimination (see Chapter 20).

Losing control of choice of method was not entirely a disadvantage to the chemistry laboratory. Standardized procedures allow for inter-laboratory comparison of results and the monitoring of within-day and day-to-day intra-laboratory reproducibility. However, automated instruments require careful adherence to the manufacturer's operating instructions to prevent introduction of machine errors. The need to have laboratory personnel with mechanical skills soon gave way to service contracts, reagent and instrument rental/lease agreements, over-the-phone troubleshooting, and 24-hour on-call service for repair or loan of a replacement instrument. Contracts for reagent purchase led to eventual ownership of the instrument, which was soon succeeded by a more versatile instrument and a new reagent contract.

Primary control of the analytical process has shifted from the clinical chemist in a hospital setting to multi-discipline teams of scientists developing closed system analyzers and working in relative obscurity behind a corporate logo of industry and subject to the restraints imposed by the need to protect proprietary information.

The aca and Ektachem heralded the trend in clinical chemistry to replace continuous flow and discrete sample batch analyzers with fully automated random access systems. As this eventually became the industry standard during the late 1980s and early 1990s, new technological features were developed to further improve the benefits of evolving technology. New capabilities rapidly coming into play are primary closed tube sampling, positive sample identification, multi-tasking software, STAT interrupt, and automatic repeat and dilution. Balanced against these features is the laboratory's insistence on ease of operation, reduced maintenance requirements, and decreased operating costs. Added to this are the numerous manufacturers who continually introduce new instrument designs to satisfy the expanding needs and preferences of clinical chemistry laboratories. Some of these companies are: Abbott, Baxter, Beckman, Boehringer Mannheim, Corning, Coulter, Instrumentation Laboratory, Bayer (Technicon), Behring, Olympus, Johnson & Johnson (Kodak), Roche, and Becton Dickinson.

Perhaps industry's greatest contribution to the clinical laboratory has been the conception and manufacture of instruments that perform tests that could not be done without mechanization. One of the more significant benefits of industry's involvement in clinical chemistry is the increasing uniformity of test results, as automated systems have reduced the variety of analytical methods.

Another avenue for narrowing the range of compared data is the National Reference Laboratory System. This is a program of reference materials and methods that makes it possible to trace procedures to a single reference method. This has improved the level of comparison of results between laboratories and within a laboratory over time. Industry's collaboration and cooperation in referencing their immunoassay products to common calibrators helps to ensure that results are comparable even when different methods are used. Because the development of new analytical systems is expensive, new systems are not likely to be produced in hospitals or university settings. Collaboration with industry becomes the practical means for applying concepts developed in medical centers [43].

Newcomers to clinical chemistry will not experience the techniques, the methods, and the instruments, that were the mainstay of the laboratory before mechanization began to make inroads in the late 1950s. Nostalgia being what it is, those who experienced the laboratory of the pre-automation era tend to remember first the romance and excitement of learning new methods, developing new skills, and using different apparatus, even as they recall how primitive and cumbersome, labor-intensive and tedious, the old technologies often were. It was hands-on from start to finish—no walk-away systems—you could see what was going on. The modern clinical laboratory may be a duller, less versatile, and more technical place of work, but it gets far more done at far less cost than anything that came before it.

NOTES AND REFERENCES

1. RALPH HOLCOMBE MÜLLER, "Instrumental Methods of Chemical Analysis," *Industrial and Engineering Chemistry. Analytical Edition,* 1941, 13: 667–754, pp. 713, 702.
2. These and other procedures and the mechanisms of their reactions are reviewed by JOHN G. REINHOLD, "Flocculation Tests and Their Application to the Study of Liver Disease," *Advances in Clinical Chemistry,* 1960, 3: 83–156; H. E. SCHULTZE and J. F. HEREMANS, "Analytical Methods in Protein Chemistry," in *Molecular Biology of Human Proteins. Nature and Metabolism of Extracellular Proteins,* vol. 1, sec. 1 (Amsterdam: Elsevier, 1966), pp. 145–150.
3. WILLIAM P. BELK and F. WILLIAM SUNDERMAN, "A Survey of the Accuracy of Chemical Analyses in Clinical Laboratories," *American Journal of Clinical Pathology,* 1947, 17: 853–861. See also F. W. SUNDERMAN, SR., "The History of Proficiency Testing/Quality Control," *Clinical Chemistry,* 1992, 38: 1205–1209.
4. ROBERT E. KOHLER, *From Medical Chemistry to Biochemistry. The Making of a Biomedical Discipline* (New York: Cambridge University Press, 1982), pp. 215–216.
5. *Ibidem,* pp. 216–219.
6. *Ibidem,* pp. 219, 225, 248–249.
7. *Ibidem,* pp. 243, 248.
8. OLIVER H. GAEBLER, "Comments on the Origins and Development of Clinical Chemistry," *Clinical Chemistry,* 1958, 4: 331–338.
9. KOHLER (1982), pp. 250–251.
10. BERNARD KLEIN, "Organization and History of the New York Metropolitan Section, AACC: Some Recollections," *Clinical Chemistry,* 1987, 33: 1486–1489; LOUIS ROSENFELD, "Clinical Chemistry in New York at the Founding of the AACC: Recollection and Remembrance," *Clinical Chemistry,* 1991, 37: 2146–2149; JOSEPH BENOTTI, "Some Reminiscences of a Charter Member," *Clinical Chemistry,* 1973, 19: 1085–1086; MIRIAM REINER, "The Growth of the AACC: 25 Years," *Clinical Chemistry,* 1973, 19: 1409–1411; "Historical Note: Naming of AACC," *Clinical Chemistry,* 1973, 19: 1094. Minutes of the AACC Organizing Committee Meetings (Dec. 15, 1948; Jan. 11, Feb. 1, 1949). Also see minutes of the meeting of the AACC Executive Committee, September 3, 1958. For his experiences with the politics of licensure and other biographical events, see JOHN G. REINHOLD, "Adventures of a Clinical Chemist," *Clinical Chemistry,* 1982, 28: 2314–2323, pp. 2317–2318, 2322–2323.
11. "Gift of 'The Clinical Chemist'—and a Historical Note," *Clinical Chemistry,* 1973, 19: 348–349; MAX M. FRIEDMAN, "*Clinical Chemistry*: The History of a Journal," *Clinical Chemistry,* 1973, 19: 1315–1317; J. STANTON KING, "Clinical Chemistry: A Fragmentary History (1969–1977)," *Clinical Chemistry,* 1994, 40: 2106–2110; J. I. ROUTH, "Training of Clinical Chemists in the United States: A Brief History," *Clinical Chemistry,* 1974, 20: 1251–1253. See also F. WILLIAM SUNDERMAN, SR., "The Foundation of Clinical Chemistry in the United States," *Clinical Chemistry,* 1994, 40: 835–842.
12. These sections were derived from the following: *IFCC Handbook 1994–1996*; Peter Broughton and John Lines, *The Association of Clinical Biochemists. The First Forty Years,* Roy Sherwood, ed. (London: ACB Venture Publications, 1996), pp. 9–20. For a review

of the organizational structure and activities of the IFCC, see DONALD S. YOUNG, "International Federation of Clinical Chemistry: Present and Future," *Clinical Chemistry*, 1988, 34: 202–207. See also C. P. STEWART, "Professor E. J. King Eulogy," *Clinical Chemistry*, 1965, 11: 194–197.

13. NORMAN D. LEE, "A History of Bio-Science Laboratories," *Clinical Chemistry*, 1994, 40: 149–157.
14. S. B. BARKER, M. J. HUMPHREY, and M. H. SOLEY, "The Clinical Determination of Protein-Bound Iodine," *Journal of Clinical Investigation*, 1951, 30: 55–62.
15. OTTO A. BESSEY, OLIVER H. LOWRY, and MARY JANE BROCK, "A Method for the Rapid Determination of Alkaline Phosphatase With Five Cubic Millimeters of Serum," *Journal of Biological Chemistry*, 1946, 164: 321–329.
16. LOUIS BERGER, "Sigma Diagnostics: Pioneer of Kits for Clinical Chemistry," *Clinical Chemistry*, 1993, 39: 902–903; OLIVER H. LOWRY, "How to Succeed in Research Without Being a Genius," *Annual Review of Biochemistry*, 1990, 59: 1–27, p. 9.
17. ARTHUR KARMEN, "A Note on the Spectrophotometric Assay of Glutamic-Oxalacetic Transaminase in Human Blood Serum," *Journal of Clinical Investigation*, 1955, 34: 131–133.
18. STANLEY REITMAN and SAM FRANKEL, "A Colorimetric Method for the Determination of Serum Glutamic Oxalacetic and Glutamic Pyruvic Transaminases," *American Journal of Clinical Pathology*, 1957, 28: 56–63.
19. ROSALYN S. YALOW, "Radioimmunoassay: A Probe for the Fine Structure of Biologic Systems," *Science*, 1978, 200: 1236–1245. For a brief biography of Yalow, see *Science*, 1977, 198: 594.
20. SOLOMON A. BERSON, ROSALYN S. YALOW, A. BAUMAN, M. A. ROTHSCHILD, and K. NEWERLY, "Insulin-I^{131} Metabolism in Human Subjects: Demonstration of Insulin Binding Globulin in the Circulation of Insulin Treated Subjects," *Journal of Clinical Investigation*, 1956, 35: 170–190.
21. ROSALYN S. YALOW and SOLOMON A. BERSON, "Immunoassay of Endogenous Plasma Insulin in Man," *Journal of Clinical Investigation*, 1960, 39: 1157–1175.
22. DAVID H. SOLOMON, JOSEPH BENOTTI, LESLIE J. DEGROOT, *et al.*, "A Nomenclature for Tests of Thyroid Hormones in Serum: Report of a Committee of the American Thyroid Association," *Journal of Clinical Endocrinology and Metabolism*, 1972, 34: 884–890.
23. F. V. LAZAREV and M. K. TRIFONOVA, "The Role of Apparatus in Cognition and Its Classification," in *Contributions to a Philosophy of Technology. Studies in the Structure of Thinking in the Technological Sciences*, Friedrich Rapp, ed. (Dordrecht, Holland: D. Reidel Publishing Company, 1974), pp. 197–209.
24. L. TONDL, "On the Concepts of 'Technology' and 'Technological Sciences,'" in *Ibidem*, pp. 1–18; see also J. BÜTTNER, "Technical Evolution in the Clinical Laboratory," *Pure & Applied Chemistry*, 1982, 54: 2011–2016, pp. 2012–2013.
25. BÜTTNER (1982), p. 2013.
26. *Ibidem*, pp. 2011–2012.
27. "Clinical Chemistry's Man on Horseback. Leonard T. Skeggs, Jr.," *Chemical & Engineering News*, 1970, 48: 54–58; see also LEONARD T. SKEGGS, "New Dimensions in Medical Diagnoses," *Analytical Chemistry*, 1966, 38: 31A–44A. For additional biographical data, other scientific work, and illustrations of early models of the AutoAnalyzer during its developmental stage, see LENA A. LEWIS, "Leonard Tucker Skeggs—A Multifaceted Diamond," *Clinical Chemistry*, 1981, 27: 1465–1468.

28. LEONARD T. SKEGGS, JR., "An Automatic Method for Colorimetric Analysis," *Clinical Chemistry*, 1956, 2: 241 (Abstract); "An Automatic Method for Colorimetric Analysis," *American Journal of Clinical Pathology*, 1957, 28: 311–322. See also LEONARD T. SKEGGS, JR., "Principles of Automatic Chemical Analysis," in *Standard Methods of Clinical Chemistry*, vol. 5, Samuel Meites, ed. (New York: Academic Press, 1965), pp. 31–42.
29. LEONARD T. SKEGGS, Jr. and HARRY HOCHSTRASSER, "Multiple Automatic Sequential Analysis," *Clinical Chemistry*, 1964, 10: 918–936, p. 919; see also LEONARD T. SKEGGS, JR., "Multiple Automatic Sequential Analysis. I. Theoretical Considerations," *Clinical Chemistry*, 1963, 9: 442–443 (Abstract).
30. ANDRÉS FERRARI and RALPH OVERMAN, "Medical Technology," *Annals of the New York Academy of Sciences*, 1969, 166: 1027–1030.
31. TONDL (1974), p. 11. The term "automation" appears to have been coined independently in 1947 by D. S. Harder of the Ford Motor Company and by J. Diebold.
32. F. L. MITCHELL, "Present and Future Trends of Automation in Clinical Chemistry" (pp. 180–190), in *Methods in Clinical Chemistry*, vol. 1, 7th International Congress of Clinical Chemistry, Geneva/Evian, September 8–13, 1969 (Baltimore: University Park Press, 1970), p. 183. Numerous automated systems, mostly of British design and manufacture are briefly described in this article.
33. "Instruments for Clinical Chemistry Labs. The Big Swing to Automation," *Chemical & Engineering News*, December 9, 1963, part one, pp. 112–128; W. C. CRAWFORD, JR., "The Robot Chemist: A New Approach to the Automation of Wet-Chemistry Analysis," *Annals of the New York Academy of Sciences*, 1968–69, 153: 655–659. For a detailed operating description of the Robot Chemist, see also WILMA L. WHITE, MARILYN M. ERICKSON, and SUE C. STEVENS, *Practical Automation for the Clinical Laboratory* (St. Louis: The C. V. Mosby Company, 1968), pp. 249–290. For a description of an early model of the Robot Chemist, and other automated systems in various stages of development, see WALTON H. MARSH, *Automation in Clinical Chemistry* (Springfield, Illinois: Charles C. Thomas, Publishers, 1963), pp. 19–34.
34. MATHEWS B. FISH and PAUL R. JENSEN, "Experience With the Robot Chemist," (pp. 65–72), in *Automation and Data Processing in the Clinical Laboratory*, Geoffrey M. Brittin and Mario Werner, eds. (Springfield, Illinois: Charles C. Thomas, Publisher, 1970); see also HOWARD J. NEWMAN, "A New Approach to the Automation of Wet-Chemistry Analysis," *Clinical Chemistry*, 1966, 12: 554 (Abstract).
35. DONALD S. YOUNG and JAMES D. GALLUP, "Hycel Mark X: Principles and Initial Operating Experience," (pp. 99–109), Brittin and Werner, eds., 1970.
36. ROBERT T. BURKHARDT and JOHN G. BATSAKIS, "An Interlaboratory Comparison of Serum Total Protein Analyses," *American Journal of Clinical Pathology*, 1978, 70 (Supplement): 508–510.
37. MERLE A. EVENSON, "Preliminary Field Evaluation of the duPont Automatic Clinical Analyzer," (pp. 89–98), Brittin and Werner, eds., 1970; R. G. NADEAU, "Development of the duPont Automatic Clinical Analyzer (ACA) System: Part I.," *Clinical Chemistry*, 1968, 14: 778–779 (Abstract).
38. J. E. ROCHTE and R. C. MEYER, "Design of the New Beckman Automatic Chemical Analyzer," *Clinical Chemistry*, 1968, 14: 780 (Abstract); NADEAU (1968), pp. 778–779 (Abstract); see also RODNEY E. WILLARD, "Evaluation of the Beckman DSA-560 Analyzer," (pp. 73–84), Brittin and Werner, eds., 1970.
39. MITCHELL (1970), pp. 186–188. For a description of the Vickers 300 and other automated systems, see DONALD S. YOUNG, "Automation" (pp. 187–211), in *Fundamentals of*

Clinical Chemistry, Norbert W. Tietz, ed., 2nd ed. (Philadelphia: W. B. Saunders Company, 1976), pp. 199–200.
40. WHITE, ERICKSON, and STEVENS (1968), pp. 241–248.
41. NORMAN G. ANDERSON, "Computer Interfaced Fast Analyzers," *Science*, 1969, 166: 317–324; DONALD W. HATCHER and NORMAN G. ANDERSON, "GeMSAEC: A New Analytic Tool for Clinical Chemistry Total Serum Protein With the Biuret Reaction." *American Journal of Clinical Pathology*, 1969, 52: 645–651. The acronym GeMSAEC is derived from the major sources of support which are the National Institute of General Medical Sciences and the United States Atomic Energy Commission.
42. TERRY L. SHIREY, "Development of a Layered-Coating Technology for Clinical Chemistry," *Clinical Biochemistry*, 1983, 16: 147–155. See also HENRY CURME and ROYDEN N. RAND, "Early history of Eastman Kodak Ektachem slides and instrumentation," *Clinical Chemistry*, 1997, 43: 1647–1652.
43. DONALD S. YOUNG, "Industry's Contribution to the Development of Clinical Chemistry," *Journal of the International Federation of Clinical Chemistry*, 1994, 6: 2–3.

CHAPTER XX

PITFALLS OF PUBLICATION

At the turn of the century, many practitioners had eagerly and incautiously embraced the new hematological, bacteriological, and chemical tests that they saw as the promise of exactness to replace the old empirical uncertainty. The first two decades of the century were a period of rapid innovation, especially in clinical biochemistry. The emergence of routine laboratory diagnosis became a controversial issue during this period.

In 1902 the assistant resident physician in charge of the clinical laboratory at Johns Hopkins Hospital wrote: "Clinical laboratories are growing in favor and influence; publishers have produced a superabundance of text-books which purport to 'make clinical chemistry easy'; medical journals accept at sight articles on almost any chemical subject, some of scientific value, some of practical value, some of no value." Every practitioner gets a free course in chemistry through the mails and is swamped with pamphlets detailing some recent chemical achievement. "The whole practical medical world, in fact, is studying chemistry." Publication in clinical chemistry, he continued, was becoming the goal of the "scientific practitioner" [1]. The preoccupation with publication was seen as a hindrance to progress.

"It is the excessive publication of half-knowledge, of doubtful fact, and of loose inquiry. The propensity for authorship is an old disease, but it has assumed a development commensurate with the prodigious development of everything else in the age in which we live. It is harmless in journalism, less so in literature, but positively dangerous in science; for it fills this with immature or false observation, and takes the time of others to remove the obstructions placed in the stream of knowledge ... It is far from being a disease only of the untried or ignorant; famed workers, too, may succumb to it, listening, perhaps, to the entreaties of friendly and sleepless editors" [2].

DANGERS OF LABORATORY DIAGNOSIS

The Mayo brothers, William James (1861–1939) and Charles Horace (1865–1939), were concerned about the general tendency by physicians to abandon the old methods and to rely too much upon laboratory analysis and too little on patient history and clinical examination. They warned against the danger of "laboratory diagnosis," and insisted that the laboratory findings only be a supplement to the clinician's personal study of the patient [3].

As early as 1908, a warning was sounded against abandoning what is good in the old because of the helpfulness of the new. That such a fear should be expressed revealed the strong hold exerted by laboratory methods on "the minds of developing clinicians" [4]. Physicians debated the relative merits of laboratory tests and bedside diagnosis. It was a symptom of the larger conflict between two generations of clinicians eyeing each other suspiciously from opposite sides of the clinical reform movement. In his 1910 report on medical education, Abraham Flexner chided both sides when he asked, who has priority—the laboratory pathologist armed with his microscope, or the bedside clinician with his stethoscope [5].

As late as 1919, the director of laboratories at New York City's Bellevue Hospital had strong words of reproach for those who

> "overemphasize the importance of laboratory procedures. This tendency, which is becoming more pronounced each year, appeals strongly to those faddists among whom any test which suggests an easy approach to the solution of any problem, or which promises a division or evasion of responsibility, is assured of a kindly reception. Whether its credentials are written in the language of science or in that of pseudoscience appears to make little difference. It panders to laziness, which is man's most easily accessible weakness" [6].

The writer was also sharply critical of entrusting nonmedical faculty with the teaching of medical subjects, e.g., physiology, physiological chemistry, pharmacology, histology, and especially anatomy.

Much of the early criticism probably had to do with the reluctance of the senior generation of medical men to unlearn the older skills of diagnosis that they were taught and to adopt the new diagnostic capabilities of laboratory technology. However, like most new ideas, the rank and file of the medical profession were also slow to change. Neither the stethoscope, electrocardiogram, sphygmomanometer, or clinical thermometer, were received with enthusiasm when they first came on the scene [7].

THE LABORATORY: POSSIBILITIES AND AN IMAGE PROBLEM

Despite these expressed reservations, it was during the period from 1910 to 1920 that hospital administrators and the medical profession came to realize the possibilities of a fully equipped and properly functioning hospital laboratory as an important and distinct administrative unit or service. This usually consisted of five or six divisions: anatomical pathology (autopsy), hematology, bacteriology, serology, immunology, and biochemistry, staffed by trained, often salaried professionals. In addition to providing routine laboratory tests to aid the physician in his diagnosis, prognosis, and treatment of the disease, the professional staffs were also expected to cooperate with the clinical staff, to instruct internes, nurses, technicians, and physicians of the community, in laboratory methods and their clinical significance, and to do research [8].

Many hospitals also set up small laboratories adjacent to individual wards, where attending physicians could perform routine tests on the blood, urine or feces of their patients to verify a diagnosis or follow drug therapy. Such ward laboratories had been established at the Johns Hopkins Hospital in 1895–96 with the encouragement of William Osler, and at Bellevue Hospital in 1899 [9]. They were valued by the clinicians of Osler's generation as a vital part of the physician's role in laboratory testing. More complex procedures were sent to the central pathological laboratory of the hospital [10]. Eventually, with increasing variety and complexity of new tests—which quickly became routine tests—research in the central laboratory fell victim to the rapid increase in volume of routine diagnostic tests. Soon there was no time for anything else, and these central laboratories gradually became merely service units. Time for research had been compromised by the very service role that had made the laboratory possible [11].

The laboratory was not always accorded the respect its importance deserves. The following comments made in 1918 are reminiscent of the long held popular misconception of "the laboratory" [12].

"Usually after the hospital has been completely erected, certain space, unsuitable for any other purpose, is assigned to the laboratory. It is thus that we find this department frequently located in basements, in out of the way nooks and corners, in outhouses or roof structures built as an afterthought. The laboratory is gloomy, the ventilation unsuitable, and the general conditions such as to make the scientist working there cognizant of a spirit of depression in his assistants and help."

"It must be definitely understood that the twentieth century hospital must have a laboratory—not a makeshift, two by four, 'urine room,' not a gloomy, unventilated, poorly cleaned cranny, but well-constructed, properly lighted, scientifically equipped quarters. A hospital has been defined as a hotel with an operating room and laboratory attached. It is just as improper to have an inadequate laboratory as a dark and dirty operating room."

SCIENTIFIC MEDICINE AND OVERUSE OF THE LABORATORY

Another problem—the pattern of ordering too many tests—was recognized long before automation began to make inroads into the laboratory during the 1960s, 1970s, and 1980s. Throughout this century there have been numerous criticisms directed at the overuse and misuse of the laboratory. In 1912, a Parisian physician, formerly from New York, while touring American hospitals, reported his surprise at the unusually large number of laboratory tests routinely requested for the patients. Blood counts were as common as temperature or pulse-taking and analysis of urine. He concluded that "the diagnosis and treatment of a given patient depended more on the result of these various tests than on the symptoms present in the case" [13].

Twenty years later, the most common problem in the usage of medical resources cited by doctors was the large number and unintelligent use of laboratory tests ordered in hospitals as a matter of course, without apparent relevance to the condition for which the patient was admitted, and without apparent understanding of the test's meaning or limitations. Clinical application of blood chemistry determinations was especially abused [14]. It is surprising that this later perception of excessive laboratory testing occurred twenty-five years before the appearance of automated chemical analysis and its by-product, mass-produced laboratory data.

The lure and fascination of machine-produced data was an outgrowth from the twentieth century's faith in science and technology. The doctor whose diagnosis and follow-up depended heavily on technology could believe that he was practicing scientific medicine. Numbers or graphs conveyed an exactness and finality that lessened the uncertainty under which the physician often had to work. However, diagnostic technology has become a double-edged sword. Its use can increase the doctor's knowledge of disease, but it also can erode his confidence in his ability to make independent judgements based on his own techniques and experiences of gathering data [15]. Consequently, fears were expressed about the growing dependence on laboratory tests as the chief means of diagnosis. These have diverted the physician from the older and more difficult methods of clinical examination by observation at the patient's bedside, viz., palpation, percussion, and auscultation. Laboratory findings can provide additional useful information to assist in, but not to determine, the diagnosis, prognosis, or treatment. They can never replace clinical observation and careful deductive reasoning [16].

The fear of malpractice suits, blamed in part for the rise in the number of laboratory tests, is not a recent phenomenon. In the 1930s, some physicians felt pressured "by public opinion—the patient, his family, his friends—to utilize every laboratory test" even when physical examination readily revealed the diagnosis [17]. During the 1950s, physicians were criticized for ordering laboratory procedures "for the record" or "for protection" [18].

The capability of automated devices to produce more test results at a lower cost gave rise to the almost universal identification of "routine" or "baseline" tests—an ever expanding category [19]—and patterns of ordering "packages," "panels," "profiles," "screens," and "surveys," or other groupings named according to the instrument and its output, e.g., "SMA-6," or "SMA-12," instead of specific single tests [20]. Data that did not fit the clinical picture was either ignored or repeated until it did conform [21]. As expected, this increased the number of laboratory tests done. Physicians who received large numbers of reports containing normal laboratory data tended to disregard the abnormal results [22]. Although chemistry profiles only occasionally lead to new diagnoses in patients who appear healthy, nevertheless, physicians continued to order test panels.

Physicians continually expressed their concern over the possible harm done to the practice of medicine and to medical education by overdependence on laboratory tests. The routine ordering of many laboratory tests was labeled "excessive diagnostic inquisitiveness" and a "shot in the dark," in the 1930s [23]; "professionally unwise and economically unsound," in the 1940s [24]; "shot-gun testing," in the 1950s [25]; "wasteful, unproductive and conducive to 'decerebrate medical practice'," in the 1960s [26]; and in the 1970s, an "unchecked drift into the technologically thorough, sometimes obsessively complete workup of our patients in our teaching hospitals," directed to all diagnostic possibilities. It was a call for technologic restraint and a patient-oriented approach to problems, rather than the problem-oriented approach to patients [27]. One physician in 1944 described the approach to patients as "a five-minute history followed by a five-day barrage of special tests in the hope that the diagnostic rabbit may suddenly emerge from the laboratory hat. ..." [28].

Concern over increasing costs due to overuse of laboratory tests led to studies in the 1970s to develop guidelines for the optimal use of laboratory data [29]. The idea of guidelines was not new. In 1928 a physician described twelve routine chemical blood tests in common use and the indications for the ordinary clinical patient, and added: "The average physician in general practice will not see more than ten or a dozen cases a year in which a chemical analysis of the blood will be of any value to him in diagnosis or treatment" [30].

The effective use of laboratory resources has remained a matter of study to this day. In a 1996 report, the low diagnostic yield and therapeutic yield of blood counts, chemistry panel, thyroid tests, and urinalysis, support the recommendations of the U.S. Preventive Services Task Force against the routine use of these tests for case-finding, i.e., testing patients who seek medical care to detect early asymptomatic disease, in the ambulatory setting. However, periodic testing for hyperlipidemia is recommended [31].

From the mid-1950s through the 1960s, some hospitals reported the number of laboratory tests ordered per patient doubled every five years, while the patient census increased only slightly. "The ever expanding role of the laboratory in clinical diagnosis and patient care is evident on the wards, in the records, in the building plans and, not least of all, in the finance office of every hospital" [32]. Hospitals in the United States, Canada, and Great Britain reported substantial increases in the number of laboratory tests ordered, ranging from 7% annually to a doubling every five to seven years [33].

We were warned in 1982 about the danger of dehumanization as a threat to the work of the clinical laboratory. We were also urged to avoid a "black box" or "pushbutton" philosophy towards equipment [34]. This unwelcome dividend of the technical approach to modern medicine was nothing new. Nearly six decades earlier a clinician wrote: "Laboratory methods tend to make one forget the patient altogether in the nicety of the scientific" [35]. Unfortunately, the warning went unheeded as the laboratory was engulfed by an onslaught of ever newer models of chemical analyzers, as manufacturers, domestic and foreign, competed aggressively in the emerging global economy.

Commenting on the changing relationship as the laboratory staff, physically and in their thinking, moves away from the patient, one writer said: "First we digitized them, then we punched them into cards, and now we have reduced them to a few spots of magnetism on a strip of tape" [36]. However, the expanding versatility of new instrumentation in recent years has generated the return of a laboratory presence at the patient's bedside and has also produced problems for the clinical laboratory. One of these problems is the rapid proliferation of instruments for *in vitro* testing at the patient's bedside. These new capabilities have driven a demand for more immediate laboratory results of "emergency procedures" in order to speed patient evaluation and thereby improve patient outcome. Testing at or

near the patient's bedside, in the operating room, and in the intensive care units, has caused considerable concern about accuracy and quality management and that it may lead to faulty clinical decisions because of discrepancies between laboratory results and point-of-care testing (POCT).
Saving of time and reduction of costs are factors in this new application. In one study, glucose analysis was available only one to two minutes sooner at the bedside than when performed in the central laboratory receiving the blood specimen by pneumatic tube. No significant adverse medical outcomes were associated with this difference in turnaround time. The cost of bedside testing was approximately twice that of the central laboratory [37]. Proof that bedside testing produces a superior medical outcome than that of central laboratory testing—with less than a five-minute difference in turnaround time—would have to factor in the costs of the consequences of the relatively lower level of accuracy and precision of bedside testing conducted by nonlaboratory personnel when compared to central laboratory testing. Discrepancies in results are related to the complexity of the instruments used [38]. More study is needed before POCT can gain widespread approval.

A NEW PARTNERSHIP—A NEW VOCABULARY

There is little evidence of a linear historical progression within any given scientific discipline toward a so-called mature state. In the past, a feeling of continuity was experienced due to the competition of attitudes, approaches, and input from the different divisions of chemistry and biology—to which we must now add industry and government.
The rapid development of sophisticated instrumentation and the accompanying diagnostic kit methodology spurred a new relationship between the clinical laboratory, industry, and the health insurance companies, with the government regulating the activities of all three through a constantly growing new vocabulary emanating from business and the government.
Federal regulatory agencies are having a significant impact on how clinical laboratories, current and future, are to be staffed, outfitted, operated, and accredited; beginning with the Clinical Laboratory Improvement Act of 1967, followed by the Clinical Laboratory Improvement Amendments of 1988 (CLIA 1988) and its implementation by the Health Care Financing Administration (HCFA). In addition, rules, regulations, and guidelines come from the Food and Drug Administration (FDA), the National Institute of Drug Abuse (NIDA), the Occupational Safety and Health Administration (OSHA), the Department of Transportation

(DOT), the Environmental Protection Agency (EPA), the Equal Employment Opportunity Commission (EEOC), and even the U.S. Postal Service because it determines postal rates for scientific publications. All are having a significant and probably long-lasting effect on the practice of clinical laboratory medicine in the United States.

If this weren't enough, there are the rigors of inspection—voluntary by the College of American Pathologists, mandatory by the individual State's Departments of Health, and by the quasi-official Joint Commission for Accreditation of Healthcare Organizations (JCAHO). An added overlay to this regulatory environment by the government, is the ever-present potential for periodic changes in guidelines and directives or even a reversal in direction as a result of perceived or imagined election mandates.

What emerges from this alphabet soup of acronyms is the need to deal with laboratory accreditation, personnel certification and licensure, quality assurance, quality control, statistical analysis, productivity, and efficiency. From the marketplace, the catch-phrases are turnaround time, bottom-line, cost-benefit ratio, cost-effectiveness, third-party payers, satellite laboratories, and reference laboratories.

A new element of regulation entered this mix in the fall of 1996. After much delay, new HCFA documentation rules took effect. Hospitals became subject to the same medical necessity documentation required of commercial and physician office laboratories as a condition of payment by the federal program. Previously, laboratories were paid for an automated profile of tests if three of the tests were medically justified. The new requirements were established to reduce test utilization. HCFA believed that many automated tests performed by laboratories were medically unnecessary and should not be reimbursed. With the new policy, all automated multi-channel testing (billing) codes will be eliminated and replaced by codes for single tests or the new organ and disease-specific panels that carry the presumption of medical necessity.

LOOKING TO THE FUTURE—COST CONTAINMENT AND MANAGED CARE

In former, simpler times, the future was likely to be very much like the present, so there was little to be gained from predicting it more precisely. However, the accelerating rate of change in the new healthcare environment—technology-driven, cost-driven, and politically-driven—is forcing us to pay more attention to trying to predict the future. And the future has come upon us quite suddenly, from an unexpected direction and in an unsettling form. The social and economic forces of an aging population and politically sensitive welfare entitlements and government benefits, will have a major influence on the resources allocated to the healthcare delivery system. Consequently, physicians and hospitals are making a transition from fee-for-service to the era of "managed care" contracts—and clinical laboratories are becoming, in some ways, a casualty of the changes. The key components of managed care are cost containment, increased efficiency, decreased utilization of services (or prevention of overuse), diagnostic effectiveness, and outcomes assessment and analysis.

The collective frustration by the hospital community and the public over the escalation of healthcare costs has been one of the factors in the rapid emergence of managed care. Its goal is to provide insurance coverage for more people at a lower cost, with a fixed premium to the subscribers on the one hand, and negotiated or capitated fees (per subscriber per month for specified services) to physicians and hospitals on the other hand. The contractual arrangements made with hospitals in particular, offer little margin over their costs and have thus compelled hospitals to seek ways to reduce their expenses. Hospitals have reduced the length of stay of many patients and have moved many former inpatient services to an outpatient setting [39].

In many countries, these costs have been rising faster than the rate of inflation, due to technological advances, patient demand, and changes in demographics. Healthcare is now seen as a commodity offered by *providers* who must compete in the market-place for *purchaser's* dollars, which they are doing by wide-ranging multi-media advertising of their services. Hospitals are already competing with one another for providing services in the drive to improve efficiency [40].

By-products of this new business orientation are mergers, takeovers, closings, consolidations, and acquisitions, in all branches of healthcare—hospitals, physicians' groups, insurance companies, health maintenance organizations (HMOs), reference laboratories, and diagnositc devices corporations. The language of industrial America now applies to the nation's health centers. Driven by changes in the organization and financing of healthcare, institutions are forming combinations of physicians and facilities so large, they cannot be ignored by the HMO-purchasers of healthcare. Even medical schools are exploring possible mergers.

Decreasing reimbursement to providers, coupled with consolidation and integration, are the major forces transforming healthcare into an industry—the "medical-industrial complex" that is managed care. As providers respond to financial pressures, they will increasingly focus on total patient care delivered by multidisciplinary teams through disease management strategies and treatment algorithms. There will also be an increased emphasis on wellness programs, disease prevention, and early diagnosis—all of which adds up to changes in the practice of medicine [41].

In further response to cost containment pressures—at the same time as demands are made for more and better services—hospital laboratories have adopted some of the business practices of the commercial laboratories: networking with other hospitals to create shared laboratory facilities; setting up external testing services; and creating joint ventures and management agreements with commercial reference laboratories. Managed care and the increasingly corporate face of medical care are driving the system toward adoption of more technology and an expected cost savings coming largely from efficiencies in information management. The clinical chemist will find that he is in the information business with his role being less chemist and more clinical. The analysis of data to improve clinical, economic, and total patient outcomes will be the key to successful laboratory operation.

Concern about the cost of healthcare is driving administrators and managers of hospitals and health services to try to control costs. One area that comes under increasing examination is laboratory services. As a percentage of the hospital's total healthcare cost, this varies from about 3% to 4% (U.K.) to nearly 10% (U.S.A.). Despite this wide variation in usage and cost of services, laboratories worldwide are under increasing pressure to control expenditures and become cost-effective. There is an overutilization of laboratory resources that is driven by medico-legal concerns, availability of technology, and lack of incentives for cost containment [42].

In the past, laboratory professionals found it difficult to affect test-ordering practices or promote appropriate use of the laboratory by clinicians. The clinical laboratory was a profit center for the hospital, with profitability resting largely on the number of tests performed. Cost per test was often a secondary concern, especially when charges could be increased as needed to cover costs. With the implementation of healthcare reform, laboratories will no longer be able to manage budgets by increasing test fees or the number of tests performed.

However, reducing laboratory costs is only part of the answer. The laboratory will have to examine the effects of utilization of its test offerings on patient care in achieving favorable patient outcome and speedier discharge. Evaluation of the diagnostic effectiveness of tests should include technical efficacy in obtaining reproducible results; clinical efficacy as measured by impact on the clinician's diagnosis; also, the effect of the test on the cost of treatment and on patient outcome in terms of quality of life, morbidity, and mortality. Tests that are ineffective should be eliminated [43]. One approach to unnecessary testing may be to establish a hierarchy of tests, singly or in combination, that is most appropriate and cost-effective for making a rapid decision in a particular clinical situation.

Medicine is currently experiencing a shift away from the principle of "do no harm" to a new model that directs it to "do what works," contain costs, and measure outcomes. Clinicians and laboratorians are advised to evaluate their current practices and then change them, before outside sources force a change first and then an evaluation of the impact of these changes [44].

Finance management, i.e., cost analysis and budgeting, appear destined to become new road markers for the successful practice of clinical chemistry. Not to be neglected are resource management and utilization allocation, e.g., personnel and materials, as well as skills in negotiation and presentation. To succeed in the rapidly changing laboratory environment, clinical chemists must fill new roles outside the former boundaries of the traditional laboratory. They must develop leadership and management skills, particularly in regard to participation in multidisciplinary teams functioning in the "total laboratory"—a new laboratory without labels but arranged according to technology [45].

"Traditional clinical chemists" who focused exclusively on test development, test interferences, quality assurance and quality control, face a bleak future in the typical health center unless they possess diversified talents. The "new" clinical chemists must become *clinical laboratory scientists* by acquiring knowledge of clinical medicine and enlarging their knowledge of the other service branches of laboratory medicine, e.g., coagulation, genetics, molecular biology, environmental toxicology, etc., in order to optimize their contributions to patient care. Diseases are not categorized by laboratory subdisciplines, and the physician would welcome helpful information that comes from a single source with knowledge across the traditional boundaries dividing the clinical laboratory [46].

An identical warning had been sounded twenty years earlier. Complaining of the growing resemblance of chemical department to "supermarkets," Poul Astrup said that chemists should *"develop this service from a static description of concentration of components"* into dynamic and interpretive descriptions of *"the diagnostic values of the results,"* and the *"biological and medical relevance* in interpreting significance for the clinicians." He added that "the discipline will grow only if the clinical chemists *think of* themselves as belonging primarily to the biological/medical sphere and only secondarily to the technical/analytical sphere" [47].

NEW CAREER OPPORTUNITIES FOR THE CLINICAL CHEMIST

In the wake of large closed system (black-box) push-button automation performing much of the analytical work of the laboratory, with its affiliated methodology wedded to manufacturer-provided reagents, with contractual repair and trouble-shooting services, and responsibility for quality assurance largely assumed by manufacturers—an arrangement no more challenging than making a phone call—there developed a growing competition from clinical pathologists for hospital positions in clinical chemistry. At the same time, new career opportunities also opened up for clinical chemists. They filled positions in the growing field of commercial reference and speciality testing laboratories; and in the research and development, production, quality control, marketing, and management functions of the industrial companies whose innovative technology created the revolution in clinical laboratory automation. Other clinical chemists were attracted to positions in city, state, and federal regulatory agencies that administered the many new clinical laboratory legislative oversight programs.

CONCLUSION

Clinical chemistry played an integral part in the unprecedented developments in medicine following the end of World War II. Medical conditions which had probably always existed were identified for the first time in scientific terms and treated on the basis of a better understanding of the biochemical mechanisms of the disease. Better methods for the investigation of such common conditions as diabetes, renal, liver, bone, and endocrine disease were devised, in many cases leading to better treatment and monitoring of the condition. Clinical chemistry became increasingly involved in a wide spectrum of disease. The dominant role has been played by the application of technological innovation to the investigation of disease. Many other medical disciplines were also developing in parallel, but clinical chemistry moved as quickly as any, and today there are few medical conditions in which the clinical chemist is not involved.

Having evolved as a branch of analytical biochemistry that was mainly engaged in the development of chemical methods to investigate and monitor diseases, clinical chemistry now encompasses once seemingly

CONCLUSION

unrelated sciences such as physiology, endocrinology, inherited metabolic diseases, immunology, and pharmacology (toxicology). Looming over the horizon are the challenges and promises of robotics, nanotechnology, microminiature analysis, and the new interface between chemistry and biology—analysis of genetic material for diagnosis of inherited and neoplastic diseases, DNA fingerprinting, testing of biomarkers, and more—adventures that will probe the mysteries of life at the molecular level. Already in operation are systems with completely automated pre-analytical functions (sorting, centrifuging, cap removal, aliquoting, bar-code labeling, and transport to analyzer).

The first U.S. installation of the Hitachi Clinical Laboratory Automation System (CLAS), marketed by the Boehringer Mannheim Corporation,[*] Indianpolis, Indiana, is operating at the South Bend (Indiana) Medical Foundation core laboratory serving three hospitals. CLAS is a turnkey, third-generation automation system that by mid-1998 was functioning in more than 100 laboratories worldwide, with most in Japanese hospitals. The system can process more than 300 primary tubes per hour; the number of results depends on the number and configuration of Hitachi analytical modules in the system. In Japan, the scope of automatic clinical laboratory applications ranges from chemistry alone to fully integrated chemistry, hematology, urinalysis, coagulation, and immunoassay.

Automation will be followed by consolidation of currently distinct laboratory disciplines, and single analytical systems will perform tests currently performed in separate laboratories. These high-throughput analyzers will decrease the demand for technologists and further drive the trend toward combining laboratory specialties into a *laboratory without partitions.*

When total electronic recording of patient medical histories becomes widely available, privacy protection schemes will have to be developed. These should include complete notification to the patient of the intended use for the information and consent from the patient for any secondary uses. Each use will need to be documented to assure complete accountability and security.

We are becoming increasingly aware of the tyranny of geometric growth. Whereas time proceeds at a linear pace, change and progress, moving at geometric speed, are sweeping us pell-mell toward the unknowns of the next millennium. Social and technological change, in and out of the laboratory, have become more rapid, unrelenting and unsettling than ever before. To keep from being swept into the future before completely

---

[*] acquired by Roche Diagnostics in 1998.

understanding the present, and to cushion against the impact of continuous change, we must seize hold of ancestral landmarks. We must look to the history of the past—there is nowhere else to look—to understand how we arrived at the present and to prepare for the future. If we should take history for granted, we would become orphans in time—castaways on a desert island called "the present"—with no idea of where we came from or where we are going.

NOTES AND REFERENCES

1. CHARLES P. EMERSON, "Some Clinical Aspects of Chemistry," *Journal of the American Medical Association*, 1902, 38: 1359–1362.
2. J. M. DACOSTA, "Tendencies in America," *Transactions of the Association of American Physicians*, 1897, 12: 1–8, p. 7.
3. HELEN CLAPESATTLE, *The Doctors Mayo* (Minneapolis: The University of Minnesota Press, 1941), pp. 425–426; see also JASPER HALPENNY, "On Clinical and Laboratory Methods of Diagnosis," *Canadian Medical Association Journal*, 1924, 14: 671–673.
4. LEWELLYS F. BARKER, "Medical Laboratories: Their Relations to Medical Practise [sic] and to Medical Discovery," *Science*, 1908, 27: 601–611, p. 608.
5. ABRAHAM FLEXNER, *Medical Education in the United States and Canada. A Report to the Carnegie Foundation for the Advancement of Teaching. Bulletin Number Four* (Boston: Merrymount Press, 1910), pp. 91–92.
6. DOUGLAS SYMMERS, "Defects in the Teaching of Pathology, and the Lay Professor," *Journal of the American Medical Association*, 1919, 73: 1651–1655, p. 1654.
7. WILLIAM DOCK, "The Place of the Gadget in the History and the Practice of Medicine," *Stanford Medical Bulletin*, 1955, 13: 138–143.
8. O. J. WALKER, "Organizing a Modern Hospital Laboratory," *Modern Hospital*, 1921, 16: 502–506; FRANCIS CARTER WOOD, "The Hospital Laboratory," *Bulletin of the American College of Surgeons*, 1917, 3: 20–24.
9. C. N. B. CAMAC, "Hospital and Ward Clinical Laboratories," *Journal of the American Medical Association*, 1900, 35: 219–227, p. 223. This article has illustrations of clinical laboratories in America, Canada, and Europe.
10. LEWELLYS F. BARKER, "The Organization of the Laboratories in the Medical Clinic of the Johns Hopkins Hospital," *Bulletin of the Johns Hopkins Hospital*, 1907, 18: 193–198.
11. ROBERT E. KOHLER, *From Medical Chemistry to Biochemistry. The Making of a Biomedical Discipline* (New York: Cambridge University Press, 1982), pp. 233, 236–237.
12. MAX KAHN, "Construction of Laboratories Ordinarily Given No Expert Consideration in Building of Hospitals—Organization, Director, Budget, Routine, Record-Keeping, Research, Etc.," *Modern Hospital*, 1918, 11: 271–274, p. 271.
13. CECIL KENT AUSTIN, "Medical Impressions of America," *Boston Medical and Surgical Journal*, 1912, 166: 799–804, p. 801. The writer was astonished at the "anti-physiological" practice among the public of chewing gum (p. 804).
14. SYDNEY R. MILLER, "Contemporary Fads and Fallacies, Therapeutic and Diagnostic, Which Reflect Dangerous Professional Credulity," *Pennsylvania Medical Journal*, 1932, 35: 347–354, pp. 348–349. The indiscriminate and unintelligent requests for X-rays and electrocardiograms were criticized as a diagnostic fad (p. 349); RALPH G. STILLMAN, "The

Significance of Laboratory Tests and Methods," *New York State Journal of Medicine*, 1935, 35: 757–766, p. 759.
15. STANLEY JOEL REISER, *Medicine and the Reign of Technology* (New York: Cambridge University Press, 1978), pp. 161–162, 173.
16. C. F. HOOVER, "The Reputed Conflict Between the Laboratories and Clinical Medicine," *Science*, 1930, 71: 491–497, pp. 492, 493; LEWIS A. CONNER, "Relation of Laboratory Aids to the Practice of Medicine and Surgery," *Journal of the American Medical Association*, 1923, 81: 871–873; JAMES B. HERRICK, "The Relation of the Clinical Laboratory to the Practitioner of Medicine," *Boston Medical and Surgical Journal*, 1907, 156: 763–768, p. 767; BARKER (1908), p. 607; H. M. MARVIN, "The Therapeutic Trial," *Connecticut State Medical Journal*, 1954, 18: 850; FRANCIS W. PEABODY, "The Physician and the Laboratory," *Boston Medical and Surgical Journal*, 1922, 18: 324–328, p. 325 and discussion by George R. Minot, p. 327; STILLMAN (1935), p. 757.
17. WILLIAM SEAMAN BAINBRIDGE, "Clinical and Laboratory Diagnoses," *Medical Journal and Record*, 1932, 136: 265–270, p. 269.
18. LOUIS K. DIAMOND and F. STANLEY PORTER, "The Inadequacies of Routine Bleeding and Clotting Times," *New England Journal of Medicine*, 1958, 259: 1025–1027, p. 1025.
19. "Laboratory Posture," *New England Journal of Medicine*, 1963, 268: 1196 (Editorial).
20. REISER (1978), pp. 159–160.
21. "Laboratory Posture," *New England Journal of Medicine*, 1963, 268: 1196 (Editorial).
22. D. M. YOUNG, NINA DRAKE, and R. J. WEIR, "The Advent of Chemical Screening Techniques," *Canadian Medical Association Journal*, 1968, 98: 868–870; ROY N. BARNETT, W. HAROLD CIVIN, and IRWIN SCHOEN, "Multiphasic Screening by Laboratory Tests—An Overview of the Problem," *American Journal of Clinical Pathology*, 1970, 54: 483–492, p. 486.
23. MILLER (1932), p. 349; STILLMAN (1935) p. 760.
24. TINSLEY R. HARRISON, "The Value and Limitation of Laboratory Tests in Clinical Medicine," *Journal of the Medical Association of the State of Alabama*, 1944, 13: 381–384, p. 382.
25. MARVIN (1954), p. 850.
26. "Routine Laboratory Tests," *New England Journal of Medicine*, 1966, 275: 56 (Editorial).
27. DAVID E. ROGERS, "On Technologic Restraint," *Archives of Internal Medicine*, 1975, 135: 1393–1397, pp. 1393, 1397.
28. HARRISON (1944), p. 382.
29. BARNETT et al. (1970); PAUL F. GRINER and BENJAMIN LIPTZIN, "Use of the Laboratory in a Teaching Hospital. Implications for Patient Care, Education, and Hospital Costs," *Annals of Internal Medicine*, 1971, 75: 157–163; CHRISTOPHER C. KORVIN, RICHARD H. PEARCE, and JOHN STANLEY, "Admissions Screening: Clinical Benefits," *Ibidem*, 1975, 83: 197–203; RICHARD H. DIXON and JOHN LASZLO, "Utilization of Clinical Chemistry Services by Medical House Staff. An Analysis," *Archives of Internal Medicine*, 1974, 134: 1064–1067.
30. REED ROCKWOOD, "Chemical Tests of the Blood. Indications and Interpretation," *Journal of the American Medical Association*, 1928, 91: 157–166, p. 157.
31. BENOIT J. BOLAND, PETER C. WOLLAN, and MARC D. SILVERSTEIN, "Yield of Laboratory Tests for Case-Finding in the Ambulatory General Medical Examination, "*American Journal of Medicine*, 1996, 101: 142–152, p. 152. See also ref. 43.
32. Editorial (1963), see ref. 21.

33. H. F. BARNARD, "Growth of Medical Laboratory Work During 1920–2000," *British Medical Journal*, 1976, 1: 383–384; H. E. VICKERS, "Is Our Pathology Really Necessary? Parkinson versus Beeching," *Lancet*, 1966, 2: 46–47; J. S. A. ASHLEY, "Demand for Laboratory Services," *British Medical Bulletin*, 1974, 30: 234–236; IRWIN M. HILLIARD, "The Threat of the Modern Laboratory to the Art and Science of Medicine," *Canadian Medical Association Journal*, 1961, 85: 483–486; DAVID SELIGSON, "Clinical Laboratory Automation," *Journal of Chronic Diseases*, 1966, 19: 509–517.
34. J. BÜTTNER, "Technical Evolution in the Clinical Laboratory," *Pure & Applied Chemistry*, 1982, 54: 2011–2016, p. 2015.
35. HALPENNY (1924), p. 672.
36. F. L. MITCHELL, "Present and Future Trends of Automation in Clinical Chemistry" (pp. 180–190), in *Methods in Clinical Chemistry*, vol. 1, 7th International Congress of Clinical Chemistry, Geneva/Evian, September 8–13, 1969 (Baltimore: University Park Press, 1970), p. 189.
37. JAMES W. WINKELMAN, DONALD R. WYBENGA, and MILENKO J. TANASIJEVIC, "The Fiscal Consequences of Central vs Distributed Testing of Glucose," *Clinical Chemistry*, 1994, 40: 1628–1630.
38. AMIN A. NANJI, RAYMOND POON, and IRWIN HINBERG, "Decentralized Clinical Chemistry Testing: Quality of Results Obtained by Residents and Interns in an Acute Care Setting," *Journal of Intensive Care Medicine*, 1988, 3: 272–277; see also BETTE SEAMONDS, "Medical, economic, and regulatory factors affecting point-of-care testing. A Report of the Conference on Factors Affecting Point-of-Care Testing, Philadelphia, PA, 6–7 May 1994," *Clinica Chimica Acta*, 1996, 249: 1–19.
39. "The Future of Clinical Chemistry and its Role in Healthcare: a Report of the Athena Society," *Clinical Chemistry*, 1996, 42: 96–101, p. 96.
40. R. SWAMINATHAN, "Health-Care Reform and the Pathology Laboratory," *Journal of the International Federation of Clinical Chemistry*, 1995, 7: 84–85.
41. AACC Task Force on the Changing Practice Environment, "The Changing Environment for the Practice of Clinical Chemistry," *Clinical Chemistry*, 1996, 42: 91–95, p. 92.
42. SWAMINATHAN (1995).
43. MARC D. SILVERSTEIN and BENOIT J. BOLAND, "Conceptual Framework for Evaluating Laboratory Tests: Case-Finding in Ambulatory Patients," *Clinical Chemistry*, 1994, 40: 1621–1627.
44. *Idem*.
45. Responses by clinical chemists to the impact of managed care are discussed in refs. 39 and 41. For a preliminary report, see also *Clinical Chemistry*, 1995, 41: 639–640.
46. *Idem*. For a review and discussion of proposed changes in postdoctoral training programs to meet the challenges of the new laboratory environment, see ALAN H. B. WU and EDWARD W. BERMES, JR., "Report of the Third Conference on Education in Clinical Chemistry," *Clinical Chemistry*, 1997, 43: 167–173.
47. POUL ASTRUP, "Clinical Chemistry—A Changing Discipline," *Clinical Chemistry*, 1975, 21: 1709–1715, pp. 1712, 1713.

BIBLIOGRAPHY

BOOKS

ACKERKNECHT, ERWIN H. *Medicine at the Paris Hospital 1794–1848*. Baltimore: The Johns Hopkins Press, 1967.
ANITSCHKOW, N. "Experimental Arteriosclerosis in Animals." In *Arteriosclerosis. A Survey of the Problem*, Edmund V. Cowdry, ed. New York: Macmillan, 1933.
ASTRUP, POUL, PETER BIE, and HANS CHR. ENGELL. *Salt and Water in Culture and Medicine*. Translated from Danish by Kirsten Skovbjerg and Andrew L. Cameron-Mills. Copenhagen: Munksgaard, 1993.
ASTRUP, POUL, and JOHN SEVERINGHAUS. *The History of Blood Gases, Acids and Bases*. Translated from Danish by Patrick Graham Jørgensen. Copenhagen: Munksgaard, 1986.
BATES, ROGER G. *Determination of pH. Theory and Practice*, 2nd ed. New York: John Wiley & Sons, 1973.
BEALE, LIONEL S. *The Microscope in its Application to Practical Medicine*, 3rd ed. Philadelphia: Lindsay and Blakiston, 1867.
-----. *The Microscope in Medicine*, 4th ed. London: J. and A. Churchill, 1878.
BECK, EUGENE A. "The Discovery of Fibrinogen and Its Conversion to Fibrin." In *Fibrinogen*, Koloman Laki, ed. New York: Marcel Dekker, Inc., 1968.
BEHRMAN, RICHARD E., ed. *Nelson Textbook of Pediatrics*, 14th ed. Philadelphia: W. B. Saunders Company, 1992.
BEN-DAVID, JOSEPH. *The Scientists's Role in Society. A Comparative Study*. Englewood Cliffs, New Jersey: Prentice-Hall, Inc., 1971.
BERZELIUS, IÖNS JACOB. *A View of the Progress and Present State of Animal Chemistry of Iöns Jacob Berzelius, M. D.* Translated from Swedish by Gustavus Brunnmark. London: 1813.
BEST, CHARLES H. "Epochs in the History of Diabetes." In *Diabetes*, Robert H. Williams, ed. New York: Paul B. Hoeber, Inc., 1960.
BIRD, GOLDING. *Urinary Deposits, Their Diagnosis, Pathology, and Therapeutical Indications*, 2nd ed. London: John Churchill, 1846.
BOWMAN, JOHN E. *A Practical Handbook of Medical Chemistry*, 4th ed. London: John Churchill, 1862.
BOYLE, ROBERT. *Memoirs For The Natural History of Humane Blood, Especially The Spirit of that Liquor*. London: Samuel Smith, 1684.
BRADBURY, S. *The Evolution of the Microscope*. Oxford: Pergamon Press, 1967.

BRIGHT, RICHARD. *Reports of Medical Cases, Selected With a View of Illustrating The Symptoms and Cure of Diseases By a Reference to Morbid Anatomy*, (2 Vols.). London: Longman, Rees, Orme, Brown, and Green, 1827, 1831.

BROCK, WILLIAM H. *The Norton History of Chemistry*. New York: W. W. Norton & Company, 1993.

BRONK, DETLEV W. "Foreward." In George W. Corner. *A History of the Rockefeller Institute, 1901–1953. Origins and Growth*. New York: The Rockefeller Institute Press, 1964.

BROUGHTON, PETER and JOHN LINES. *The Association of Clinical Biochemists. The First Forty Years*, Roy Sherwood, ed. London: ACB Venture Publications, 1996.

BROWN, JOHN. "Locke and Sydenham." In *Horae Subsecivae*, 1st series, 4th ed. Edinburgh: David Douglas, 1882.

BUERGER, JANET E. *French Daguerreotypes*. Chicago and London: The University of Chicago Press, 1989.

BÜTTNER, JOHANNES. "Johann Joseph von Scherer (1814–1869). A commentary on the early history of clinical chemistry." In *History of Clinical Chemistry*, J. Büttner, ed. Berlin and New York: W. de Gruyter, 1983. Translated from German in *Journal of Clinical Chemistry and Clinical Biochemistry*, 1978, 16: 478–483.

BÜTTNER, JOHANNES and CHRISTA HABRICH. *Roots of Clinical Chemistry*. Darmstadt, Germany: Git Verlag Gmbh, 1987.

CAMERON, H. C. *Mr. Guy's Hospital 1726–1948*. London: Longmans, Green and Co., 1954.

CASTIGLIONI, ARTURO. *A History of Medicine*, 2nd ed. Translated from Italian and edited by E. B. Krumbhaar. New York: Alfred A. Knopf, 1947.

CHESNEY, ALAN M. *The Johns Hopkins Hospital and The Johns Hopkins University School of Medicine. A Chronicle*, (3 Vols.), Vol. 2, *1893–1905*. Baltimore: The Johns Hopkins Press, 1958.

CHILD, ERNEST. *The Tools of the Chemist. Their Ancestry and American Evolution*. New York: Reinhold Publishing Corporation, 1940.

CHITTENDEN, RUSSELL H. *The Development of Physiological Chemistry in the United States*. New York: The Chemical Catalog Company, Inc., 1930.

-----. *The First Twenty-five Years of the American Society of Biological Chemists*. Baltimore: Waverly Press, 1945.

CLAPESATTLE, HELEN. *The Doctors Mayo*. Minneapolis: The University of Minnesota Press, 1941.

CUSHING, HARVEY. *The Life of Sir William Osler*, (2 Vols.), Vol. 2. Oxford: Oxford University Press, 1925.

DACOSTA, JR., JOHN C. *Clinical Hematology. A Practical Guide to the Examination of the Blood With Reference to Diagnosis*, 2nd ed. Philadelphia: P. Blakiston's Son & Co., 1905.

DAVIES, RICHARD. *Essays to Promote the Experimental Analysis of the Human Blood*. Bath: J. Leake, 1760.

DIXON, KENDAL. "Some Biochemical Signposts in the Progress of Neurology." In *The Chemistry of Life. Eight Lectures on the History of Biochemistry*, Joseph Needham, ed. London: Cambridge University Press, 1970.

DIXON, MALCOLM. "The History of Enzymes and of Biological Oxidations." In *The Chemistry of Life. Eight Lectures on the History of Biochemistry*, Joseph Needham, ed. London: Cambridge University Press, 1970.

ELLIS, HAROLD. *A History of Bladder Stone*. Oxford: Blackwell Scientific Publications, 1969.

FIESER, LOUIS F. and MARY FIESER. *Natural Products Related to Phenanthrene*, 3rd ed. New York: Reinhold Publishing Corporation, 1949.

FISH, MATHEWS B. and PAUL R. JENSEN. "Experience With the Robot Chemist." In *Automation and Data Processing in the Clinical Laboratory*, Geoffrey M. Brittin and Mario Werner, eds. Springfield, Illinois: Charles C. Thomas, Publisher, 1970.

FISHBERG, ARTHUR M. *Hypertension and Nephritis*, 5th ed. Philadelphia: Lea & Febiger, 1954.

FLEXNER, ABRAHAM. *Medical Education in the United States and Canada. A Report to the Carnegie Foundation for the Advancement of Teaching*, Bulletin Number Four. Boston: Merrymount Press, 1910.

-----. *An Autobiography*. New York: Simon and Schuster, 1960.

FLORKIN, MARCEL and ELMER H. STOTZ, eds. Vol. 30, *A History of Biochemistry* (Marcel Florkin). Amsterdam: Elsevier Publishing Company, 1972.

FORTUNE, W. B. "Color Comparimeters." In *Analytical Absorption Spectroscopy. Absorptimetry and Colorimetry*, M. G. Mellon, ed. New York: John Wiley & Sons, Inc., 1950.

FOSTER, W. D. *A Short History of Clinical Pathology*. Edinburgh and London: E. & S. Livingstone Ltd., 1961.

FREEMAN, JAMES A. and MYRTON F. BEELER. *Laboratory Medicine—Clinical Microscopy*. Philadelphia: Lea & Febiger, 1974.

FRUTON, JOSEPH S. *Molecules and Life. Historical Essays on the Interplay of Chemistry and Biology*. New York: Wiley-Interscience, 1972.

GAMBLE, JAMES L. *Chemical Anatomy Physiology and Pathology of Extracellular Fluid. A Lecture Syllabus*, 6th ed. Cambridge, Massachusetts: Harvard University Press, 1954.

GARRISON, FIELDING H. *An Introduction to the History of Medicine*, 4th ed. Philadelphia: W. B. Saunders Company, 1929.

GERMAN, WILLIAM MCKEE. *Doctors Anonymous. The Story of Laboratory Medicine*. New York: Duell, Sloan and Pearce, 1941.

GILLISPIE, CHARLES COULSTON, ed. *Dictionary of Scientific Biography*. New York: Charles Scribner's Sons, (14 Vols.), 1970-1976.

GRADWOHL, R. B. H. *Clinical Laboratory Methods and Diagnosis. A Textbook on Laboratory Procedures With Their Interpretation*, 3rd ed. (2 Vols.), Vol. 1. St. Louis: C. V. Mosby Company, 1943.

HALLER, JR., JOHN S. *American Medicine in Transition. 1840–1910.* Urbana: University of Illinois Press, 1981.
HARAWAY, DONNA JEANNE. *Crystals, Fabrics, and Fields. Metaphors of Organicism in Twentieth-Century Developmental Biology.* New Haven: Yale University Press, 1976.
HARRISON, G. A. *Chemical Methods in Clinical Medicine. Their Application and Interpretation With Techniques of Simple Tests,* 3rd ed. New York: Grune and Stratton, 1947.
HASSALL, ARTHUR HILL. *The Urine in Health and Disease: Being an Exposition of The Composition of The Urine, and of the Pathology and Treatment of Urinary and Renal Disorders,* 2nd ed. London: John Churchill and Sons, 1863.
HAWK, PHILIP B., BERNARD L. OSER, and WILLIAM H. SUMMERSON. *Practical Physiological Chemistry,* 13th ed. New York: McGraw-Hill Book Company, Inc., 1954.
HEILENZ, SIEGFRIED. *The Liebig-Museum in Giessen. Guide Through the Museum and a Liebig-Portrait From a Current Viewpoint.* Darmstadt, Germany: E. Merck, 1986.
HENRY, RICHARD J. *Clinical Chemistry: Principles and Technics.* New York: Harper & Row, 1964.
HERRMANN, ROLAND and C. T. J. ALKEMADE. *Chemical Analysis by Flame Photometry,* 2nd ed. Translated from German by Paul T. Gilbert, Jr. New York: Interscience Publishers, 1963.
HODGKIN, THOMAS. *Medical Reform. An Address Read to the Harveian Society, at the Opening of its Seventeenth Session, October 2, 1847.* London: John Churchill, 1847.
HOLMES, FREDERICK L. "Introduction." In Justus Liebig. *Animal Chemistry or Organic Chemistry in its Application to Physiology and Pathology,* William Gregory, ed. New York and London: Johnson Reprint Corporation, 1964. (Facsimile of the Cambridge Edition of 1842.)
JACOB, FRANÇOIS. *The Logic of Life. A History of Heredity.* New York: Random House (Pantheon Books), 1973.
JAKSCH, RUDOLF V. *Clinical Diagnosis: The Bacteriological, Chemical, and Microscopical Evidence of Disease,* 4th ed. Translated from German by James Cagney. London: Charles Griffin and Company, Limited, 1899.
JOHNSON, W. B. *History of the Progress and Present State of Animal Chemistry,* (3 Vols.), Vol. 1. London: J. Johnson, 1803.
JONES, HENRY BENCE. *On Animal Chemistry in its Application to Stomach and Renal Diseases.* London: John Churchill, 1850.
-----. *Lectures on Some of the Applications of Chemistry and Mechanics to Pathology and Therapeutics.* London: John Churchill and Sons, 1867.
KASS, AMALIE M. and EDWARD H. KASS. *Perfecting the World. The Life and Times of Dr. Thomas Hodgkin, 1798–1866.* Boston: Harcourt Brace Jovanovich, 1988.

KATZ, LOUIS N. and JEREMIAH STAMLER. *Experimental Atherosclerosis.* Springfield, Illinois: Charles C. Thomas, 1953.
KELLY, HOWARD A. and WALTER L. BURRAGE, eds. *Dictionary of American Medical Biography.* New York and London: D. Appleton and Company, 1928.
KING, E. J. and I. D. P. WOOTTON. *Micro-Analysis in Medical Biochemistry,* 3rd ed. New York: Grune & Stratton, 1956.
KINGZETT, CHARLES THOMAS. *Animal Chemistry or The Relations of Chemistry to Physiology and Pathology. A Manual for Medical Men and Scientific Chemists.* London: Longmans, Green, and Co., 1878.
KIRK, PAUL L. *Quantitative Ultramicroanalysis.* New York: John Wiley & Sons, Inc., 1950.
KISSKALT, K. *Max von Pettenkofer.* Stuttgart: Wis. Verlagsges, 1948.
KNIGHT, DAVID. *Humphry Davy. Science & Power.* Oxford: Blackwell Publishers, 1992.
KNIGHTS, JR., EDWIN M., RODERICK P. MACDONALD, and JAAN PLOOMPUU. *Ultramicro Methods for Clinical Laboratories.* New York: Grune & Stratton, 1957.
KOCH, FREDERICK C. and MARTIN E. HANKE. *Practical Methods in Biochemistry,* 4th ed. Baltimore: Williams & Wilkins Company, 1943.
KOHLER, ROBERT E. *From Medical Chemistry to Biochemistry. The Making of a Biomedical Discipline.* New York: Cambridge University Press, 1982.
KRITCHEVSKY, DAVID. *Cholesterol.* New York: John Wiley & Sons, Inc., 1958.
LAITINEN, HERBERT A. and GALEN W. EWING, eds. *A History of Analytical Chemistry.* Washington, D.C.: American Chemical Society, 1977.
LANGRISH, BROWNE. *The Modern Theory and Practice of Physic.* London: A. Bettesworth and C. Hitch, 1735.
LAZAREV, F. V. and M. K. TRIFONOVA. "The Role of Apparatus in Cognition and Its Classification." In *Contributions to a Philosophy of Technology. Studies in the Structure of Thinking in the Technological Sciences,* Friedrich Rapp, ed. Dordrecht, Holland: D. Reidel Publishing Company, 1974.
LEICESTER, HENRY M. *Development of Biochemical Concepts From Ancient to Modern Times.* Cambridge, Massachusetts: Harvard University Press, 1974.
MAGATH, THOMAS B. "Minimal Albuminuria and Tests for Albumin in the Urine." In *The Kidney in Health and Disease,* Hilding Berglund and Grace Medes, eds. Philadelphia: Lea & Febiger, 1935.
MAJOR, RALPH H. *Classic Descriptions of Diesease. With Biographical Sketches of the Authors.* Springfield, Illinois: Charles C. Thomas, 1932.
MARCET, ALEXANDER. *An Essay on The Chemical History and Medical Treatment of Calculous Disorders,* 2nd ed. London: Longman, Hurst, Rees, Orme, and Brown, 1819.
MARSH, WALTON H. *Automation in Clinical Chemistry.* Springfield, Illinois: Charles C. Thomas, Publishers, 1963.
MAULITZ, RUSSELL C. *Morbid Appearances. The Anatomy of Pathology in the Early Nineteenth Century.* New York: Cambridge University Press, 1987.

MCCOLLUM, ELMER VERNER. "Stanley Rossiter Benedict, 1884–1936." In *Biographical Memoirs of the National Academy of Sciences*. Washington, D.C.: National Academy of Sciences Press, 1952.
MEITES, SAMUEL. *Otto Folin. America's First Clinical Biochemist*. Washington, D.C.: American Association for Clinical Chemistry, Inc., 1989.
Merck Index, 5th ed. Rahway, New Jersey: Merck & Co., Inc., 1940.
MITCHELL, F. L. "Present and Future Trends of Automation in Clinical Chemistry." In *Methods in Clinical Chemistry*, Vol. 1. 7th International Congress of Clinical Chemistry, Geneva/Evian, September 8–13, 1969. Baltimore: University Park Press, 1970.
MOORE, F. J. *A History of Chemistry*. New York: McGraw-Hill Book Co., Inc., 1918.
MORGAGNI, GIOVANNI BATTISTA. *De Sedibus et Causis Morborum per Anatomen Indagatis, libri quinqui*, (On the Seats and Causes of Disease Investigated by Anatomy, 5 Vols.). Venice: 1761.
MYERS, VICTOR CARYL. *Practical Chemical Analysis of Blood. A Book Designed as a Brief Survey of This Subject for Physicians and Laboratory Workers*, 2nd ed. St. Louis: C. V. Mosby Company, 1924.
NAITO, HERBERT K. "Cholesterol." In *Methods in Clinical Chemistry*, Amadeo J. Pesce and Lawrence A. Kaplan, eds. St. Louis: C. V. Mosby Company, 1987.
NATELSON, SAMUEL. *Microtechniques of Clinical Chemistry*, 2nd ed. Springfield, Illinois: Charles C. Thomas, Publisher, 1961.
NEUBAUER, CARL and JULIUS VOGEL. *A Guide to the Qualitative and Quantitative Analysis of the Urine, Designed Especially for The Use of Medical Men*, 4th ed. Translated from German by William Orlando Markham. London: The New Sydenham Society, 1863.
OBER, WILLIAM B., ed. *Great Men of Guy's*. Metuchen, New Jersey: Scarecrow Reprint Corp., 1973.
PACKARD, FRANCIS R. *Some Account of the Pennsylvania Hospital From Its First Rise to the Beginning of the Year 1938*. Philadelphia: Engle Press, 1938.
PAGEL, WALTER. *Paracelsus*, 2nd ed. Basel: Karger, 1982.
PANTON, PHILIP NOEL. *Clinical Pathology*. London: J. & A. Churchill, 1913.
PARTINGTON, J. R. *A Short History of Chemistry*, 2nd ed. London: Macmillan and Co., Limited, 1951.
-----. *A History of Chemistry*, (4 Vols.), Vols. 2, 3, 4. London: Macmillan & Co. Ltd., 1961, 1962, 1964.
PETERS, JOHN P. and DONALD D. VAN SLYKE. *Quantitative Clinical Chemistry*, (2 Vols.), Vol. 2, *Methods*. Baltimore: The Williams & Wilkins Company, 1932.
PETERSON, M. JEANNE. *The Medical Profession in Mid-Victorian London*. Berkeley: University of California Press, 1978.
POYNTER, F. N. L. "Medical Education in England Since 1600." In *The History of Medical Education*, C. D. O'Malley, ed. Berkeley: University of California Press, 1970.

BIBLIOGRAPHY 525

PROUT, WILLIAM. *An Inquiry Into the Nature and Treatment of Diabetes, Calculus, and Other Affections of the Urinary Organs*, 2nd ed. Philadelphia: Towar and Hogan, 1826.
----. *Chemistry, Meteorology, and the Function of Digestion, Considered With Reference to Natural Theology*. Philadelphia: Carey, Lea & Blanchard, 1834.
----. *On the Nature and Treatment of Stomach and Renal Diseases; Being an Inquiry Into the Connexion of Diabetes, Calculus, and Other Affections of the Kidney and Bladder, With Indigestion*, 5th ed. London: John Churchill, 1848.
PURDY, CHARLES W. *Practical Uranalysis and Urinary Diagnosis. A Manual for the Use of Physicians, Surgeons, and Students*, 4th ed. Philadelphia: The F. A. Davis Company, Publishers, 1898.
RALFE, CHARLES HENRY. *Clinical Chemistry*. Philadelphia: Henry C. Lea's Son & Co., 1883.
REES, G. O. *On the Analysis of the Blood and Urine, in Health and Disease. With Directions for the Analysis of Urinary Calculi*. London: Longman, Orme, Brown, Green, & Longmans, 1836.
----, G. OWEN. *On the Analysis of the Blood and Urine, in Health and Disease; and on The Treatment of Urinary Diseases*, 2nd ed. London: Longman, Brown, Green, and Longmans, 1845.
REISER, STANLEY JOEL. *Medicine and the Reign of Technology*. New York: Cambridge University Press, 1978.
ROLLO, JOHN. *Cases of the Diabetes Mellitus; ...*, 2nd ed. London: C. Dilly, 1798.
ROSENFELD, LOUIS. *Origins of Clinical Chemistry. The Evolution of Protein Analysis*. New York: Academic Press, 1982.
SAHLI, HERMANN. *A Treatise on Diagnostic Methods of Examination*, 4th ed., Francis P. Kinnicutt and Nath'l Bowditch Potter, eds. Translated from German. Philadelphia and London: W. B. Saunders Company, 1907.
SCHAFER, EDWARD. *An Introduction to the Study of the Endocrine Glands and Internal Secretions*. California: Stanford University, 1914.
SCHULTZE, H. E. and J. F. HEREMANS. *Molecular Biology of Human Proteins. Nature and Metabolism of Extracellular Proteins*, (2 Vols.), Vol. 1. Amsterdam: Elsevier, 1966.
SHAFFER, PHILIP ANDERSON. "Otto Folin, 1867–1934." In *Biographical Memoirs of the National Academy of Sciences*. Washington, D.C.: National Academy of Sciences Press, 1952.
SHARPEY-SCHAFER, E. *The Endocrine Organs. An Introduction to the Study of Internal Secretion*, 2nd ed. (2 Vols.), Vol. 2. London: Longmans, Green and Co. Ltd., 1926.
SHRYOCK, RICHARD HARRISON. *The Development of Modern Medicine. An Interpretation of the Social and Scientific Factors Involved*. Philadelphia: University of Pennsylvania Press, 1936.
SNELL, FOSTER DEE and CORNELIA T. SNELL. *Colorimetric Methods of Analysis, Including Some Turbidimetric and Nephelometric Methods*, 3rd ed. Vol. 1, *Theory-Instruments-pH*. New York: D. Van Nostrand Company, Inc., 1948.

STEPHENS, HARRISON. *Golden Past. Golden Future: The First Fifty Years of Beckman Instruments, Inc.* Claremont, California: Claremont University Center, 1985.
SZABADVÁRY, FERENC. *History of Analytical Chemistry.* Translated from Hungarian by Gyula Svehla. London and New York: Pergamon Press Ltd., 1966. (Reprinted in 1992. Langhorne, Pennsylvania: Gordon and Breach Science Publishers S. A.)
TEMKIN, OWSEI. *Galenism. Rise and Decline of a Medical Philosophy.* Ithaca, New York: Cornell University Press, 1973.
THACKRAY, ARNOLD and JEFFREY L. STURCHIO. "The Education of an Entrepreneur: The Early Career of Arnold Beckman." In *The Beckman Symposium on Biomedical Instrumentation* (April 12, 1985, New York City), Carol L. Moberg, ed. New York: The Rockefeller University in association with Beckman Instruments, Inc., 1986.
TROUSSEAU, A. *Lectures on Clinical Medicine*, 3rd ed. Translated from French by John Rose Cormack and P. Victor Bazire, (2 Vols.), Vol. 1. Philadelphia: Lindsay & Blakiston, 1873.
VAN SLYKE, DONALD D. and J. PLAZIN. *Micromanometric Analyses.* Baltimore: The Williams & Wilkins Company, 1961.
VAN SLYKE, DONALD DEXTER. *Oral History.* Bethesda, Maryland: National Library of Medicine, 1971.
VARLEY, HAROLD. *Practical Clinical Biochemistry*, 3rd ed. London: William Heinemann Medical Books, Ltd., 1963.
VAUGHAN, VICTOR C. *Hand-Book of Chemical Physiology and Pathology With Lectures Upon Normal and Abnormal Urine*, 3rd ed. Ann Arbor: Ann Arbor Printing and Publishing Company, 1880.
WARNER, JOHN HARLEY. *The Therapeutic Perspective. Medical Practice, Knowledge, and Identity in America, 1820–1885.* Cambridge, Massachusetts: Harvard University Press, 1986.
WASHBURN, FREDERICK A. *The Massachusetts General Hospital. Its Development, 1900–1935.* Boston: Houghton Mifflin Company, 1939.
WASSON, TYLER, ed. *Nobel Prize Winners.* New York: The H. W. Wilson Company, 1987.
WESSELOW, O. L. V. DE. *The Chemistry of the Blood in Clinical Medicine.* London: Ernest Benn Ltd., 1924.
WHITE, WILMA L., MARILYN M. ERICKSON, and SUE C. STEVENS. *Practical Automation for the Clinical Laboratory.* St. Louis: The C. V. Mosby Company, 1968.
WILLIAMS, CHARLES J. B. *Memoirs of Life and Work.* London: Smith, Elder, & Co., 1884.
WILLIS, THOMAS. Treatise III (Of Urines) and Treatise IX, Part I (Pharmaceutice Rationalis: or, an Exercitation of the Operations of Medicines in Humane Bodies: shewing The Signs, Causes, and Cures of most Distempers incident thereunto), in *Practice of Physick.* Translated from Latin by S. Pordage. London: T. Dring, C. Harper, and J. Leigh, 1684.

WINKELMAN, JAMES, DONALD C. CANNON, and S. LAWRENCE JACOBS. "Liver Function Tests, Including Bile Pigments." In *Clinical Chemistry. Principles and Technics*, 2nd ed., Richard J. Henry, Donald C. Cannon, and James W. Winkelman, eds. Hagerstown, Maryland: Harper & Row, 1974.
WITH, TORBEN K. *Bile Pigments. Chemical, Biological, and Clinical Aspects.* Translated from Danish by J. P. Kennedy. New York: Academic Press, 1968.
WOOD, F. C., KARL M. VOGEL, and L. W. FAMULENER. *Laboratory Technique. The Methods Employed at St. Luke's Hospital, New York*. East Stroudsburg, Pennsylvania: The Press Publishing Company, 1922.
YOUNG, DONALD S. "Automation." In *Fundamentals of Clinical Chemistry*, 2nd ed., Norbert W. Tietz, ed. Philadelphia: W. B. Saunders Company, 1976.

MISCELLANEOUS

BAUSCH & LOMB. "A History of Innovation." (company publication)
KIMBLE KLIPPINGS. Memorial issue for Col. Kimble. (company publication)
The Echo. Franklin Lakes, New Jersey: Becton Dickinson and Company, September 1991.

JOURNALS

Acta Medica Scandinavica
American Journal of Cancer
American Journal of Clinical Pathology
American Journal of Diseases of Children
American Journal of Insanity
American Journal of Medical Sciences
American Journal of Physiology
American Laboratory
American Scientist
Analyst
Analytical Biochemistry
Analytical Chemistry
Annalen auf Chemie und Pharmacie
Annalen der Chemie und Pharmacie
Annalen der Physik
Annalen der Physik und Chemie
Annales de Chimie et de Physique
Annali Universale di Medicina e Chirurgia (Milano)
Annals of Clinical and Laboratory Science
Annals of Clinical Biochemistry
Annals of Internal Medicine
Annals of Medical History
Annals of Medicine and Surgery

Annals of the New York Academy of Sciences
Annals of Science
Annals of Tropical Medicine and Parasitology
Annual Review of Biochemistry
Applied Scientific Research
Archiv für Experimentell Pathologie und Pharmakologie
Archives of Internal Medicine
Archives of Pathology
Archives of Pathology and Laboratory Medicine

Beitrage zur Chemische Physiologie und Pathologie
Biochemische Zeitschrift
Biochimica et Biophysica Acta
Biological Chemistry Hoppe-Seyler
Boston Medical and Surgical Journal
British and Foreign Medical Review
British Journal of Experimental Pathology
British Medical Bulletin
British Medical Journal
Bulletin de la Societe Chimique de Paris
Bulletin of the American Academy of Medicine
Bulletin of the American College of Surgeons
Bulletin of the History of Medicine
Bulletin of the Johns Hopkins Hospital
Bulletin of the New York Academy of Medicine
Bulletin of the Scientific Instrument Company
Bulletin of the U.S. Medical Department
Bumed News Letter

Canada Medical and Surgical Journal
Canadian Medical Association Journal
Cancer Research
Chemical Abstracts
Chemical & Engineering News
Chemical News (London)
Chemical Reviews
Chymia
Clinica Chimica Acta
Clinical Biochemistry
Clinical Chemistry
Clio Medica
Comptes Rendus des Séances de la Societe de Biologie et de Ses Filiales
Comptes Rendus Hebdomadaires des Séances de l'Académie des Sciences

Comptes Rendus Societe de Biologie
Connecticut State Medical Journal

Deutsches Archiv für Klinische Medizin
Dublin Journal of Medical Science

Edinburgh Medical and Surgical Journal
Endeavour
Enzymologia
Ergebnisse der Physiologie
Experimentia

Febs Letters
Federation Proceedings

Gastroenterology
Geriatrics
Guy's Hospital Reports

Harvey Lectures
Helvetica Chimica Acta
History
Hoppe-Seyler's Zeitschrift für Physiologische Chemie

Industrial and Engineering Chemistry. Analytical Edition
Interstate Medical Journal
Investigative Urology
Isis

Journal für Practische Chemie
Journal of Biological Chemistry
Journal of Chemical Education
Journal of Chronic Diseases
Journal of Clinical Chemistry and Clinical Biochemistry (European)
Journal of Clinical Endocrinology and Metabolism
Journal of Clinical Investigation
Journal of Clinical Pathology
Journal of Experimental Medicine
Journal of General Physiology
Journal of Industrial and Engineering Chemistry
Journal of Intensive Care Medicine
Journal of Laboratory and Clinical Medicine
Journal of Nutrition
Journal of Physiology

Journal of the American Chemical Society
Journal of the American Medical Association
Journal of the History of Biology
Journal of the History of Medicine and Allied Sciences
Journal of the International Federation of Clinical Chemistry
Journal of the Iowa State Medical Society
Journal of the Medical Association of the State of Alabama
Journal of the Optical Society of America (and Review of Scientific Instruments)
Journal of the Royal Society of Medicine

Lancet
Liebig's Annalen der Chemie und Pharmacie
London Medical Gazette
London Medical Journal

Medical and Surgical Reporter
Medical History
Medical Journal and Record
Medical Journal of Australia
Medical Life
Medical News
Medical Observations and Inquiries (London)
Medical Record
Medical Times
Medico-Chirurgical Review, and Journal of Practical Medicine
Medico-Chirurgical Transactions
Microchemical Journal
Middlesex Hospital Journal
Mikrochemie
Modern Hospital
Muenchener Medizinische Wochenschrift

Nature (London)
Naturwissenschaften
New England Journal of Medicine
New York Medical Journal
New York State Journal of Medicine
Notes and Records of the Royal Society of London

Pediatrics
Pennsylvania Medical Journal
Perspectives in Biology and Medicine
Philippine Journal of Science
Philosophical Magazine

Philosophical Transactions of the Royal Society of London
Physiological Reviews
Proceedings of the Royal Australian Chemical Institute
Proceedings of the Royal Society of London
Proceedings of the Royal Society of Medicine
Proceedings of the Society for Experimental Biology and Medicine
Pure & Applied Chemistry

Quarterly Journal of Medicine

Report of the British Association for the Advancement of Science
Revue Hebdomadaire de Chimie Scientifique et Industrielle

Scandinavian Journal of Clinical & Laboratory Investigation
Science
Scientific American
Skandinavisches Archiv für Physiologie
Spectrochimica Acta
Stanford Medical Bulletin

Tractrix
Transactions of the American Association of Genito-Urinary Surgeons
Transactions of the Association of American Physicians
Transactions of the Clinical Society of London
Transactions of the Faraday Society
Transactions of the Medical Society of London

Urology

Wiener Klinische Wochenschrift

Yale Journal of Biology and Medicine

Zeitschrift für Analytische Chemie
Zeitschrift für Klinische Medizin

AUTHOR INDEX

Portrait pages are in italics.

Abbe, Ernst 435
Abderhalden, Emil 295, 298
Abel, John J. 179, 341, 342
Abell, Liese L. 380, 381
Actuarius, Johannes 7
Adams, W. 452
Addison, Thomas 84
Alexander I. 153
Alkemade, C. T. J. 371
Allen, William 115, 116, 128 (note 29)
Alsberg, Carl L. 247
Ambrosioni, Felice 159
Amend, Bernard G. 235–36, 256
Andersen, E. 387
Anderson, Norman 495
Andral, Gabriel 92, 93, 95, 122, 439
Ångström, Anders J. 365
Aretaesus 156
Argand, Aimé 363
Aristotle 12, 16, 24, 25, 437
Armstrong, A. Riley 404
Arrhenius, Svante A. 203
Ashwood, Edward R. 477
Astrup, Poul 513
Autenrieth, W. 379, 380, 422
Avicenna 157
Ayer, Frederick F. 226
Ayer, Josephine M. 226

Babcock, Stephen 237, 238
Babington, William 114, 115
Babo, Lambert H. C. K. von 200
Baglivi, Giorgio 18
Bang, Ivar C. 202, 301, *302*, 303, 320, 321, 325

Banting, Frederick G. 177–79, 183 (note 58)
Barker, Fordyce 306
Barker, Lewellys F. 467n
Barker, S. B. 475
Barlow, George H. 90
Barnes, R. Bowling 367, 368
Barreswil, Louis C. 160, 321
Baruch, Hans 491
Bausch, John J. 240, 256, 258, 259
Beale, Lionel S. 219, 220, 225
Beaumont, William 46, 47
Becker, Christopher 72
Becker, Christopher A. 72
Beckman, Arnold O. 207, 208, *209*, 210, 214 (note 49), 459, 461, 462
Becquerel, Alfred 92
Becton, Maxwell W. 310
Beeler, Myrton F. 147
Beer, August 260–62
Behre, Jeanette A. 329, 330
Belk, William P. 467
Bell, Richard D. 357
Benedict, Stanley R. 252, 269, 275, 287, *288*, 289, 300, 305, 319, 320, 321, 322, 324, 329, 330, 334–36, 357, 445, 468, 469, 470
Bennett, J. Hughes 219, 225
Benotti, Joseph 471
Berg, Wilhelm 452
Bergeim, Olaf 355
Berkman, Sam 475
Bernard, Claude 46, 108, 135, 136, 143, 159, 163, *164*, 165, 204, 249, 398
Bernard, Félix 261

AUTHOR INDEX

Berson, Solomon A. 478, *479*, 481
Berthelot, Marcellin 69, *70*, 71, 377
Berzelius, Jöns J. 41, 49, 63, *64*, 65, 67, 92, 107, 113, 122, 190, 395–96, 398, 418n
Bessey, Otto A. 405
Best, Charles H. 177, 178, 179, 183 (note 58), 321
Bidder, Friedrich 47
Biot, Jean B. 162, 263
Bird, Golding 48, 50–51, 93, 119
Bjerrum, Niels 206
Black, Joseph 15, *31*, 32, 39, 114, 115, 116
Blackall, John 86
Blix, Magnus G. 196
Bloor, Walter R. 284, 301, 379–80, 381
Blum, E. 370
Bodansky, Aaron 404, 470
Boedeker, C. H. D. 49
Boerhaave, Hermann 40, 41, 49
Bois-Reymond, Emil du 108
Booth, James C. 162
Borelli, Giovanni A. 18
Bostock, John 86, 88, 89, 90, 93, 114
Böttger, Wilhelm C. 140, 203
Bouguer, Pierre 260, 261
Bourget, Louis 139
Bowers, Jr., George N. 405
Bowman, John E. 134–35
Boyle, Robert 23, *24*, 25–28, 117, 122, 185, 186, 351, 483
Brewster, David 263
Brian, Thomas 10–11
Briggs, A. P. 357
Bright, Richard 43, 50, 84–86, *87*, 88–91, 93, 99, 101, 102, 106, 140
Brodie, Benjamin 114
Broida, Dan 477, 478
Brønsted, Johannes N. 206
Brown, John 227
Brown, Michael S. 377
Brunner, Johann C. 158

Buchanan, Andrew 438
Buchner, Eduard 399, 400, 401, 413 (note 12)
Buchner, Hans 399
Bunge, Gustav P. A. von 200
Bunsen, Robert W. E. 187, 188, 190, 236, 337, 362–65, 371
Burchard, H. 378–81
Burtis, Carl A. 477

Camac, C. N. B. 230
Cameron, F. K. 360
Campbell, Walter R. 329, 421
Cannizzaro, Stanislao 211 (note 2)
Cantani, Arnoldo 154
Caraway, Wendell T. 336
Carr, Julius J. 470
Cawley, Thomas 157
Celsus, Aulus C. 4, 5, 18
Chaney, Albert 475
Cherry, Ian S. 403
Chevreul, Michel E. 157, 329, 377
Chiene, John 221
Chittenden, Russell H. 247, *248*, 249–51, 275, 276, 305, 443, 448
Clanny, William R. 151–53
Clark, C. P. 146
Clark, E. P. 355
Clark, William M. 207
Clausen, S. W. 360
Cole, Rufus 292, 294, 295
Collardeau, F. 254
Collip, James B. 178, 179, 183 (note 58), 355
Consden, R. 428
Cotlove, Ernest 353
Cotugno, Domenico 85
Count Rumford 116
Courtois, Bernard 358
Cowles, Edward 278
Crandall, Jr., Lanthan A. 403
Cremer, Max 207
Cromwell, Oliver 37

Cronstedt, Alexander F. von 76 (note 17)
Crookes, William 365
Cruickshank, William 40, 50, 85, 89, 157, 158
Cullen, Glenn E. 173, 295, 300, 338, 354
Cullen, William 115, 157

DaCosta, Jr., John C. 307
Daguerre, Louis J. 263
Daland, Judson 196, 198, 232
Dalton, John 358
Dam, Henrik 283
Darwin, Charles 159
Darwin, Charles R. 159
David, Louis 33, 34
Davidsohn, Israel 320, 449
Davies, Richard 29
Davy, Humphry 114, 116, 186, 357, 358, *359*, 395
Day, P. 452
Dekkers, Fredericus 85, 132
Denigès, G. 357
Denis, Prosper S. 377, 418, 438
Denis, Willey G. 260, 284, 286, 315 (note 29), 333, 383
Desaga, Peter 236, 363
Descroizilles, François A. H. 73, 76 (note 24)
Dickinson, Sr., Fairleigh S. 310
Döbereiner, Johann W. 117
Dobson, Matthew 157, 161
Dock, George 228, 229, 243 (note 40)
Doisy, Edward A. 283, 357
Dole, Malcolm 207
Dole, Vincent P. 300
Donleavy, John J. 352
Donné, Alfred 218
Dotti, Louis B. 470, 471
Duboscq, Jules 254, 255-60, 262-63, *264*, 276, 279, 286, 304, 448, 451
Dubowski, Kurt 327
Dubský, J. V. 352

Duclaux, Émile P. 398
Duhamel du Monceau, Henri L. 361
Dumas, Jean B. A. 57, *58*, 59, 101, 113, 142, 165, 298
Dunlop, D. M. 449
Durrum, Emmett L. 429

Ehrlich, Paul 141, 192, 385
Eijkman, Christiaan 346 (note 46)
Eimer, Carl 236, 256
Einhorn, Max 161, 228, 229
Eisenman, George 370
Elevitch, Franklin R. 433
Elliotson, John 52 (note 18)
Elster, J. 452
Enenkel, H. J. 429
Enriques, Eugen 387
Esbach, Georges H. 139, 146, 147, 228
Evelyn, Kenneth A. 387, 455
Everett, Mark R. 451

Failyer, G. H. 360
Faraday, Michael 53, 116, 116n, 358, 424
Fearon, William R. 339
Fehling, Hermann 124, 140, 143, 160, 161, 319
Fine, Joseph 422
Fine, Morris S. 444
Fischer, Edgar 429
Fischer, Emil 165, *166*, 167, 295, 401
Fischer, Hans 211 (note 4)
Fisher, Chester G. 236
Fishman, William H. 406
Fiske, Cyrus H. 284, 357
Flexner, Abraham 269, *270*, 271, 272, 313 (note 1), 504
Flexner, Simon 226, 242 (note 32), 271, 289, *291*, 292, 294, 295
Flint, Austin 139
Folin, Otto K. 147, 169, 252, 255, 269, 271, 275, 276, *277*, 278-89, *290*, 299, 304, 305, 307, 316 (note 46),

321–25, 332–36, 337, 338, 445, 447, 468, 469, 470
Forbes, John 80
Foster, G. L. 321
Foucault, Léon 263
Fourcroy, Antoine F. de 37, 38, 40, 49, 56, 77, *78*, 79, 140, 217
Frabot, C. 333
Franklin, Benjamin 37
Fraunhofer, Joseph 362, 364
Free, Alfred 388
Free, Helen M. 388
Freeman, James A. 147
Fresenius, Carl R. 360
Fridericia, L. S. 169
Friedman, Max M. 470, 471
Friedman, Sydney M. 370
Froesch, E. Rudolf 327
Fuld, E. 419
Funk, Albert 379, 380

Gaebler, Oliver H. 469
Galat, A. 326
Galen 5, 14, 16, 18, 49, 156, 185
Galilei, Galileo 483
Gamble, James L. 175
Gammeltoft, S. A. 206
Garrod, Alfred B. 40, 93, 138, 139, 143, 232, 332
Garrod, Archibald E. 40
Gates, Frederick T. 289
Gavarret, Jules 92
Gay-Lussac, Joseph L. 41, 73, 108, 358
Geiger, Phillip L. 112
Geitel, N. 452
Geraghty, J. T. 341
Gerhardt, Carl 142, 168
Gibson, K. S. 452
Gies, William J. 252, 422
Gmelin, Leopold 46, 137, 141, 159, 232, 356, 384, 388
Goethe, Johann W. von 117
Goldstein, Joseph L. 377

Golub, Orville 475
Gomberg, Moses 292
Gortner, Ross A. vii
Göttling, Johann F. A. 117
Gowers, William R. 192, 193, 195
Graaf, Regnier de 305
Gradwohl, R. B. H. 234, 244 (note 57)
Greenberg, David M. 333
Greiner, Emil 236
Grigaut, A. 379
Gróf, P. 387
Grunbaum, B. W. 431
Gutman, Alexander B. 406
Gutman, Ethel B. 406

Haber, Fritz J. 207, 214 (note 46)
Haldane, John S. 169, 195
Haldimand, Jane 53 (note 33)
Hales Stephan 30
Hallervorden, Eugen 167
Hallwachs, Wilhelm 452
Halverson, John O. 355
Hammarsten, Olof 276, 301, 438
Hanna, Marion I. 421
Harrison, G. A. 449
Harrow, Benjamin 451
Hartree, E. F. 327
Harvey, William 27, 37, 185, 305
Hassall, Arthur H. 133, 134
Hasselbalch, Karl A. 206
Hassenfratz, Jean H. 165
Hastings, A. Baird 300
Hawk, Philip B. 142, 202, 252, 448, 470
Hedin, Sven 196, 197
Heidelberger, Michael 300
Heine, Carl 253
Heller, Johann F. 120, *121*, 124, 139, 140
Henderson, Lawrence J. 169, 204, 206, 247
Henry VIII 10
Henry, John B. 320, 449
Henry, Richard J. 320, 475–77
Henseleit, Kurt 299

Herapath, Thornton J. 253
Herbert, Freda K. 406
Herter, Christian A. 251
Hewson, William 438
Hildebrand, Joel H. 203
Hiller, Alma E. 300, 422
Hippocrates 2, 4, 18, 37, 156
Hitchings, George H. 283
Hjertén, Stellan 433
Höber, Rudolf O. A. 203
Hodgkin, Thomas 84, 95, 218
Hoffman, Friedrich 351
Hoffman, William S. 451, 458
Hofmeister, Franz 166, 204, 401, 422
Home, Everard 114
Home, Francis 157
Hooke, Robert 185, 215
Hope, Thomas C. 115
Hopkins, F. Gowlaud 332
Hoppe-Seyler, Felix 188, *189*, 190, 191, 249, 255
Horsford, Eben N. 235, 236
Houghton, Arthur A. 239
Houghton, Sr., Amory 239
Houton de Labillardière, Jacques J. 254
Howe, Paul E. 419–21
Howland, John 155
Hüfner, Carl G. von 195
Huggett, A. St. G. 328
Huggins, Charles 405
Humboldt, Alexander von 110
Hünefeld, Friedrich L. 245
Hunter, Andrew 329
Hunter, Charles 306

Israeli, Isaac 10

Jackson, Charles T. 162
Jacobs, Walter A. *296*
Jacquelain, Augustin 253
Jaffe, Max 141, 255, 280, 280n, 345 (note 39), 330, 331
Jaksch, Rudolf von 139–44

Jaquet, Jules A. 173
Jendrassik, L. 387
Johnson, Lyndon B. 301
Johnson, W. B. 86
Jones, Henry Bence 43, 47, 50, 93, 102, *103*, 104–08, 119, 127 (note 9), 138, 140, 162, 219, 365
Jones, Walter 252
Joseph, Glenn 208, 209
Judaeus, Isaac 10

Kahn, Joseph 470
Kantor, J. L. 422
Kaplan, Lawrence A. 477
Kauder, Gustave 418
Kay, Herbert D. 404
Keilin, D. 327
Kekulé, Friedrich A. 71
Kendricks, Arthur 202
Keston, Albert S. 328
Kimble, Evan E. 238
King, Earl J. 404, 405, 474
Kingsbury, F. B. 145, 146
Kingsley, George R. 421, 423
Kingzett, Charles T. 135–36
Kirchhoff, Gottlieb S. 395
Kirchhoff, Gustave R. 190, 362–65, 371
Kjeldahl, Johan G. C. T. 59, 60, *61*, 62, 63, 141, 142, 283, 417, 435–36
Kleiner, Israel S. 451
Kleiner, Joseph J. 310–12
Klemensiewicz, Zygmunt 207
Kletzinsky, V. 251
Klobusitzky, D. von 428
Knoop, Franz 295
Koch, Frederick C. 451
Koch, Robert 231
Kohn, J. 431
Kolbe, Adolph W. H. 69, 71
Kolthoff, I. M. 360
König, P. 427–29
Kossel, Albrecht 276, 292
Kramer, Benjamin 355, 360

Krebs, Hans 299
Kühne, Wilhelm F. 250, 250n, 292, *397*, 398
Kulz, Eduard 167
Kunitz, Moses 402
Kussmaul, Adolf 167, 168
Kuttner, Theodore 357

Laennec, René T. H. 80, 81
Lagrange, Joseph L. 163
Lambert, Johann H. 260, 261
Lampadius, Wilhelm A. 253
Lang, Konrad 352
Langrish, Browne 28, 29, 122, 219
Latta, Thomas 152, 177
Laval, Gustav de 200
Lavoisier, Antoine L. 32, 33, *34*, 35 (note 14), 67, 71, 115, 163, 358
Leeuwenhoek, Anton van 215, 218
Legal, Emmo 141, 142, 168
Lemery, Nicolas 55
Lenard, Philipp 452
Lengyel, B. 370
Leopold I. 37
Lerner, F. 406
Leube, Wilhelm O. 145, 228
Levene, Phoebus A. T. 292, *294*, 295
Lewis, Gilbert N. 207
Liebermann, Carl T. 378–81
Liebig, Justus 38, 49, 67, 75, 105–08, *109*, 110–13, 114, 117, 119, 120, 122, 124, 133, 137, 142, 153, 163, 165, 249, 279, 298, 329, 337, 352, 396, 398–400, 413 (note 10), 418
Linacre, Thomas 10
Lister, Joseph 39, 216, 219, 307
Lister, Joseph J. 216, 218, 241 (note 2)
Locke, John 1, 27
Loeb, Jacques 276
Lomb, Henry 240, 256, 258, 259
Longsworth, Lewis G. 425
Louis I of Hesse 109, 110
Louis XIV 37

Lower, Richard 186
Lowry, Oliver H. 405, 477, 478
Lowry, Thomas M. 206–07
Ludwig, Carl F. W. 187, 188, 249
Luer, Hermann W. 306, 310
Lundegårdh, Henrik G. 366–68
Lundsgaard, Christen 206, 300

Macallum, A. B. 333
Macdonald, Dean 342
MacInnes, Duncan A. 207
Maclean, John 116
Macleod, John J. R. 178, 179, 183 (note 58)
Magath, Thomas B. 147
Magendie, François 46, 57, 163
Magnus, Heinrich G. 187, 188
Maimonides, Moses 5
Malloy, Helga T. 387
Malpighi, Marcello 437, 438
Marcet, Alexander *48*, 49, 50, 53 (note 33), 86, 93, 114, 115, 119
Marggraf, Sigismund A. 361
Marriott, William M. 155
Martin-Solon, Fernand 85
Mathews, Albert P. 252, 451
Mayo, Charles H. 503
Mayo, William J. 503
Mayow, John 186
McComb, Robert B. 405
McCrudden, Francis H. 293–94
Mckenna, Mary H. 470
McLean, Frederick C. 300
Méhu, Camille 418
Mendel, Lafayette B. 250, 252, 287, 443, 448
Mendeleev, Dimitri I. 211 (note 2)
Mène, Charles 255
Menten, Maud L. 408
Mering, Joseph F. von 177
Mettler, Erhard 72
Meulengracht, E. 385
Meyer, Gustave M. *296*, 298, 299

AUTHOR INDEX

Meyer, Lothar 187, 188, 190, 211 (note 2)
M'Gregor, Robert 101, 159
Michaelis, Leonor 403, 408
Miescher, Johann F. 292
Milatz, J. M. W. 371
Millon, Auguste N. E. 124, 132
Mink, Frieda 422
Minkowski, Oscar 167–68, 177
Mirsky, I. Arthur 480, 481
Mitscherlich, Eilhardt 160
Mohr, Karl F. 73, *74*, 75, 132, 351
Moore, N. S. 436
Morgagni, Giovanni B. 1
Mulder, Gerardus J. 140, 142, 245, 418, 418n
Müller, Alexander 254
Müller, D. 327
Müller, Ralph H. 457
Myers, Victor C. 195, 252, 305, 330, 340, 354, 361, 403, 443, *444*, 445, 446, 470, 485

Nakashima, Miyoshi 370
Napoleon I 37, 79, 358
Napoleon III 37
Natelson, Samuel 175, 176, 470
Naunyn, Bernhard 167, 168
Needham, Joseph 402
Nessler, Julius 60, 62, 253, 255, 279, 283, 337, 338, 419
Neubauer, Carl 131, 132, 141
Neuner, A. 306
Newton, Isaac 37, 361, 483
Nixon, D. A. 328
Noble, Edward C. 178
Northrop, John H. 402
Nydick, A. J. 470, 471
Nylander, E. 328

Oertling, L. 72
Offer, Th. R. 333
Ohm, Georg S. 424

Ohmori, Yoshihisa 405
Opie, Eugene L. 177
Ormsby, Andrew A. 340
Oser, Bernard L. 142, 448
O'Shaughnessy, William B. 151–53, 156
Osler, William 200, 222–24, 231, 232, 289, 505
Osmond, F. 357
Ostwald, Wilhelm 203

Page, Irvine 300
Palmer, Walter W. 169, 300
Panton, P. N. 232
Panum, Peter L. 418
Paracelsus 4, 7, 12, 16, 24, 25, 49, 117
Pasteur, Louis 399, 400
Pauling, Linus 211 (note 4)
Pavy, Fredrick W. 145, 161
Payen, Anselme 253, 395
Péligot, Eugène M. 59
Pellin, Ph. 255
Pelouze, Theophile J. 163
Pepper, Jr., William 224
Pepys, Samuel 37
Persoz, Jean F. 395
Perutz, Max 211 (note 4)
Pesce, Amadeo J. 477
Peter the Great 37
Peters, John P. 147, 300, 436, 449, 450, 476, 477
Pettenkofer, Max J. von 124, *125*, 129 (note 44), 137, 279, 329
Petters, Wilhelm 142
Pflüger, Eduard F. W. 188, 191
Philip, Wilson 45
Plateau, Joseph A. 263
Pliny the Elder 26, 253
Porges, Otto 419
Pott, R. 422
Poulletier de la Salle, François P. L. 377
Pravaz, Charles G. 306
Prévost, Jean L. 101
Priestley, John G. 169

Priestley, Joseph 39
Pritchett, Henry S. 269, 271
Prout, William 19, 41, *42*, 43–50, 52 (notes 18, 24), 93, 114, 119, 154
Purdy, Charles W. 144, 145, 147, 200–02, 448

Ralfe, Charles H. 137–39
Rappleye, W. C. 352
Réaumur, René de 19
Rees, George O. 90, 91, 93, 99, *100*, 101, 102, 119, 140, 159
Rehberg, Poul B. 331
Reimann, Stanley P. 453
Reiner, Miriam 470
Reinhold, John G. 471
Reiss, Emil 435
Remsen, Ira 246
Renold, Albert E. 327
Reuss, Ferdinand F. 424
Riegler, E. 422
Rittenberg, David 280, 281
Ritthausen, H. 422
Roberts, William 135
Robertson, T. Brailsford 435
Robison, Robert 404
Rockefeller, John D. 289, 292
Rogers, Leonard 155
Rokitansky, Karl F. von 122
Rollo, John 158
Rona, Peter 403
Rose, Ferdinand 139, 422
Rosenthal, Sanford M. 342
Rouelle, Hilaire M. 40
Rowntree, L. G. 341, 342
Rush, Benjamin 116
Rynd, Francis 306

Sahli, Hermann 139, 193
Saifer, Abraham 328
St. Martin, Alexis 46
Saliceti, Guglielmo 85
Salkowski, Ernst L. 276, 332, 378

Sanford, Arthur H. 448–49
Schales, Otto 352
Schales, Selma S. 352
Scheele, Carl W. 37–38, 358
Scherer, Johann J. 120, 122, *123*
Schiff, Hugo 423
Schmidt, Alexander 154, 418
Schmidt, Carl E. H. 47, 153–55, 200, 354
Schmiedeberg, Oswald 200, 378
Schoenheimer, Rudolf 280, 281, 380, 381
Schoenlein, Johann L. 120
Schuhknecht, Wolfgang 367
Schwann, Theodor A. H. 396
Schwarzenbach, Gerold 356
Segalove, Milton 475
Sellards, Andrew W. 155
Sendroy, Julius 300
Shaffer, Philip A. 252, 284
Shakespeare, William 10
Sharpey-Schäfer, Edward A. 177
Shinowara, George Y. 405
Simon, Johann F. 120
Sivó, Rudolf 387
Skeggs, Jr., Leonard T. 484, 485, *486*
Snapper, J. 385
Snel, Willebrord 435
Sobel, Albert E. 470
Sobotka, Harry H. 470
Soleil, Jean B. F. 104, 262, 263
Solomon xviii
Somogyi, Michael 325–26, 403
Sondern, Frederick 233
Sørensen, Søren P. L. 148, 168, 204, *205*
Soret, C. 366
Spallanzani, Lazzaro 19
Sperry, Warren M. 380, 381
Spiro, K. 419
Stadelmann, Ernst 167
Stahl, Georg E. 66
Stanley, Wendell M. 402
Steel, Matthew 451

AUTHOR INDEX

Stenbeck, Thor 196–98, 199
Stevens, William 151, 152
Stewart, C. P. 449
Stieglitz, Julius 276
Strubell, Alexander 435
SubbaRow, Yellapragada 284, 357
Sumner, James B. 283, 401, 402
Sunderman, Sr., F. William 209, 436, 467
Suzuki, Umetaro 404
Sydenham, Thomas 2, 37
Sylvius, Franciscus de le Boë 16, 17, 18

Takeuchi, T. 338
Talalay, Paul 405
Tatum, Whitall 237, 238
Taylor, Norman B. 179
Teller, Joseph D. 328
Thannhauser, J. S. 387
Theis, Ruth C. 357
Thenard, Louis J. 108–09, 157
Thompson, Benjamin 116
Thomson, Thomas 114
Thudichum, Johann L. W. 247, 249, 265 (note 11)
Thurneisser, Leonhardt 7
Tiedemann, Friedrich 46, 159
Tiemann, George 306
Tietz, Norbert A. 320, 477
Tisdall, Frederick F. 355, 357, 360
Tiselius, Arne W. K. 424–27, 429
Todd, James C. 448–49
Trinder, P. 380
Trommer, Carl A. 134, 135, 140, 143, 159, 160
Trousseau, Armand 94, 95, 138
Trtílek, J. 352
Turba, F. 429

Van den Bergh, A. A. Hymans 385, 387
Van Helmont, Johannes B. 12, *13*, 14, 15, 16, 18, 32, 49

Van Slyke, Donald D. 147, 169–75, 178, 188, 195, 252, 269, 275, 285, 292, *293*, 295, *296*, 297–301, 305, 330, 335, 338, 339, 341, 352, 354, 436, 445, 449–51, 465, 468, 469, 470, 474, 476, 477, 483
Van Slyke, Lucius L. 292
Varrentrapp, Franz 57, 59, 60
Vaughan, Victor C. 136–37
Vauquelin, Nicholas L. 37, 38, 40, 49, 140
Vesalius, Andreas 186
Vierordt, Karl von 192, 365
Vinci, Leonardo da 483
Virchow, Rudolf 188, 219, 225, 438
Vogel, Heinrich A. 160
Vogel, Julius 131, 132
Volhard, Jacob 351–52
Volta, Alessandro 424

Waibel, Ferdinand 367
Wakley, Thomas 52 (note 24)
Walsh, Alan 371
Walter, B. 261
Walter, Friedrich 167
Wang, M. C. 406
Wanklyn, James A. 60
Weichselbaum, T. E. 423
Welch, William H. 225, 289
Wells, Benjamin B. 449
Wells, William C. 86
Wesselow, O. L. V. de 447
West, John B. 179
Wheatstone, Charles 263, 365
White, Edwin C. 342
Whitehorn, J. C. 352
Widal, Fernand 231, 341, 354
Wiedemann, Gustave 422
Wieland, Theodor 429
Will, Heinrich 57, 59, 60
Williams, Charles J. B. 94
Willis, Thomas 10, 12, 157, 180 (note 16)
Wilson, Woodrow 310

Windaus, Adolf 378, 379, 380
Winkler, Lajos 62
Wöhler, Friedrich 38, 41, 49, 67, *68*, 69, 71, 108, 120, 162, 396
Wohlgemuth, Julius 402–03
Wollaston, William H. 39, 40, 49, 50, 114, 158, 159, 361
Wood, Alexander 306
Wren, Christopher 305

Wu, Hsien 284, 285, 300, 322, 323, 333, 334, 338

Yalow, Rosalyn S. 478, *480*, 481
Young, John R. 19

Zak, Bennie 380, 381
Zeiss, Carl 451, 484
Ziemssen, Hugo von 225

SUBJECT INDEX

Aberration
 chromatic 215, 216
 spherical 215, 216
aca DuPont 494, 497
Acetoacetic acid 142, 168
Acetone, tests 141–142, 168
Acetylene, synthesis of 69, 71
Acid hematin 193
Acid intoxication 168
Acid-base balance 155, 172, 297
Acidmeter 209
Acidosis
 definition 169, 297
 diabetic 156, 167, 168, 297
 nephritic 168, 282
 origin of term 168
Actuarius 7
Ad eundem 79
Advances in Clinical Chemistry 472
Aesculapian quacks 124
Air-pump 167, 168, 169, 186, 187, 188, 190
Albumin
 binding reagents 421–22
 biuret reaction 139
 color reactions in urine 137
 Esbach's estimation of 139
 gravimetric procedure 132, 137
 origin of term 85, 418
Albuminometer, Esbach 146–47
Albuminuria 85, 145
 Bright's disease 84–91
 edema 85, 86, 88, 89
 false-positive 89–90
 insurance companies 145–46
 Kingsbury-Clark standards 145–46

nitric acid ring test 120, 124, 139, 140
 origin of term 85
 orthostatic 145
 postural 145
 test strips 144
Albustix 148
Alchemists laboratory Frontispiece
Alchemy
 kitchen chemistry 3
 philosopher's stone 23
 quintessential spirit 3
 secrecy 3
 symbols 33
 transmutation 3, 4, 23
Alcohol spirit lamp 50, 122, 127
 (note 9), 135
Alcoholic intoxication 265 (note 11)
Alkali reserve 155, 173
Alkaline tide 104–05
Alveolar air 169
Amateur scientists 23
American Association of (for) Clinical Chemistry 287, 467, 470, 472
 organizational meetings 470–71
 publications of 472
American Chemical Journal 246
American Chemical Society 236, 251
 Division of Biological Chemistry 251
American College of Surgeons 234
American Expeditionary Forces 233
American Journal of Physiology 251
American Journal of Science and Arts 250
American Medical Association 272, 273, 443, 469
 chemistry demonstrations 443

543

American Physiological Society 251
American Society of Biological
 Chemists 251, 252, 286, 287, 322,
 472
American Society of Clinical
 Pathologists 234, 467
 Board of Registry 234–35
American Thyroid Association
 thyroxine iodine 482
Amino acids 250
 deamination, site of 299
 formation of urea 299
 plasma levels 298, 299
 resynthesis to protein 298, 299
Ammonium cyanate, rearrangement
 of 67
Amylase 402–03
Analytical balance 7, 32, 71–72, 235
 extinction, threat of 497
Analytical techniques 252
 colorimetry 253–55
 gasometry 169–175, 176
 gravimetry 101, 132, 143, 148, 332, 417
 titrimetry 72–75
Analyzers
 batch analysis 490–96
 continuous flow 485–88
 random access 494, 496, 497
Anatomical furnace 7, 8
Anatomy, investigation of 2
 chemical, of urine 14
 dissection, objection to 80
 pathology, local 1, 83
 postmortem 1, 2, 18, 80, 218, 220, 224
  clinical correlation with 83, 84, 91
 Warburton Anatomy Act of 1832 80
Animal chemistry 55, 56, 245
Anima, directive forces of 66
Animal Chemistry 112–113, 119
 critique of 113
 impact of 113
Animal Chemistry Society 114
Animal materials 56

Annalen, Liebig's 112, 250, 351
Antigen-antibody reaction 482
 non-isotopic labels 482
Antisepsis 39, 216, 307
Aphorisms, medical 5
Aplanatic foci, microscope 216
Apocrypha xviii
Apothecaries' Act of 1815 50, 117
Apparatus, chemical 135, 136
 innovative devices 73
 portable 43, 49–50, 359
Arc and spark 366, 368
Archeus 4
Aristotelian four elements 12, 16, 24, 25
Army laboratories 233, 458
Ashing. *See* Calcination
Aspartate aminotransferase 407
Association of Clinical Biochemists 473,
 474
Association of Clinical Pathologists
 234, 473
Astra, Beckman 494
Asu 5
Atherosclerosis 382–84
Atomic absorption spectrometry 371–72
 calcium, analysis 372
Auscultation, stethoscope 80–81
AutoAnalyzer 257, 338, 340, 354, 387,
 457, 458, 465, 484, 485, 487, 488,
 490, 492, 497
 design configuration 486–88
 industrial applications 488
 Skeggs, Leonard 485
 training facilities 489–90
AutoChemist, Swedish 494
Automation 483
 continuous flow 485–88
 batch analysis 490–96
 random access 494, 496, 497
Autopsy. *See also* Postmortems
 assisting at 223
 clinical correlation with 80, 83, 84, 91,
 220

SUBJECT INDEX

Azobilirubin 141, 385, 388
  direct reaction 385–87
  indirect reaction 385, 386, 389
Azote 33

Babcock bottle 237, 244 (note 65)
  hand-operated centrifuge 237
Bacteriology 221, 229–31
Balance
  analytical 7, 32, 71–72, 235
  anion-cation 175
  humoral 2, 9
Bar graphs 175
Barreswil's reagent 160
Baseline tests 507
Bausch & Lomb Optical 240, 256
Beckman Instruments
  Astra 494
  DSA 494
  DU spectrophotometer 458–62
  pH meter 207–11, 462
Becton Dickinson Company 152, 309–13
Bedside medicine 2, 18, 79, 84, 120, 506, 508–09
Beer's law 260–62
Bence Jones protein 102–04, 138
Benign tumor, Folin's 281
Berichte, Berzelius's 65
Best's sugar methods 178, 321
Beta hydroxybutyric acid 167, 168
Bicarbonate level 169, 173
Bilirubin 384–90
  analysis, methods of 387, 388
  azobilirubin 385, 388
  Ehrlich's diazo reaction 141, 385, 386, 387
  fractions 384–87
  froth test 143
  Gmelin's test 141
    modifications of 141, 142
  icterus index 385
  jaundice 384, 389, 390

  kernicterus 389
  erythroblastosis fetalis 389, 393 (note 37)
  phototherapy 389
  spectrophotometric, direct assay 390
  standards
    artificial 387–88
    reference 388, 468
  urine
    dipstick 388
    tablet test 388–89
Bilirubinometer 390
Binocular vision 263
Biochemical Journal 251
Biochemical Society 473
Biochemisches Centralblatt 251
Biochemistry 251–52
  academic development of 251, 274–75
  clinical connection to 468–70
  separation from physiology 191, 251–52
Biochemists 251
  context of practice 251–52
Biological chemistry 251, 252
  at Harvard University 275
Bio-Science Laboratories 474–77
Biuret reaction 139, 422–23
  biuret as standard 422–23
  carbamyl groups 423
  reagent stability 423
Bladder stone 37–38, 39
Bleeding
  fluid depletion 2
  transfusion 186
  exchange 389
Blistering 2, 307
Blood analysis. *See also* Books, blood
  and urine increasing use of 304
Blood clotting
  buffy layer 27
  inflammatory crust 27, 437
Blood gas pump 167–69, 186–88, 190

SUBJECT INDEX

blood gases
  early studies 185–88
  equilibrium and transport 296–97
  manometric 173–75
  volumetric 169–73
Blood group incompatibility 389, 390
Bloodletting 279, 305
Blood volume 343
Blowpipe 39, 43, 50, 76 (note 17), 132, 235, 364
Blutfarbstoff 190
Blutschmidt 154
Board of Registry, ASCP 234
Body fluids
  depletion of 2, 14
  renewed interest in 92, 95
Body, as machine 18
Books, blood and urine 26–29, 43–44, 49–50, 99–102, 105, 107, 122, 131–144, 232
Böttger's test 140
Boyle's chemical tests 25–26
  analysis of blood 26–28
Brain chemistry 265 (note 11)
Bright's disease 84–91, 295
British Vickers Analyzer 495
Bromsulphalein retention 342
Brookhaven National Laboratory 300
Buffer systems 204
Bureau of Standards, National 239, 384, 452, 468
Burette 73
Burner
  Bunsen-Desaga 187, 236, 363–64
  Fisher 236
Butterfat in milk 237

Calcination 27, 55, 185
calcium analysis 354–56
  ashing 355
  atomic absorption 372
  complex formation 355–56
  direct photometry 356

flame photometry 372
  oxalate precipitation 355
  potassium permanganate titration 355
Calculi
  books on disorders of 43, 44, 49, 50, 107
  bladder 37–40
  concept of formation 107–08
  constituents of 14, 38, 39, 49
  electrolysis of 108
  lithontriptic remedies 31, 38–39
  lithotomy 37–39
  reagents for analysis 43, 49–50
  urinary 31, 37–40, 49–50, 78
Cambridge University 79, 126
CAP survey 340, 345 (note 39), 347 (note 67), 354, 370, 372, 382, 387, 422
Capillary blood 139, 279, 301, 303, 307, 320, 321, 446
Carbohydrate, origin of term 154
Carbon dioxide
  capacity of plasma 172
  carbonated water 39
  chaos 15
  combining power 156, 173, 195, 285, 483
  content in plasma 172, 297
  diabetic coma, in 167
  discovery of 15, 31–32
  fixed air 15, 32, 39
  gas sylvestre 15
  manometric analysis 173–75
  vacuum extraction 171
  volumetric analysis 170–73
Carbonated water 39
Career opportunities 514
Carnegie Foundation 269, 313 (note 1)
Casein 56
Catalysis, origin of term 396
Catalytic force 395–96, 398
Cation-anion balance 175

SUBJECT INDEX 547

Cell, first use 215
Cell-free extract 399–400, 401
Centrifugal analyzers 495–96
Centrifuge 196–202, 237, 301
  centrifugals 201
  electric 200–202
  hand-operated 237
  hematocrit 196
  urine sediments 44, 50, 119, 120, 132, 133, 135, 137, 196, 200, 201, 202, 218, 219, 448
Cephalin-cholesterol flocculation 466
Cereal seeds, pregnancy test 5
Cesium, discovery of 364. *See* Spectroscope
Chainomatic balance 72
Chaos 15
Chemical
  anatomy of urine 14
  diagnostic signs 124
  diseases 40, 105
  dissection 92
  glassware 237–39
  medicines 4, 16, 44, 52 (note 18)
  nomenclature
    according to Lavoisier 33
    authority for 63–65
  portable apparatus 43, 49–50, 135, 136
  suppliers 235–36
  tests in 1914 232
  transformation 3
Chemical News 255
Chemist
  internal 4
  pathological 120, 265 (note 11)
Chemist-Microscopist 223
Chemistry
  animal 55, 56, 112–13, 119
  clinical laboratories 222–26
  image problem 231, 232, 444, 505
  diagnostic signs 124, 126
  founder of modern 117
  inorganic 55, 66, 69, 71

  laboratory course, first 16, 110–11, 117
  lectures, popular 114–17
  medical curriculum 106, 274–75
  pathological 126
  private courses 120, 162, 163, 219–220
  quantitative phase, start 31
  rejection by Trousseau 94–95
  science of observation 42
  separation from medicine 117
Chloride analysis 351–54
  amperometric titration 353
  automated 354
  chloride shift 354
  classical method, gravimetric 351
  iodometric titration 352
  mercuric nitrate titration 352
  origin of name 358
Cholera 151–56
  chloride level in 354
  epidemic 151–53
  fluid-electrolyte therapy in 151–56
  reduced mortality 155–56
  government hoax 153
Cholesterol
  color reactions 378, 379–80
  coronary heart disease
    atherosclerosis, role in 377, 382–84
  education program 384
  digitonin, reaction with 378–79
  enzymatic 381–82
  fractions 379–81
  gravimetric 379
  origin of name 377
  reference methods 380–81
  saponification of 379–80
  standardization program 384
Chromatic aberration 215, 216
Circulation of blood 27
Clinic 242 (note 18)
Clinica Chimica Acta 287
Clinical chemical laboratory
  analytical techniques 252
  basic apparatus 465

first use of term 122
in Great Britain 447
revival in Germany 126
revival in Germany 126
Clinical Chemistry 287, 472
  Advances in 472
  Standard (Selected) Methods of 472
Clinical Chemistry: Principles and Technics 320, 476
Clinical Chemistry: Theory, Analysis, and Correlations 477
Clinical laboratory 221–26
  bacteriology 221, 223, 229, 231
  hematology 92, 231
  lectures 18, 79, 120
    in United States 222–26, 445–447
  wards 90
  microscopy 218, 221, 223, 225, 230
  pathology 88, 218–26, 230–31, 242 (note 18), 443
  sciences, applied 230
  scientific luxuries 230
  scientific medicine 226–28
    lack of application 232
Clinical Laboratory Improvement Act of 1967 472
  Amendments of 1988 509
Clinical picture, reliance on 2, 82
Clinician, origin of term 242 (note 18)
Clinistix 327
Clinitest tablet 326–27
Clotting of blood 437–38
  catalytic action 438
  diagnostic significance 437
  fibrinogen 438–39
College of American Pathologists 467, 510. *See also* CAP survey
College of New Jersey 116
College of Physicians and Surgeons 250, 281, 448
College of Physicians of London 10
Colloidal gold 466
Color indicators 25

Colorimeter, photoelectric
  Cenco-Sheard-Sanford 454, 458
  Coleman Jr 458–60
  Evelyn 455–56
  Klett-Summerson 455, 457
  Kober-Klett 257
Colorimeter, visual
  colorimeter-nephelometer 256, 257
  Duboscq light path 256, 258–60
Color matching
  early applications 253–54
  techniques of 254
Combustion 16, 30, 32, 33, 185, 186, 187, 188
Commercial laboratories 232–34
  Bio-Science 474–77
Competition, friendly
  Folin and Benedict 287–89, 322, 324
Conservation of matter 15, 32–33
  transmutation 3, 15, 23
Continuous flow analysis 485–88
Conversations on Chemistry 53 (note 33)
Cooking technology 3
Corning glass 239
Coronary heart disease 377, 382–84
Cost-containment 411, 511, 512
Creatine 138, 329
Creatinine 329–32
  blood, colorimetric method 282
  clearance 331
  discovery of 129 (note 44), 279, 329
  doubts of existence 329–30
  enzymatic method 330
  gravimetric analysis 132
  Jaffe's reaction 141, 330, 331, 345 (note 39)
  24-hour volume, verification of 331–32
Crotonic acid 167
Cupping
  dry 317 (note 68)
  wet 307
Curriculum, medical 106, 274–75

SUBJECT INDEX 549

Cystic oxide 40
Cystinuria 39–40

Daguerreotypes 263–64
Dalmatian dog
   purine metabolism in 288
Department of Agriculture 236
De Sedibus 1
Depletion of body fluids 2, 14
Diabetes
   adult-onset 480
   acid intoxication 168
   acidosis 156, 167, 168, 297
   association with pancreas 157–58
   coma 167
   excretion of ammonia 167, 168
   glycosuria 50, 156, 157, 161, 304
   honey urine 156, 157
   hyperglycemia 156, 158
   hyperpnoea in 167–69
   ketone bodies 168
   organic acid in urine 167
   polyuria 156
   removal of pancreas 177
   theory of 104
Diacetyl monoxime, urea analysis 339–40
Diagnosis, laboratory *vs* bedside 503–04, 506, 507
Diagnostic chemistry 105, 126
   qualitative 230
   signs 124, 126
   yield of tests 508
Diarrhea of urine 156
Diarrheal disease 155–56
Diastase 395, 396, 398
Diazo reaction 385, 386, 387
Diffusion of medication 365
Digestion, gastric
   chemical 15
   mechanical 18
   products of 250, 298
Digitonin, reaction with 378–79

Dipstick, glucose 327, 328
   multiple test 328
   protein 148
Discovery of elements 63, 357–58, 364
Discrete sample analyzers 490–96
Disease, humoral imbalance 2, 14
   chemical 40, 105
Dissection
   chemical 92
   stigma of 80
Distillation
   destructive dry 25, 27–29, 55, 122
   steam 301
Dorpat University 47, 153, 154, 179 (note 7)
Dow Chemical Company 476
Dropsy 85, 86, 88, 89
Dry-cupping 317
Dry combustion 59
Dry-slide chemistry 496
Duboscq colorimeter 255–59, 304, 448, 449, 457, 458
   calculations on 260, 262
   discovered by Folin 276
   early uses of 255
   Folin's first use 255, 279, 304
   light path 256, 258–60
   turbidimetry with 260
Duboscq's optical instruments 262–63
DuPont Analyzer, aca 494
Dye-binding, albumin 421–22
Dynamic equilibrium 281, 400

Eastern cultures 5
Edema 85, 86, 88, 89, 93
Edinburgh university 79, 84, 115
Education, medical 106, 272–75
Effervescence, physiological 16
Ehrlich's test, bilirubin 141
Einhorn tube 161, 228, 229
Eimer & Amend 236, 256
Ektachem analyzer 496
Electrolysis, calculi 108

Electronic voltmeter 208–09
Electroendosmosis 424
Electrophoresis, protein
 moving boundary 420, 423–27, 428
 agarose gel 433–34
 amino acids 428–29
 cellulose acetate 431–34
 filter paper 427–31, 434, 466
Elementary principles, three 4, 12, 16, 24, 25
Elements, four 12, 16, 24, 25
 classification of 211 (note 2)
 detection *in vivo* 281, 365
 three principles 4, 12, 16, 24, 25
 two by Van Helmont 12
Eli Lilly and Company
 purification of insulin 179
 Tes-tape 328
Emission spectrum 361–65
Emulsin 396
Enclosure movement 57
Endogenous metabolism 280
Enzymatic methods
 cholesterol 381–82
 creatinine 330
 glucose 327–28
 urea 338, 340
 uric acid 336
Enzyme
 catalytic force 395–96, 398
 clinical units 407
 colloidal carrier 401–02
 ferments 396–400
 origin of term 397
 protein identity of 401–02, 413 (note 14)
 spontaneous changes 395
 theory of life 400–01
Equation, Henderson–Hasselbalch 204, 206
Erythroblastosis fetalis 389, 393 (note 37)
Esbach's method 139, 146–47, 228

Eudiometer 59, 187
Euglobulin 419
Evacutainer tube 312. *See also* Vacutainer tube
Evelyn photometer 455, 456
Evolution of technology 483–84
Exchange transfusion 389
Exogenous metabolism 280
Extract, cell-free 399–400
Extractives 99, 101, 122, 135, 138

Factory teaching 489–90
False-positive, proteinuria 89–90
Fat in blood 122, 143, 284, 301
Federation of Clinical Chemistry, International 474
Fehling's solution 124, 140, 143, 160, 161, 319
Fermentation
 alcoholic 71, 158, 159, 396, 399, 400
 cell-free extract 399–400, 401
 chemical 399, 400, 401
 contact catalysis 398, 399
 Einhorn tube 161
 mechanical 399
 microbial 399, 400
 physiological 16, 19
 putrefaction 399
 specific gravity tables 104, 162
 sugar in urine 134, 137, 157, 161
 in blood 135–36
 vitalist 398–400
 yeast 137, 396–400
Ferments 395, 396–400
 catalytic force 395–96
 organized 396, 400
 unorganized 396–97, 400
 vital 15
Fertilizers, artificial 214 (note 46)
Fibrin 29, 56, 90, 93, 100, 135, 137, 138, 298, 333, 418, 439
 enzymatic process 154, 438
Fibrinogen 438–39

SUBJECT INDEX

Filtrate, protein-free 284–86
Finger-tip blood 139, 279, 301, 303, 307, 320, 321, 446
First International Congress of Clinical Chemistry 474
Fisher Scientific Company 235–36, 455
  burner 236
Fistula, pancreatic 398
Fixed air 15, 32, 39
Flame photometry 366–70
  calcium analysis 372
  Perkin–Elmer 368–69
  spectrography 367, 368
Flexner Report 269–73
  laboratory sciences, neglect of 271
  medical education, deficiencies in 272
  role for clinical chemistry 271
Fluid-electrolyte therapy in cholera 151–56
Folin-Wu tube 322–23
Foodstuffs
  animal materials 56
  classification 44, 154
  extraction, chemical 56
  nutritional value 298
  performed assimilation 165, 298
  rearrangement 107
  shortage 55–57
  solubilization 298
Foreign societies 473–74
Formalin titration 168, 204
Formol gel 466
Four elements, Aristotelian 12, 16, 24, 25
Fraunhofer lines 362, 364
French medicine 84
Free radicals, organic 292
Fundamentals of Clinical Chemistry 320, 477
Fusible calculus 39
Future developments 498, 511–13, 515

Galatest, urine 326
Galenical medicine 14, 16

Gallnuts 26, 253, 266 (note 22)
Garrod's thread test 138, 139, 232, 332
Gas analysis
  manometric 173–75
  volumetric 169–73
Gas carbonum 15
Gas sylvestre 15, 32
Gaseous state 29–32
Gastric analysis 228, 232
Gastric digestion
  chemical 15
  digestive principle 396
  outside the stomach 396
  pepsin 250, 396, 398
  products of 250, 298
  trituration 15, 18, 19
  vital ferment 15, 18
Gastric fistula 46, 47
Gastric juice 19, 46–47
  Beaumont's research 46
  hydrochloric acid, discovery of 46–7
  response to food 46
  solvent action 19
General practitioner 118
Genetic defect 40
Gerhardt's test 142
Giessen, laboratory at 108–12, 122, 249, 329
Glass electrode 203, 207–08, 209
Globulin
  classification 142, 418–19
  electrophoretic components 420
  euglobulin 419
  fractionation 418–21
  Howe's method 419–20
  pseudoglobulin I, II 419
Glassware, chemical 237–39
Glucose analysis, blood 102, 135–36, 143, 320–24
  arterial-venous difference 136, 321
  Bang's method 320
  Bernard's method 135, 143
  enzymatic 327–28

Fehling's solution 143, 160–61
ferricyanide 325
Folin-Wu 322–24
iodometric titration 178, 301, 320, 325
literature review 343 (note 3)
method modifications 321–24
ortho-toluidine 327
picrate 319–22
yeast fermentation 324
glucose analysis, urine 134, 137, 144, 319–20
Benedict's method 319–20
Clinistix 327
Clinitest tablet 326
enzymatic 327–28
detection 144
Fehling's solution 143, 160–61
fermentation 140, 157, 159, 161–62
decrease in specific gravity 162
Galatest powder 326
Moore's test 134
taste 44
Trommer's test 134, 135, 140, 159–60
glycogenic function of liver 165
glycosuria 50, 156, 157, 158, 165, 304
abolish the term 322
Gmelin's test 50, 137, 141, 232
Guaiac test for blood 137
Guy's Hospital 49, 50, 84, 88, 90, 101, 115, 218, 220
clinical wards 90, 221
microscopy department 218
chemistry course at 115
Gymnasia, Germany 246

Halle University 66, 245
Harvard University 169, 175, 273, 275, 276, 283, 284
catalog 247
Harvey Society 271
Healing, Hippocratic 2, 4
Health Maintenance Organizations 511

Heat and acid test, proteinuria 140, 146, 148
false-positive 89–90
Heidelberg University 46, 187, 188, 249, 250, 292, 362, 398
Heller's ring test 120, 124, 139, 140, 146
Hematocrit 196
Hematology 231
Hemoglobin 190, 194, 390
acid hematin 193
analysis 190, 192
carboxyhemoglobin 190, 195
cyanmethemoglobin 195
isolation of 190
oxygen capacity of 195
specific gravity method for 297, 436–37
spectral absorption of 137, 139, 142, 190, 192, 232, 448
standards 193–96, 468
Hemoglobinometer 193, 195, 228
Hemometer 193
Henderson–Hasselbalch equation 204, 206
Hippocratic healing 2, 4
bedside observation 2
self-healing 4
writings on urine 5
Homogentisic acid 49
Honey urine 156, 157
Hospital medicine 84
Howe's protein fractionation 419–20
compared to electrophoresis 420
Humoral imbalance 1, 2, 14, 82
Humoral pathology 2, 9, 83, 95
four humors 9
return to 92, 95
Hydrochloric acid, discovery of 46–7
Hydrogen electrode 202–03, 204, 206, 207
Hydrogen ions 203–04, 207, 208
activity of 213 (note 40)
proton donors 207

SUBJECT INDEX 553

Hygeine, development of 129 (note 44)
Hyperbilirubinemia 384, 389
Hyperglycemia 102, 156, 158, 165
Hyperpnoea 155, 167, 168
Hypodermic syringe
   Becton Dickinson 309–13
   disposable 312
   evolution of 305–07
   glass 306, 310
   Multifit, interchangeable 310–12
   needles 168, 313
   origin of term 306
   sterile technique 307
   Vacutainer tube 312–13

Iatrochemistry 4, 10, 16, 23, 94
Iatromechanist 18
Iatrophysicist 18
Icterus index 385
Ictostix 388
Ictotest 388
International Federation of Clinical Chemistry 474
Image problem for chemistry 231, 232, 444, 505
Inborn errors 40
Incompatibility, blood group 389
Industrial Revolution 57, 115, 483, 490
Indicators, color 25
Inflammatory crust 27, 437
Insanity
   chemical basis, toxins 265 (note 11)
   McLean Hospital 278
Instruments, physical 83, 226, 228
Insulin 156
   controversy over isolation 183 (note 58)
   discovery of 177–79
   effect on biochemical testing 304
   origin of term 177
   radioimmunoassay of 478–81
Insurance coverage 145
Internal chemist 4

International
   authority for nomenclature 63–65
   enzyme unit 407
   Federation of Clinical Chemistry 474
Iodine, identification of 358
Iodometric titration 60, 178, 301, 320, 325, 352
Ion selective electrodes 370
Islands of Langerhans 177
Isoenzymes 405–06
Isomerism 67
Isotope label
   radioactive 478–81
   stable 281

Jack bean 338
Jaffe's test 141, 255
Jaundice 384, 385, 389
Johns Hopkins Hospital 222, 223, 225, 230, 273, 291, 292, 503, 505
Journal of Biological Chemistry 251, 269, 286, 287, 300, 335
Journal of Experimental Medicine 291
Journal of Laboratory and Clinical Medicine 287, 445
Journal of the American Chemical Society 246
Julius Hospital 122

Kernicterus 389
Ketone bodies 168
Kimble Glass Company 238
Kingsbury–Clark standards 145–46
Kidney function tests
   creatinine clearance 331
   phenolsulphonephthalein excretion 341–42
   urea clearance 297, 341
Kit methods
   hemoglobin 192
   LaMotte 478
   radioimmunoassay 481
   Sigma 477–78

554 SUBJECT INDEX

Kjeldahl method 59–63, 141, 142, 283, 301, 417, 419, 435–36
Klett Manufacturing Company 256, 424
Klett–Summerson photoelectric colorimeter 455, 457

laboratory
 Army 233
 assistants 132, 232–33, 446
 bacteriology 221
 commercial 232–34, 278
 course, first 16, 110–11, 117
 divisions 505
 endowment 222–23, 224–26
 furniture 235
 image problem 231, 232, 444, 505
 integrated 513
 low cost 230
 medicine, practice of 84, 233, 467n 471, 475
 Pepper 224–25
 pharmacist's 77
 portable 43, 49–50
 private 120, 162, 163, 219–220
 technician's school 234, 244 (note 57)
 testing at bedside 508–09
 ward 77, 90, 221, 223, 230, 505
Laboratory diagnosis 504
 chemical signs 124
 dependence on 503–04, 506, 507
 expanding role of 508
 laboratory vs bedside 503–04, 506–08
 overuse of testing 506, 507
 scientific medicine 506
Laboratory Manual, Folin's 447
Laboratory medicine 84, 233, 467n 471, 475, 510, 513
Lancet, The 44, 52 (note 24), 79, 118, 152, 153
 book review 136
 letter to 151
Lectures, popular 114–17
Legal's test 141, 142

Lens-makers, trial and error 216
Leyden University 79, 85
 bedside teaching at 18
Library medicine 84
Liebig's
 Annalen 112, 250, 351
 conflicts 75, 113
 urea method 124
Liqnum nephriticum 25
Lime water 32, 39
Lipase 403–04
lithic acid 37, 39
lithontriptic remedies 31, 38–39
 carbonated water 39
 electrolysis 108
Lithotomy 37–39
Litmus paper 25
Liver function tests 342, 466
London Medical Gazette
 running debate in 45
Lovibond comparator 193, 194

Machine
 body as 18
 laboratory 483
Malpractice, fear of 507
Managed care 511, 512
Manometric gas analysis 173–75
 microgasometer 175, 176
Manual of Chemical Physiology 247
Marburg University 276
 Folin discovers colorimetry 276
Mark 10 Analyzer, Hycel 492–94
Marketing terms 497
Massachusetts General Hospital 223, 284, 315
Matière savonneuse 40
Matula, uroscopy 5, 7
McGill University
 microscope course 222
McLean Hospital
 mental diseases 278
 urine analysis 279
Mechanics' Institutions 115

Mechanistic rationalism 119
Mecolab Analyzer 495
Medical aphorisms 5
Medical chemistry 245, 274
Medical education
  deficiencies 106
  Flexner Report 271–73
  innovation 225
Medical-industrial complex 511
Medical leadership 84
medical papyrus 5
Medical schools
  Flexner Report 271–74
  proprietary 272
  reform of 273–75, 313 (note 1)
Medical technologists 232–34
  national association of 234
  schools for 234, 244 (note 57)
  women technicians 233–234
Medicasters 6, 12
Medicine, scientific 119, 226–28, 289
  change in 84
Melanic acid 48–49
Metabolic pool 281
Metabolism
  dynamic 281
  endogenous 280
  exogenous 280
  inborn errors of 40
  wear and tear, theory of 280–81
Methemoglobin 139, 169, 195
Method development 497, 498
Methods in Clinical Chemistry 477
Metpath Laboratory 495
Metropolitan Life Insurance Company 145, 289
Microgasometer 175, 176
Micromethods 182 (note 53), 301–03
  317 (note 63), 320
Microscope 2, 83, 126, 215–19, 221–25,
  227, 230, 231, 235
  achromatic lenses 216
  aplanatic foci 216
  Chemist-Microscopist 223

  compound 215, 216, 218
  red cells, description of 218
  Guy's Hospital 218
  lecture course 219
  photomicrographs 131, 218, 448
  room, private 219
  single lens 215, 216
Middle Ages, uroscopy 5, 9, 10, 12
  blood analysis 26
  outmoded terminology 33
Millon's reaction 124, 132
Mineral analysis 63
Modern chemical textbook, first 33
Mofette 33
Molecular movements 95
Molisch's test 140
Moore's test 134
Morbid anatomy, emphasis on 83, 84, 91
Moving boundary electrophoresis 423–27
Mulberry calculus 40
Multifit syringe 311
  Japanese import 310–11
Murexide 38, 143, 356
Muriatic acid 46

National Bureau of Standards 239, 384, 452, 468
National Health Service 473
National Committee for Clinical Laboratory Standards 468
National Institute of Standards and Technology 239, 384, 388, 392 (note 22)
Natural philosophy 26, 109–10
Naturphilosophie 119
Nnephelometer-colorimeter 256, 257
Nephrotic urine, protein in 89
Nessler reagent 60, 62, 253, 255, 279, 283, 337, 338, 340, 419
Nitric acid ring test 120, 124, 139–40, 146

Nitrogen analysis 57, 59–60, 62–63
   Dumas method 57, 59
   Kjeldahl method 60–62
   Nessler reagent 62
   organic compounds 57, 59
   protein conversion factor 142
   Wanklyn procedure 60
   Will-Varrentrapp method 59, 60
Nitrogen-free diet 57
Nobel Prize 167, 211 (note 4), 214
   (note 46), 283, 292, 346 (note 46),
   377, 379, 401, 402, 424, 480
Nomenclature
   authority for 63–65
   by Lavoisier 33, 67, 71
Nonprotein nitrogen 115, 282–83, 284,
   285, 286
Nonspecific tests 466
Nonisotopic labels 482
Normal system 74
Normal *vs* natural 81–82
Nucleic acids 292, 293
Nucleoproteins 292
Nutrition 55, 275, 298
Nylander's test 140

Optical instruments
   Bausch & Lomb 240
   Duboscq's 262–63
Optical laws 260–62
Organic chemistry 55, 245, 275
Organic compounds
   definition of 71
   nitrogen analysis of 57–63
   synthesis of 68–71
Orthostatic albuminuria 145
Oxford University 79, 126, 169
Oxygen capacity 169, 195
Oxyhemoglobin 169, 190

Packages, test 507
Pancreas 157–58, 177
   acinous tissue 178
   fistula 398
   islands of Langerhans 177
   ligation of ducts 177
Panels, test 507
Papyrus 156
   equal arm balance 72
   medical 5
Paracelsian three elements 4, 12, 16, 24,
   25
Parenteral therapy in cholera 151–56
Pathological cabinet 223
   anatomy 80, 1–2
   chemist 120, 265 (note 11)
   chemistry 120, 122, 126, 265 (note 11)
Pathology
   clinical 88, 218–26, 230–31, 242
   (note 18), 443
   humoral 2, 9, 83, 92, 95
   localized 1
Peking Union Medical College 300
Pennsylvania Hospital 226, 242
   (note 31)
Pepsin 250, 396, 398
Peptones 138, 140, 147, 250, 298
Peptonuria 140
Pettenkofer's test 124, 137
Pflügers Archiv 191
pH 204, 206, 207
   operational definition of 213 (note 40)
pH meter 207, 208–11, 462
   electronic voltmeter in 208
Pharmacist's laboratory 77
Pharmaceutical Society of Great Britain
   128 (note 29)
Phenakistoscope 263
Phenolsulphonephthalein excretion
   341–42
Phenylhydrazine 165–66
   osazone formation 143, 162, 324
Philadelphia General Hospital 223, 242
   (note 31)
Philadelphia Medical College 116
Philosopher's stone 23

Philosophy, natural 26, 109–10
  speculative 119, 126
Phlebotomy 26, 305–09, 312
Phlogiston 114
Phosphatase
  acid 406
  alkaline 404–06
Phosphorus 356–57
Phosphocreatine, discovery of 284
Photelometer, Cenco-Sheard-Sanford 454
Photoelectric Clinical Chemistry 458
Photoelectric photometry
  advantages of 455
Photoelectric effect 452, 453, 483
Photometer, filter
  Cenco-Sheard-Sanford 454, 458
  Evelyn 455–56
  Klett–Summerson 455, 457
  Kober–Klett 257
  visual 451–52
Photometry, direct
  filter flame 366–70
  internal standards 368
  reflectance 496
Photomicrographs 131, 218, 448
Phototherapy 389
Physical bedside examination 80, 84
  physician's practice 118
Physiological chemistry 190, 245–48,
  249–52, 274–75
  books 448, 451
  branch of chemistry 247
  development in U.S. 246
  separation from physiology 191, 252
  teaching, beginning of 247
  traditional subservience of 252, 275
Physiology 163, 191, 231, 245, 247, 249,
  251, 275, 515
Pipette, double-stopcock 171, 172, 173
Pisse-Prophet 6, 10, 11
Plough Court 116
Pneumatic trough 30
Point-of-care testing 508–09

Polariscope 50, 104, 141, 162, 235, 263,
  319
Polypetides, synthesis of 167
Polyuria 156–58
Popular lectures 114–17
Post-Graduate Hospital 305, 443, 445,
  447
  blood drawing at 308, 446
  chemical tests at 445–47
Postmortems 1, 2, 18, 80, 218, 220, 223,
  224
  clinical correlation with 80, 83, 84,
  91, 220
Postural albuminuria 145
Potassium analysis
  atomic absorption 371–72
  clinical utility 360–61, 369–70
  colorimetric 360
  discovery of 357–58
  flame photometry 360, 367–70
  gravimetric 359–60
  ion-selective electrode 370
  titrimetric 360
Potassium permanganate
  bilirubin standard 388
  calcium analysis 355
Practical Chemical Analysis of Blood
  445–47
Practitioner, general 118
Pregnancy, tests for 5, 482
Practical Physiological Chemistry 448
  Duboscq colorimeter in 448
  Purdy's centrifuge in 448
Principles, three 4, 12, 16, 24, 25
Proficiency testing 234, 466–67
Profiles, test 507
Proposal to hospitals, Folin's 286
Prostatic specific antigen 406
Proteolytic digestion 250
protein
  albumin binding reagents 421–22
  analysis 122, 417–439
  biuret color reaction 422–23

SUBJECT INDEX

chemistry, beginning of 55
classical separation 417-19
cleavage products of 250
  fate of 298-99
coagulation 417
electrophoresis
  moving boundary 423-27
  photographic record 425, 427, 434
  solid media 427-34
error of indicators 146, 148
fractionation 418-21
globulin, origin of term 418
Howe's fractionation 419-20
Kjeldahl analysis 142, 417, 419, 436
metabolism 297-99
nitrogen conversion factor 142
nutritional value of 298
origin of term 418n
peptide theory 166, 401
proteios 418n
refractive index 435-36
salt precipitation 418-21
solubilization of 298
specific gravity 297, 436-37
structure, theory of 166, 401
sulfate-sulfite reagent 421
Protein-bound iodine 475, 482
Protein-free filtrate 283, 284-86, 325-26, 355
Proteinuria
Bence Jones protein 102-04
Bright's disease 84-91
qualitative procedures
  dipstick 148
  Esbach's 147
  heat and acid 140, 146, 148
  false-positive 89-90
  Heller's ring test 120, 124, 139, 140, 146
quantitative procedures
  biuret 422-23
  gravimetric 147-48
  turbidimetric 286

Pseudoglobulin I, II 419
Publication pitfalls 503
Pure Food and Drug Act of 1906 238
Purging, physiologic 2
Purpurate of ammonia 38
Putrefaction 16, 19, 399

Quality assurance 467
Quality control 467
Quantitation in medicine 82-83
Quantitative Clinical Chemistry 300, 449-51
Quintessential spirit 3

Radioimmunoassay 478-82
Random access analyzers 494, 496
Red blood cell counts 192
  microscopic examination of 218
  sedimentation of 27, 437
Reference material 384, 388, 392 (note 22), 468
Referral chemistry 474-77
Reflectance photometry 496
Refractive index
  moving boundary electrophoresis 425
  protein 435
Regulatory agencies 509-10
Regurgitation, gastric 19
Remedies, chemical 16, 44, 52 (note 18)
Renaissance artists 3, 6, 10, Frontispiece
Renal disease. *See* Bright's disease
Respiration 16, 30, 32, 33, 185-86, 188
  artificial 185
  compensated acidosis 206
  intratracheal 185
Revolution
  French 77, 79, 80, 83
  Industrial 57, 115, 483, 490
Ring test, proteinuria 120, 124, 139, 140
Robot Chemist 490-92
Rocella, as color indicator 25

SUBJECT INDEX

Rockefeller Institute Hospital 292, 293, 295, 297, 300, 436
  Van Slyke's research at 295–299
Rockefeller Institute for Medical Research 226, 271, 289, 291, 295, 296, 300, 450
Routine tests 507
Royal College
  of Physicians 44
  of Surgeons 37, 118
Royal Institute of Chemistry 473
Royal Institution 115, 116
Royal Society 46, 114, 126, 185
Rubidium, discovery of 364. *See* Spectroscope
Rubner's test 140

Saccharimeter (Saccharometer). *See also* Polariscope Biot's instrument 104, 162, 263
St. Andrew's University 51
St. Bartholomew's Hospital 40, 220
St. Luke's Hospital 230, 447
  laboratory tests at 447
St. George's Hospital 102, 105, 127 (note 9)
St. Thomas's Hospital 220, 221, 265 (note 11), 447
  laboratory tests at 447
Scandinavian Journal of Clinical & Laboratory Investigation 287
Scholastics
  ancient texts 1
  traditional medicine 14, 18
Scientific medicine 119, 226–28, 289, 506
  change in 84
Screening groups 507
Standard (Selected) Methods 472
Sheffield Scientific School 247, 250, 287, 448
  Laboratory of Physiological Chemistry 251, 443
Sigma, kit methods 477–78

SI units 408–12
  objections to 408–09, 410–12
  origins of 408, 409–10
Sequential multiple analysis 488, 489
Societies, clinical chemistry 470, 473–74
Society of Apothecaries 117, 118
Sodium analysis
  atomic absorption 371–72
  clinical utility 360–61, 369–70
  colorimetry 360
  discovery of 357–58
  flame photometry 360, 367–70
  gravimetry 359–60
  ion-selective electrode 370
  titrimetry 360
Specific gravity 7, 14. *See also* Urine
  protein analysis by 297, 436–37
  sugar analysis by 104, 162
Spectrochemical analysis 361–65
Spectrography 367, 368
Spectrophotometer
  Beckman DU 458, 459–62
  Coleman Jr 458–60
  visual 365–66
Spectroscope 190, 361, 362–63, 365–66
  arc and spark 366, 368
  hemoglobin derivatives 365
  Fraunhofer lines 362, 364
Spectrum analysis
  Bunsen burner 362–64
  diffusion of drugs 365
  emission and absorption spectra 362, 364
  new elements 364
  priority claims 364–65
Spherical aberration 215, 216
Spirit lamp, alcohol 50, 122, 127 (note 9), 135
Standards
  artificial 254, 387–88, 455–56
  certified 388
  reference material 384, 388, 392 (note 22), 468, 498

SUBJECT INDEX

Stethoscope 80–81, 83, 94, 220
Stereoscope 263
Strassbourg University 190, 191, 249
Students
   Folin's 283–84
   Van Slyke's 300
Sugar, polariscope 162
   structural formulas 165–66
Sulfosalicylic acid 145–46, 147, 148, 260, 286
Surgeon-apothecary 118
Sweating 2
Swedish AutoChemist 494–95
Synthetic fats 71
Syringe, hypodermic
   Becton Dickinson 309–313
   disposable 312
   evolution of 305–07
   interchangeable 310–12
   Luer glass 306, 310
   Multifit 311
   origin of term 306
   sterile technique 307
   Vacutainer tube 312–13

Tablet test, urine
   acetone 142
   bilirubin 388
   glucose 326–27
Tariff Act of 1883, 162
Tariff Act of 1922, 239
Tartu 179 (note 7)
Technicon Corporation 257, 465, 485, 488
Technology
   basic laboratory (1925–1960) 465
   evolution of 483–84, 498
   restraint of 507
Tes-Tape 328
Test ordering packages 507
Tests, increasing volume of 506–08
   obsolete 466

Textbooks
   blood and urine analysis 131–44
   clinical chemistry 137–38, 445–51, 458, 476–77
Theory of life, enzyme 400–01
Therapy, changing role 81–82
Thread test 138, 139, 232, 332
Thymol flocculation 466
Thymol turbidity 466
Thyroid function tests 475, 482
Titrimetric analysis 72–75
   burette 73
   pipette 73
Tools, laboratory 483
Torricellian vacuum 173, 187
Transaminase enzymes 407, 466, 478
Transfusion of blood
   exchange 389
   first 186
Transmutation 3, 23
Tree experiment 14
Trephination 37
Tribute to Ivar Bang 301
Trichloroacetic acid 285, 286, 355
Triple phosphate calculus 39
Trituration, gastric 15, 18, 19
Trommer's test 134, 135, 140, 143, 159–60
Trypsin, pancreatic 250, 396, 398
Tube, Folin-Wu 322–23
Tübingen University 188, 190, 192
Tungstic acid filtrate 284–85, 355
Turbidimetric methods 145–46, 286

Unguentum Aegyptiacum 159
University chemical laboratory, first 16
University of Pennsylvania Hospital 224
Urea
   ammonium cyanate 67
   analysis 41, 101–02, 142, 336–40
   aeration of ammonia 338–39, 340
   conductivity 340

diacetyl monoxime 339–40
Folin's method 337–38
gasometric 133, 139, 141, 144, 337
gasometric 133, 139, 141, 144, 337
Liebig's method 133, 137, 144, 337
urease 338, 340
Van Slyke's method 338–39
clearance 297, 341
insoluble salts of 133
isomerism of 67
metabolic source of 299
preparation of 40, 41
synthesis of 41, 67–69
Urethanes 276
Uric acid 347 (note 67)
  analysis
    Benedict's methods 334–35
    colorimetric 333–36
    Folin's methods 333–35
    Garrod's thread test 138, 139, 232, 332
    gravimetric 332
    Hopkin's method 332–33
    uricase 336
  Dalmation dog, metabolism in 288
  derivatives of 108
  discovery of 37–38
  Folin's interest in 276
  murexide reaction 143
Uril 108
Urine
  acetone in 141–42
  albumin, discovery of 132
    detection of 137, 139
  anatomization of 7–8
  bile in 50
  books on 43–44, 49–50, 99, 102, 107, 122, 131–44, 232
  calculi 14, 37–38, 49
  chemical anatomy of 14
  color reactions 139
  creatinine 142, 279–80, 281
    gravimetric 132

  dipstick analysis, protein 148
  distillation of 7
  early cultures 5
  glucose, detection of 134, 137, 144
    fermentation 134, 137
  medical aphorisms 5
  nephrotic protein patterns 89
  Pisse-Prophet 6, 10, 11
  protein, gravimetric 147–48
  routine tests on 44
  sediment in 8, 44, 50, 119, 120, 131, 132, 133, 135, 137, 196, 200, 201, 202, 218, 219, 448
    photomicrographs of 131, 218, 448
  specific gravity of 14, 50, 88–89, 90, 104, 120, 132, 140, 162
  timed collections of 92, 132, 134, 331, 341
  urea, insoluble salts of 133
  visual examination of. See Uroscopy
Uroscopy 5, 9, 10, 12
  color wheel 9
  matula 5, 7
  Medicasters 6, 12
  uromancy 9, 10
  water-caster 10
Urous acid 49

Vacutainer tube 152, 312–13
  development of 311–13
  polyester gel insert 313
Vacuum, Torricellian 173, 187
Van Helmont
  tree experiment 14
  two elements 12
Venesection 142, 305, 320, 321
Venipuncture 139, 255, 305–09, 312, 317 (note 70), 320 into general use 304
Verdigris 159
Vickers Analyzer 495
Visual filter photometers 451–52
Visual spectrophotometer 365–66

Vital ferment
  force 65–67, 69, 106–07, 413 (note 10)
  power 398
Vital phenomena, chemical terms 16
Vitalism, concept of 66, 66–69
  challenge to 95, 119
Vitalists 43, 45, 66–67, 94
Vivisection 163
Vocabulary, new 509–10
Volumetric analysis 72–75
  burette 73
  pipette 73
Vomiting, fluid depletion 2

Warburton Anatomy Act of 1832 80
Ward laboratories 77, 90, 220, 221, 222, 224, 230, 505
Wasserschmidt 154
Water-caster 10
Water and salt therapy 151–56
Water, primary substance 14
Wear and tear, theory of 280–81
Western Reserve University 443, 485
Wet-cupping 307
Whitall Tatum & Company 237, 238

William Pepper Laboratory 224–25
Wissenschaft 246
World War I 232, 233, 238, 255, 303, 310, 313, 413 (note 12), 474
World War II 111, 235, 240, 311, 355, 423, 436, 458, 465, 467, 475, 476, 485, 514
Würzburg University 120, 121, 122

Xanthic oxide 49
Xanthine 49, 166
Xanthoproteic reaction 132, 140

Yeast
  cell-free extract 399–400
  fermentation 396–400
Young Ladies Academy of Philadelphia
  lectures to 117

Zeitschrift für Physioloqische Chemie 191, 246, 255, 276
Zymase 399, 400
Zymogens 398
Zinc turbidity 466